Lecture Notes in Mechanical Engineering

Lecture Notes in Mechanical Engineering (LNME) publishes the latest developments in Mechanical Engineering—quickly, informally and with high quality. Original research reported in proceedings and post-proceedings represents the core of LNME. Volumes published in LNME embrace all aspects, subfields and new challenges of mechanical engineering. Topics in the series include:

- Engineering Design
- Machinery and Machine Elements
- Mechanical Structures and Stress Analysis
- Automotive Engineering
- Engine Technology
- Aerospace Technology and Astronautics
- Nanotechnology and Microengineering
- Control, Robotics, Mechatronics
- MEMS
- Theoretical and Applied Mechanics
- Dynamical Systems, Control
- Fluid Mechanics
- Engineering Thermodynamics, Heat and Mass Transfer
- Manufacturing
- Precision Engineering, Instrumentation, Measurement
- Materials Engineering
- Tribology and Surface Technology

To submit a proposal or request further information, please contact the Springer Editor of your location:

China: Ms. Ella Zhang at ella.zhang@springer.com
India: Priya Vyas at priya.vyas@springer.com
Rest of Asia, Australia, New Zealand: Swati Meherishi at swati.meherishi@springer.com
All other countries: Dr. Leontina Di Cecco at Leontina.dicecco@springer.com

To submit a proposal for a monograph, please check our Springer Tracts in Mechanical Engineering at http://www.springer.com/series/11693 or contact Leontina.dicecco@springer.com

Indexed by SCOPUS. All books published in the series are submitted for consideration in Web of Science.

More information about this series at http://www.springer.com/series/11236

José Machado · Filomena Soares ·
Justyna Trojanowska · Erika Ottaviano
Editors

Innovations in Mechanical Engineering

Springer

Editors
José Machado
Department of Mechanical Engineering
University of Minho
Guimarães, Portugal

Filomena Soares
Department of Industrial Electronics
University of Minho
Guimarães, Portugal

Justyna Trojanowska
Poznan University of Technology
Poznan, Poland

Erika Ottaviano
Dept.of Civil and Mechanical Engineering
University of Cassino and Southern Lazio
Cassino, Italy

ISSN 2195-4356 ISSN 2195-4364 (electronic)
Lecture Notes in Mechanical Engineering
ISBN 978-3-030-79164-3 ISBN 978-3-030-79165-0 (eBook)
https://doi.org/10.1007/978-3-030-79165-0

This Springer imprint is published by the registered company Springer Nature Switzerland AG
The registered company address is: Gewerbestrasse 11, 6330 Cham, Switzerland

Preface

This volume of Lecture Notes in Mechanical Engineering gathers selected papers presented at the First International Scientific Conference (ICIE'2021), held in Guimarães, Portugal, on 28–30 June 2021. The conference was organized by the School of Engineering of University of Minho, throughout MEtRICs and Algoritmi Research Centres.

The aim of the conference was to present the latest engineering achievements and innovations, and to provide a chance for exchanging views and opinions concerning the creation of added value for the industry and for the society. The main conference topics include (but are not limited to):

- Innovation
- Industrial Engineering
- Mechanical Engineering
- Mechatronics Engineering
- Systems and Applications
- Societal Challenges
- Industrial Property

The organizers received 213 contributions from 24 countries around the world. After a thorough peer review process, the committee accepted 126 papers written by 412 authors from 18 countries for the conference proceedings (acceptance rate of 59%), which were organized in three volumes of the Springer Lecture Notes in Mechanical Engineering.

This volume, with the title "Innovations in Mechanical Engineering", specifically reports and presents methods and technologies for modelling, simulation and design of mechanical systems, with a special focus on research spanning from engineering design and testing of medical devices, evaluation of new materials and composites, for different industrial applications, fatigue and stress analysis of mechanical structures, and application of new tools such as 3D printing and CAE 3D models, and decision support systems. Last but not least, it analyses important issues proposing a good balance of theoretical and practical aspects. This book consists of 44 chapters, prepared by 156 authors from 11 countries.

Extended versions of selected best papers from the conference will be published in the following journals: Sensors, Applied Sciences, Machines, Management and Production Engineering Review, International Journal of Mechatronics and Applied Mechanics, SN Applied Sciences, Dirección y Organización, Smart Science, Business Systems Research and International Journal of E-Services and Mobile Applications.

A special thank to the members of the International Scientific Committee for their hard work during the review process.

We acknowledge all that contributed to the staging of ICIE'2021: authors, committees and sponsors. Their involvement and hard work were crucial to the success of ICIE'2021.

June 2021

José Machado
Filomena Soares
Justyna Trojanowska
Erika Ottaviano

Contents

Machinability of the 18Ni300 Additively Manufactured Maraging Steel Based on Orthogonal Cutting Tests

Tiago E. F. Silva[1], Pedro A. R. Rosa[2], Ana R. Reis[1], and Abílio M. P. de Jesus[1](✉)

[1] INEGI/FEUP, University of Porto, Rua Dr. Roberto Frias, 4200-465 Porto, Portugal
ajesus@fe.up.pt

[2] IDMEC, Instituto Superior Técnico, Av. Rovisco Pais 1, 1049-001 Lisbon, Portugal

Abstract. Metallic additive manufacturing is a trending topic of manufacturing, being nowadays intensively investigated due to its innumerous advantages, such as design freedom. Some challenges remain, namely the need to perform post-processing operations of the parts towards improved surface finishing, which in some cases may involve machining operations. In addition, in some industries, the compatibility of additively manufactured inserts is assured by machining operations. Therefore, understanding the machinability of additively manufactured materials leads to timely research. This paper presents research on metal cutting supported by orthogonal cutting operations, aiming at investigating the machinability of the additively manufactured 18Ni300 maraging steel. Material build direction and tool rake angle were investigated. In addition, conventional material was tested for comparison purposes. Cutting loads, specific cutting pressure, shear angle, friction and chip geometry are evaluated according to Merchant theory. Despite the higher flow stress and anisotropic behaviour of the additively manufactured steel, their specific cutting pressure is lesser influenced by the metallurgical condition than the geometric effect of the cutting tool (rake angle).

Keywords: Additive manufacturing · LPBF · Maraging steel · Machinability · Specific cutting pressure

1 Introduction

Formerly known as rapid prototyping, additive manufacturing (AM) has shifted its focus towards industrial applications rather than the development of isolate non-functional prototypes, over the last decade. Originally limited to a group of polymers, it quickly evolved to metal processing. The current range of metallic material options combined with the possibility of attaining intricate geometrical features (some of them impossible, otherwise) has attracted the industry.

The compelling advantages of AM and its development as manufacturing process have contributed to the fact that most of the fabricated parts will serve as a functional

J. Machado et al. (Eds.): icieng 2021, LNME, pp. 1–13, 2022.
https://doi.org/10.1007/978-3-030-79165-0_1

component or be integrated in an assembly. Figure 1a displays the most common application purposes of AM, while Fig. 1b shows the substantial increase of metallic additive manufacturing (MAM) portrayed by the number of sold machines over time. It is, however, important to note the need for post-processing of metallic printed parts. These are fabricated on a build plate, sometimes with support structures that are required to be removed. Heat treatment for stress relief and hot isostatic pressing (HIP) may also be required for some parts. Moreover, the surface quality of the generated near-net shapes is low, imposing the need for finishing machining operations [1]. In order to ensure functionality and allow for assemblage, metallic AM parts require demanding dimensional tolerances and high surface quality, which is not yet compatible with the additive process. Features such as precision and threaded holes as well as reference contact surfaces are exclusive of subtractive processes, which has resulted in the appearance of hybrid manufacturing [2]. In conclusion, the benefit of AM is highly significant in areas that deal with expensive materials and high complexity parts, where a considerable amount of machining is required [3].

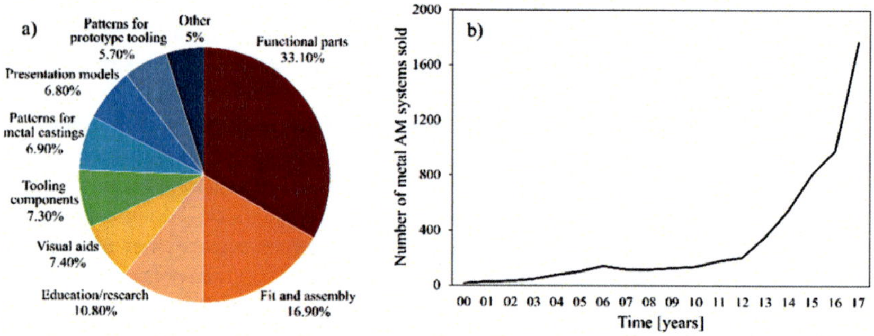

Fig. 1. Application purpose of AM (a) and number of metallic AM systems since 2000 (b), adapted from [4].

For the exposed reasons, to understand the machinability of additively manufactured metals is mandatory in order to allow the best post-processing techniques of such metallurgical condition of the alloys. It is of great value to investigate if conventional wrought/rolled/laminated metallurgical conditions exhibit significant differences with respect to the additive manufactured ones. Therefore, the present work presents an experimental research composed of orthogonal cutting tests performed on conventional (CM) and additively manufactured (AM) maraging steels (18Ni300). The results of instrumented orthogonal cutting tests are analyzed based on Merchant orthogonal cutting model [5]. One of the main outputs of the cutting tests is the specific cutting pressure defined as:

$$K_c = \frac{F_c}{bt_0}, \tag{1}$$

where K_c is the specific cutting pressure, F_c is the cutting force, b is the cutting width and t_0 is the uncut chip thickness. Also, knowing the cutting and feed force components, one may compute the friction coefficient between the chip and the tool rake angle:

$$\mu = \tan\ \beta = \frac{F_c \tan\alpha + F_f}{F_c - F_f \tan\alpha}, \quad (2)$$

where μ is the friction coefficient, β is the friction angle, F_c is the cutting force component, F_f is the feed force component, and α is the tool rake angle. Resorting to the measured chip thickness ratio, one may compute the shear angle:

$$\varphi = \tan^{-1}\frac{r_c \cos\alpha}{1 - r_c \sin\alpha}, \quad (3)$$

where r_c is chip thickness ratio defined as:

$$r_c = \frac{t_0}{t_c}, \quad (4)$$

and t_0 is the uncut chip thickness and t_c is the actual chip thickness. One of the main contributions of Merchant work was to propose an alternative to Eq. (3) to compute the shear angle, based on minimum energy theory:

$$\varphi = \frac{\pi}{4} - \frac{1}{2}(\beta - \alpha) \quad (5)$$

Besides cutting tests, this study compares the compression behavior of the alloy in its different metallurgical variations. Compression tests are very often used to identify constitutive models for metal cutting simulation.

2 Experimental Details

2.1 Materials

The material selected for this research is the 18Ni300 maraging steel in both conventional and additive manufactured forms. The AM material was processed using Laser Powder-Bed Fusion (LPBF). An ytterbium laser with a spot size of 70 μm was used to process the material samples with the following parameters: laser power, P_L = 400 W; scan speed, v_c = 0.86 m/s; hatch spacing, h_s = 95 μm; and average powder size, d_s = 40 μm; powder size distribution, PSD (d_{10}/d_{90}) = 15/45 μm. As-built material was used in the tests. Figure 2 displays the 18Ni300 build plate, where different groups of specimens (randomly distributed) can be seen. Blocks with the red dots were used for orthogonal cutting tests and cylinders with blue dots were used for compression tests aiming at characterizing the flow stress behavior of the material under quasi-static conditions. Table 1 summarizes the chemical composition of the 18Ni300 steel and reference values for conventional material according to MIL-S-46850D [6]. This steel is a Ni-Co-Mo iron alloy with capabilities of age hardening by precipitation.

2.2 Orthogonal Cutting Tests

Orthogonal cutting tests are often used to portray more intricate real machining operations such as turning or milling. It is important to note that this setup is closer to an experimental methodology for characterizing materials towards metal cutting operations (more representative than, for example, compression tests) rather than an actual machining operation. In the present work, orthogonal cutting tests were performed with the aim of measuring cutting loads and infer on its eventual difference among the distinctly manufactured maraging steels. In the AM samples, two distinct cutting directions were contemplated: parallel and perpendicular to build direction. Friction at the tool-chip interface in orthogonal cutting conditions are subject of study. The influence of tool geometry influence (namely rake angle) has also been a studied parameter in the present experimental procedure. There are multiple possible setups for performing orthogonal cutting tests. Facing turning operations of a circular workpiece [7] and cutting off a pre-machined workpiece with ribs/disk [8] are the most common setups on a lathe. Similar configurations have been developed in milling setup [9, 10]. In these setups, the uncut chip thickness is developed in a continuous way as a result of a predefined feed value.

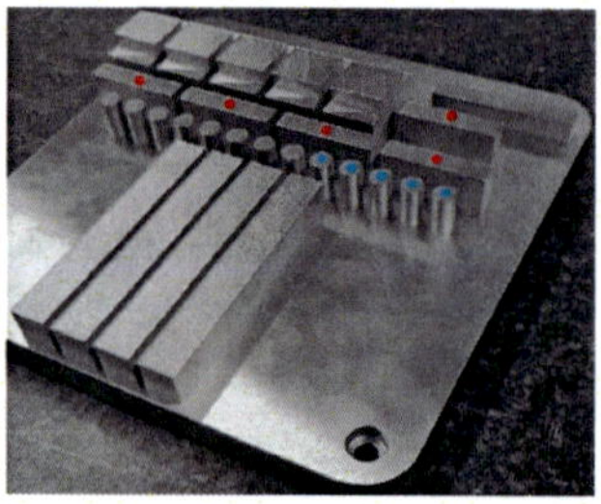

Fig. 2. 18Ni300 as-built material samples attached to the build plate.

Table 1. Chemical composition (wt%) of 18Ni300 steel with respective reference values [6].

	Ni	Co	Cr	Mo	Ti	Si	Mn	C	P	S
Reference, min	18.0	8.5	–	4.6	0.5	–	–	–	–	–
Reference, max	19.0	9.5	–	5.2	0.8	0.10	0.10	0.03	0.01	0.01
AM material	18.80	8.84	–	5.15	0.65	0.05	0.03	0.02	< 0.001	< 0.001

In this study, orthogonal cutting instrumented experiments were performed on a mechanical shaping machine. Despite being industrially outdated, this type of machine tool is quite relevant towards the experimental representation of the orthogonal cutting phenomenon due to its robustness and high rigidity that translates into the capability of sustaining high cutting loads in close speed conditions to cutting. In addition, it presents some advantages when compared to the turning and milling setups for orthogonal cutting due to the simplified kinematics and possibility to apply fixed uncut chip thicknesses. On the limitations side, one may refer the cutting speed limits and impossibility to cover a wide range of cutting speed values. Figure 3 illustrates the mechanical shaping machine, which is essentially a quick-return mechanism. The machine allows different ram strokes with the cutting speed varies within the stoke. Nevertheless, the cutting speed for larger stokes could be stable in the central part of the stroke around 25.8 m/min, which as used in the cutting tests.

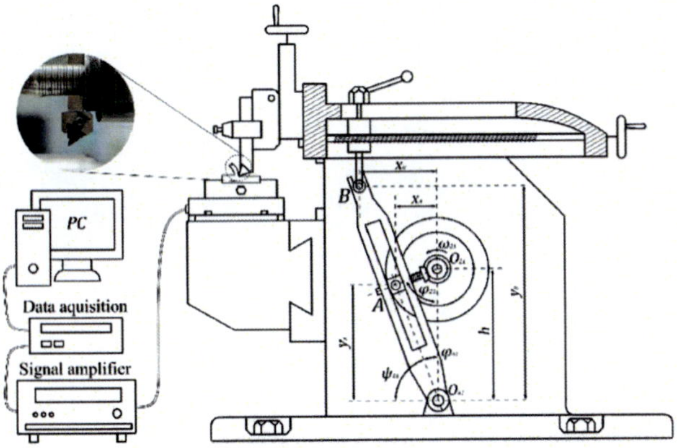

Fig. 3. Schematic representation of the shaping machine orthogonal cutting setup, its kinematics and nomenclature.

Load measurement was performed through a three-component piezoelectric dynamometer (Kistler 9257B). A charge amplifier (Kistler 5070A) and a data acquisition system (Advantech 4711A) enabled signal conversion and data collection, which was further analyzed using typical processing software. The printed raw geometries intended for orthogonal cutting tests were cut through wire-EDM in their longitudinal and transverse directions. This way, a single printed block results in eight orthogonal cutting specimens. All final specimens were ground to 1mm thickness. Two cutting directions were tested: perpendicular to build direction (with a cutting length of 24 mm) and parallel to build direction (cutting length of 19 mm). Identical specimens were prepared for the conventional maraging steel. Figure 4 shows the initial printed raw geometry and the 8 final specimens. Specially-designed inserts with three available cutting edges of 4 mm in length were used in the orthogonal cutting tests, manufactured by Palbit S.A. The inserts were mounted on a tool holder with a V pocket with their inverse geometry (ensuring rigidity), which in turn was fixed to the shaping machine ram. Figure 5 illustrates the two tested tool geometries, which differ solely in rake angle. A physical vapor deposition (PVD) AlTiN layer was applied as coating (PH7920 grade), as it improves tool life, workpiece surface finish and decreases cutting loads when compared to uncoated carbides, mostly due to the enhanced friction conditions at the tool-workpiece interface [11, 12].

In order to ensure the suitability of the orthogonal cutting tools, preliminary tests were performed to check the integrity of the cutting edge, rake angle and overall structural strength. For that purpose, an optical microscopy inspection of the tool after each successive test was conducted. No changes in the geometry were observed, especially the cutting edge that remains without any wear signals. However, as shown in Fig. 6 chip material seems to accumulate on the rake face after the first cut. In order to avoid the influence of build-up material on the tool's rake face, an unused rake surface was used for every test.

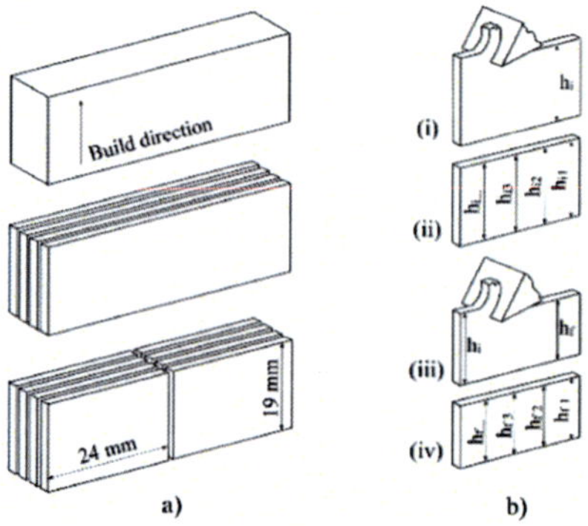

Fig. 4. Orthogonal cutting specimen preparation thru wire-EDM cutting (a) and uncut chip thickness measurement procedure from height measurements before and after cut (b).

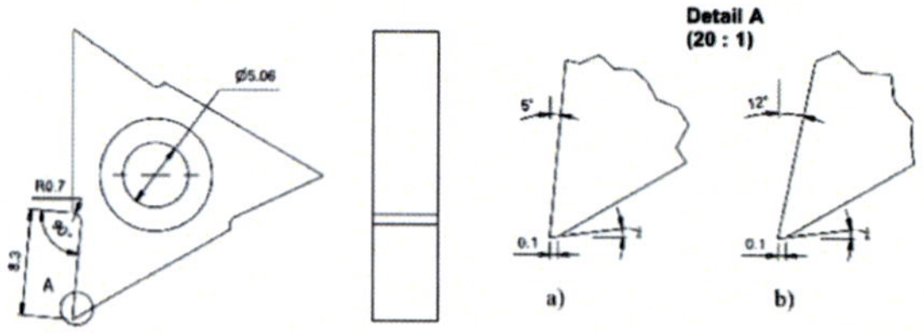

Fig. 5. Insert geometry used in the orthogonal cutting tests with (a) 5° and (b) 12° rake angle.

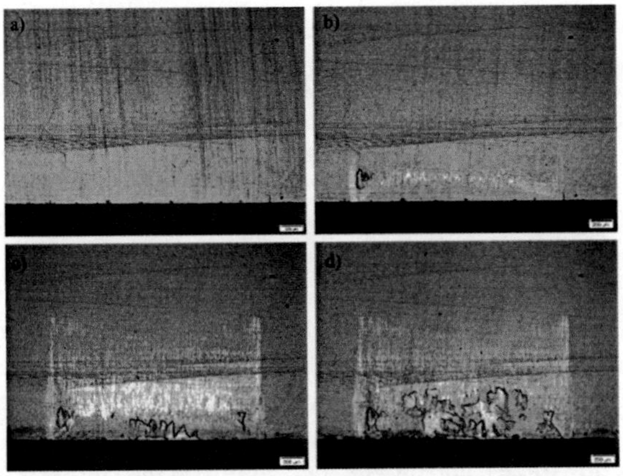

Fig. 6. Tool rake face ($\alpha = 5°$) optical microscopy after successive orthogonal cutting tests (t_0 ~ 0.1 mm): (a) untested geometry; (b) after 1 test (c) after 2 tests and (d) after 3 tests.

3 Results and Discussion

3.1 Flow Stress Characterization Based on Compression Test Data

The flow stress was evaluated for different metallurgical conditions of the 18Ni300 steel. The obtained results from the instrumented compression tests in quasi-static conditions are displayed in Fig. 7. The influence of metallurgical condition and build direction are evidenced on the stress-strain response. The higher mechanical strength of the AM metallurgical condition is consistent with the typically finer microstructure of these alloys. The 18Ni300 maraging steel presents identical work hardening (almost perfect plastic behavior) and a smaller yield stress difference (CM is approximately 10% lower than AM). A slight anisotropic behavior is also noticeable in the AM material. The highest mechanical strength occurs for compression in perpendicular direction to build, which can be explained by the higher grain boundary density in that specific loading path.

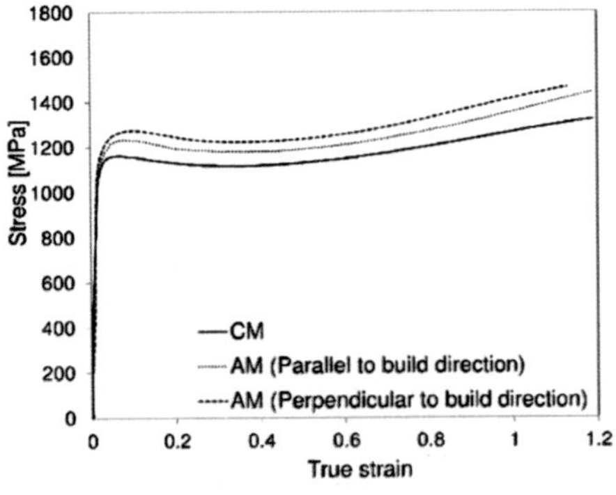

Fig. 7. Influence of metallurgical condition on the stress vs. strain compression behavior.

3.2 Orthogonal Cutting Test Results

The developed experimental setup successfully enabled orthogonal cutting conditions for a wide range of uncut chip thicknesses. Figure 8 shows an example of the obtained cutting loads of an orthogonal cutting test for an average uncut chip thickness of 0.165 mm. Both transient at starting of chip formation (for cutting length < 5 mm) and steady-state metal cutting conditions can be clearly identified. The latter were used for the analysis of cutting load on the distinct maraging steel samples. While the cutting force component, F_c, achieves a stabilized plateau, the feed force, F_f, shows some trend to increase along the test which may be related with friction coefficient evolution. Figure 9a shows the experimental measurements of cutting load (F_c), using a cutting tool with a 5° rake angle. A markedly linear increase of the cutting force with the increase of uncut chip thickness is found, regardless of the metallurgical conditions or cutting direction (CD). Differences in cutting force can be found, depending on the cutting direction (relatively to build direction). Despite its higher significance for bigger uncut chip thicknesses, the orthogonal cutting tests performed with the tool movement perpendicular to build direction resulted in bigger cutting forces. In contrast, lower loads seem to result from the orthogonal cutting tests performed in conventionally manufactured maraging steel. This trend shows the strong compatibility between F_c magnitude and the mechanical strength of the cut material. The influence of tool geometry on the measured cutting loads (illustrated in Fig. 9b) is in accordance with the expected higher loads for smaller rake angle [13]. Interestingly, it has a higher influence on the cutting load than the material's build direction.

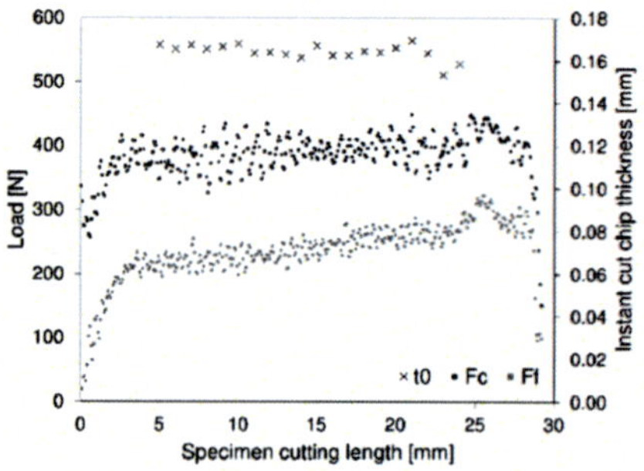

Fig. 8. Experimental cutting loads for orthogonal cutting with $t_0 = 0.165$ mm.

As discussed in literature review, specific cutting pressure (K_c) is commonly used for load estimation purposes, according to the selected cutting parameters, and these curves are usually obtained from experimental procedures. Figure 10 presents the K_c curves obtained from orthogonal cutting experiments. The results show a significant increase in K_c for a decrease in the chip section and K_c stabilization for t_0 increase. These are typical trends associated with the size effect in metal cutting. The occurrence of defects in metals (e.g. grain boundaries and missing atoms creating vacancies in gains' structure) is generally acknowledged and the chance of encountering those stress-reducing defects decreases for smaller sizes of removed material [14]. Taniguchi [15] has conducted a

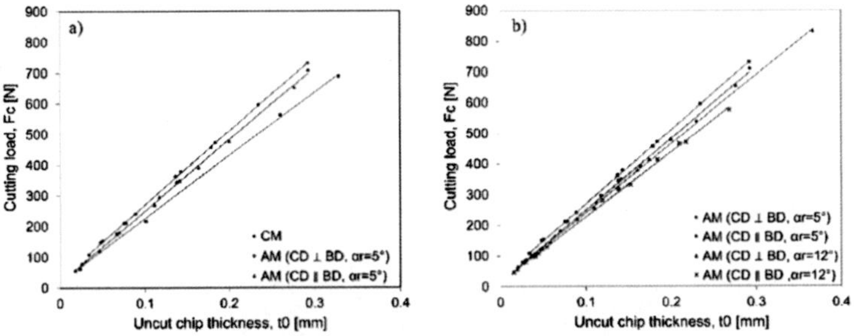

Fig. 9. Influence of metallurgical condition (a) and rake angle (b) on the cutting force for maraging steel in orthogonal cutting conditions.

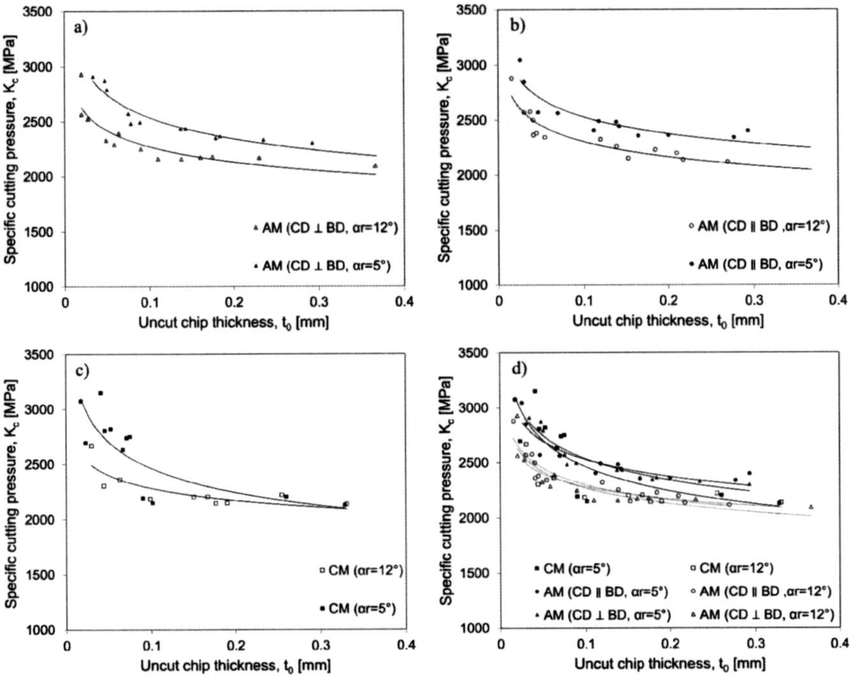

Fig. 10. Influence of rake angle on the specific cutting pressure of AM 18Ni300 cut in perpendicular direction to BD (a), parallel direction to BD (b) and the CM 18Ni300 (c). Comparison of all K_c results, showing prevailing influence of rake angle (d).

series of experimental tests for increasing size of cut (or deformed) material, finding that the value of K_c tends to the materials' mechanical strength, typically obtained through tensile testing. The same effect seems to be identified for the maraging steel as the maximum strength values obtained from compression testing (in simultaneous speed and temperature) are approximately 1900 MPa, regardless of metallurgical condition.

Moreover, K_c values seem to be very similar whether the material is additively or conventionally manufactured, as illustrated in Fig. 10d. No obvious distinction can be identified for different cutting directions relatively to build direction. A more notorious distinction is noticed from rake angle variation. Tool geometry shows a significant influence on the specific cutting pressure of all tested maraging steel conditions, as depicted from Fig. 10a to Fig. 10c.

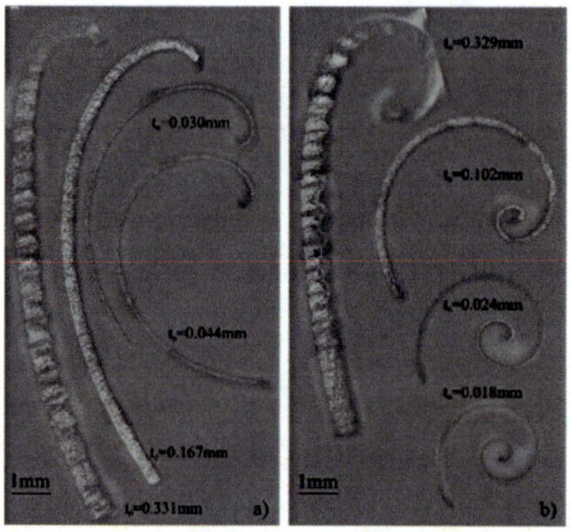

Fig. 11. Chip morphology of CM maraging steel obtained from orthogonal cutting tests with a 12° (a) and 5° (b) tool rake angle, for distinct uncut chip thickness.

Concerning chip morphology, no substantial differences were found for the distinctly manufactured maraging steels. On the other hand, a smaller tool rake angle seems to promote a higher chip curvature (smaller chip radius), which goes in line with literature [16]. Moreover, increasing uncut chip thickness also promotes an increase of chip curvature radius. It has been observed that for $t_0 \sim 0.3$ mm chip geometry shifts from continuous to lamellar morphology, likely due to a localized shearing zone promoted by the high strains, strain rates, temperature and, mostly, the low thermal conductivity of maraging steel, since for smaller t_0 there is higher heat diffusion. Figure 11 shows the chip morphology for different t_0 and both tool rake angles on the conventionally manufactured maraging steel. These images were obtained through optimal microscopy, enabling the measurement of chip thickness and calculation of compression chip ratio.

Friction coefficient was calculated through Eq. (2) for all types of maraging steel. Results show a slight tendency for higher friction coefficient as t_0 decreases (refer to Fig. 12a). A possible reason for such effect might be a progressive change of the friction mechanism, for smaller t_0. It is well known from literature that, in the metal cutting process, there are three basic components of sliding friction: adhesion, ploughing and plastic deformation [17]. For smaller t_0 the relative size of the cutting edge becomes more significant, which may result in an increased contribution of that friction component, thus increasing the overall friction values. It is also important to emphasize the very high friction values (higher than the $\mu = 0.577$ theoretical limiting value) obtained from cutting force measurement methods. Despite decades of research, friction in metal cutting is still a controversial topic, and despite its apparently correct theoretical calculation

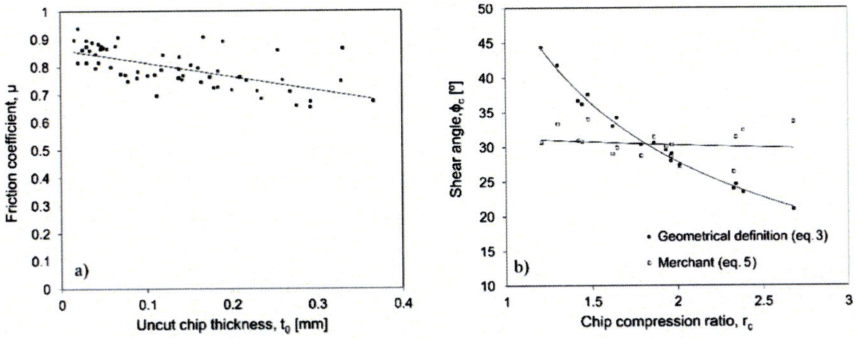

Fig. 12. Friction coefficient results calculated through load ratio of Eq. (2), for all metallurgical conditions (a) and shear angle predictions using geometrical and Merchant's metal cutting theory (b).

through the measured cutting loads, the experimental values keep raising challenging questions to researchers. Shear angle was calculated through the geometrical description of metal cutting, depicted in Eqs. (3) and (4) and Merchant's metal cutting theory, Eq. (5). Results are illustrated in Fig. 12b. Higher values of chip compression ratio occurred for chips with smaller uncut chip thickness. This is in accordance with the theory that a more intense plastic deformation and ploughing effect develop in such t_0 conditions, meaning that the measured chip thickness is potentially influenced by those effects. Shear angle evolution is rather distinct, depending on the calculation method. Merchant's model seems to underestimate φ values for realistic chip compression ratios. Despite the popularity of the model, the mismatch of Merchant's theory has been noticed by several authors [18].

4 Concluding Remarks

The following concluding remarks are outlined from this work: i) 18Ni300 additively manufactured steel shows higher flow stress values (as-built condition) than the conventionally produced material; ii) The strain hardening of the material is reduced. A small degree of anisotropic behavior was verified for the AM material, with the perpendicular to build direction showing higher strength; iii) As regards the cutting tests, those performed with the tool movement perpendicular to build direction resulted in higher cutting forces. In contrast, lower loads seem to result from the orthogonal cutting tests performed in conventionally manufactured maraging steel which is consistent with their lower flow stresses; iv) Concerning the specific cutting pressure values, the tool rake angle had the major role. Scale effects on specific cutting pressures were observed for the smaller uncut chips thicknesses; v) Serrated chips were observed mostly for the highest tested chip thicknesses with is typical of these materials; vi) The computed friction coefficients were always above the limit theoretical value for this parameter of 0.577, which shows limitations of the Merchant theory. Therefore, the computed values should be regarded as apparent values and not true ones; vii) As regards the experimental shear angle evaluations, it demonstrated the inadequacy of the Merchant equation, showing a strong correlation with the chip compression ratio.

Summarizing, the as-build AM material shows higher strength properties than the conventional counterpart, which make it a good candidate for engineering components. Its machinability is not significantly affected, besides the strength increase; nevertheless, the possibility of generating serrated chips may endanger the surface finish of post-processing operations.

Acknowledgments. This work has been conducted under the scope of MAMTool (PTDC/EME-EME/31307/2017) and AddStrength (PTDC/EME-EME/31307/2017) projects, funded by Programa Operacional Competitividade e Internacionalização, and Programa Operacional Regional de Lisboa funded by FEDER and National Funds (FCT). Support of PALBIT SA is also fully acknowledged. This work was also supported by FCT, through IDMEC, under LAETA, project UIDB/50022/2020.

References

1. Fox, J.C., Moylan, S.P., Lane, B.M.: Effect of process parameters on the surface roughness of overhanging structures in laser powder bed fusion additive manufacturing. Procedia CIRP **45**, 131–134 (2016)
2. Flynn, J.M., Shokrani, A., Newman, S.T., Dhokia, V.: Hybrid additive and subtractive machine tools – research and industrial developments. Int. J. Mach. Tools Manuf. **101**, 79–101 (2016)
3. Froes, F., Boyer, R.: Additive Manufacturing for the Aerospace Industry, 1st edn Elsevier, Amsterdam, Netherlands (2019)
4. Campbell, I., Diegel, O., Kowen, J., Wohlers, T.: Wohlers report 2018: 3D printing and additive manufacturing state of the industry: annual worldwide progress report. Wohlers Associates. Technical report (2018)
5. Merchant, M.E.: Mechanics of metal cutting process. I – orthogonal cutting and type 2 chip. J. Appl. Phys. **16**(5), 267–275 (1945)
6. U.S. Department of Defense. MIL-S-46850D – Steel: Bar, Plate, Sheet, Strip, Forgings, and Extrusions, 18 Percent Nickel Alloy, Maraging, 200 KSI, 250 KSI, 300 KSI, and 350 KSI, High Quality (1991)
7. Markopoulos, A.P.: Finite Element Method in Machining Processes. Springer Science & Business Media, Berlin (2012)
8. Outeiro, J.C., Costes, J.-P., Kornmeier, J.R.: Cyclic variation of residual stress induced by tool vibration in machining operations. Procedia CIRP **8**, 493–497 (2013)
9. Daoud, M., Jomaa, W., Chatelain, J.F., Bouzid, A.: A machining-based methodology to identify material constitutive law for finite element simulation. Int. J. Adv. Manuf. Technol. **77**(9–12), 2019–2033 (2014)
10. Wang, B., Liu, Z., Su, G., Ai, X.: Brittle removal mechanism of ductile materials with ultrahigh-speed machining. J. Manuf. Sci. Eng. 137(6), 061002 (2015)
11. Varghese, V., Akhil, K., Ramesh, M.R., Chakradhar, D.: Investigation on the performance of AlCrN and AlTiN coated cemented carbide inserts during end milling of maraging steel under dry, wet and cryogenic environments. J. Manuf. Process. **43**, 136–144 (2019)
12. Endrino, J.L., Fox-Rabinovich, G.S., Gey, C.: Hard AlTiN, AlCrN PVD coatings for machining of austenitic stainless steel. Surf. Coat. Technol. **200**(24), 6840–6845 (2006)
13. Günay, M., Aslan, E., Korkut, I., Şeker, U.: Investigation of the effect of rake angle on main cutting force. Int. J. Mach. Tools Manuf. **44**(9), 953–959 (2004)

14. Shaw, M.C.: The size effect in metal cutting. Sadhana **28**(5), 875–896 (2003)
15. Taniguchi. N.: The art of nanotechnology. PhD thesis, Seattle (1993)
16. Devotta, A., Beno, T., Löf, R., Espes, E.: Quantitative characterization of chip morphology using computed tomography in orthogonal turning process. Procedia CIRP **33**, 299–304 (2015)
17. Melkote, S.N., et al.: Advances in material and friction data for modelling of metal machining. CIRP Ann. **66**(2), 731–754 (2017)
18. Uhlmann, E., et al.: An extended shear angle model derived from in situ strain measurements during orthogonal cutting. Prod. Eng. Res. Devel. **7**(4), 401–408 (2013)

Machining Process Time Series Data Analysis with a Decision Support Tool

Katarzyna Antosz[1](✉), Dariusz Mazurkiewicz[2], Edward Kozłowski[3], Jarosław Sęp[1], and Tomasz Żabiński[4]

[1] Faculty of Mechanical Engineering and Aeronautics, Rzeszow University of Technology, Powstańców Warszawy 8, 35-959 Rzeszów, Poland
{katarzyna.antosz,jsztmiop}@prz.edu.pl

[2] Mechanical Engineering Faculty, Lublin University of Technology, Nadbystrzycka 36, 20-618 Lublin, Poland
d.mazurkiewicz@pollub.pl

[3] Faculty of Management, Lublin University of Technology, Nadbystrzycka 38, 20-618 Lublin, Poland
e.kozlovski@pollub.pl

[4] Faculty of Electrical and Computer Engineering, Rzeszow University of Technology, W. Pola 2, 35-959 Rzeszów, Poland
tomz@prz.edu.pl

Abstract. Dynamic industrial data growth necessitates the development of several new concepts of these data analysis that will allow to select not only the right data, but also to apply appropriate methods in order to extract knowledge from them. For this purpose, the possible use of decision trees as a decision support tool for a machining process data analysis was discussed in this article. With the use of the generated decision rules, we identify parameters that affect the state of a blade (blunt, sharp). In consequence that makes it possible to predict its future state at specific values of the identified parameters. Decision trees enable the analyses of the importance of each variable for the dependent variable. This makes it possible to analyse how each individual parameter and the relationships between them affect the condition of a cutter blade. The results of variables importance for a decision tree analysis can be used to determine the most important input variables, while rejecting those which do not affect the condition of a cutter blade. The study offers some promising results. It is confirmed by the achieved prediction model quality indicators.

Keywords: Decision trees · Identification of cutter state · Machining process analysis

1 Introduction

It is well known that the data analysis and use of Big Data play a major role in the current Industry 4.0 challenges and applications. In Industry 4.0 vision, data processing is expected to bring major changes in manufacturing in general, and the spread of novel

J. Machado et al. (Eds.): icieng 2021, LNME, pp. 14–27, 2022.
https://doi.org/10.1007/978-3-030-79165-0_2

technologies will enable a stepwise increase of productivity in manufacturing companies. However, barriers to successful digital transformation still exist [1–5]. Among several of them, one can point to heterogeneous data streams that cannot be well processed in order to realize the automated decision support, due to the lack of strong analytic capabilities. Sophisticated tools, extensive knowledge and experience required pose an important challenge. Despite the promising results [1, 6–9], there are still many factors which prevent an effective implementation of expert systems that support maintenance services or system operators. An effective analysis of the collected industrial data can significantly improve a decision making process in enterprises. Using artificial intelligence systems is an effective tool for parameter identification of complex manufacturing systems [10]. In order to cope with dynamic data growth, it is necessary to develop concepts of their analysis that will allow to select not only the right data, but also to apply appropriate methods in order to extract knowledge from them. For this purpose, the possible use of decision trees as a decision support tool for a machining process data analysis was discussed in this article. Time series data from the real industrial milling machine were used to verify this modelling method effectiveness while identifying the state of a cutting tool.

2 Research Problem Formulation

The primary aim of the presented research is to identify the cutter state based on the data (instrument readings, signal readouts) obtained from sensors (accelerometers, microphones). In our case the identification consists in determining whether the cutter is sharp or not. The outcome variable has a categorical value in the case considered. Thus, the presented task is related to a classification problem. Numerous techniques (classifiers) might be used to solve a classification problem. In the research presented below for the cutter state identification, the classification trees were applied. In order to create a classification tree, the domain of input variables (a set of possible realizations of observations) should be divided into k separated regions. Each region corresponds to an appropriate categorical value (from a set of possible outcomes) of a response variable. The definition of region consists in the creation of conditions which can be satisfied by input variables and it is called a rule. A set of rules determines a classification tree.

3 Decision Trees

Decision trees are a method of data modeling that can be used to solve both classification and regression problems. This method allows to perform analyses that lead to finding a set of logical rule conditions, type, "if, then", helping to classify the examined objects clearly. In data mining and machine learning, decision trees are considered as predictive models. This is one of the most popular and effective methods of data mining which is very often used for prediction. Classification trees are created when a dependent variable is qualitative, and regression trees – with a continuous form of a dependent variable. Classification trees are used to determine the belonging of objects to classes, based on the measurements of one or more describing variables, determining their impact on a qualitative dependent variable – the forecast (predicted) variable.

Prediction can be understood as a model which can be used to estimate (calculate) the value (or a range of values) of an attribute. The value of this attribute may be, in particular, a class label. Classification trees help to generate the rules in very complex multidimensional cases as opposed to human abilities. The received rules are usually presented in the form of a tree, are transparent even in the case of large trees [9]. The most commonly partition algorithms used in decision trees are: the TDIDT, CART, CHAID and C.4.5 algorithms. However, the most advanced method of building classification and regression trees is the CART algorithm (*Classification and Regression Trees*) proposed by Breiman [12]. A characteristic feature of this method is excessive tree growth and pruning of individual nodes to reduce the description of leaves (with a slight increase of classification error). This algorithm allows to compare the extended model and the model with a reduced number of nodes. The construction of a decision tree is carried out by conducting an in-depth search of all available variables and all possible splits in a data set for each decision node (t) by selecting an optimal split [13]. $\{(y_i, x_i)\}_{1 \leq i \leq n}$ denotes an analysed data set, where $y_i \in \{c_1, c_2, \ldots, c_s\}$ and $x_i = (x_{i1}, x_{i2}, \ldots, x_{ik}) \in R^k$. The values $c_1, c_2, \ldots, c_s$ denote possible classes for the feature y. The classification task consists in dividing space R^k on q separated regions, where each region corresponds to an appropriate class. Based on the observation feature $x_i = (x_{i1}, x_{i2}, \ldots, x_{ik})$ we need to classify the analysed object.

The authors of the algorithm suggest to use the Gini index, also called the node pollution measure (or impurity measure). The entire space R^k is divided into q separated regions, $R_1 \cup R_2 \cup \ldots \cup R_q = R^k$. For the node m, $1 \leq m \leq q$, representing region R_m, the Gini index is determined as follows (1):

$$Q_G(m) = \sum_{j=1}^{s} p_{mi}(1 - p_{mi}) = 1 - \sum_{j=1}^{s} p_{mi}^2, \tag{1}$$

where p_{mi} is a conditional probability for j-th class in a node, s – a number of classes. In node m with n_m observations the conditional probability for j-th class is equal (2):

$$p_{mi} = \frac{\#\{y = c_i : x \in R_m\}}{n_m}. \tag{2}$$

The qualitative assessment of the generated decision trees is carried out through a k-fold cross validation and confusion matrix. A confusion matrix shows classification errors by class. Based on the confusion matrix, numerical indicators can be calculated. They are determined as follows:

$$Accuracy = \frac{TP + TN}{TP + TN + FP + FN}, \tag{3}$$

$$True\ Positive\ Rate = Sensivity = \frac{TP}{TP + FN}, \tag{4}$$

$$Specificity = 1 - False\ Positive\ Rate = \frac{TN}{TN + FP}, \tag{5}$$

$$Positive\ Predictive\ Value = \frac{TP}{TP + FP}, \tag{6}$$

$$Negative\ Predictive\ Value = \frac{TN}{TN + FN}, \tag{7}$$

$$Prevalence = \frac{TP + FN}{TP + TN + FP + FN}, \tag{8}$$

$$Detection\ Rate = \frac{TP}{TP + TN + FP + FN}, \tag{9}$$

$$Detection\ Prevalence = \frac{TP + FP}{TP + TN + FP + FN}, \tag{10}$$

$$Balanced\ Accuracy = \frac{Sensivity + Specificity}{2}, \tag{11}$$

$$False\ Alarm\ Rate = \frac{FP}{TP + FP}. \tag{12}$$

The most important is *Accuracy*, which determines the prediction ability of a decision tree. These indicators were presented in detail and discussed in the works of [14–16].

Decision trees are widely used in the analysis of a large number of significant problems – from a signal analysis to a medical analysis. The results of their applications in different areas show their many advantages such as:

- they are understandable, they create a transparent and easy to understand data representation,
- they are more effective than numerically oriented methods when used to analyse the categorized variables,
- they identify the importance of variables and relationships between them,
- they do not require special data preparation for the analysis,
- they can effectively (relatively quickly) work even on large data sets.

4 Numerical Example

For the formulated research problem, a decision tree was created. The main goal was to obtain:

- a model that explains the changes in the condition of a cutter blade depending on the values of the measured parameters,
- a set of decision rules that can be used to predict the state of a cutter for the given values of the measured parameters,
- ranking of the importance of variables, presenting information about the parameters which mostly affect the cutter blade state.

In the algorithm creating a binary decision tree, the criteria that affect the size of a tree and its adaptation to learning data were adopted. The decision tree was created in the R *(rpart)* program. The following tree construction criteria were used: classification tree (method = 'class') and the following parameters that control the tree building procedure: a complexity parameter (*cp* = 0.001), a minimum number of observations that have to exist on the node in order to be able to attempt a split (*minsplit* = 5), a number of variables competing at the output (*maxcompete* = 50), a number of surrogate variables (*maxsurrogate* = 10), a method of determining which surrogate variables will be used (*usesurrogate* = 2), a number of cross- validation (*xval* = 10) and a maximum depth of a tree (*maxdepth* = 7). A decision tree was generated for such defined parameters. Additionally, for every observation the elements of a learning dataset were determined: for a blunt cutter $y = 1$ and for a sharp cutter (no warnings) y = 0 were adopted.

The experiments were conducted in the laboratory testbed consisting of Haas VM-3 CNC machine equipped with an inline direct-drive spindle and a set of sensors: 7 accelerometers integrated with temperature sensors, 1 acoustic emission sensor and 1 force and torque sensor. A four-blade, solid carbide milling cutter with TiAln coating (Aluminum Titanium Nitride) and a spindle speed of 860 rpm were used for the tests. Two signals, AccSignal1 and P2y collected during milling experiments conducted on Haas VM-3 CNC machine, were analysed. An acceleration signal (AccSignal1) was collected from the acceleration sensor mounted on a lower bearing of the machine in-line direct-drive spindle. P2y is a force signal parallel to the Y-axis which was collected from the force and torque sensor mounted in the chuck. During the experiments, the data from the sensors were collected using a platform for rapid prototyping of intelligent diagnostic systems [17] composed of Beckhoff Industrial PC (IPC C6920), TwinCAT 3, Matlab/Simulink projects and a distributed input/output system based on EtherCAT protocol. IPC was used for signal acquisition and communication with a PC computer equipped with Matlab/Simulink system, which performed data collection in External Mode. The PC computer received data from IPC and stored them as binary Matlab files (mat) on the hard drive. The real-time PLC module for data acquisition operated with the main sampling interval of 2 ms. The signal duration stored in one file was equal to 640 ms. The sampling interval for accelerometers and acoustic sensors was equal to 40 μs (25 kHz sampling frequency, 16 000 samples in one file). The sampling interval for force and torque signals was equal to 2 ms (500 Hz sampling frequency, 320 samples in one file). In the analog input modules EL3632 from Beckhoff, used to connect acceleration and acoustic sensors, an oversampling factor (defined as a number of probes per one main sampling interval) was set to 50. For force and torque signals the analog input modules EL3104 from Beckhoff were used. For AccSignal1, stationary properties were tested using the Augmented Dickey-Fuller (ADF) test [18, 19]. At the level of significance $\alpha = 0.01$, the working hypothesis regarding the existence of a unit root was rejected. In addition, the significance of the autocorrelation {r_τ}_(1 $\le \tau \le$ 200), using the Ljung-Box test [20, 21] was made. Using theorems of Herglotz ranks, AccSignal1 was identified by the value of an autocorrelation function. Therefore, for AccSignal1 signal to a training set, the values of the autocorrelation function for shifts $1 \le \tau \le 200$ were identified and marked as $x_1, x_2, \ldots, x_{200}$. Based on ADF test, it was shown for some P2y signals that a unit root exists. These signals were identified by means of ARIMA class models, where the autoregression order and the moving average order do not exceed 5: integration degree x_{201}, values of autoregression coefficients $x_{202}, \ldots, x_{206}$, coefficient values of the moving average $x_{207}, \ldots, x_{211}$, the value of constant term x_{212} and the variance of external disturbances x_{213}.

The first of the assessed parameters of a decision tree which provides the quality of classification trees is the size of the tree. It comes directly from the degree of tree pruning, which is responsible for the elimination of overfitting of a classifier. Less complex classifiers (with a fewer decision nodes and leaves) are more preferred. Figure 1 shows the most-developed decision tree consisting of 20 splits nodes and 22 terminal nodes. For each node, the probability value of the appearance of a result in a given class, and the percentage of the result in a data set were given. It can be concluded that according to the defined parameters, only 20 variables (out of 213) were used for the tree construction: x_1, $x_{10}, x_{12}, x_{124}, x_{133}, x_{154}, x_{159}, x_{165}, x_{175}, x_2, x_{213}, x_{28}, x_{34}, x_{38}, x_4, x_{47}, x_{75}, x_{81}, x_{82}$ and x_{89}. These variables have the largest discrimination ability, i.e. they define the split of a data set into individual segments (terminal nodes). When the structure of the presented decision tree is analysed, it should be noted that its first and most important split (into classes marked with blunt and sharp) was carried out according to parameter $x213$. In the blunt class 34% of the results of a data set with the probability of appearing in this class is at the level of 0.01. However, in the sharp class 66% of the results of a data set with the probability of appearance of the result in this class is 0.86. Analysing further splits of the right part of the tree, it can be seen that the next parameters differentiating further splits are $x81$, then $x2$ and $x154$. The analysis of all splits presented in the analysed tree allowed to generate 22 decision rules (22 leaf nodes) for the recognition of the state of a cutter. The chosen decision rules for class marked as sharp, are presented in Table 1. Decision rules show how each individual parameter and the relationships between them affect the condition of a cutter blade.

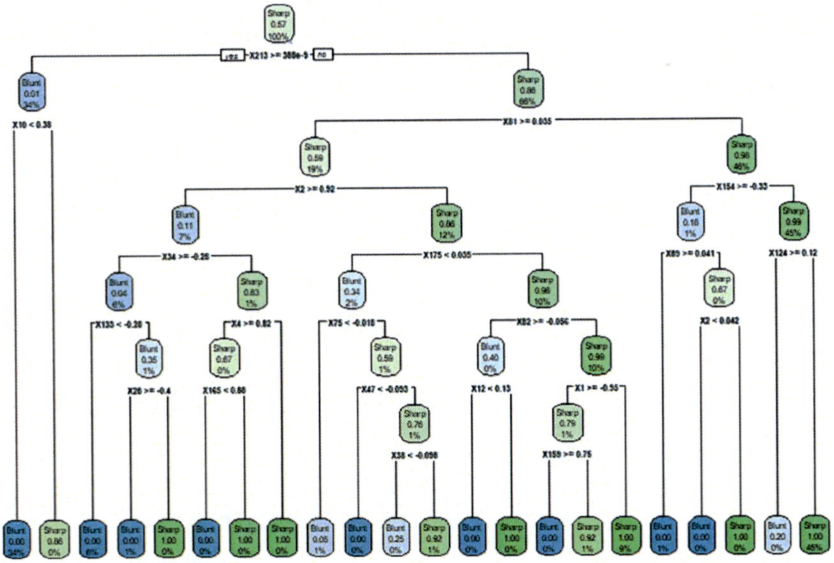

Fig. 1. Decision tree for cutter state recognition for $cp = 0.001$.

Table 1. Set of chosen rules generated for the decision tree for $cp = 0.001$

No.	Decision rules
1.	Sharp when X213 > = 3.9e−07 & X10 > = 0.38
2.	Sharp when X213 < 3.9e−07 & X81 > = 0.035 & X2 < 0.919 & X175 > = 0.035 & X82 < −0.056 & X1 > = −0.95 & X159 < 0.75
3.	Sharp when X213 < 3.9e-07 & X81 > = 0.035 & X2 < 0.919 & X175 < 0.035 & X75 > = −0.018 & X47 > = −0.093 & X38 > = −0.098
4.	Sharp when X213 < 3.9e−07 & X81 < 0.035 & X154 < −0.33 & X124 < 0.12
5.	Sharp when X213 < 3.9e−07 & X81 > = 0.035 & X2 > = 0.919 & X34 > = −0.28 & X133 > = −0.28 & X28 < −0.4
6.	Sharp when X213 < 3.9e−07 & X81 > = 0.035 & X2 > = 0.919 & X34 < −0.28 & X4 > = 0.82 & X165 > = 0.88
7.	Sharp when X213 < 3.9e−07 & X81 > = 0.035 & X2 > = 0.919 & X34 < −0.28 & X4 < 0.82
8.	Sharp when X213 < 3.9e−07 & X81 > = 0.035 & X2 < 0.919 & X175 > = 0.035 & X82 > = −0.056 & X12 > = 0.13
9.	Sharp when X213 < 3.9e−07 & X81 > = 0.035 & X2 < 0.919 & X175 > = 0.035 & X82 < −0.056 & X1 < −0.95
10.	Sharp when X213 < 3.9e−07 & X81 < 0.035 & X2 > = 0.042 & X154 > = −0.33 & X89 < 0.041

The practical use of such a model is very difficult. Therefore, in order to assess the complexity of a decision tree, cross-validation was used. Table 2 shows the results of cross-validation for all the sequences of the generated classification trees: a complexity parameter (*cp*), a number of internal nodes of the tree (*nsplit*), an error for the training data (*rel error*), a classification error of cross-validation (*xerror)* and a standard deviation (*xstd*).

Table 2. Cross-validation results for all sequences of the generated classification trees.

No	cp	nsplit	rel error	xerror	xstd
1.	0.7844184	0	1.0000000	1.0000000	0.0246382
2.	0.0640342	1	0.2155816	0.2177161	0.0145100
3.	0.0149413	3	0.0875133	0.0949840	0.0098067
4.	0.0138741	4	0.0725720	0.0917823	0.0093683
5.	0.0085379	5	0.0586980	0.0789755	0.0083472
6.	0.0053362	6	0.0501601	0.0747065	0.0081518

(continued)

Table 2. (*continued*)

No	cp	nsplit	rel error	xerror	xstd
7.	0.0048026	7	0.0448239	0.0747065	0.0084747
8.	0.0032017	9	0.0352188	0.0704376	0.0081518
9.	0.0021345	14	0.0192102	0.0672359	0.0079511
10.	0.0010672	17	0.0128068	0.0618997	0.0079511
11.	0.0010000	21	0.0085379	0.0640342	0.0086619

An optimal model size is the one which gives the smallest cross-validation error, i.e. the *xerror* value. For the analysed tree, the cross-validation error was *xerror* = 0.0640342. The graphic presentation of the cross-validation results is shown on Fig. 2.

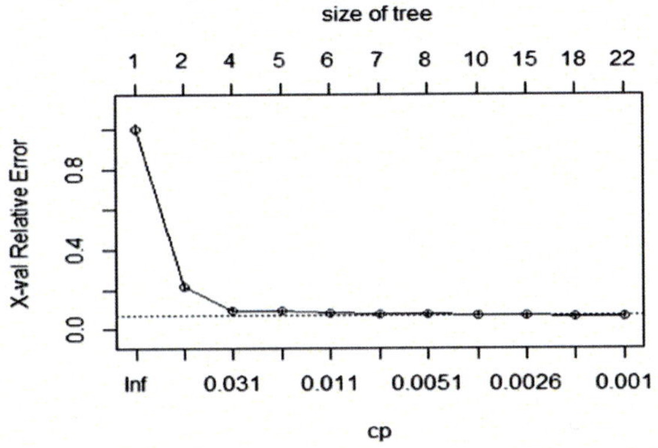

Fig. 2. The relationship of the prediction error on cp.

The analysis of the results presented in Fig. 2 allows to select the value of the parameter *cp* that the pruned tree has the best classification ability for. In order to choose *cp* parameter value, the suggested by Breiman [12] 1 SE rule was used. This rule proposes for an optimal tree to choose a tree from among those of which the total fraction of incorrect classifications is distant by no more than one standard deviation from the minimum of a fraction of incorrect classifications. Function *Plotcp(rpart) Rpacket* default defines a line at the height of one standard deviation (the minimum value of the obtained standard deviation). In the analysed case, it is a tree with *nsplit* = 7 and *cp* = *0.0048026*. The last element of the generated decision tree analysis was a confusion matrix. The generated confusion matrix is presented in Table 3, with the sharp cutter assumed as a negative case (N), while the blunt cutter is a positive case (P). The confusion matrix requires the following values to be determined: TP (True Positive) means a number of cases for which the blunt cutter was correctly recognized, TN (True Negative) – a number of cases for which the sharp cutter was correctly recognized, FP (False Positive) – a number of cases for which the blunt cutter was recognized as sharp (false alarm), FN (False Negative) – a number of cases for which the sharp cutter was recognized as blunt.

Table 3. Confusion matrix for cp = 0.001.

	Reference		
	State	Blunt	Sharp
Prediction	Blunt	932	3
	Sharp	5	1233

The confusion matrix showed that 8 (5 + 3) from n = 2173 of the analysed variants were incorrectly classified It means that the prediction error was (8/2173) ≈ 0.37%. The analyses of the indicator values of prediction model quality confirmed the prediction error value (Table 4). The Accuracy value is 0.9963, what means that the value of the prediction error is 1 − Accuracy = 0.0037(0.37%). This value indicates a very high predictive ability of the analysed tree.

Table 4. The values of indicators of prediction model quality for cp = 0.001.

Indicator	Value
Accuracy	0.9963
Sensitivity	0.9946638
Specificity	0.9975728
Positive predictive value	0.9967914
Negative predictive value	0.9959612
Precision	0.9967914
Recall	0.9946638
F1	0.9957265
Prevalence	0.4312011
Detection rate	0.4289001
Detection prevalence	0.4302807
Balanced accuracy	0.996

A characteristic feature of binary trees is their excessive growth, and the pruning ability to reduce the number of leaf nodes (with a slight increase of a classification error). This allows to compare the extended model and the model with a reduced number of nodes. Less complex classifiers (with a fewer decision and leaf nodes) are more preferred.

Using the analysis results of the relationship of the prediction error and the value of *cp* parameter (see Fig. 2), the analysed decision tree was pruned for $cp = 0.005$. On Fig. 3, the definition of the tree for $cp = 0.005$ is shown.

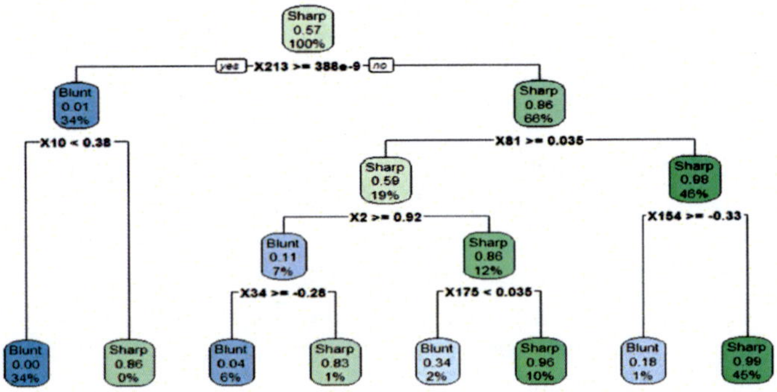

Fig. 3. Decision tree for the cutter state recognition (cp = 0.005).

It can be concluded that with the defined parameters only 7 variables (discriminating variables) from among 213: x_{10}, x_{154}, x_{175}, x_2, x_{213}, x_{34} and x_8 were used for the tree structure. This is the effect of pruning a tree according to the inherent parameter of cp, whose analysis was presented in Fig. 2. The generated tree has 7 split nodes and 8 terminal nodes (leaf nodes), and thus it generates 8 decision rules (see Table 5).

Table 5. Set of rules generated for the pruned decision tree (cp = 0.005).

No.	Decision rule
1.	Blunt when X213 > = 3.9e−07 & X10 < 0.38
2.	Blunt when X213 < 3.9e−07 & X81 > = 0.035 & X2 > = 0.92 & X34 > = −0.28
3.	Blunt when X213 < 3.9e−07 & X81 < 0.035 & X154 > = −0.33
4.	Blunt when X213 < 3.9e−07 & X81 > = 0.035 & X2 < 0.92 & X175 < 0.035
5.	Sharp when X213 < 3.9e−07 & X81 > = 0.035 & X2 > = 0.92 & X34 < −0.28
6.	Sharp when X213 > = 3.9e−07 & X10 > = 0.38
7.	Sharp when X213 < 3.9e−07 & X81 > = 0.035 & X2 < 0.92 & X175 > = 0.035
8.	Sharp when X213 < 3.9e−07 & X81 < 0.035 & X154 < −0.33 A

In Table 6 the confusion matrix for the pruned tree is presented.

Table 6. Confusion matrix for cp = 0.005.

	Reference		
	State	Blunt	Sharp
Prediction	Blunt	919	24
	Sharp	18	1212

The confusion matrix showed that 42 (18 + 24) from n = 2173 of the analysed variants were incorrectly classified. It means that the prediction error was (42/2173) ≈ 1.93%. The analyses of the indicator values of prediction model quality confirmed the prediction error value (Table 7). The Accuracy value is 0.9807, which means that the value of the prediction error is 1 − Accuracy = 0.0193 (1.93%). This value indicates a decrease of the predictive ability of the analysed tree. However, pruning the tree reduced its complexity, and thus eliminated "overfitting".

Table 7. The values indicators of prediction model quality for cp = 0.005.

Indicator	Value
Accuracy	0.9807
Sensitivity	0.9807898
Specificity	0.9805825
Positive predictive value	0.9745493
Negative predictive value	0.9853659
Precision	0.9745493
Recall	0.9807898
F1	0.9776596
Prevalence	0.4312011
Detection rate	0.4229176
Detection prevalence	0.4339623
Accuracy balanced	0.9806861

Decision trees enable the analyses of the importance of each variable on the dependent variable. The importance of a variable determines the variable's contribution in the created decision tree. Importantly, the specified importance of a variable applies only to the decision tree for which it was determined. Hence, a variable, which is important for one decision tree, cannot be important for another tree, even if both are created from the same dataset. In addition, a decision tree pruning changes not only the size of the tree, but also the importance of individual variables. Moreover, high significance of a variable does not necessarily mean that the variable is included in one of the decision tree nodes. It depends on the specifics of a data set – if another input variables contain relevant information, then the lack of even the most important input variable on the tree will have no impact on the quality of the prediction. In the analysed case, the analysis of variable importance allowed to identify the parameters that affect the condition of a cutter blade. The ranking of variable importance for a pruned tree ($cp = 0.005$) is shown on Fig. 4. The importance ranking was limited from 213 to 20 variables.

The highest values indicate the largest variable influence on the cutter state. From the 20 most important variables affecting the condition of a cutter blade, the largest influence has variable x_{213} (value 715.90), and the lowest variable x_3 (78.32). Not all variables are included in a decision tree, but they play a significant role in determining the splits in the node. The variables x_{10}, x_{154} and x_{34}, which were used to construct the tree for $cp = 0.005$, were not included in the 20 most important variables. These variables obtained the values of 10.19, 28.38 and 13.81, respectively.

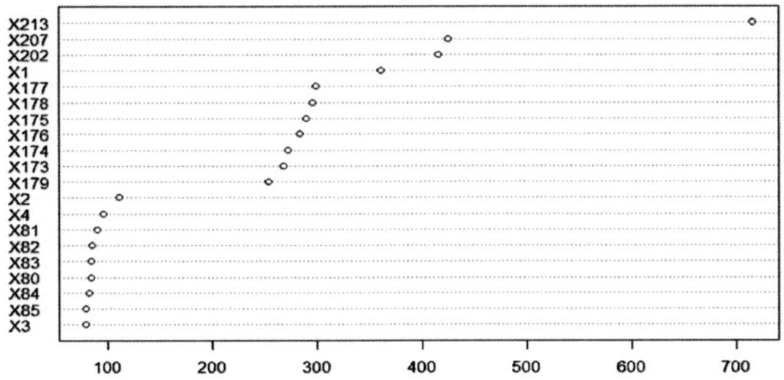

Fig. 4. Ranking of variables importance for the pruned tree (cp = 0.005).

The results of analysing the variable importance for a decision tree can be used to determine the most important input variables, while rejecting those which do not affect the condition of a cutter blade.

5 Summary

Several challenges of machining process durability and reliability can be considered as one of the most important issues, also in the context of Computerized Maintenance Management Systems development, predictive maintenance and real-time monitoring of production processes. Particularly, it is due to the growing role of diagnostic systems, in which every typical, numerically controlled machine tool is equipped. Therefore, the authors undertook the task of developing a solution that would enable effective identification of the condition of a cutting tool with the use of a decision support tool such as decision trees. The presented method offers a simple way to identify the cutter state. The study gives promising results which are confirmed by the prediction model quality indicators. Sensitivity equals 0.98 and the false alarm rate is equal to 0.0193. The presented analytical solution, verified on the basis of the real data from an industrial machine tool, can be used as part of the system for recognizing a wear rate of a cutting tool during the production process, based on the analysis of an acoustic signal.

In future work, the analyzes will be extended in order to identify the possibility of linking the physical aspects of the process with the autocorrelation coefficients contained in the tree, to parametrize the data, as well as to search for the possibility of generalizing the proposed solution to make it more universal. In addition, the other methods of selecting variables for the decision tree (e.g. PCA methods) will be included in future work.

References

1. de Jonge, B.: Maintenance Optimization Based on Mathematical Modeling. University of Groningen, SOM Research School, Groningen, Holland (2017)
2. Yan, J., Meng, Y., Lei, L., Li, L.: Industrial big data in an industry 4.0 environment: challenges, schemes, and applications for predictive maintenance. IEEE Access **5**, 23484–23491 (2017)
3. Valis, D., Mazurkiewicz, D., Forbelska, M.: Modelling of a transport belt degradation using state space model. In: Proceedings of the 2017 IEEE International Conference on Industrial Engineering & Engineering Management, pp. 949–953. IEEE, Singapore (2017)
4. Vališ, D., Mazurkiewicz, D.: Application of selected Levy processes for degradation modelling of long range mine belt using real-time data. Arch. Civil Mech. Eng. **18**(4), 1430–1440 (2018)
5. Varela, M.L.R., Putnik, G.D., Manupati, V.K., Rajyalakshmi, G., Trojanowska, J., Machado, J.: Integrated process planning and scheduling in networked manufacturing systems for I4.0: a review and framework proposal. Wireless Netw. **27**(3), 1587–1599 (2019)
6. Jasiulewicz-Kaczmarek, M., Żywica, P.: The concept of maintenance sustainability performance assessment by integrating balanced scorecard with non-additive fuzzy integral. Eksploatacja i Niezawodnosc – Maintenance Reliab. **20**(4), 650–661 (2018)
7. Kozłowski, E., Mazurkiewicz, D., Żabiński, T., Prucnal, S., Sęp, J.: Assessment model of cutting tool condition for real-time supervision system. Eksploatacja i Niezawodnosc - Maintenance Reliab. **21**(4), 679–685 (2019)
8. Borucka, A., Grzelak, M.: Application of logistic regression for production machinery efficiency evaluation. Appl. Sci. **9**, 4770 (2019)
9. Antosz, K., Paśko, Ł, Gola, A.: The use of intelligent systems to support the decision-making process in Lean Maintenance management. IFAC PapersOnLine **52**(10), 148–153 (2019)
10. Pavlenko, I., Trojanowska, J., Ivanov, V., Liaposhchenko, O.: Parameter identification of hydro-mechanical processes using artificial intelligence systems. Int. J. Mechatron. Appl. Mech. **5**, 19–26 (2019)
11. Gareth, J., Witten, D., Hastie, T., Tibshirani, R.: An introduction to statistical learning with applications with R. Springer, London (2013)
12. Breiman, L., Friedman, J.H., Olshen, R.A., Stone, C.J.: Classification and Regression Trees. Chapman & Hall, New York (1984)
13. Larose, D.T.: Discovering Knowledge From Data. Introduction to Data Mining. Scientific Publisher PWN, Warsaw (2013)
14. Costa, E.P., Lorena, A.C., Carvalho, A.C.P.L.F., Freitas, A.A.: A review of performance evaluation measures for hierarchical classifiers. In: Evaluation Methods for Machine Learning II: Papers from the AAAI-2007 Workshop, pp. 182–196. AAAI Press (2007)
15. Provost, F., Fawcett, T., Kohavi, R.: The case against accuracy estimation for comparing classifiers. In: Proceedings of the ICML-1998, pp. 445–453. Morgan Kaufmann, San Francisco (1998)
16. Powers, D.: Evaluation: from precision, recall and F-score to ROC, unforcedness, nakedness & correlation. J. Mach. Learn. Technol. **2**, 37–63 (2011)

17. Żabiński, T. Mączka, T., Kluska, J.: Industrial platform for rapid prototyping of intelligent diagnostic systems trends. In: Mitkowski, W., Kacprzyk, J., Oprzędkiewicz, K., Skruch P. (eds.) Advanced Intelligent Control, Optimization and Automation Polish Control Conference, Kraków, Poland, pp. 712–21. Springer, Heidelberg (2017) https://doi.org/10.1007/978-3-319-60699-6_69
18. Charemza, W.W., Syczewska, E.M.: Joint application of the Dickey-Fuller and KPSS tests. Econ. Lett. **61**(1), 17–21 (1998)
19. Dickey, D.A., Fuller, W.A.: Distribution of the estimators for autoregressive time series with a unit root. J. Am. Stat. Assoc. **74**, 427–431 (1979)
20. Box, G.E.P., Pierce, D.A.: Distribution of residual autocorrelations in autoregressive-integrated moving average time series models. J. Am. Stat. Assoc. **65**(332), 1509–1526 (1970)
21. Ljung, G.M., Box, G.E.P.: On a measure of lack of fit time series models. Biometrika **65**(2), 297–303 (1978)

Stainless Steel Deep Hole Drilling with EDM

Jan Hošek(✉)

Faculty of Mechanical Engineering, Czech Technical University, Technická 4, 16607 Praha 6, Czech Republic
Jan.Hosek@fs.cvut.cz

Abstract. The electro discharge machining is a promising production technique allowing to overcome the limits of standard drilling process especially in case of small and deep holes production in tough materials. A drawback of the EDM technique is the necessity to find optimal machining conditions depending on materials and dimensions of the workpiece and the electrode. Even in case of an optimal set of machining parameters, the machining process still shows instability mainly related to the flushing and debris removal mechanism. The aim of this study is to find how the machining flushing level, the machining feed rate, and the machining process stability relate to the machined hole's depth for a combination of the sinking process with a hollow electrode with inner flushing. We machined out series of holes that show that inner flushing increases the machining depth compared to machining with no electrode inner flushing. Experiments also showed the machining feed rate increases with a decreasing dielectric oil flushing pressure. There was not found any limit in the hole depth up to the depth of 48 mm in stainless steel, even for a low-level flushing corresponding to dropping of individual drops at the electrode tip. It was found that decreasing of the flushing pressure increases the standard deviation of the feed rate measured along the hole machining depth. There was also found an effect of strong drop of the feed rate in a small hole depth related to the increases of the machining area caused by workpiece inner peak formation inside the hollow electrode.

Keywords: Electro discharge machining · Stainless steel · Process stability

1 Introduction

Deep hole drilling is a difficult task in tough materials such as stainless steel. Standard drilling techniques usually fail due to the low stiffness of the cutting tool for a small hole diameter D for depths bigger than $10D$. An alternative technique usable for deeper holes drilling is the electro discharge machining (EDM). The EDM technique is a complex machining procedure employing simultaneous effects of electrical, thermal, hydrodynamic and chemical treatment of the workpiece. The EDM is based on high frequency sparking in a gap between the tool electrode and machined material, which generates plasma. The thermal effect of sparks melts the surfaces, the hydrodynamic effects remove the molten material and debris solidify from melt and vapors.

While the volume removed by a single spark is small this technique is suitable for precise machining. On the other hand, with increases of the machining precision, the

J. Machado et al. (Eds.): icieng 2021, LNME, pp. 28–33, 2022.
https://doi.org/10.1007/978-3-030-79165-0_3

material removal rate (MRR) is decreasing. MRR depends on many factors such as melting temperature, thermal and electrical conductivity, thermal capacity of both tool and workpiece materials. Appropriate machining conditions as discharge current, voltage, pulse on-time, duty factor and other need to be carefully selected and optimized with respect to the workpiece and electrode materials, electrode shape, dimensions, and other factors [1–4]. Even in case of optimal set of basic machining parameters, the machining process still shows instability mainly related to the flushing and debris removal mechanism [5]. The most common technique of the EDM machining is the sinking process, where the electrode periodically jumps to allow for an exchange of the dielectric fluid media in the discharge gap. This technique employs turbulent flows and bubbles generation for the debris removal during electrode lifting. It works well up to a certain depth of the hole where the electrode jump speed is not able to remove the debris anymore and the debris is trapped and accumulated within the gap [6, 7]. Another common technique of the debris treatment is the electrode rotation and inner flushing. A comparative study of different flushing techniques [8] showed that the most efficient flushing techniques are forced pump internal flushing under the electrode rotation and, even better, the forced pump internal flushing under the electrode rotation and its vibration. Under the electrical discharge drilling the electrode is rotating and the flushing pressure reaches different levels up to the 25 MPa [9], but the effect of the dielectric flushing effect on MRR is not reported uniformly. Munz et al. [9] reported an optimal flow rate, where higher and lower flow rates decrease the machining feed rate. Lee [10] reported increases of MRR when flushing pressure decreases.

This study aims to investigate the influence of flushing pressure in case of non-rotating hollow electrode with employing the electrode jumping.

2 Experimental Procedure

The experiments were carried out on the electrical discharge sinking machine Sodick AP1L filled with dielectric oil total Diel MS 5000. A copper electrode 1/0.5 mm was used and the flushing pressure was varied at levels 0.02, 0.04, 0.1, 0.14, and 0.2 MPa. Material parameters of the electrode, the workpiece and the dielectric fluid are summarized in Table 1.

Individual holes were machined down to different depth levels. The aim of the study was to find a relation between the machining flushing level, the machining feed rate, and the machining process stability along the machined hole depth. For this reason, the hole sinking was interrupted every 0.5 mm and the actual electrode wear was measured by electro-contact approach. The data were used for evaluation of the actual hole depth, machining feed rate and electrode wear ratio (EWR).

Table 1. Material properties.

Workpiece	EN 1.4301/AISI 304/X5CrNi18-10	Stainless steel
Melting temperature	1400–1450	°C
Thermal conductivity	15/20 °C–22/500 °C	W/m·K
Electrical resistivity	0.73/20 °C–1.25/500 °C	$\Omega \cdot mm^2/m$
Specific heat capacity	500	J/kg·K
Density	7920	kg/m^3
Tensile strength	500–700	N/mm^2
Electrode	99.9% copper	Copper
Melting temperature	1083	°C
Thermal conductivity	401/20 °C–361/500 °C	W/m·K
Electrical resistivity	0.017/20 °C–0.03/500 °C	$\Omega \cdot mm^2/m$
Specific heat capacity	383/20 °C–438/500 °C	J/kg·K
Density	8933	kg/m^3
Tensile strength	224–314	N/mm^2
Dielectric fluid	Diel MS 5000	Total
Flash point	>130	°C
Thermal conductivity	0.126	W/m·K
Electrical resistivity	$4 \cdot 10^{12}$/20 °C–10^{11}/100 °C	Ω·cm
Specific heat capacity	1860	J/kg·K
Density	843	kg/m^3
Kinematic viscosity	6.5/20 °C–3.9/40 °C	mm^2/s

The machining was performed under fixed machining conditions summarized in Table 2.

Table 2. Fixed machining conditions.

Parameter	Value	Units
ON time duration	45	μs
OFF time duration	24	μs
Main power voltage	90	V
Main power current	3	A
Jump speed	4	m/min
Down time	0.16	s

3 Results and Discussion

An electrode sinking copper tube electrode 1/0.5 mm without any flushing under condition summarized in Table 2 was taken as reference machining. Next sinking was tested under different inlet pressure levels: 0.2 MPa two holes, 0.14 MPa one hole, 0.1 MPa one hole 0.04 MPa six holes, 0.02 MPa two holes, and 0.01 MPa one hole. Measured sinking data are visualized in next figures. Figure 1 shows trend of increasing hole depth depending on machining time. Coordinate system origin is at the surface.

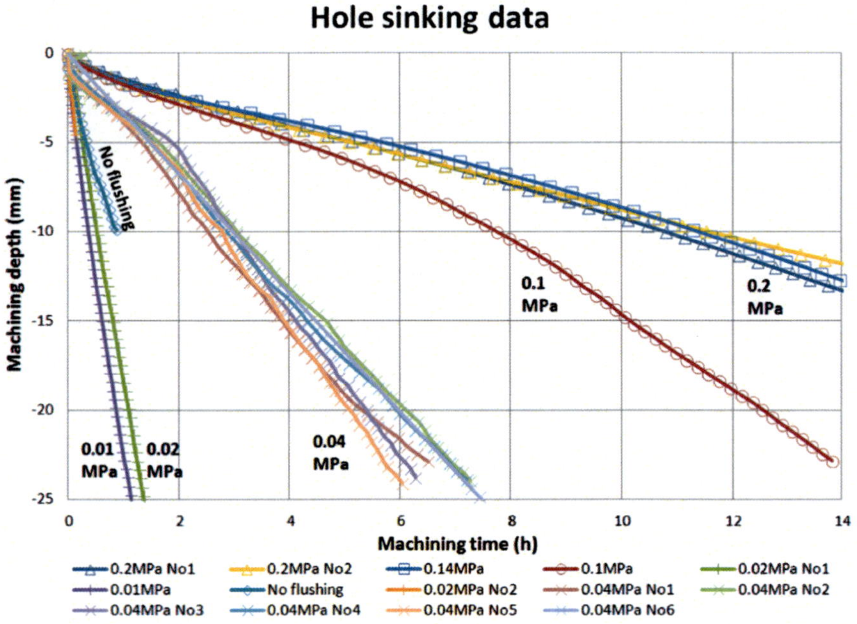

Fig. 1. A sinking holes depth evolution in time for different flushing pressures.

Computed values of the Feed rate, the EWR, ratio of the feed rate standard deviation to its mean value, and the total machining depth are summarized in Table 3.

The Fig. 1 and Table 3 clearly visualize the differences for various flushing pressures. The lowest machining feed rate was achieved under the highest applied flushing pressure. There is almost no difference between flushing under pressures 0.2 and 0.14 MPa. Differences fall within uncertainties of the process. The only meaningful difference is a drop of the EWR of the 0.14 MPa flushing compared to 0.2 MPa flushing pressure. 0.1 MPa flushing pressure shows significant increases of the machining feed rate and decreases of the EWR. For the 0.04 MPa flushing pressure the trend continues and the maximum feed rate about 20 mm/hour with minimum EWR about 3% was achieved for the flushing pressure about 0.01 to 0.02 MPa. Such values are even better than for machining with no flushing, which becomes to jam at depth 9.86 mm. Based on measured data it is possible to point out another outcome. With decreasing the flushing pressure, the process instability increases as it is shown in Fig. 2.

Table 3. Machining data.

Pressure (MPa)	Feed rate (mm/h)	EWR (%)	Std/mean (%)	Depth (mm)
0.2	0.95	48.8	8.3	25.47
0.2	0.74	53.3	13.1	31.24
0.14	0.92	44.1	13.9	20.70
0.1	1.82	29.1	25.8	22.85
0.04	3.62	8.4	22.6	22.90
0.04	3.37	8.1	18.4	23.96
0.04	4.36	7.5	29.1	23.79
0.04	3.39	7.5	15.5	23.87
0.04	4.32	6.5	33.9	24.10
0.04	4.95	5.5	26.6	48.54
0.02	18.45	2.4	11.4	29.22
0.02	27.08	3.4	26.9	4.57
0.01	20.59	2.9	16.8	29.15

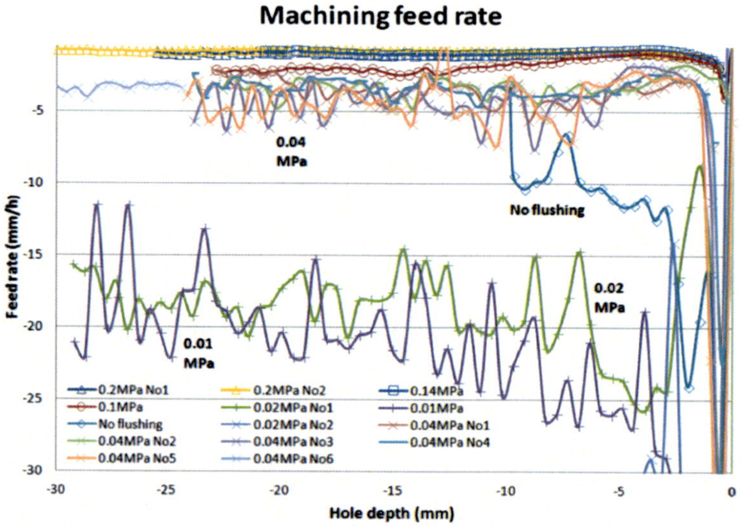

Fig. 2. A machining feed rate as function of the hole depth.

The Fig. 2 shows an important effect of strong decreases of machining feed rate for a small hole depth and stabilization of the process or slight increases of the feed rate for the hole depth more than 5 mm. It may explain the highest measured feed rate for 0.02 MPa pressure while the machined hole was low in depth, below 5 mm. It is clearly visible in Fig. 1 that no flushing sinking follows similar feed rate as the low pressure flushing but at a depth of few millimeters it slows down, and it finally jams at a depth of about 10 mm. The change of the machining trends for the depth of more than 5 mm can be explained by the development of an inner cone of workpiece material inside the hollow electrode.

4 Conclusions

The aim of the study was to find the influence of the machining flushing level to the machining feed rate and the machining process stability in case of the sinking process with a hollow electrode provided with inner flushing. Machining of holes series was performed. The data showed a limit of the no-flushing machining depth of about 10 mm and an increase of the machined depth with electrode inner flushing. There was not found any limit in the hole depth up to the depth of 48 mm in stainless steel. It was found that the machining feed rate increases with a decreasing dielectric oil flushing pressure. Despite the low pressure flushing corresponding to dropping of individual drops at the electrode tip, the machining feed rate reached the highest values with the lowest electrode wear ratio of about 3%. It was found that decreasing the flushing pressure increases the standard deviation of the feed rate measured along the hole depth. There was also found an effect of a strong drop of the feed rate in small hole depth related to the increase of the machining area caused by workpiece inner peak formation inside the hollow electrode. The effects related to the low level of flushing in case of hollow electrode sinking will be the next topic for the study to find the optimal dielectric fluid flux maximizing the machining feed rate and minimizing the electrode wear.

References

1. Kuppan, P., Rajadurai, A., Narayanan, S.: Influence of EDM process parameters in deep hole drilling of Inconel 718. Int. J. Adv. Manuf. Technol. **38**, 74–84 (2007)
2. Joshi, S.N., Pande, S.S.: Thermo-physical modeling of die-sinking EDM process. J. Manuf. Process. **12**, 45–56 (2010)
3. Ferraris, E., Castiglioni, V., Ceyssens, F., Annoni, M., Lauwers, B., Reynaerts, D.: EDM drilling of ultra-high aspect ratio micro holes with insulated tools. CIRP Ann. Manuf. Technol. **62**, 191–194 (2013)
4. D'Urso, G., Ravasio, C.: Material-Technology Index to evaluate micro-EDM drilling process. J. Manuf. Process. **26**, 13–21 (2017)
5. Yılmaz, V., Sarikaya, M., Dilipak, H.: Investigation of deep-drilled micro-hole profiles in Hadfield steel. Mater. Test. **58**(3), 224–230 (2016)
6. Cetin, S., Okada, A., Uno, Y.: Electrode jump motion in linear motor equipped die-sinking EDM. J. Manuf. Sci. Eng. **125**, 809–815 (2003)
7. Liao, Y.S., Wu, P.S., Liang, F.Y.: Study of debris exclusion effect in linear motor equipped die-sinking EDM process. Procedia CIRP **62**, 123–128 (2013)
8. Ni, H., Gong, H., Dong, Y.H., Fang, F.Z., Wang, Y.: A comparative investigation on hybrid EDM for drilling small deep holes. Int. J. Adv. Manuf. Technol. **95**(1–4), 1465–1472 (2017). https://doi.org/10.1007/s00170-017-1282-1
9. Munz, M., Risto, M., Haas, R.: Specifics of flushing in electrical discharge drilling. Procedia CIRP **6**, 83–88 (2013)
10. Lee, S.H., Li, X.P.: Study of the effect of machining parameters on the machining characteristics in electrical discharge machining of tungsten carbide. J. Mater. Process. Technol. **115**, 344–358 (2001)

Check for updates

Experimental Research of the Tribological Properties of D-Gun Sprayed WC – Co Coatings

Yuriy Kharlamov, Volodymyr Sokolov, Oleg Krol(✉), and Oleksiy Romanchenko

Volodymyr Dahl East Ukrainian National University, 59-a Central Pr, Severodonetsk 93400, Ukraine

Abstract. The results of experimental researches of the tribological properties of D – Gun coatings from powder of alloy VK8 (WC-8%Co) are given. The main studies were carried out under conditions of ring – on – ring test without lubrication (dry friction). For testing in conditions of ring – on – ring test the device for metal – cutting machine tools is developed. Coatings were spraying on steel samples immediately after turning without jet – abrasive preparation of surfaces for spraying. The influence of the sliding velocity on friction coefficient, comparative wear resistance and temperature of friction pairs are studied. All friction pairs are characterized by a decrease of friction coefficient and an increase of temperature in friction zone with increasing of sliding velocity. It is shown that in the range of sliding velocities from 0,104 to 0,6 m/s there is a more intensive decrease of friction coefficient and with a further increase of sliding velocity its rate decreases and at 3,14 m/s reaches 0,145. In studied range of sliding velocities, when use steel samples as counterbody, the friction coefficient has smaller values.

Keywords: D–gun spaying coating · Tribological properties · Wear rate · Coefficient of friction · Sliding speed · Dry friction

1 Introduction

The application of protective coatings allows finding solution of various problems associated with reducing of material and energy intensity of production and operation of machines, increasing of their reliability, creating and mastering of new products of modern technique [1–5]. Perspective methods are thermal spraying of coatings from powder materials and among them D – Gun spraying [1, 6–10]. The main advantages of D – Gun sprayed coatings are: the ability to obtain durable coatings during spraying; reduced quality requirements for preparation of sprayed surface; wider possibilities of regulation of thermal cycle of coating and product formation; high growth rate of coating thickness; relative simplicity of designs and high reliability of technological equipment; high density of received coatings, etc.

J. Machado et al. (Eds.): icieng 2021, LNME, pp. 34–45, 2022.
https://doi.org/10.1007/978-3-030-79165-0_4

2 Literature Review

The creation and implementation of effective technological processes and equipment [1, 6, 11–14] for getting protective coatings with increased strength and low porosity requires complex research of the physico-mechanical and tribological properties of D – Gun spraying coatings [1, 6, 15–20]. One of the major problems of tribology is the development of scientific basis of materials compatibility in friction units, as well as practical recommendations for creation of wear – resistant antifriction and friction materials, new technologies of hardening and coating, new lubricants and additives to them, including for biological friction units and parts [7, 15, 21–25]. The technique and technology of tribological coatings is field of science and technology that rapidly developing. New types of coatings are constantly being developed and their application is expanding in various fields of science and technology, ranging from parts and machine couplings, as well as consumer goods to medical instruments and prostheses [6, 26–30].

Among the materials of D–Gun spraying coatings particular interest is tungsten carbide, which is used as a solid component of various compositions with metals and other refractory compounds [1, 6–9, 25, 29]. However, tribological properties of detonation gas coatings have not been sufficiently researched, especially in conditions of dry friction.

Objective. The objective of this work is research of tribological properties of D–Gun spraying coatings from powder of alloy VK8 (WC-8%Co).

3 Research Methodology

Coatings from standard powder of alloy VK8 (WC-8%Co) were applied to steel samples after turning on the equipment of Volodymyr Dahl East Ukrainian National University. The jet – abrasive preparation of surfaces for spraying was not carried out. Also, the detonating mixture of gases of acetylene – oxygen in a ratio of 1:1 was applied, and for purging of detonation combustion chamber was applied nitrogen.

The microhardness of obtained coatings at a load of 100 g exceeded 1600 kg/mm^2. Typical feature of D – Gun spraying coatings is the low porosity (less than 1%) and high adhesion with base material. The coatings have good resistance to shock loads. The high strength of detonation coatings is explained by realization of mechanisms that occur during process of coating layer formation from individual particles that associated with their high – speed collision: plastic deformation; increasing the density of dislocations, etc.

Tribological tests of coatings were carried out in conditions of ring – on – ring test without lubrication on a specially designed device (Fig. 1) for universal metal – cutting machine tools with a vertical spindle. In this work, vertically – milling and vertically – drilling machine tools were used. The specific load on the samples during the test was kept constant and was equal to 1,0 MPa. Continuous record of friction torque and temperature was performed using a thermo-pair mounted in bottom stationary sample at a distance of 1mm from the friction surface. Coated samples were used as the bottom and samples from normalized, hardened steel 45 (0,42–0,5% C) and with coating were used as upper sample with rotation. Coated samples were grinded by circles of synthetic diamond and running – in before testing. The thickness of coating layer on samples after machining was 0,1–0,15 mm.

In the device, the upper rotating sample (5) is mounted on the mandrel (10), which through the spherical heel (11) contacts the spherical head of the finger (9) pressed into the shank (8) mounted on the spindle (7) of machine tool. The rotation from the shank is transmitted by pins (6). the spherical heels and the finger provide self – alignment of the samples, making it easier to fit the samples over the friction surface. Adjustment of the rotation speed of the upper sample is ensured by changing the speed of the machine spindle. The lower fixed sample (12) is mounted on a cylindrical mandrel, which is fixed to a spindle (4) mounted on thrust bearings in a floating movable sleeve (3). The latter is placed in the guide holes of the housing (13) mounted on the table (22) of the machine. The axial loading on samples is carried out by a lever (14) with a load (16) through a clamp (2) fixed on the movable sleeve.

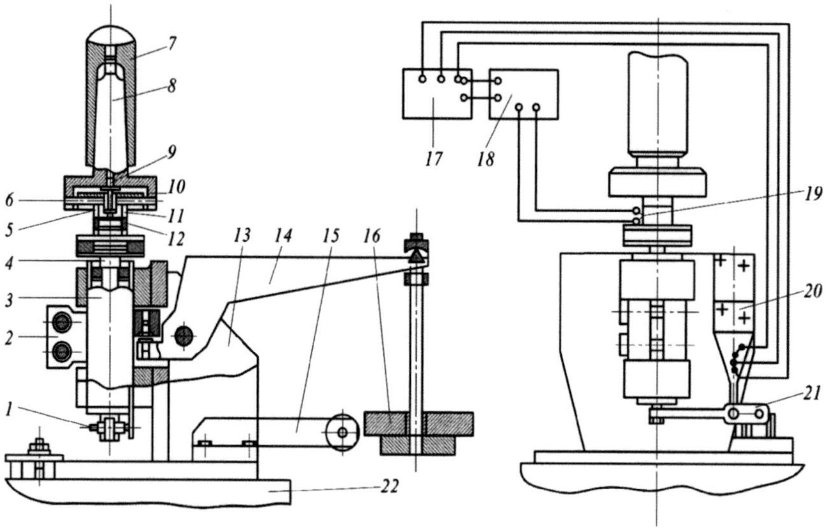

Fig. 1. Device for testing of coatings

For measurement of torque at the lower end of shaft mounted a lever (21) with a pressure screw (1), that have contacts with elastic beam (20) on which wire resistance sensors are glued. Continuous record of friction torque is performed by an automatic potentiometer EPR – 09 (18) through an amplifier (17). The calibration of power measuring system is carried out by weight. For this purpose, a bracket (15) with a roller for tipping of cable on which the weights are suspended is mounted on body.

The criterion of samples comparative wear resistance rate was determined by their periodic weighing and taking into account the mass loss of samples materials.

For measurement of the temperature in the process of friction in a fixed sample at a distance of 1 mm from friction surface thermo-pair (19) with wire diameter 0,5 mm was mounted. The thermopair is welded to the samples by application of capacitor welding method. Continuous record of temperature during friction is also carried out by potentiometer EPR – 09. In process of wear resistance tests of detonation coatings the sliding velocity (determined by spindle turnover number of used machine tool), specific pressure, combination of materials of friction pair, coefficient of mutual overlap of samples, thickness of coatings, technological modes of their spraying and other parameters can be different. The change of coefficient of mutual overlap of the samples is achieved by the manufacture of a fixed sample with grooves of appropriate sizes.

Thus, this device and methodology allow to make comprehensive researches of the processes of friction and wear of detonation coatings in conditions of dry friction. This is necessary to select the optimum combination of materials and coatings for friction units. In addition, the methodology allows to evaluate the influence of technological parameters of spraying process on wear resistance of obtained coatings.

4 The Main Results

Friction of solid bodies is related to three successive and interrelated steps [25]: (1) interaction of surfaces with regard to the influence of environment; (2) change of surface layers as a result of interaction; (3) destruction of surface layers due to the two previous steps. The main types of friction interaction are: a slice of material; plastic stamping; elastic displacement; grasping of films; grasping of surfaces. In these dry – friction tests, the atmospheric air was used as working environment. The frictional heating that occur during test of friction surfaces results to their oxidation and formation of oxide films. The adhesion interaction of friction surfaces is also not excluded.

The results of tribological tests of friction pair are presented in Fig. 2: coating of powder of alloy VK8 (WC-8%Co) (lower sample) – normalized steel 45 (0,42–0,5% C), t – temperature (°C), f – friction coefficient, K – comparative wear resistance. This case is characterized by a decrease of controlled values of friction coefficient in researched range of sliding velocities from 0,39 to 0,145 and an increase of temperature of friction surfaces with increasing sliding velocity (Fig. 2, curves 1, 2, respectively) up to 480 °C. Moreover, in the range of sliding velocities from 0,104 to 0,6 m/s there is a more intense decrease of friction coefficient (curve 1) – from 0,39 to 0,27, with a further increase of sliding velocity, the rate of its decrease decreases and at 3,14 m/s reaches reach 0,145. The temperature of friction surfaces is characterized by a smoother increase with increasing sliding velocity. The observed decrease of decrease rate of friction coefficient at a sliding velocity about 0,6 m/s can be explained by changes of interaction mechanisms of friction surfaces when their temperature increases (curve 2).

Significant differences are observed in values and character of change of wear intensity criterion with increasing of sliding velocity of samples with and without coating. For bottom coated sample with increasing of sliding velocity the comparative wear resistance criterion decreases (Fig. 2, curve 3) and gradual increase in range from 0,45 to 1 m/s. With increasing of velocity up to 1,7 m/s the comparative wear resistance criterion of coating gradually decreases and then retains a low value. Even more sharp changes of wear intensity criterion have upper sample from steel 45 (0,42–0,5% c) (curve 4). In range of velocities 0,104–0,65 m/s it decreases from 2,64 to 0,12, that is 22 times. Then in range of velocities 0,65–1,42 m/s the wear intensity criterion increases to 0,74 and retains in future about the same value.

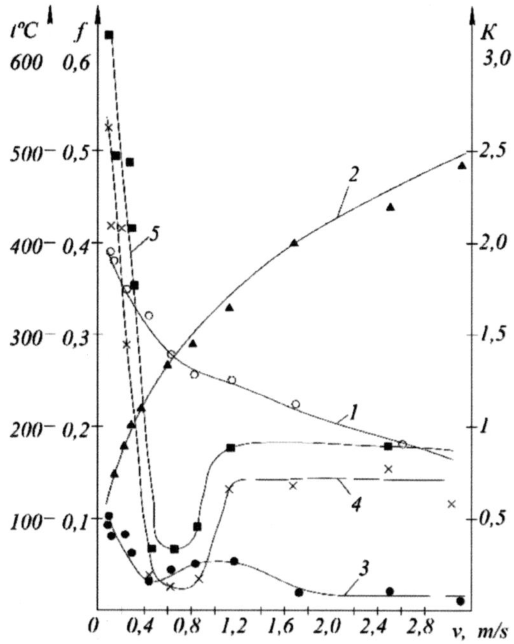

Fig. 2. Effect of sliding velocity on parameters of friction and wear of friction pair: coating of alloy VK8 (WC-8%Co) – steel 45 (0,42–0,5% C): 1 – coefficient of friction; 2 – temperature; 3 – comparative wear resistance of sample with coating; 4 – comparative wear resistance of sample from steel; 5 – total comparative wear resistance of samples

The friction surfaces of samples after testing were researched by optical metallography to evaluate features of friction interaction of their structures under conditions of dry sliding friction (Fig. 3). Increase of sliding velocity leads to changes of wear mechanism and its intensity.

Wide and rather deep grooves are observed on friction surfaces of both samples. The friction process, especially at low sliding velocities, is characterized by abundant release of wear products in form of a brown color powder. The sample from steel 45 (0,42–0,5% C) has a larger number of grooves and their dimensions are larger than the coating. These grooves are formed during rolling of emerging wear products.

In the entire researched range of sliding velocities the comparative wear resistance of sample without coating (steel 45 (0,42–0,5% C) exceeds comparative wear resistance of coating of alloy VK8 (WC-8%Co), especially at low sliding velocities up to 0,6 m/s, which confirms the significant influence of sliding velocity on wear mechanism. The comparative wear resistance decreases sharply as sliding velocity increases up to 0,4 m/s. Moreover, the minimum wear intensity of steel samples is observed at a sliding velocity of 0,62 m/s, which corresponds to inflection point on curve of friction coefficient. But already at a sliding velocity of 0,82 m/s begins a sharp increase of comparative wear resistance up to a sliding velocity of 1,2 m/s, and with a further increase of sliding velocity comparative wear resistance remains at the same level.

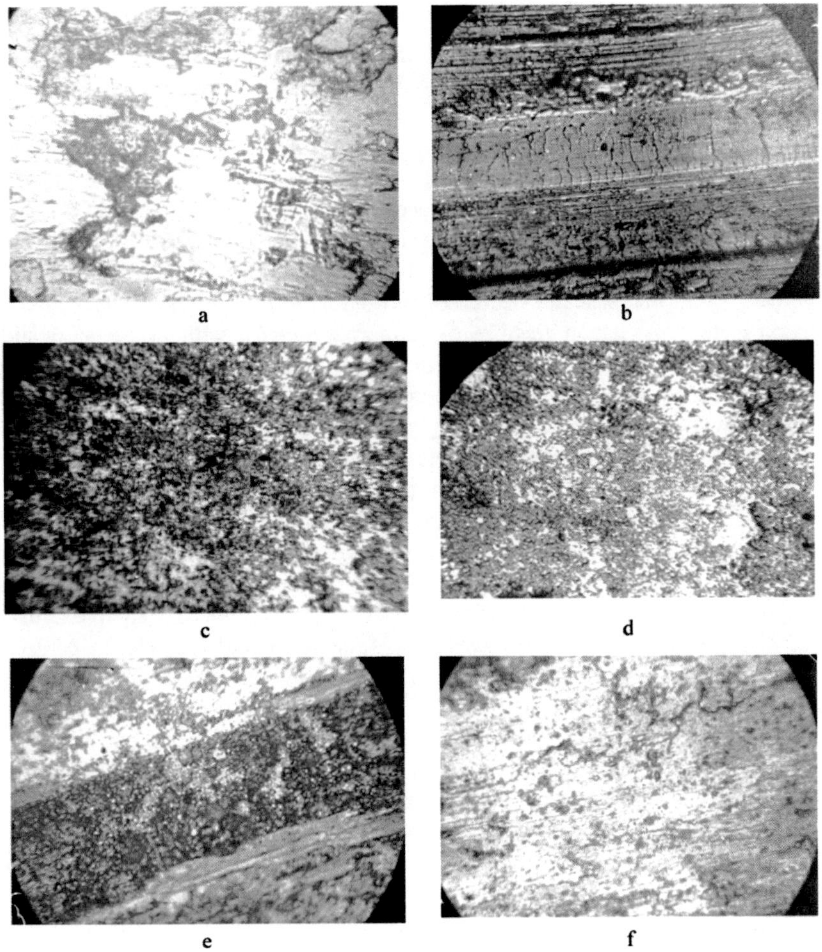

Fig. 3. Microphotos of friction surfaces (×240): (a) – steel 45 (0,42–0,5% C), V = 0,83 m/s (530 rpm); (b) – steel 45 (0,42–0,5% C), V = 2,5 m/s (1600 rpm); (c) –VK8 (WC-8%Co), V = 0,32 m/s (195 rpm); (d) – VK8 (WC-8%Co), V = 0,83 m/s (530 rpm); (e) – VK8 (WC-8%Co),V = 3,14 m/s (2000 m / min); (f) – VK8 (WC-8%Co), V = 3,14 m/s (2000 rpm)

Compare wear features of contacting samples with coating and steel 45 (0,42–0,5% C). The wear rates of both samples decrease with increasing of sliding velocity to values 0,42–0,62 m/s, and for coating it decreases up to 3 times, and for steel 45 (0,42–0,5% C) – more than 25 times. With further increase of sliding velocity up to 1 m/s, the comparative wear resistance of coating increases only 1,6 times, and the steel 45 (0,42–0,5% C) – 8 times. Thus, the coincidence of ranges of sliding velocities at which a change of samples comparative wear resistance is observed indicates the phenomena and processes that affect on their interaction. Despite intense frictional heating in this and other combinations of friction pairs exfoliation of coatings during testing process was not observed, that indicating the high strength of obtained coatings.

Figure 4 presents the results of tribological tests of friction pair: coating of alloy VK8 (WC-8%Co) (lower sample) – hardened steel 45 (0,42–0,5% C; HRC 40…45). For this friction pair, with increasing sliding velocity from 0,104 to 2,5 m/s, there is also a characteristic decrease of controlled values of friction coefficient from 0,4 to 0,15 and an increase of temperature of friction surfaces (Fig. 4, curves 1, 2, respectively) to 410 ^{0}C. Moreover, in range of sliding velocities from 0,104 to 0,4 m/s there is a more intense decrease of friction coefficient (curve 1) – from 0,4 to 0,26, with a further increase of sliding velocity, its decrease rate decreases and at 2,5 m/s reaches 0,15. The temperature of friction surfaces is characterized by more sharper increase with increasing of sliding velocity. At a value of about 1,4 m/s, the temperature value stabilizes and remains almost unchanged. The observed decrease of decrease rate of friction coefficient at a sliding velocity about 0,4 m/s can be explained by changes of mechanisms of friction surfaces interaction with increasing of their temperature (curve 2).

For this friction pair, there are also significant differences of values and character of change of wear intensity with increasing of sliding velocity of coated and uncoated samples is observed. But character of change of wear intensity is significantly different from the friction pair of alloy VK8 (WC-8%Co) – normalized steel 45 (0,42–0,5% C). For the lower sample with coating with increasing sliding velocity a sharp decrease of comparative wear resistance is observed (Fig. 4, curve 3), and at a velocity of 0,4 m/s, its smooth increase begins, which stabilizes at a sliding velocity about 1,2 m/s. Thus, only one extreme (minimum) value of wear intensity is observed.

Even more distinctive changes in the intensity of wear have the upper sample of hardened steel 45 (0,42–0,5% C), (curve 4). At a velocity range of 0,104–0,4 m/s, the comparative wear resistance increases sharply, and then decrease more sharply and at a sliding velocity 0,6 m/s reaches a minimum value, and then it begins gradually increase.

In a comparative analysis of curves of dependence of sliding velocity from comparative wear resistance of both samples and friction coefficient, it is easy to identify the critical values of sliding velocities at which mechanisms of friction and wear are changing. Thus, at a sliding velocity about 0,4 m/s, there is inflection on curve of friction coefficient – the minimum wear intensity of coating and maximum wear intensity of hardened steel 45 (0,42–0,5% C). With the increase of sliding velocity in range from 0,104 to 0,4 m/s wear intensity of steel sample increase and comparative wear resistance of coating decreases. At a sliding velocity range of 0,4–0,8 m/s a sharp decrease of comparative wear resistance of steel sample is accompanied by a smooth increase of comparative wear resistance of coating. As the sliding velocity increase further, a sharp increase of comparative wear resistance of hardened steel is accompanied by a rapid slight increase of comparative wear resistance of coating to an almost stable value.

In Fig. 5 presented microphotos of friction surfaces of coating and working with it in a pair sample from hardened steel 45 (0,42–0,5% C) at a sliding velocity of 3,14 m/s. There are no obvious thermal cracking on wear paths of steel sample the following.

In Fig. 6 presented the results of tribological tests of friction pair: coating of VK8 (WC-8%Co) – coating of VK8 (WC-8%Co). For this friction pair, with increase of sliding velocity from 0,104 to 3,14 m/s, characteristic decrease of controlled values of

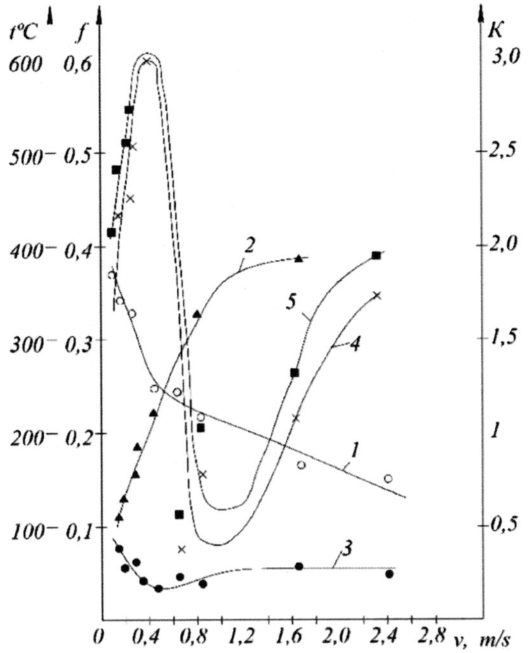

Fig. 4. Effect of sliding velocity on parameters of friction and wear of friction pair: coating of VK8 (WC-8%Co) – steel 45 (0,42–0,5% C; HRC 40…45): 1 – coefficient of friction; 2 – temperature; 3 – comparative wear resistance of sample with coating; 4 – comparative wear resistance of sample from steel; 5 – total comparative wear resistance of samples

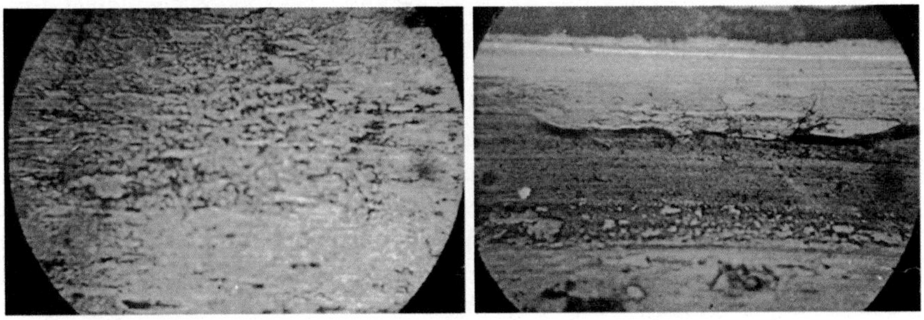

a b

Fig. 5. Microphoto of friction surfaces (× 240): (a) – coating of VK8 (WC-8%Co), $V = 3,14$ m/s; (b) – steel 45 (0,42–0,5% C; HRC 40…45), $V = 3,14$ m / s.

friction coefficient from 0,48 to 0,14 and an increase of temperature of friction surfaces up to 460 °C is observed (Fig. 6, curves 1, 2, respectively). Moreover, in range of sliding velocities from 0,104 to 0,64 m/s there is a more intense decrease of friction coefficient (curve 1) – from 0,48 to 0,317, with a further increase of sliding velocity its rate decreases and at 3,14 m/s reaches value of 0,14. The temperature of friction surfaces of this combination of materials is characterized by the most dramatic increase rate with increase of sliding velocity from 0,104 to 0,64 m/s. Then the increase rate of temperature decreases, but the temperature continues to rise. The points of inflection on curves for friction coefficient and temperature are quite close by value to sliding velocity of 0,64 m/s, which clearly indicates a change of interaction mechanisms of friction surfaces when their temperature reaches certain critical value.

This friction pair has significant differences in values and character of change of wear rate criterion with increase of sliding velocity. First, the comparative wear resistance criterion here is much lower compared to friction pairs that previously considered. Secondly, the curves of comparative wear resistance criterion have several extremes. Third, despite uniformity of sample materials, the top of them exhibits a reduced comparative wear resistance in entire range of sliding velocity (curve 4).

The comparative wear resistance criterion of both samples is minimal with a minimum sliding velocity of 0,104 m/s and gradually increase as sliding velocity increases up to 0,3 m/s. Then the comparative wear resistance also gradually decreases as sliding velocity increases up to 0.45 m/s and then gradually increases as sliding velocity increases for lower sample up to 0,64 m/s and to 0,82 m/s for upper sample. Further increase of sliding velocity up to 1,2 m/s leads to a gradual decrease of comparative wear resistance of coatings on both samples to a minimum value. As the sliding velocity increases, the comparative wear resistance of upper sample does not change and the lower sample increases up to 2,2 m/s. Then the comparative wear resistance criterion of lower sample is gradually reduced. In general, the comparative wear resistance of coated samples for a given friction pair (coating – coating) is the lowest, except for lower sample in sliding velocity range of 1,6–3 m/s.

In a comparative analysis of curves of dependence of sliding velocity from comparative wear resistance of both samples, the temperature and friction coefficient it is not difficult to identify critical values of sliding velocities at which the friction and wear mechanisms are changing. Thus, at a sliding velocity of 0,6 m/s, there is a inflection on curve of friction coefficient, which would also correspond to extreme of comparative wear resistance of coating on lower sample. Increase of comparative wear resistance of coatings at low sliding velocities up to 0,3 m/s may be related to process of running in of surface friction. Other extremums on comparative wear resistance curves require a more detailed analysis of friction processes and wear of solid surfaces without lubrication. Testing by scheme roller – insert on friction machine with lubricant are also presented. The coated insert and roller from steel 45 (0,42–0,5% C) was used. As the load increased from 1,2 MPa to 8 MPa, friction coefficient decreased from 0,07 to 0,03. When the load increase, friction coefficient gradually increases (Fig. 7), keeping value below 0,05 up to loads of 40 MPa at low comparative wear resistance of samples. On the same machine, in conditions of dry friction and specific loading of 3–6 MPa, a good running–in ability of coating with a steel roller at the reciprocating movement of the upper carriage with the pad is established. Both rubbing surfaces become mirror – like. In comparative tests on wear of lower rotating roller from steel 45 (0,42–0,5% C; HRC 45) about upper stationary application of coating increased wear resistance more than 10 times, compared with hardened steel 45 (0,42–0,5% C), despite the fact that dimensions of hole were smaller, and therefore, and above specific pressure in friction area.

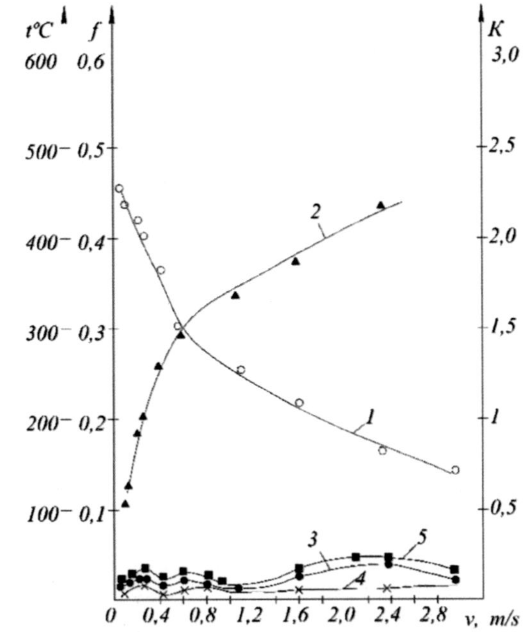

Fig. 6. Effect of sliding velocity on parameters of friction and wear of friction pair: coating of VK8 (WC-8%Co) – coating of VK8 (WC-8%Co): 1 – coefficient of friction; 2 – temperature; 3 – comparative wear resistance of lower sample; 4 – comparative wear resistance of upper sample; 5 – total comparative wear resistance of samples

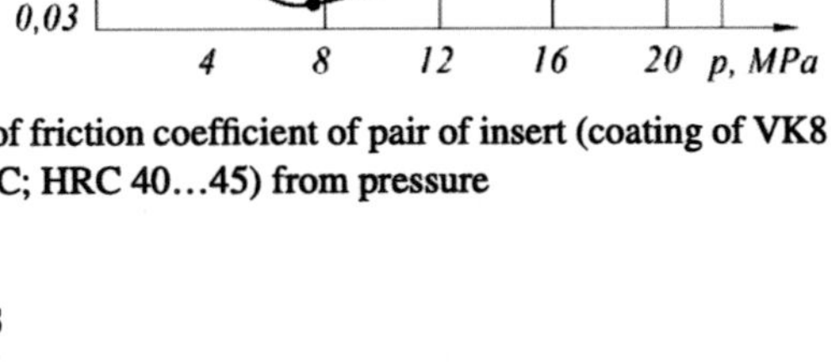

Fig. 7. Dependence of friction coefficient of pair of insert (coating of VK8 (WC-8%Co)) – roller (steel 45, 0,42–0,5% C; HRC 40…45) from pressure

5 Conclusions

D – Gun spraying method allows to obtain high – strength coatings from fine powders of solid alloys of tungsten carbide – cobalt. The coatings were obtained by spraying standard powder of alloy VK8 (WC-8%Co). The main studies were carried out under conditions of ring – on – ring test without lubrication (dry friction). For testing in conditions of ring – on – ring test the device for metal – cutting machine tools is developed. Coatings were spraying on steel samples immediately after turning without jet – abrasive preparation of surfaces for spraying.

The influence of the sliding velocity up to 3 m/s on friction coefficient, comparative wear resistance and temperature of friction pairs are studied: coating – normalized steel 45 (0,42–0,5% C); coating – hardened steel 45 (0,42–0,5% C) and coating – coating. All friction pairs are characterized by a decrease of friction coefficient and an increase of temperature in friction area with increase of sliding velocity (Fig. 1). Moreover, in range of sliding velocities from 0,104 to 0,6 m/s there is a more intense decrease of friction coefficient is observed, with a further increase of sliding velocity its decrease rate decreases and at 3,14 m/s reaches 0,145.

In researched range of sliding velocities, when steel sample used as a counterbody, the friction coefficient has slightly smaller values.

References

1. Kharlamov, Y.: The development of D-Gun spraying technologies. Visnik. V. Dahl EUNU **7**(237), 114–132 (2017)
2. Krol, O., Sokolov, V.: Parametric Modeling of Gear Cutting Tools. In: Gapiński, B., Szostak, M., Ivanov, V. (eds.) MANUFACTURING 2019. LNME, pp. 3–11. Springer, Cham (2019). https://doi.org/10.1007/978-3-030-16943-5_1
3. Karpus, V., Ivanov, V., Dehtiarov, I., Zajac, J., Kurochkina, V.: Technological assurance of complex parts manufacturing. In: Ivanov, V., et al. (eds.) DSMIE 2018. LNME, pp. 51–61. Springer, Cham (2019). https://doi.org/10.1007/978-3-319-93587-4_6
4. Pavlenko, I., Simonovskiy, V., Demianenko, M.: Dynamic analysis of centrifugal machines rotors supported on ball bearings by combined application of 3D and beam finite element models. J. Phys. Conf. Ser. Mater. Sci. Eng. **233**, 012053 (2017)
5. Krol, O., Sokolov, V.: Parametric modeling of transverse layout for machine tool gearboxes. In: Gapiński, B., Szostak, M., Ivanov, V. (eds.) Manufacturing 2019. LNME, pp. 122–130. Springer, Cham (2019). https://doi.org/10.1007/978-3-030-16943-5_11
6. Kharlamov, Y., Budag'janc, N.A.: D-Gun processes in industry. Publishing house of VUGU, Lugansk (1998)
7. Yushchenko, K.A., Borisov, Y.S., Kuznetsov, V.D., Korzh, V.M.: Surface Engineering. Scientific Thought, Kiev (2007)
8. Anciferov, V.N., Bobrov, G.V., Druzhinin, L.K.: Powder metallurgy and spray coatings. In: Mitin, B.S. (eds.). Metallurgy, Moscow (1987)
9. Davis, J.R. (ed.): Handbook of Thermal Spray Technology. ASM International, Ohio (2004)
10. Pawlowski, L.: The Science and Engineering of Thermal Spray Coatings, 2nd edn. John Wiley & Sons Ltd, Chichester, England (2008)
11. Sokolov, V., Krol, O., Stepanova, O.: Automatic control system for electrohydraulic drive of production equipment. In: International Russian Automation Conference (RusAutoCon) 2018. IEEE (2018)
12. Fesenko, A., Basova, Y., Ivanov, V., Ivanova, M., Yevsiukova, F., Gasanov, M.: Increasing of equipment efficiency by intensification of technological processes. Periodica Polytechnica Mech. Eng. **63**(1), 67–73 (2019)
13. Rogovyi, A., Khovanskyy, A.: Application of the similarity theory for vortex chamber superchargers. J. Phys. Conf. Ser. Mater. Sci. Eng. **233**, 012011 (2017)
14. Sokolov, V., Krol, O., Stepanova, O.: Nonlinear simulation of electrohydraulic drive for technological equipment. J. Phys. Conf. Ser. **1278**, 012003 (2019)

15. Movshovich, A.Y., Chernaja, Y.A., Ishchenko, G.I.: Investigation of influence of physico–mechanical characteristics of D-Gun sprayimg coatings on wear resistance of cutting elements of reworkable dies. Vestnik NTU HPI: Collect. Sci. Papers Them. Release New Solutions Mod. Technol. **45**, 22–28 (2011)
16. Faradzhallah, M.A., Panarin, V.E., Bys', S.S.: The problem of technological residual stresses in improving the quality of D-Gun spraying coatings. Visnik Khmelnytskyi Natl. Univ. **2**, 14–18 (2011)
17. Kundrák, J., Mitsyk, A., Fedorovich, V., Morgan, M., Markopoulos, A.: The use of the kinetic theory of gases to simulate the physical situations on the surface of autonomously moving parts during multi – energy vibration processing. Materials **12**(19), 3054 (2019)
18. Korobov, Y.S.: Analysis of Properties of Thermal Coatings Part 2 Evaluation of Coating Parameters. Publishing House of Ural University, Yekaterinburg (2016)
19. Rjahovskij, A.V., Kosenko, V.V., Vlasenko, V.N.: Features of estimation of adhesion strength of D-Gun spraying coatings. Weapons Syst. Milit. Equip. **3**(31), 215–217 (2012)
20. Ilyushchenko, A.F., Okovityj, V.A., Kundas, S.P., Formanek, B.: Formation of Thermal Sprauing Coatings Theory and Practice. Bestprint, Minsk (2002)
21. Pierre, L., Fauchais, J., Heberlein, V., Maher, R., Boulos, I.: Thermal Spray Fundamentals: From Powder to Part. Springer Science + Business Media, New York (2014)
22. Lin, C.K., Berndt, C.C.: Measurement and analysis of adhesion strength for thermally sprayed coatings. J. Therm. Spray Technol. **3**(1), 75–104 (1994)
23. Sokolov, V., Krol, O., Stepanova, O.: Choice of correcting link for electrohydraulic servo drive of technological equipment. In: Ivanov, V., et al. (eds.) DSMIE 2019. LNME, pp. 702–710. Springer, Cham (2020). https://doi.org/10.1007/978-3-030-22365-6_70
24. Sokolov, V.: Diffusion of circular source in the channels of ventilation systems. In: Fujita, H., Nguyen, D.C., Vu, N.P., Banh, T.L., Puta, H.H. (eds.) ICERA 2018. LNNS, vol. 63, pp. 278–283. Springer, Cham (2019). https://doi.org/10.1007/978-3-030-04792-4_37
25. Chichinadze, A.V., Braun, J.D., Bushe, N.A.: Fundamentals of tribology (friction, wear, lubrication). In: Chichinadze, A.V. (eds.) Mashinostroenie, Moscow (2001)
26. Heintze, G.N., McPherson, R.: Fracture toughness of plasma sprayed zirconia coatings. Surface Coating Technol. **134**, 15–23 (1988)
27. Fedorovich, V., Mitsyk, A.: Mathematical simulation of kinematics of vibrating boiling granular medium at treatment in the oscillating reservoir. Key Eng. Mater. **581**, 456–461 (2014)
28. Rogovyi, A.: Energy performances of the vortex chamber supercharger. Energy **163**, 52–60 (2018)
29. Maksimovich, G.G., Shatinskij, V.F., Kopylov, V.I.: Physicochemical Processes During Plasma Spraying and Destruction of Coated Materials. Nauk Dumka, Kiev (1983)
30. Fedorov, V.V.: Kinetics of Damage and Destruction of Solids. Fan, Tashkent (1985)

Ball Milled Al Spheres for the Manufacturing of Casting-Based Al-CNT Composites

Hélder Puga[1], Vitor Hugo Carneiro[1(✉)], and Manuel Vieira[2]

[1] CMEMS-UMinho, University of Minho - Campus de Azúrem, Guimarães, Portugal
d5425@dem.uminho.pt
[2] CEMMPRE - Centre for Mechanical Engineering, Materials and Processes, University of Porto, R. Dr. Roberto Frias, 4200-465 Porto, Portugal

Abstract. Carbon nanotube reinforced aluminum matrix composites are considered a promising solution for applications that require high specific mechanical properties. Even though there are numerous methods for their manufacturing, these are frequently based on powder metallurgy approaches, limiting the fabrication of components with significant volumes and complex shapes. Casting, as a manufacturing technique, is regarded as the most appropriate route to obtain complex-shaped components with a relative high microstructural quality. These techniques to obtain Al-CNT cast components is still challenging due to the agglomeration, lack of dispersion, reduced bonding and density of the CNTs in Al. To address these issues, CNTs are usually pre-processed by ball-milling with Al powder to promote bonding and disperse the reinforcement, however, these techniques are not really beneficial to casting approaches due to the increase of Al_2O_3 content that do not disperse within the Al alloy melts. This study proposes the use of Al spheres (~1 mm) in these ball milling techniques to prevent significant plastic deformation, the formation of Al flakes and the increase in Al_2O_3 content. It is shown that CNTs may be dispersed and bonded to the Al sphere surfaces. Results suggest that this is a promising novel technique to allow a successful implementation of casting-based routes to fabricate high-volume and complex-shaped Al-CNT components.

Keywords: Aluminum · CNT · MMC · Ball milling · Casting

1 Introduction

Since their discovery in the early 1990s [1], carbon nanotubes (CNTs) have been considered a promising reinforcement for matrix composites due to their remarkable mechanical properties (e.g. Young's modulus ~ 1TPa and strength in the range of 30–150 GPa) [2–5]. It is expected that even in low volume fractions, CNT reinforcements can significantly improve the mechanical properties of a matrix.

CNTs have successfully been introduced in both polymer- and metal-based matrix composites [6–8]. However, due to the possibility of being dissolved in a liquid surfactant (e.g. ethanol) and ultrasonicated with CNTs, polymer-based matrix composites are

J. Machado et al. (Eds.): icieng 2021, LNME, pp. 46–56, 2022.
https://doi.org/10.1007/978-3-030-79165-0_5

relatively easier to synthesize than metal-based MMCs [9]. This is especially true for Al alloys, that due to their affinity to the resultant hydrogen from the water decomposition (H2O) in CNT pre-processing may promote exothermal reactions, as CNTs are especially prone to hydrogen storage in their interior and tube walls [10].

Carbon nanotube reinforced aluminum matrix (Al-CNT MMCs) composites are considered a promising route to obtain lightweight bulk materials with a wide range of applications due to their high specific strength, hardness and corrosion resistance [4, 11–17], especially in transportation industries to reduce fuel and energy consumption. The inclusion of CNTs in an Al matrix is reported to enhance load transfer capacity [2, 18], promoting an Orowan strengthening dislocation mechanism [19]. CNT addition is also known to generate grain refinement [20], which is beneficial to enhance mechanical properties by a Hall-Petch effect.

During the last three decades there was a significant effort to develop routes to efficiently integrate CNTs into an Al-based matrix [21], including: (i) ball milling [5, 13], rolling processes [22, 23], friction-stir [24, 25], powder metallurgy [26, 27], spaying [28] and the combination of the referred techniques (e.g. [29, 30]).

Synthesizing Al-CNT MMCs is, however, a complex procedure due to the difficulty of dispersing the reinforcement due to strong Vander Walls interaction, i.e. CNTs tend to agglomerate and promote defects such as voids and pores [31, 32]. This is especially relevant due to the affinity of both Al and CNT for hydrogen entrapment [10] which combined may cause premature crack nucleation. There are additional difficulties caused by the formation of brittle Al_4C_3 due to the Al-CNT surface reaction [33, 34] and the existence of the Al_2O_3 surface layer [35, 36], both of which compromise the direct Al-CNT bonding.

It has been shown that ball milling successfully disperses CNTs and improves the mechanical properties of an Al matrix [13, 19, 32, 37]. The processing parameters in these techniques are rather important since they directly influence the homogeneity and final properties of MMCs [10]. Ostovan et al. [38] stated that milling time has a relevant impact on the dispersion of multi-walled carbon nanotubes (MWCNTs) and, consequently, in the mechanical properties of Al-CNT MMCs. These authors also suggested that CNT content has a crucial influence in these properties, showing that high mass fraction higher than 2%wt may actually be detrimental in terms of physical and mechanical properties. Liu et al. [39] have shown a uniform CNT dispersion into the Al matrix after 6 h of ball milling, while further increasing this time seriously damages the CNTs and promotes hydrogen entrapment [10]. It is known that mechanical property enhancement in Al-CNT nanocomposites is not only compromised by the interfacial reaction and bonding [32, 34], but also due to the damage of CNTs [40].

Ball milling has been performed using low to high energy approaches. Liao et al. [9] proposed that high energy ball milling has better CNT mixing within the Al matrix than low energy milling. These authors, however, have also shown that high energy milling damages the tubular structure of the CNTs. Liu et al. [19] have suggested that medium rotation (i.e. a compromise between low and high energy ball milling) may be an efficient approach to avoid some of these issues while achieving sufficient reinforcement dispersion.

Recently, ball milling with Al powders and CNTs was used as a precursor technique to implement reinforcements in cast components, by adding their ball-milled morphology into Al melts that are then mechanically stirred [21, 41, 42], ultrasound treated [43, 44] or induction melted [45] to promote CNT dispersion. The introduction of ball milled CNT-Al powder in Al melts presents some challenges due to the CNT poor wettability, tendency to agglomerate and density mismatch [9, 44]. There is still a need to develop efficient techniques to promote a good dispersion and well-bonded CNTs in cast products. Additionally, ball milling approaches permanently deform the Al powder into flake-like morphologies [26, 39, 46]. Although this may be appropriate in powder metallurgy approaches, it generates a very high surface area and significant volume fraction of Al_2O_3. Our initial experiments have shown that the Al_2O_3 thin layer in these flakes inhibits their full melting and, thus, prevents them from really being integrated in the melt.

To address these issues, ball milling with CNT is performed with Al spheres (1 mm) to attempt to uniformly distribute and bond CNTs to an Al matrix, while preventing the significant formation of Al_2O_3 during the pre-processing (i.e. ball milling) stage.

2 Methodology

The mixture between aluminum powders and CNTs was performed by a twostep process: (i) an initial slurry-based dispersion; and (ii) ball milling. These processes were performed by the adoption of two approaches (according to Table 1) that contemplate the analysis of using Al powders and spheres. Details on the specific physical properties of the referred materials are also detailed in Table 1.

Table 1. Properties of used materials and manufacturing processes.

Materials		Process step	
Aluminum	CNT	Slurry dispersion	Ball milling
Powder (1 μm, 99.8% Al)	>95% purity Diameter~30–50 nm Length~0.5–2 mm	– Mechanical Stirring in solution (50% deionized water + 50% ethyl alcohol); – Vacuum chamber dried	– Argon atmosphere; – 6 mm stell balls; – 400 rpm for 8 h
Ball (1 mm, 98.0% Al)			

For the first step, a slurry based dispersion was performed by the addition of 2%wt of CNT and the reminiscent mass fraction of Al powder/balls (see Table 1) in a solution (50% deionized water + 50% ethyl alcohol). The slurry was mechanically stirred and dried in a vacuum chamber.

The dried slurry was then ball milled using a planetary ball milling machine under an Argon inert atmosphere with 6 mm stainless steel balls (20:1 ball-to-powder/sphere weight ratio). According to the suggestion of Liu et al. [19], an intermediate milling energy was performed by maintaining a 400 rpm rotation rate for 8 h to avoid CNT damage and still promote an their uniform dispersion.

The final products from both Al powder and sphere approaches were observed by scanning electron microscopy (JEOL JSM 6301F/Oxford INCA Energy 350) to detail CNT distribution and the final configuration of the Al products to predict their impact when added to an Al alloy melt. Energy Dispersive Spectroscopy (EDS) was also used to characterize the reagents, estimate their composition and assess the inclusion of contaminants in the final ball milled product. The initial powder and spheres were also characterized by the referred techniques to determine their initial dimensions and morphology.

3 Results and Discussion

Figure 1 shows the SEM micrographs of the initial Al powders (Fig. 1(a)) and spheres (Fig. 1(b)). Further detailing both Al configurations by EDS, shows that while the powders are fundamentally composed by Al (Fig. 1(c)), the spheres are alloyed with Mg (Fig. 1(d)). This is expected, given that spheres are composed by a $98\%_{wt}$ of Al, thus, they have a $2\%_{wt}$ composition of other elements. In these other elements, there is a predominance of Mg. Traces of elemental oxygen and Au were also captured by the EDS, being attributed to: (i) the thin alumina (Al2O3) layer in the surface of the powders (~5 nm [47]); and (ii) the detection of the Au/Pd coating that was used in the preparation of the SEM samples.

Relatively to the Al powder and sphere dimensions, Fig. 2 shows that they reveal an average diameter of, respectively, 1.25 ± 0.24 μm and 1.13 ± 0.03 mm. These results imply that there is a small deviation to their nominal diameters (respectively, 1 μm and 1 mm), having an overall larger size than expected. Still the difference between the sizes of sample type, i.e. powder and sphere, tends a ~10^3 magnitude, appropriate to test the influence of Al substrate size in the dispersion and bonding of CNTs for MMC casting purposes.

Figure 3(a) displays the post-ball milled Al-powder, showing severe plastic deformation, promoted by the steel balls, and assume a flake configuration, a common feature in powder metallurgy Al/CNT composites already reported by other authors [37]. The magnification in Fig. 3(b) and the presence of carbon in the Z3 EDS spectrum (Fig. 3(c)) suggest that the surface of these Al flakes are rich in CNTs. This shows that the two step, i.e. slurry dispersion and ball milling, were able to disperse the nanotubes. The oxygen content in the Z3 EDS spectrum (Fig. 3(c)), also identifies the Al2O3 layer in the surface of the flakes.

Due to the flake configuration of the post-processed samples, however, it is observable (Fig. 3(a)) that this shape promotes the increase of Al_2O_3 volume fraction. According to the model for plastic deformation in Fig. 4, as the initial spherical shape of the powders is deformed, first into thick plates and finally to the flake morphology, there is a relevant increase in the overall surface that contacts with oxygen, motivating this increase in Al_2O_3 content. This is further evidenced by the model presented in Eq. 1, where the volume fraction of Al_2O_3 ($V_{\%,Al2O3}$) is related to the total volume of the flake (V_T) and the volume of Al (V_{Al}). It may be observed that for a given simplified model of a

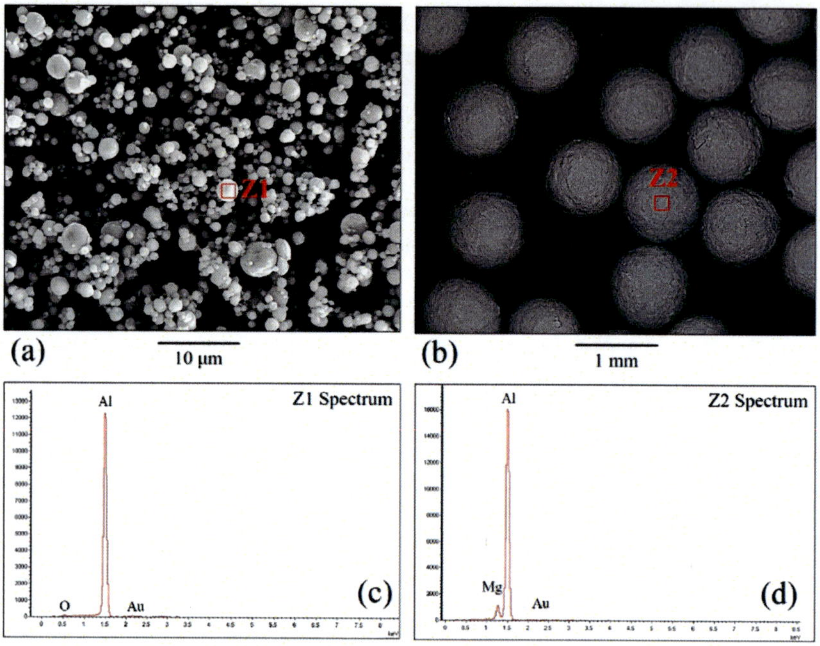

Fig. 1. SEM micrographs of Al (a) powders and (b) spheres; EDS spectra in (c) Z1 and (d) Z2, respectively in a powder and sphere samples.

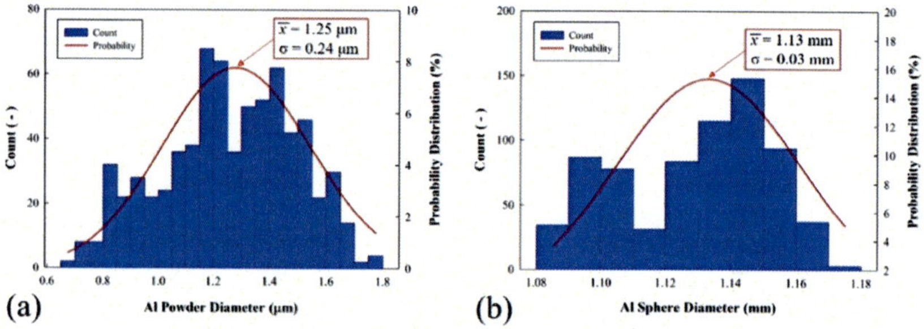

Fig. 2. Al sample dimension characterization: (a) powders and (b) spheres.

flake, that its V_T and V_{Al} will depend on their thickness (t_i), length (l_i) and width (w_i), respectively for the total (i = T) and Al (i = Al) volumes. Recurring to Eq. 2, it may be theoretically inferred that as the thickness (t_{Al}) of Al is reduced due to the isochoric plastic deformation and, eventually, approximates a zero value, the flake also tends to be composed only by Al_2O_3.

$$V_{\%,Al_2O_3} = 1 - \frac{V_{Al}}{V_T} = 1 - \frac{t_{Al}w_{Al}l_{Al}}{t_T w_T l_T} \tag{1}$$

$$\lim_{t_{Al} \to 0} \left(V_{\%,Al_2O_3} \right) \tag{2}$$

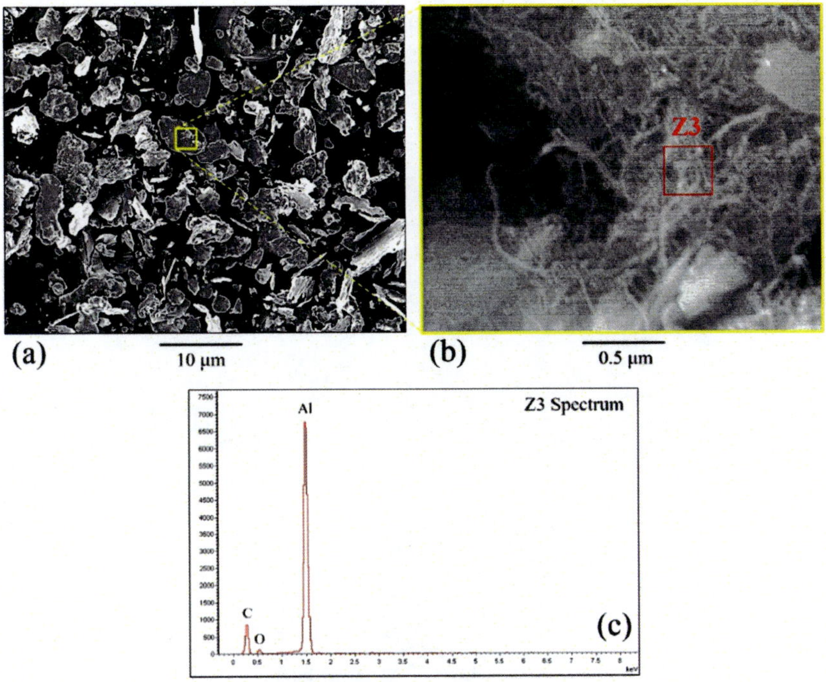

Fig. 3. SEM micrographs of Al flakes with CNT dispersed in the surface: (a) shape characterization and (b) magnified surface detailing CNTs, where (c) is the Z3 EDS spectrum.

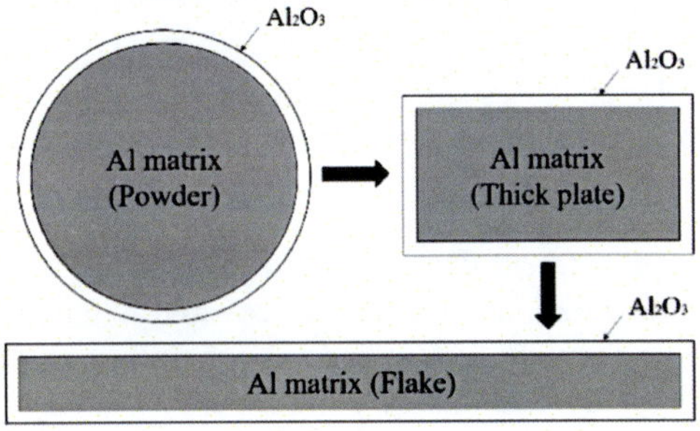

Fig. 4. Model for the powder plastic deformation during the ball milling stage.

This high increase in Al_2O_3 content is not beneficial in a MMC casting process. Although the Al enclosed in the Al flakes may change phase, the reduced Al volume fraction (i.e. as the volume fraction of Al_2O_3 increases, effectively limits the pressure to break the alumina layer and connect to an Al alloy melt. This means that the flakes are not really dispersed in an Al alloy melt and, therefore, release the CNTs for them to be dispersed within the Al matrix.

Figure 5 details the morphology of the post-milled Al spheres. It is shown that, even though there is evidence of plastic deformation by the observation of dints in the surface of the spheres, their sphericity is not compromised. This implies that there are no significant changes to the Al_2O_3 content and they maintain their ability to completely melt. Therefore, the Al content in the sphere is able to be fully dissolved in an Al alloy melt and release the CNTs in its surface.

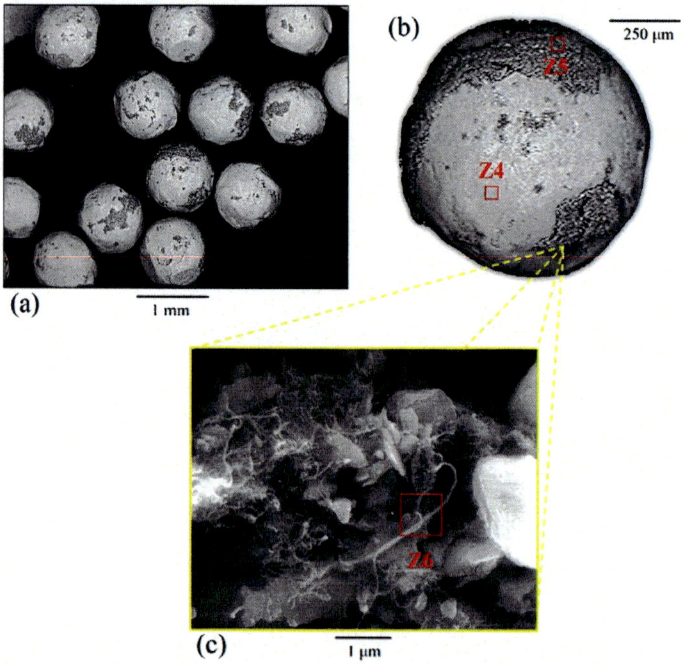

Fig. 5. SEM micrographs of Al spheres with CNT dispersed in their surface: (a) shape characterization and (b) unitary sphere with (c) magnified surface detailing CNT distribution.

Figure 5(b) shows a different contrast between different areas of the Al spheres, being suggested that the dark colored areas are covered with dispersed CNTs. This is attested by the EDS spectra of Figs. 6(a) and (b), where it is shown that the light gray areas (e.g. Figure 5(b) – Z4) are mainly composed by the Al matrix, the previously identified alloying Mg (Fig. 1(d)) and residual elemental carbon. The latter is suggested to be a trace element from the ball milling process. The presence of carbon in the dark areas of Fig. 5(b) is much higher, as shown in the EDS spectrum of Fig. 6(b), proving that these are in fact CNTs that were dispersed in the surface of the Al spheres. This is further evidenced by the observation of Fig. 5(c), in which CNTs are clearly visible, and the EDS spectrum in Fig. 6(c), correspondent to Z6 in Fig. 5(c). It is worth mentioning that the concentration of the ion beam in the Z6 EDS spectrum was able to identify other trace element from the sphere alloy (e.g. Si and K), the thin surface layer of Al_2O_3 (i.e. presence of oxygen) and the Au/Pd coating that was used in the preparation of the SEM samples.

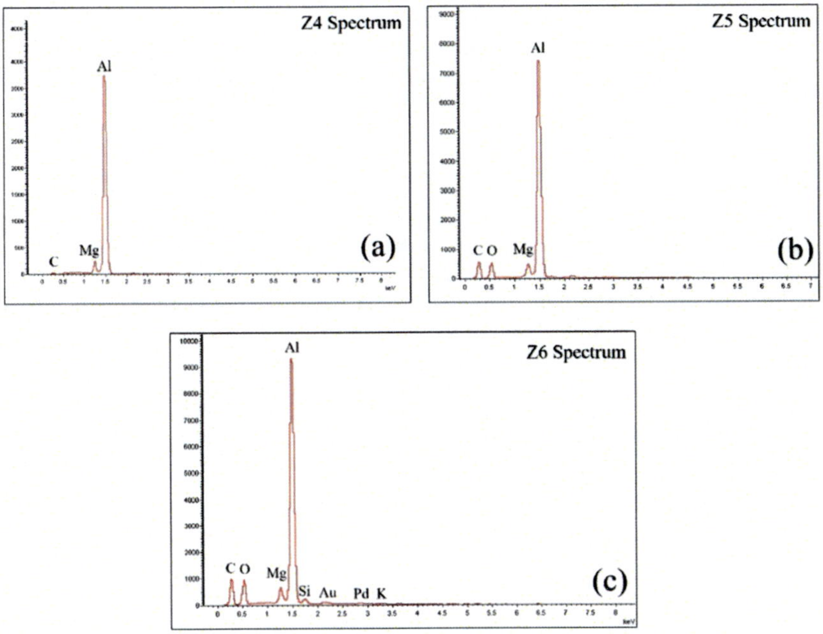

Fig. 6. EDS spectra from areas (a) Z4, (b) Z5 and (c) Z6.

4 Conclusions

This study details a novel approach that uses a slurry based and ball milling process to disperse CNTs in an Al substrate, using spheres (1 mm) instead of powders (1 μm) to allow their use in MMC casting-based manufacturing methods. The objective of the proposed technique is avoiding the formation of Al flakes that, although are appropriate for powder sintering methods, are not beneficial in casting-based methods due to their high Al_2O_3 content. The following conclusions were drawn:

(i) The use of Al spheres instead of Al powders, prevents severe plastic deformation and the formation of flakes. Therefore, the proposed method is able to prevent a high increase in particle content;
(ii) Ball-milling is able to disperse CNTs in the surface of the Al spheres, although, this dispersion is less efficient relatively the Al powders (i.e. in the surface of Al flakes);
(iii) Al spheres are known to be easily melt, relatively to Al flakes due to their high Al_2O_3 content, thus, given their bonding to dispersed CNTs are a promising solution for the manufacturing of casting-based MMC components.

Acknowledgements. This work was supported by PTDC/EMEEME/30967/2017 and NORTE-0145-FEDER-030967, co-financed by the European Regional Development Fund (ERDF), through the Operational Programme for Competitiveness and Internationalization (COMPETE

2020), under Portugal 2020, and by the Fundação para a Ciência e a Tecnologia – FCT I.P. national funds. Also, this work was supported by Portuguese FCT, under the reference project UIDB/04436/2020, Stimulus of Scientific Employment Application CEECIND/03991/2017.

References

1. Iijima, S.: Helical microtubules of graphitic carbon. Nature **354**, 56–58 (1991)
2. Tjong, S.C.: Recent progress in the development and properties of novel metal matrix nanocomposites reinforced with carbon nanotubes and graphene nanosheets. Mater. Sci. Eng. R Rep. **74**, 281–350 (2013)
3. Coleman, J.N., Khan, U., Blau, W.J., Gunko, Y.K.: Small but strong: A review of the mechanical properties of carbon nanotube–polymer composites. Carbon **44**, 1624–1652 (2006)
4. Bakshi, S.R., Lahiri, D., Agarwal, A.: Carbon nanotube reinforced metal matrix composites - a review. Int. Mater. Rev. **55**, 41–64 (2010)
5. Jargalsaikhan, B., Bor, A., Lee, J., Choi, H.: Al/CNT nanocomposite fabrication on the different property of raw material using a planetary ball mill. Adv. Powder Technol. **31**, 1957–1962 (2020)
6. Hou, Y., Tang, J., Zhang, H., Qian, C., Feng, Y., Liu, J.: Functionalized few-walled carbon nanotubes for mechanical reinforcement of polymeric composites. ACS Nano **3**, 1057–1062 (2009)
7. Thostenson, E.T., Ren, Z., Chou, T.-W.: Advances in the science and technology of carbon nanotubes and their composites: a review. Compos. Sci. Technol. **61**, 1899–1912 (2001)
8. Peigney, A., Laurent, C., Flahaut, E., Rousset, A.: Carbon nanotubes in novel ceramic matrix nanocomposites. Ceram. Int. **26**, 677–683 (2000)
9. Liao, J., Tan, M.-J.: Mixing of carbon nanotubes (CNTs) and aluminum powder for powder metallurgy use. Powder Technol. **208**, 42–48 (2011)
10. Liu, F., et al.: Preparation of short carbon nanotubes by mechanical ball milling and their hydrogen adsorption behavior. Carbon **41**, 2527–2532 (2003)
11. Bradbury, C.R., Gomon, J.-K., Kollo, L., Kwon, H., Leparoux, M.: Hardness of Multi Wall Carbon Nanotubes reinforced aluminium matrix composites. J. Alloy. Compd. **585**, 362–367 (2014)
12. George, R., Kashyap, K.T., Rahul, R., Yamdagni, S.: Strengthening in carbon nanotube/aluminium (CNT/Al) composites. Scripta Mater. **53**, 1159–1163 (2005)
13. Choi, H., Shin, J., Min, B., Park, J., Bae, D.: Reinforcing effects of carbon nanotubes in structural aluminum matrix nanocomposites. J. Mater. Res. **24**(8), 2610–2616 (2009). https://doi.org/10.1557/jmr.2009.0318
14. Sridhar, I., Narayanan, K.R.: Processing and characterization of MWCNT reinforced aluminum matrix composites. J. Mater. Sci. **44**(7), 1750–1756 (2009). https://doi.org/10.1007/s10853-009-3290-5
15. Esawi, A.M.K., Morsi, K., Sayed, A., Taher, M., Lanka, S.: Effect of carbon nanotube (CNT) content on the mechanical properties of CNT-reinforced aluminium composites. Compos. Sci. Technol. **70**, 2237–2241 (2010)
16. Popov, V.N.: Carbon nanotubes: properties and application. Mater. Sci. Eng. R Rep. **43**, 61–102 (2004)
17. Casati, R., Vedani, M.: Metal matrix composites reinforced by nano-particles - a review. Metals **4**, 65–83 (2014)

18. Kim, K.T., Eckert, J., Menzel, S.B., Gemming, T., Hong, S.H.: Grain refinement assisted strengthening of carbon nanotube reinforced copper matrix nanocomposites. Appl. Phys. Lett. **92**, 121901 (2008)
19. Liu, Z.Y., Xiao, B.L., Wang, W.G., Ma, Z.Y.: Modelling of carbon nanotube dispersion and strengthening mechanisms in Al matrix composites prepared by high energy ball milling-powder metallurgy method. Compos. A Appl. Sci. Manuf. **94**, 189–198 (2017)
20. Prabhakar, G.V.N.B., Pavan Kumar, Y.V.R.S.N., Dileep Kumar, P., Prasanna Kumar, B., Gopi Raju, M., Naseema, S., Ravi Kumar, N., Jagannatham, M., Ratna Sunil, B.: Producing Al5083-CNT composites by friction stir processing: influence of grain refinement and CNT on mechanical and corrosion properties. Mater. Today Proc. **15**, 44–49 (2019)
21. Khanna, V., Kumar, V., Bansal, S.A.: Mechanical properties of aluminium-graphene/carbon nanotubes (CNTs) metal matrix composites: advancement, opportunities and perspective. Mater. Res. Bull. **138**, 111224 (2021)
22. Bi, S., et al.: Enhancing strength-ductility synergy of carbon nanotube/7055Al composite via a texture design by hot-rolling. Mater. Sci. Eng. A **806**, 140830 (2021)
23. Ramkumar, K.R., Dinaharan, I.: Accumulative roll bonding route for composite materials production. In: Reference Module in Materials Science and Materials Engineering. Elsevier (2020)
24. Alladi, A., Aluri, M., Maddela, N., Abbadi, C.R.: Recent progress of CNTs reinforcement with metal matrix composites using friction stir processing. Mater. Today Proc. (2021, in press)
25. Liu, Z.Y., Xiao, B.L., Wang, W.G., Ma, Z.Y.: Singly dispersed carbon nanotube/aluminum composites fabricated by powder metallurgy combined with friction stir processing. Carbon **50**, 1843–1852 (2012)
26. Chen, M., et al.: Design of an efficient flake powder metallurgy route to fabricate CNT/6061Al composites. Mater. Des. **142**, 288–296 (2018)
27. Trinh, P.V., et al.: Microstructure, microhardness and thermal expansion of CNT/Al composites prepared by flake powder metallurgy. Compos. A Appl. Sci. Manuf. **105**, 126–137 (2018)
28. Zhang, Y., Wang, Q., Chen, G., Ramachandran, C.S.: Mechanical, tribological and corrosion physiognomies of CNT-Al metal matrix composite (MMC) coatings deposited by cold gas dynamic spray (CGDS) process. Surf. Coat. Technol. **403**, 126380 (2020)
29. Guo, B., et al.: Microstructures and mechanical properties of carbon nanotubes reinforced pure aluminum composites synthesized by spark plasma sintering and hot rolling. Mater. Sci. Eng. A **698**, 282–288 (2017)
30. Toozandehjani, M., Ostovan, F., Jamaludin, K.R., Amrin, A., Matori, K.A., Shafiei, E.: Process−microstructure−properties relationship in Al−CNTs−Al2O3 nanocomposites manufactured by hybrid powder metallurgy and microwave sintering process. Trans. Nonferrous Metals Soc. China **30**, 2339–2354 (2020)
31. Ma, P.-C., Siddiqui, N.A., Marom, G., Kim, J.-K.: Dispersion and functionalization of carbon nanotubes for polymer-based nanocomposites: a review. Compos. A Appl. Sci. Manuf. **41**, 1345–1367 (2010)
32. Bakshi, S.R., Agarwal, A.: An analysis of the factors affecting strengthening in carbon nanotube reinforced aluminum composites. Carbon **49**, 533–544 (2011)
33. Rodríguez-Reyes, M., Pech-Canul, M.I., Rendón-Angeles, J.C., López-Cuevas, J.: Limiting the development of Al4C3 to prevent degradation of Al/SiCp composites processed by pressureless infiltration. Compos. Sci. Technol. **66**, 1056–1062 (2006)
34. Fan, G., et al.: Enhanced interfacial bonding and mechanical properties in CNT/Al composites fabricated by flake powder metallurgy. Carbon **130**, 333–339 (2018)

35. Lalet, G., Kurita, H., Miyazaki, T., Kawasaki, A., Silvain, J.-F.: Microstructure of a carbon fiber-reinforced aluminum matrix composite fabricated by spark plasma sintering in various pulse conditions. J. Mater. Sci. **49**(8), 3268–3275 (2014). https://doi.org/10.1007/s10853-014-8032-7
36. Kurita, H., et al.: Load-bearing contribution of multi-walled carbon nanotubes on tensile response of aluminum. Compos. A Appl. Sci. Manuf. **68**, 133–139 (2015)
37. Esawi, A.M.K., Morsi, K., Sayed, A., Taher, M., Lanka, S.: The influence of carbon nanotube (CNT) morphology and diameter on the processing and properties of CNT-reinforced aluminium composites. Compos. A Appl. Sci. Manuf. **42**, 234–243 (2011)
38. Ostovan, F., et al.: Effects of CNTs content and milling time on mechanical behavior of MWCNT-reinforced aluminum nanocomposites. Mater. Chem. Phys. **166**, 160–166 (2015)
39. Liu, Z.Y., Xu, S.J., Xiao, B.L., Xue, P., Wang, W.G., Ma, Z.Y.: Effect of ball-milling time on mechanical properties of carbon nanotubes reinforced aluminum matrix composites. Compos. A Appl. Sci. Manuf. **43**, 2161–2168 (2012)
40. Hassan, M.T.Z., Esawi, A.M.K., Metwalli, S.: Effect of carbon nanotube damage on the mechanical properties of aluminium–carbon nanotube composites. J. Alloy. Compd. **607**, 215–222 (2014)
41. Soni, S.K., Thomas, B., Kar, V.R.: A comprehensive review on CNTs and CNT-reinforced composites: syntheses. Charact. Appl. Mater. Today Commun. **25**, 101546 (2020)
42. Li, G., Qu, Y., Yang, Y., Zhou, Q., Liu, X., Li, R.: Improved multi-orientation dispersion of short carbon fibers in aluminum matrix composites prepared with square crucible by mechanical stirring. J. Mater. Sci. Technol. **40**, 81–87 (2020)
43. Yan, H., Huang, Z.-X., Qiu, H.-X.: Microstructure and mechanical properties of CNTs/A356 nanocomposites fabricated by high-intensity ultrasonic processing. Metall. Mater. Trans. A **48**(2), 910–918 (2016). https://doi.org/10.1007/s11661-016-3872-1
44. Li, Q., Rottmair, C.A., Singer, R.F.: CNT reinforced light metal composites produced by melt stirring and by high pressure die casting. Compos. Sci. Technol. **70**, 2242–2247 (2010)
45. Mansoor, M., Shahid, M.: Carbon nanotube-reinforced aluminum composite produced by induction melting. J. Appl. Res. Technol. **14**, 215–224 (2016)
46. Yuan, C., et al.: Enhanced ductility by Mg addition in the CNT/Al-Cu composites via flake powder metallurgy. Mater. Today Commun. **26**, 101854 (2020)
47. Saif, M.T.A., Zhang, S., Haque, A., Hsia, K.J.: Effect of native Al2O3 on the elastic response of nanoscale Al films. Acta Mater. **50**, 2779–2786 (2002)

Reliability Estimation of the Solid Lubricated Bearing Based on a General Wiener Process and Its Experimental Validation

Rentong Chen[1,2](✉), Shaoping Wang[1,2], Chao Zhang[2,3], and Mileta Tomovic[4]

[1] School of Automation Science and Electrical Engineering, Beihang University, Beijing 100191, People's Republic of China
rentongchen@buaa.edu.cn, shaopingwang@vip.sina.com
[2] Beijing Advanced Innovation Center for Big Data-Based Precision Medicine, Beihang University, Beijing 100191, People's Republic of China
cz@buaa.edu.cn
[3] Research Institute of Frontier Science, Beihang University, Beijing 100191, People's Republic of China
[4] Engineering Technology Department, Old Dominion University, Norfolk, VA 23529, USA
mtomovic@odu.edu

Abstract. Conducting accurate reliability estimation plays an important role for solid lubricated bearings, because they are widely used in highly reliable space applications. This paper proposes a general Wiener process to describe the degradation process for solid lubricated bearings. The performance indicator for solid lubricated bearings is a fusion indicator from several time domain statistics of vibration signal during operation. Mahalanobis Distance (MD) is utilized to fuse these time-domain statistics indicators into one indicator representing the degradation process. Two different time scale functions in drift term and fluctuation term of Wiener process are used to capture the nonlinearity characteristics of degradation process. Bayesian MCMC method is used to estimate all the unknown parameters in the proposed degradation model. Degradation data from experiment is used to validate the model. Results show that general Wiener process degradation model is fit well for the degradation data.

Keywords: Solid lubricated bearings · Mahalanobis Distance · Wiener process · Reliability estimation

1 Introduction

Solid lubricated bearings are widely used in space shuttles, rockets, satellites and other space drive mechanisms [1]. Once the failure occurs on the solid lubricated bearings, unpredictable consequences or economic loss cannot be manageable. Therefore, it is essential to provide accurate lifetime prediction for solid lubricated bearings used in space.

When degradation data are available, it is very important to establish appropriate degradation model which can describe the underlying degradation process accurately.

J. Machado et al. (Eds.): icieng 2021, LNME, pp. 57–67, 2022.
https://doi.org/10.1007/978-3-030-79165-0_6

According to Si [2], the degradation modeling and reliability estimation approaches in degradation testing can be clarified into three approaches which are knowledge-based approaches, physical model-based approaches and data-driven approaches. Knowledge-based approaches typically require experts' empirical knowledge on the relevant components with corresponding available degradation or failure data [3, 4]. Physical model-based approaches perform the reliability estimation by a series of mathematical and physical models based on deep understanding of physical behavior [5, 6]. Data-driven approaches use the degradation data collected by the sensors, analyze and transform them into corresponding models to represent the degradation behavior or process. At present, there are mainly two types of data-driven methods, which are based on learning methods and stochastic models, respectively. As for solid lubricated bearings, failure mechanism and degradation behavior are complex, and the operating conditions are usually harsh when they are in operation in space [7]. Therefore, it might be complicated to build an accurate physical model in order to describe the degradation process the solid lubricated bearings. There is usually a lot of degradation information in vibration signal of bearings, and time-domain features of it is widely used [8]. Stochastic process degradation model can reasonably identify the random variation in degradation process [2], and it is effective approach in capturing the degradation process of bearings. The failure mechanism of solid lubricated bearings is different from that of the bearings lubricated by oil or grease. However, experimental show that vibration signal can still be used to reflect its degradation process [8, 9]. Therefore, the degradation process of solid lubricated bearings could be described by time-domain statistics from vibration signal, and it could also be described by stochastic process model.

As for Wiener process, linear time scale function can be used when the degradation is linear. The distribution of failure time is inverse Gaussian distribution, and it has the advantage to obtain analytical solution of failure time [10, 11]. As for nonlinear degradation process, an appropriate nonlinear structure needs to be considered. The time scale transformation is one of the approaches to capture the nonlinearity of degradation process [12–14]. However, the same time scale function is used in the drift term and fluctuation term of Wiener process, and this might not be suitable for some real engineering cases. Si et al. [15] and Wang et al. [16] extended Wiener process into a degradation model in which time scale functions in drift term and fluctuation term are different, and the approximate analytical probability density function (PDF) and cumulative density function (CDF) of failure time distribution of nonlinear degradation model were derived and related theoretical results were obtained.

Based on the discussion above, a general Wiener process degradation model is used to conduct the reliability estimation of solid lubricated bearings in space. Firstly, based on the application of Mahalanobis Distance (MD) fusion of multiple statistics on bearings [17], similar method is utilized to the feature fusion of solid lubricated bearings in our study. Secondly, the degradation process is described by a general Wiener process. Three cases, including (1) linear time scale function, (2) nonlinear and the same time scale function in drift term and fluctuation term and (3) nonlinear but different time scale functions in drift term and fluctuation term, are discussed. Thirdly, the reliability estimation is conducted by Monte Carlo method instead of using the approximation function. Unknown parameters in degradation model is estimated by Bayesian MCMC method, and the model we proposed is validated by using experimental data from the degradation test.

The rest of paper is organized as follows. The failure mechanism for solid lubricated bearings is introduced, and degradation model is established based on a general Wiener process in Sect. 2, Bayesian approach is utilized to conduct unknown parameter estimation. In Sect. 3, degradation test is conducted to verify the proposed model. Estimation of unknown parameters and reliability function are derived. Misspecification errors corresponding to different degradation models are also analyzed. The paper is concluded in Sect. 4.

2 Failure Mechanism and Model Development

2.1 Failure Mechanism

Typically, a solid lubricated bearing contains outer race, inner race, rolling element and retainer, as shown in Fig. 1. No outer lubrication is needed, which is different from oil lubricated bearings. Both races are placed with solid lubricated film. The retainer is made of self-lubrication material. Rolling elements are typically made by steel, which is the non-lubrication material.

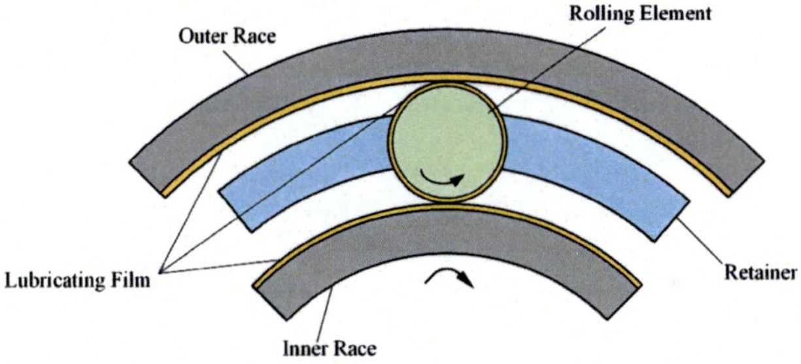

Fig. 1. Schematic of solid lubricated bearings

During the working operation of a solid lubricated bearing, two lubrication sources, including the lubrication film between the inner and outer races and the transformation from the lubrication materials of the retainer. If the transfer rate is small, the lubricant will be consumed gradually. If the transfer rate is big, the transferred film will accumulate both on the inner and outer races. It will result in blocking or heavy vibration and noise [18].

Vibration signal contains information of the degradation process of solid lubricated bearings. Related studies illustrate that the change of state of lubrication film might result in the change of the statistical characteristics of vibration signal for solid lubricated bearings [9, 19]. Time domain statistics obtained from vibration signal might be suitable to reflect degradation process. In the next subsection, fusion of multiple time-domain statistics will be analyzed in detail.

2.2 Degradation Feature Fusion

Statistical features of the vibration signal can be utilized to refelct the degradation process of bearings. These statistics features is that they are easy to implement and need low computation time. Maximum value, root mean square, kurtosis and crest factor are four types of statistical features which are extracted from vibration signal and are commonly used to reflect the working status

To conduct the reliability estimation of solid lubricated bearings, a proper degradation model for the degradation signal is necessary. One feature might not be adequate suitable to describe the degradation process. All the features can be fused into one performance indicator by calculating the MD. MD has been widely used in degradation analysis of mechanical and electrical components [20–22]. Assuming that W_{pj} represents the observation of p th ($p = 1, 2, \cdots, Q$) performance indicator (time-domain feature) on the j th ($j = 1, 2, \cdots, M$) measurement. Then the ($Q \times 1$) data vector for the normal group can be denoted by $\mathrm{W}_j (j = 1, 2, \cdots, M)$. After normalization, MDs for i th ($i = 1, 2, \cdots, N$) bearing can be obtained by

$$X_i(t_j) = \frac{1}{Q} \mathrm{Z}_{ij}^T \mathrm{C}_i^{-1} \mathrm{Z}_{ij} \tag{1}$$

where $X_i(t_j)$ denotes the fusion degradation performance indicator for the i th bearing on j th measurement, Z_{ij} is the normalized vector based on $\mathbf{W}_j$, and $\mathbf{Z}_{ij}^T$ is the transpose of $\mathbf{Z}_{ij}$. C_i is the correlation matrix which can be given by

$$\mathrm{C}_i = \frac{1}{M-1} \sum_{j=1}^{M} \mathrm{Z}_{ij} \mathrm{Z}_{ij}^T \tag{2}$$

Then, all the features are fused to one performance indicator for assessment.

2.3 Wiener Process Degradation Model

Wiener process has been commonly utilized in describing the degradation process of mechatronics components [2]. In this study, the degradation process of bearings is modeled by Wiener process. It can be given as Eq. (3)

$$M_0 : X(t) = \mu \Lambda(t; \gamma) + \sigma B(\Lambda(t; \theta)) \tag{3}$$

where μ is the drift parameter representing the degradation rate, and σ is a constant parameter used to describe the short term fluctuations. $B(\cdot)$ is the standard Brownian motion. $\Lambda(t; \gamma)$ and $\Lambda(t; \theta)$ are the time-scale transformation functions. In particular, the power law function $\Lambda(t; \gamma) = t^\gamma$ and $\Lambda(t; \theta) = t^\theta$ is widely used to characterize the nonlinear degradation path. When $\gamma = \theta = 1$, the degradation path is linear. When $\gamma, \theta > 1$, the degradation path is concave, and when $0 < \gamma, \theta < 1$, the degradation path is convex. Given D as the threshold level for the performance indicator, the failure time T can be expressed as

$$T = \inf\{t : X(t) \geq D\} \tag{4}$$

With regarding to the different values and different relationships between γ and θ, the degradation model M_0 can be extended to the following three forms.

(i) If $\gamma = \theta = 1$, and $t^{\gamma} = t^{\theta} = t$, M_0 will become a Wiener process degradation model with a linear drift, and M_0 will reduce to

$$M_1 : X(t) = \mu t + \sigma B(t) \tag{5}$$

It is known that the failure time T follows inverse Gaussian distribution with the PDF $f_{M_1}(t)$ and CDF $F_{M_1}(t)$, given by Eqs. (6) and (7), respectively [2].

$$f_{M_1}(t) = \frac{D}{\sqrt{2\pi\sigma^2 t^3}} \exp\left(-\frac{(D-\mu t)^2}{2\sigma^2 t}\right) \tag{6}$$

$$F_{M_1}(t) = \Phi\left[\frac{\mu t - D}{\sigma\sqrt{t}}\right] + \exp\left(\frac{2\mu D}{\sigma^2}\right)\Phi\left[\frac{-\mu t + D}{\sigma\sqrt{t}}\right] \tag{7}$$

(ii) If $\gamma = \theta \neq 1$, and $t^{\gamma} = t^{\theta}$, M_0 will become

$$M_2 : X(t) = \mu t^{\gamma} + \sigma B(t^{\gamma}) \tag{8}$$

In this case, the PDF $f_{M_2}(t)$ and CDF $F_{M_2}(t)$ of failure time T can be obtained as Eqs. (9) and (10), respectively.

$$f_{M_2}(t) = \frac{\gamma t^{\gamma-1} D}{\sqrt{2\pi\sigma^2 t^{3\gamma}}} \exp\left(-\frac{(D-\mu t^{\gamma})^2}{2\sigma^2 t^{\gamma}}\right) \tag{9}$$

$$F_{M_2}(t) = \Phi\left[\frac{\mu t^{\gamma} - D}{\sigma\sqrt{t^{\gamma}}}\right] + \exp\left(\frac{2\mu D}{\sigma^2}\right) \cdot \Phi\left[\frac{-\mu t^{\gamma} + D}{\sigma\sqrt{t^{\gamma}}}\right] \tag{10}$$

(iii) If $\gamma \neq \theta$, M_0 will be transformed into

$$M_3 : X(t) = \mu t^{\gamma} + \sigma B(t^{\theta}) \tag{11}$$

As for the nonlinear model M_3, no closed-form function of PDF $f_{M_3}(t)$ and CDF $F_{M_3}(t)$ could be derived. Therefore, the exact failure time distribution is difficult to be obtained in the closed-form. Based on Si [15] and Wang [16], the PDF $f_{M_3}(t)$ of failure time T can be approximated by

$$f_{M_3}(t) \cong \frac{1}{K}\frac{\theta t^{\theta-1}}{\sqrt{2\pi t^{\theta}}}\left(\frac{(D-\mu t^{\gamma})}{\sigma t^{\theta}} + \frac{\gamma\mu t^{\gamma-\theta}}{\theta\sigma}\right)\exp\left(-\frac{(D-\mu t^{\gamma})^2}{2\sigma^2 t^{\theta}}\right) \tag{12}$$

where $K = \int_0^{\infty} \frac{\theta t^{\theta-1}}{\sqrt{2\pi t^{\theta}}}\left(\frac{(D-\mu t^{\gamma})}{\sigma t^{\theta}} + \frac{\gamma\mu t^{\gamma-\theta}}{\theta\sigma}\right)\exp\left(-\frac{(D-\mu t^{\gamma})^2}{2\sigma^2 t^{\theta}}\right)dt$

Then, CDF of $F_{M_3}(t)$ can be obtained as

$$F_{M_3}(t) \cong \int_0^t f_{M_3}(u)du \tag{13}$$

Based on the Eq. (13), the reliability function can be given by

$$R_{M_i}(t) = 1 - F_{M_i}(t)(i = 1, 2, 3) \tag{14}$$

As for M_3, it can be found that $F_{M_3}(t)$ and corresponding $R_{M_3}(t)$ might be difficult to calculate based on Eqs. (13) and (14). Monte Carlo based method is utilized in our work to conduct the reliability estimation in this case. The main idea for this method is that the reliability function $R(t)$ at the exact time t equals the probability that the degradation value $X_{\max}(t)$ during time interval $[0, t]$ is less than the threshold D. This probability can be obtained by Monte Carlo method. A similar application can be found in Ref [23].

2.4 Bayesian Inference

As discussed above, the fusion degradation performance indicator for the ith $(i = 1, 2, \cdots, N)$ bearing at the jth $(j = 1, 2, \cdots, M)$ measurement during the test is $X_i(t_j)$, and the degradation increment can be denoted as $\Delta X_i(t_j) = X_i(t_j) - X_i(t_{j-1})$. Then the degradation model based on the basic model M_0 for solid lubricated bearings can be showned as Eq. (15)

$$\begin{aligned} &\Delta X_i(t_j) \sim N\left(\mu\Delta\Lambda(t_j; \gamma), \sigma^2\Delta\Lambda(t_j; \theta)\right) \\ &X_i(t_j) = \sum_{l=0}^{j} \Delta X_i(t_l) \\ &(i = 1, 2, \cdots, N; j = 1, 2, \cdots, M) \end{aligned} \tag{15}$$

With regarding to different degradation model including M_1, M_2 and M_3, there are different unknown parameters set, i.e. $\mathbf{\Theta}_{M_1} = (\mu, \sigma)$, $\mathbf{\Theta}_{M_2} = (\mu, \sigma, \gamma)$, $\mathbf{\Theta}_{M_3} = (\mu, \sigma, \gamma, \theta)$. All of these unknown parameters need to be estimated to obtain the reliability function. The likelihood function for degradation model $M_k(k = 1, 2, 3)$ can be given by

$$\ln L(\mathbf{\Theta}_{M_k}) = \sum_{i=1}^{N}\sum_{j=1}^{M} \ln\left[f_{M_k}(\Delta X_i(t_j))\right] \tag{16}$$

Due to the high dimensional parameter space in the degradation model, Bayesian Markov Chain Monte Carlo (MCMC) method is used to estimated unknown parameters. The whole Bayesian MCMC process is conducted in OpenBUGS software. All the prior information of unknown parameters is selected as non-informative. Then, the posterior probability of unknown parameters vector $\mathbf{\Theta}_{M_k}(k = 1, 2, 3)$ can be expressed as

$$\pi\left(\mathbf{\Theta}_{M_k}|\mathrm{X}\right) \propto \ln L(\mathbf{\Theta}_{M_k}) \cdot \pi\left(\mathbf{\Theta}_{M_k}\right) \; (k = 1, 2, 3) \tag{17}$$

where X is the degradation data set, $\pi\left(\mathbf{\Theta}_{M_k}\right)$ and $\pi\left(\mathbf{\Theta}_{M_k}|\mathrm{X}\right)$ are the prior distribution and posterior distribution of unknown parameters vector $\mathbf{\Theta}_{M_k}$, respectively. The Gelman-Rubin Ratio is utilized to evaluate MCMC convergence by analyzing the difference among different Markov chains.

3 Experimental Validation

3.1 Introduction to Degradation Test

Degradation test of solid lubricated bearings was carried out. The test rig is shown in Fig. 2(a), following the schematic of the test rig in Fig. 2(b). Two sets of solid lubricating bearings (one set includes two bearings) were used in our test.

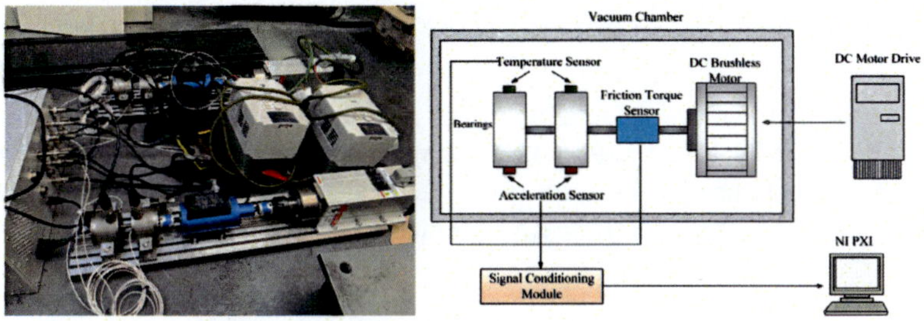

(a) Degradation test rig (b) Schematic of the Test Rig

Fig. 2. Test rig of solid lubricated bearings

The whole test rig was placed in the vacuum chamber. DC brushless motor was controlled by exclusive motor driver. Vibration sensors, temperature sensors and friction force sensors were installed on each test bearing. After conditioned by a conditioning module, signal from three sensors were collected by data collection system. Data collection system was programmed by National Instruments LabWindows CVI®.

Failure of bearings was manually judged by test operators based on vibration sensors, temperature sensors and friction force sensors. Once a test sample fails, a new sample will be installed to keep the pair running. The degradation data of newly installed was not taken into consideration. The rotation speed in our degradation test was 8000 rpm, and axial load was 50 N.

3.2 Model Validation and Analysis

MD is calculated based on the Eqs. (1) and (2) using four statistics features of vibration signal i.e. maximum value, RMS, kurtosis and crest factor. As discussed above, unknown parameters $\mathbf{\Theta}_{M_1} = (\mu, \sigma)$, $\mathbf{\Theta}_{M_2} = (\mu, \sigma, \gamma)$, $\mathbf{\Theta}_{M_3} = (\mu, \sigma, \gamma, \theta)$ in Wiener process degradation model M_1, M_2 and M_3 need to be estimated. Through Bayesian MCMC, posterior distributions of these unknown parameters can be obtained. Take model M_3 for example. Figure 3 shows the Gelman-Rubio ratio and posterior distributions of $\mathbf{\Theta}_{M_3}$. The total iterations are 200000. Gelman-Rubio ratio becomes stable after 75000 iterations. Iterations from 75001 to 200000 are chosen to calculate the mean value of $\mathbf{\Theta}_{M_3}$. Similar methods can also be used for model M_1 and M_2.

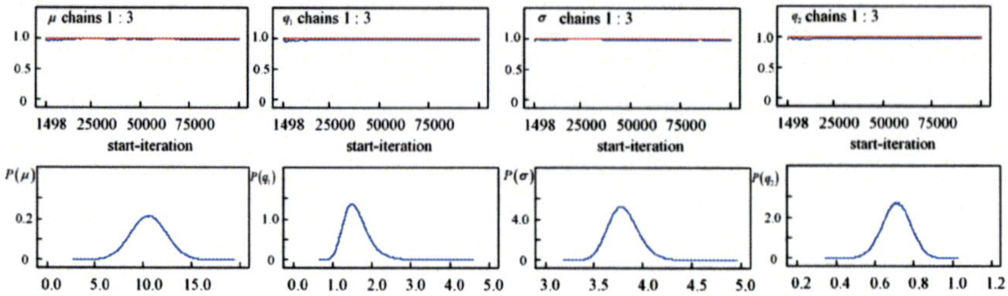

Fig. 3. Converge results and posterior distributions of unknown parameters for model M_3

Table 1 shows the mean value of unknown parameters for each degradation model, respectively. In model M_1, $\gamma = \theta = 1$, and they do not need to be estimated. In model M_2, $\gamma = \theta$, we only need to estimate one of them. These mean values will be used to conduct the reliability function for each degradation model.

Table 1. Unknown parameters estimation results

Model	μ	γ	σ	θ
M_1	12.22	–	3.649	–
M_2	12.79	0.8237	3.736	–
M_3	12.55	1.61	3.8	0.7073

Akaike information criterion (AIC) values based on the degradation data of solid lubricated bearings are conducted to compare degradation model M_1, M_2 and M_3. Results show that model M_3 has the lowest AIC value, and it indicates that model M_3 fits the degradation data better than M_1 and M_2 (Table 2).

Table 2. AIC value comparisons for degradation model

	AIC	Ranking
M_1	396.9514	3
M_2	389.9514	2
M_3	**375.7167**	**1**

Figure 4 shows the reliability function for the three different degradation models with the failure threshold 142. It can be seen that the reliability plot for model M_2 is much higher than that of M_1 and M_3, and the reliability plot for model M_1 is the second higher one. Given the same degradation data, there are large differences among three degradation models. It might affect the accuracy of reliability and lifetime estimation and cannot be ignored.

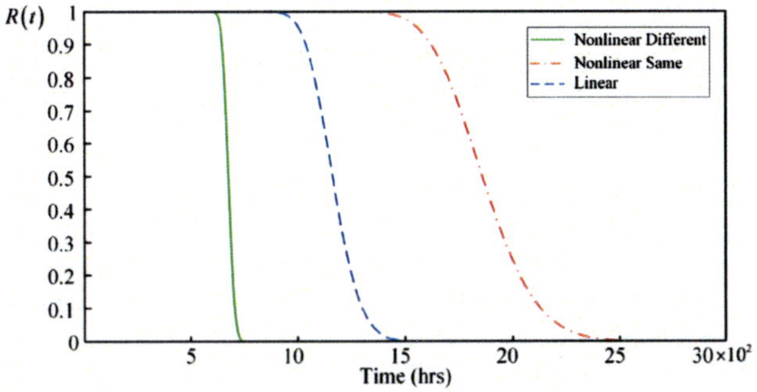

Fig. 4. Reliability estimation comparisons for three different models for solid lubricated bearings

MTTFs for different degradation models of the test bearings under specific level of stress are displayed in Table 3. It can be found that the value of MTTF based on degradation model M_3 is much closer than the other two models to the real lifetime under pre-defined stress. It can also demonstrate that it is important to use different time scale functions for drift term and fluctuation term in Wiener process for the reliability estimation for the solid lubricated bearings.

Table 3. MTTF comparisons for three degradation models

	M_1	M_2	M_3	Ture life time
MTTF (hrs)	1165.98	1868.78	675.31	696.3

4 Conclusion

In this paper, reliability estimation of solid lubricated bearings based on a general Wiener process is constructed. The performance indicator for solid lubricated bearings are fusion indicator calculated by MD considering the multiple statistics of vibration signal. Unknown parameters in each degradation model is estimated by Bayesian MCMC method. Degradation test was conducted, and the proposed degradation model is verified based on corresponding degradation data. Results show that Wiener process with two different time scale functions might be necessary in reliability estimation. Future work may address other forms of nonlinear functions for time scale function, and multivariate performance indicators for solid lubricated bearings, such as friction force, temperature, might also be considered.

References

1. Vanhulsela, A., Velsasco, F., Jacobs, R., Eersels, L., et al.: DLC solid lubricant coatings on ball bearings for space applications. Tribol. Int. **40**(7), 1186–1194 (2006)

2. Zhang, Z., Si, X., Hu, C., et al.: Degradation data analysis and remaining useful life estimation: a review on Wiener-process-based methods. Eur. J. Oper. Res. **271**(3), 775–796 (2018)
3. Alamaniotis, M., Grelle, A., Tsoukalas, L.H.: Regression to fuzziness method for estimation of remaining useful life in power plant components. Mech. Syst. Signal. Pr. **48**(1–2), 188–198 (2014)
4. Zhang, Q., Tse, P.W., Wan, X., et al.: Remaining useful life estimation for mechanical systems based on similarity of phase space trajectory. Expert Syst. Appl. **42**(5), 2353–2360 (2015)
5. Jin, G., Matthews, D., Fan, Y., et al.: Physics of failure-based degradation modeling and lifetime prediction of the momentum wheel in a dynamic covariate environment. Eng. Fail. Anal. **28**, 222–240 (2013)
6. Feng, Q., Jiang, L., Coit, D.W.: Reliability analysis and condition-based maintenance of systems with dependent degrading components based on thermodynamic physics-of-failure. Int. J. Adv. Manuf. Tech. **86**, 913–923 (2016)
7. Singh, H., Mutyala, K.C., Evans, R.D., et al.: An investigation of material and tribological properties of Sb_2O_3/Au-doped MoS_2 solid lubricant films under sliding and rolling contact in different environments. Surf. Coat. Tech. **284**, 281–289 (2015)
8. Zhang, C., Wang, S.: A new method for performance degradation assessment of solid lubricated bearings. In: 2011 6th IEEE Conference on Industrial Electronics and Applications, pp. 387–390. IEEE Press, New York (2011)
9. Zhang, C., Wang, S.: Experimental analysis of performance degradation of solid lubricated bearings with vibration and friction torque signal. In: IEEE 10th International Conference on Industrial Informatics, pp. 617–620. IEEE Press, New York (2012)
10. Peng, C., Tseng, S.: Mis-specification analysis of linear degradation models. IEEE T. Reliab. **58**(3), 444–455 (2009)
11. Tang, S., Guo, X., Zhou, Z.: Mis-specification analysis of linear Wiener process–based degradation models for the remaining useful life estimation. P. I. Mech. Eng. O-J. Ris. **228**(5), 478–487 (2014)
12. Whitmore, G.A.: Estimating degradation by a Wiener diffusion process subject to measurement error. Lifetime Data Anal. **1**(3), 307–319 (1995)
13. Huang, Z.Y., Xu, Z., Wang, W., et al.: Remaining useful life prediction for a nonlinear heterogeneous Wiener process model with an adaptive drift. IEEE Trans. Reliab. **64**(2), 687–700 (2015)
14. Zhang, C., Chen, R., Wang, S., et al.: Reliability estimation of reciprocating seals based on multivariate dependence analysis and its experimental validation. IEEE Access. **7**, 130745–130757 (2019)
15. Si, X., Wang, W., Hu, C., et al.: Remaining useful life estimation based on a nonlinear diffusion degradation process. IEEE Trans. Reliab. **61**(1), 50–67 (2012)
16. Wang, X., Balakrishnan, N., Guo, B.: Residual life estimation based on a generalized Wiener degradation process. Reliab. Eng. Syst. Eng. **124**, 13–23 (2014)
17. Wang, Y., Peng, Y., Zi, Y., et al.: A two-stage data-driven-based prognostic approach for bearing degradation problem. IEEE Trans. Ind. Inform. **12**(3), 924–932 (2016)
18. Zhang, C., Wang, S., Bai, G.: An accelerated life test model for solid lubricated bearings based on dependence analysis and proportional hazard effect. Acta Astronaut. **95**, 30–36 (2014)
19. Zhang, C., Wang, S.: Performance degradation analysis of solid lubricated bearings based on Laplace wavelet filter and self-organizing map. P. I. Mech. Eng. C-J. Mec. **228**(10), 1743–1753 (2014)
20. Kumar, S., Vichare, N.M., Dolev, E., et al.: A health indicator method for degradation detection of electronic products. Microelectron. Reliab. **52**(2), 439–445 (2012)
21. Niu, G., Singh, S., Holland, S.W., et al.: Health monitoring of electronic products based on Mahalanobis distance and Weibull decision metrics. Microelectron. Reliab. **51**(2), 279–284 (2011)

22. Wu, J., Wu, C., Cao, S., et al.: Degradation data-driven time-to-failure prognostics approach for rolling element bearings in electrical machines. IEEE T. Ind. Inform. IEEE T. Ind. Electron. **66**(1), 529–539 (2019)
23. Zhang, C., Chen, R., Bai. G., et al.: Reliability estimation of rotary lip seal in aircraft utility system based on time-varying dependence degradation model and its experimental validation. Chinese J. Aeronaut. **33**(8), 2230–2241 (2019)

Study of Thermostable Polyurethane Material Produced by Robotic Milling Machining

Alejandro Pereira[1(✉)], Maria Teresa Prado[1], María Fenollera[1], Michal Wieckzorowski[2], and Thomas Mathia[3]

[1] EEI, Campus Lagoas Marcosende, Universidad de Vigo, 36310 Vigo, Spain
apereira@uvigo.es
[2] Faculty of Mechanical Engineering and Management, Poznan University of Technology, Pl. M. Sklodowskiej-Curie 5, 60965 Poznan, Poland
[3] LTDS, Ecole Centrale de Lyon, 69134 Ecully, Cedex, France

Abstract. The thermostable polyurethane is a material used in the modeling sector and is currently replacing structural parts in control models of the automobile and aeronautics sector, for its ease of machining, in addition to providing good mechanical properties. In this work a robotic machining of a commercial polyurethane material type Necuron® 651 has been experienced.

Robotic machining offers a wide possibility of machining and modeling materials with the advantage of increasing the accessibility to pieces, compared to conventional machining centers of three or more axes. Machining has been carried out with a system consisting of a ABB® 6640-235 robot with a Peroni® high speed head and ABB rotary table. The disadvantage of robotic machining, caused by the lack of rigidity of the robot's morphology, are macro and micro-dimensional errors. Simple geometries with high speed steel tools have been made. The CAM programming has been obtained with the Software Powermill® of the company DELCAM, taking into account two types of strategies for limiting the freedom degree of the robot, to verify geometric results. Finally it has been measured and analyzed the surface generated by interferometer autofocus Alicona and has been measured the geometry of the samples in a coordinate measuring machine (CMM). The results show that both micro and macro errors are significantly reduced by making fixed one of the seven axes of the robotic system.

Keywords: Robotic machining · Dimensional metrology · Surface metrology

1 Introduction

The automotive and aeronautics industry uses polymer-based materials, such as thermosetting polyurethanes, replacing structural parts in quality inspection devices. These polymeric materials are processed in order to maximize their mechanical properties by lightening the weight, with techniques based on the foaming of plastic parts or on the optimization of their designs using advanced simulation tools. These new techniques require a mastery of these materials (formulation, processability, knowledge of their mechanical properties, microstructure …), in order to meet both aesthetic and functional criteria [1–3].

J. Machado et al. (Eds.): icieng 2021, LNME, pp. 68–81, 2022.
https://doi.org/10.1007/978-3-030-79165-0_7

Thermostability is the quality of a substance to resist irreversible changes in its physical or chemical structures by resisting decomposition or polymerization, at a high relative temperature reencountered in machining process. A thermostables polymers, an uncommon and unconventional term, is likely to refer to a thermosetting plastic that cannot be remodeled when heated, than to a thermoplastic that can be melted again and reshaped. There is still luck of precise technical knowledge of machining of specific class of Thermostable PolyUrethane (TPU) due to its rheological and tribological variability despite growing industrial interest. Following paper is devoted to one of them Necuron®

Necuron® materials have a high quality surface structure, extremely homogeneous after machining. They are dimensionally stable materials, with high edge stability and little hygroscopic. Its extensive range makes it a suitable material for various uses: application in bioengineering as a modeling material and mechanical prototyping of bones (Necuron 600 and 1300) [4, 5]; the manufacture of tools for benders with inserts of Necuron 1050, 1150 and 1300 [6]; the manufacture of functional components of the car with Necuron 1020 and 1300 [7]; and even manufacture of patterns and molds.

The improvement of productivity in the advanced manufacturing industry both in aeronautics and in the automobile field, is achieved by selecting a robotic machining to reduce cycle time, increase flexibility, as well as facilitate accessibility to the parts. These are the main advantages of robotic machining over conventional machining centers with three or more axes.

However, one of the considerable limitations is the rigidity of the robot, directly related to the hardness of the material to be machined. The application of robots is usually applied to easily mechanized materials and not to hard materials due to the strong influence of the stiffness factor in the tool-material interaction. A large investment is required for the development of a robotic cell and is easy to implement in high tolerance applications using a variety of materials, generally soft, as is the case of this research based on the study of Necuron® 651 [7].

The robotized machining of large prototypes has the disadvantage of the dimensional tolerance of the pieces. Chen et al. studied different strategies using the robot for rough machining. This process was designed with different tolerances, depending on the type of robot, spindle, tool holder and tool. As a verification of the proposed algorithm, several prototypes have been produced that have demonstrated the viability and advantages of robotic machining [8].

The surface quality is related to the cutting conditions of the finishing operations. This is the reason why it is necessary to quantify the relationship between surface roughness and parameters such as cutting speed, feed rate and depth of cut [9, 10]. In addition to these parameters, other inherent factors must be taken into account: the tool itself, the material to be machined, the cutting strategy, as well as the stability and rigidity of the machine tool-part system [11, 12].

On the other hand, in previous works, in order to study the geometric conditions and the topography of the surfaces generated by machining, the relationships between the different machining conditions and the geometric and topographic results have been analyzed, with different materials [13–17].

2 Methodology

In this work, the specimens have been designed and developed in a commercial thermostable polyurethane material called Necurom 651®. The proposed objective is based on measuring machining errors with a robotic cell consisting of an ABB 6640-235 robot with a Peroni spindle capable of speed rate of up to 60,000 rpm [8, 13, 14]. As shown in Fig. 1, it has a seventh rotary axis that constitutes the ABB rotary table, with the fixturing system. The fixing system of the tool to the hydraulic spindle head have been choosen by high speed cone HSK -50 [15–17].

Fig. 1. Robot ABB 6640-235, with seven rotary axis and Peroni® spindle.

In the specimens, simple straight geometry machining consisting of grooves of the selected tool diameter is made, in order to measure the linear interpolation error of the robot. The Necurom material has been machined, with different cutting conditions. The obtained geometry has been measured and the surface topography has been obtained. CAM programming has been carried out with the Powermill® software of DELCAM company, taking into account two types of strategies with the objective of verifying geometric results: with all degrees of freedom, and making the axis of the rotary table fixed.

2.1 Material

Necuron® 651 are modeled plates cast and injected in formulated polyurethane that can be machined with CNC. Its main properties are: low specific gravity, excellent dimensional stability, it is not altered by moisture, good resistance to abrasion, compression and flexion. Regarding machining, it has great advantages such as being a neat tool cut, without generating dust or vibrations; It does not require cutting coolant and prolongs the life of the tool. The slats can be glued together with an epoxy adhesive to generate complex volumes and optimize material, you can thread and fix inserts. Its main application is in tooling, foundry, models and prototypes, gauges and devices. Its technical characteristics are shown in Table 1.

Table 1. Technical data of Necuron® 651 (Technical data sheet Necuron® 651. NECUMER GmbH · Industriestraße 26 · D-49163 Bohmte).

Characteristics	Values
Colour	Brown
Coefficient of thermal expansion	55 × 10 – 6 K-1
Temperature resistance (ISO 75)	65°C
SHORE D	D70
Compressive strength	25 Mpa
Flexural strength	27 Mpa
Density	0.70 g/cm^3
Thermal conductivity	0.11 W/m.K
Fire protection classification	B2

The dimensions of the Necuron® 651 specimens to be tested are dimensions: 100 × 50 × 30 mm.

2.2 Machining Process and Cutting Kinetics Conditions and Settings

The tests carried out on the samples of Necuron® 651 material, have been carried out following two machining strategies, using in both a high speed steel tool (HSS) with two edges (Z), but of different diameter (H). In the first strategy, any fixed axes, unique cutting speed (Vc) and unique depth (Ap) have been considered. Five different values of the feed rate per edge (fz) between 0.05 and 0.25 mm/edge have been taken. And in the second strategy with 1 fixed axis, the unique parameters have been Ap and fz, varying the cutting speed (Vc) taking 4 values between 500 and 2,000 m/min. The cutting parameters for each test in the two specimen are shown in Table 2 (The design of experiments are showed in Table 3).

Table 2. Machining strategy.

Conditions	Necuron 651-1	Necuron 651-2
Tool, HSS	H = Ø 18 mm Z = 2 flutes	H = Ø 20 mm. Z = 2 flutes
Cutting speed	Unique Vc = 500 m/min	Variable Vc (m/min) = 500/1000/1500/2000
Axial depth	Ap = 10 mm	Ap = 5 mm
Cutting strategy	0 fixed axis. Interpolation 7 axis.	1 fixed axis. Interpolation 6 axis.
Feed rate	Variable f_z(mm/rev) 0.25/0.20/0.15/0.10/0.05	Unique f_z = 0.15 mm/filo

Table 3. Design of experiments.

	Sample	Tool H Ø(mm)	Vc (mm/min)	fz (mm/rev)	Ap (mm)	Axis	Geo	Topography
Necuron 651-1	1.1	18	500	0,25	10	0 fixed axis	Straighness	Pz, Sa, Sz, Ssk, Sku
	1.2			0,20				
	1.3			0,15				
	1.4			0,10				
	1.5			0,05				
Necuron 651-2	2.1	20	500	0,15	5	1 fixed axis	Straighness	Pz, Sa, Sz, Ssk, Sku
	2.2		1.000					
	2.3		1.500					
	2.4		2.000					

In Figs. 2 and 3, the Necuron 651-1 specimen is shown, respectively showing the 3D view, indicating the topographic study surface, and the physical specimen with the machining zones, each of them with their respective feed rate. The same procedure manufacture Necuron 651-2 specimen indicating in the physical specimen the zones, but in this case, mechanized with different cutting speeds.

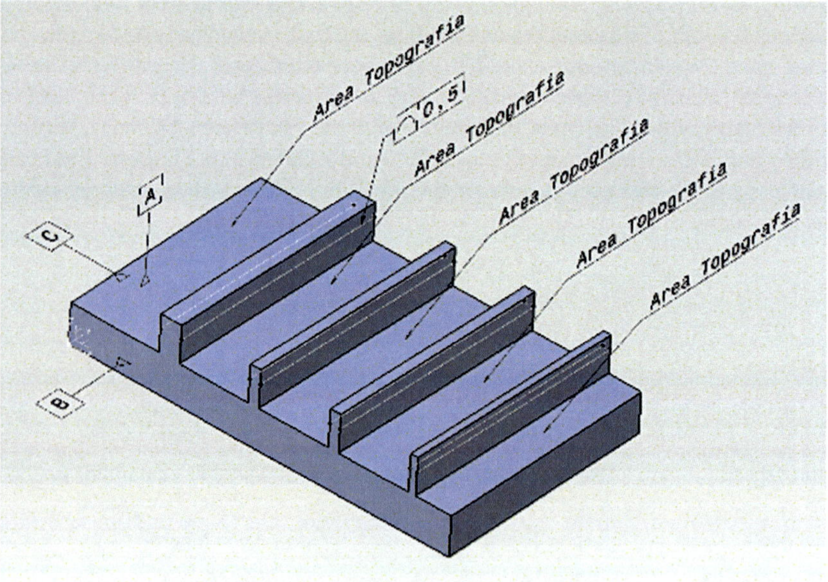

Fig. 2. Necuron® 651-1 Specimen 3D.

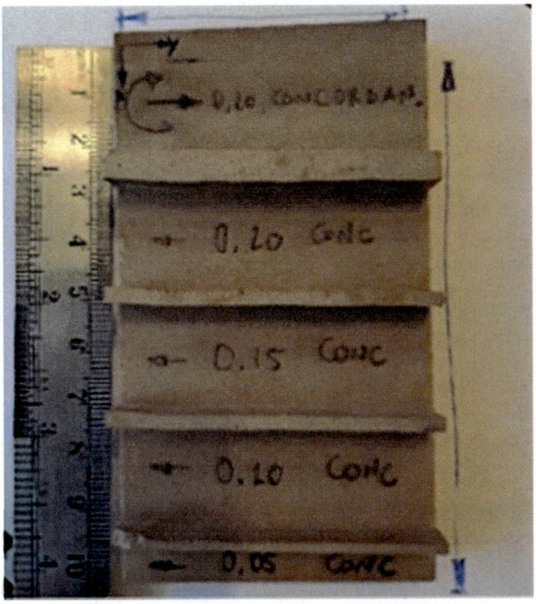

Fig. 3. Necuron® 651-1 Specimen.

2.3 Measurements

The quantification, analysis and evaluation of the micro and macro geometric errors produced by the lack of rigidity of the morphology of the robot is carried out with different techniques, differentiating between geometric and topographic measurements.

The geometry measurements have been made in Coordinate Measuring Machine (CMM), Nikon Altera 1500.700.600, programming in CAMIO 8.4® measurement software. The machine is equipped with Renishaw TP20/PH10T motorized head. The maximum permissible error (MPE) according to ISO 10360-2 is calculated by the Eq. (1) expressed in microns. The manufacturer of the CMM, ensures a maximum error less than 5.75 μm. The geometric measurements have been the straightness of the mechanized sides (Yaxis), measured by contact scanning.

$$\mathrm{MPE}(\mu\mathrm{m}) = 2 + \left(\mathrm{L}(\mathrm{mm})/400\right) \tag{1}$$

where L is the measurement length in millimeters.

$$\mathrm{MPE} = 2 + \left(1500/400\right) = 5.75\ \mu\mathrm{m}$$

In the case of study the flat areas topography, it is measured at certain points that are illustrated by an Alicona® infinite focus 3D topometer, in order to know the variation of 3D surface parameters according to ISO 25168 in the Necuron material depending on the conditions of machining (Figs. 4 and 5).

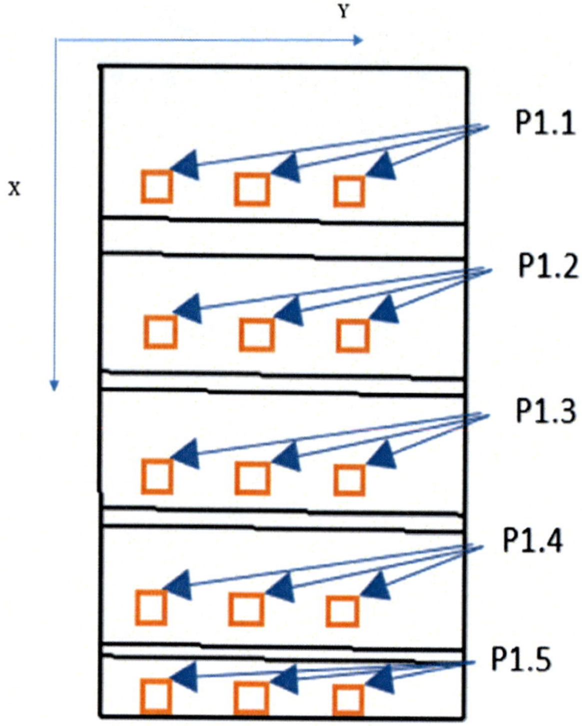

Fig. 4. Specimen 651-1. Topographic measurement

The measurements obtained have been processed with MountainMaps 7.4 software. A robust Gaussian filter according to ISO 16610-71 has been applied, with a cutoff of 0.8 mm, to separate layer of waviness and roughness [18].

The surface parameters used in the analysis of the measures, according ISO 25178 are:

- Amplitude parameter P: Maximum peak-valley height, P_z
- Amplitude parameters S: Arithmetic average surface deviation, S_a; Total peak-valley height (sampling length), S_z; Asymmetry factor (<0 valleys predominate, >0 peaks predominate), S_{sk}; and Kurtosis (Increases for higher peak acuity, central value = 3), S_{ku}.

The proposed experiment design is indicated in Table 3.

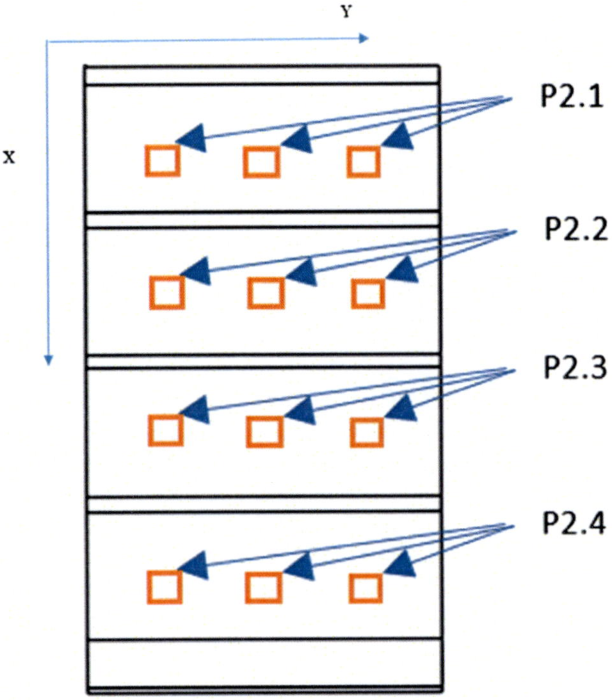

Fig. 5. Specimen 651-2. Topographic measurement

3 Results

The measurements made in the coordinate measuring machine have given the straightness defect results on Y axis, obtained from the scanning of the marked measurement zones [19]. Table 4 shows the results as they have been determined from a number of points for each shape measurement.

Table 4. Results of geometrical measurements.

Sample	Dia Ø (mm)	Vc (mm/min)	fz (mm/filo)	Ap (mm)	Ae (mm)	Axis (fixed)	Straightness (mm)
1.1	18	500	0.25	10	18	0	0.96
1.2	18	500	0.2	10	18	0	0.90
1.3	18	500	0.15	10	18	0	0.57
1.4	18	500	0.1	10	18	0	0.27

(continued)

Table 4. (*continued*)

Sample	Dia Ø (mm)	Vc (mm/min)	fz (mm/filo)	Ap (mm)	Ae (mm)	Axis (fixed)	Straightness (mm)
1.5	18	500	0.05	10	18	0	0.27
2.1	20	500	0.15	5	20	1	0.34
2.2	20	1000	0.15	5	20	1	0.30
2.3	20	1500	0.15	5	20	1	0.26
2.4	20	2000	0.15	5	20	1	0.27

It can be noted that there are significant differences in straightness. This defect are larger in specimen Necurom 651.1, than in specimen Necurom 651.2. Only the samples 1.4 and 1.5 with a feed of 0.1 mm/rev and 0.05 mm/rev are lower than samples 2.x

Figure 6, illustrates the result of the line-shaped defect (straightness) measurement of the Necuron 651-2 specimen. A straightness defect is defined as the distance between two lines, parallel to the reference line, calculated by adjustment for least squares, and that include the study profile. The significant differences in the straightness of the linear interpolations can be seen in the graph of measured points.

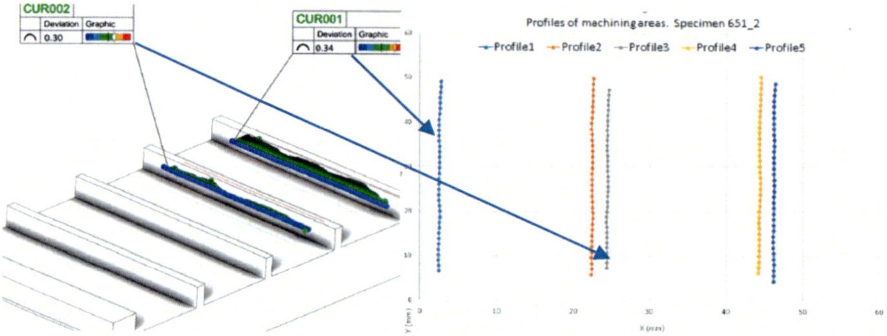

Fig. 6. Straightness (Yaxis) and profiles of Specimen Necurom 651.2

Figure 7 shows the behavior of the straightness defect with respect to the variation of the feed rate, and the discrete variation of the straightness defect, in relation to the variation of the cutting speed, with a strategy of interpolation with a fixed robot axis.

In the case of fixing an axis of the robot, linear interpolation is considerably better because the straightness defect is reduced by 40%. In the case of machining the Necuron 651-1 specimen with a feed rate of 0.15 mm, and the straightness defect of the specimen Necurom 651.2 of 0.57 mm has been obtained, with the same feed rate, a result of straightness 0.34 mm was obtained.

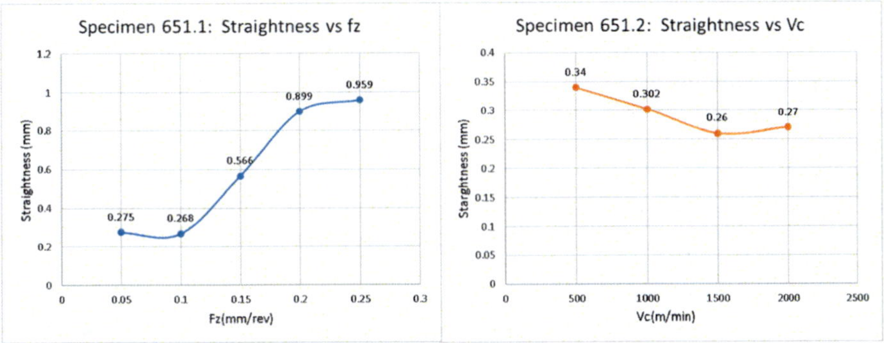

Fig. 7. Straightness versus feed rate and straightness versus cutting speed

Another interesting result is the proportionality between the straightness defect and the feed rate. The higher the feed rate is, the greater is the straightness defect. The case of greater straightness, close to the millimeter, occurs for the highest feed rate 0.25 mm/rev.

It can be concluded that in the case of specimen Necuron 651.2, a slight improvement in the straightness defect can be observed as the cutting speed increases. The error variation has been 0.08 mm in the full variation of the speed range, which indicates that a direct proportionality cannot be assigned.

Regarding the topography measurements, the results are shown in Table 5. The selected parameters have been:

Table 5. Topography measurements

N	Vc (mm/min)	fz (mm/rev)	Pz (μm)	Sa (μm)	Sz (μm)	Ssk	Sku
1.1	500	0,25	143	11,8	132	−1,12	5,1
1.2	500	0,2	158	10,8	150	−1,6	6,78
1.3	500	0,15	165	11,2	160	−1,41	5,64
1.4	500	0,1	144	10,9	142	−1,42	5,56
1.5	500	0,05	156	8,87	143	−1,84	8,1
2.1	500	0,15	154	12,6	156	−1,33	5,38
2.2	1.000	0,15	134	10,5	129	−0,992	4,93
2.3	1.500	0,5	151	11,5	149	−1,14	5,21
2.4	2.000	0,15	149	11,3	142	−0,619	4,32

- The maximum primary Profile Pz (μm), that is to say height difference between peak and valley on the measured surface, (This parameter can be associated with the flatness of the measured surface).
- The average surface area Sa (μm) has also been measured, after applying a
- robust filter to separate the roughness layer and the waveness layer, with a cut-off of 0.8 mm.
- The maximum surface area Sz (μm).
- The dimensionless parameters of asymmetry Ssk and kurtosis Sku have also been measured [20].

If we take into account the parameter Pz of the primary profile obtained without filters we can deduce that there is an intrinsic variability due to interpolation of the robot in the Z plane with a variability in the Necurom 651.1 specimen of 22 μm (165 μm–143 μm). In the case of the Necuron 651-2 specimen, taking into account that the seventh axis of the table has been fixed and that axis does not influence the Z axis, as the work object (Base reference) has been placed, the variation is 20 μm (154 μm–134 μm), similar to the case of the Necuron 651-1 specimen.

As can be seen in Figs. 8 and 9 corresponding respectively to the topography of maximum Sa for zone 2.1 of Necuron 651-2 and minimum Sa, corresponding to zone 5 of Necuron 651-1, no significant differences can be seen in surface morphology.

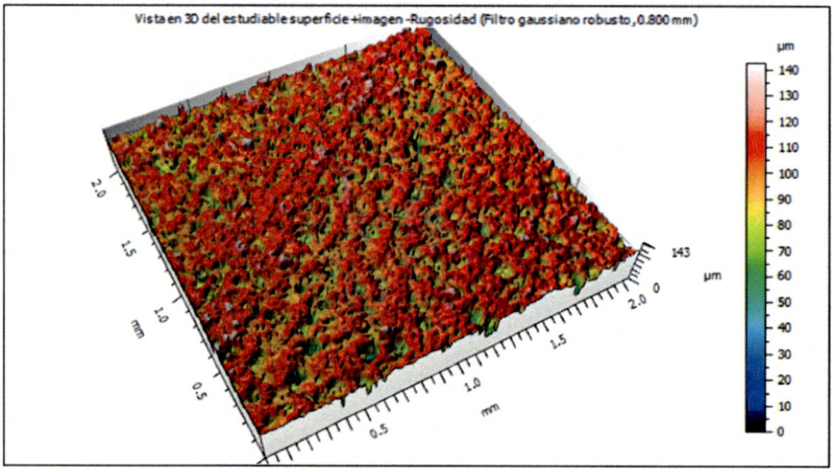

Fig. 8. Necuron 651-2, sample 2.1, Sa = 12,6 μm.

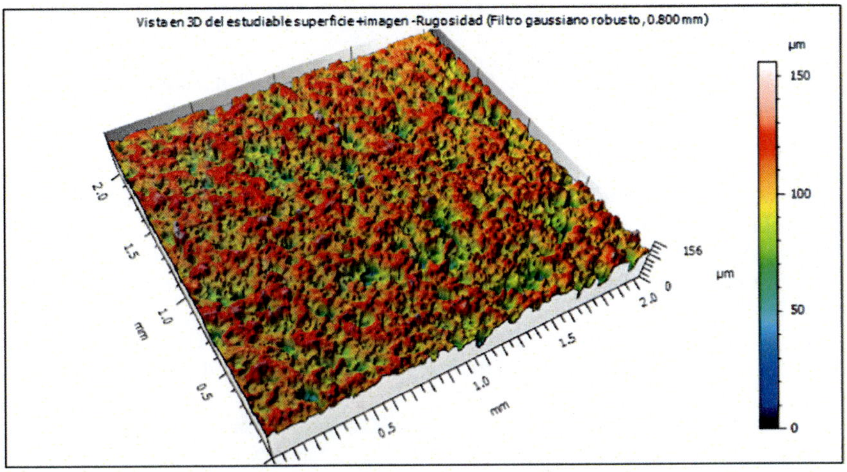

Fig. 9. Necuron 651-1, sample 1.5, Sa = 8,87 µm.

4 Conclusions

Based on the research carried out, it can be concluded that the correctly optimized robotic milling machining of soft materials has important applicability in the processing of 3D models.

One of the advantages of the application of robotic machining is the high flexibility and speed of modeling, as well as the possibility of achieving large models.

As a disadvantage of the application of robotic machining to prototyping, and as a result of the low rigidity of the system, there are certain accuracies that will obviously depend on the configuration of the system. In this work, straightness errors for high advances of ±0.5 mm have been verified for the tested cell. The results show that the geometric macro errors decrease significantly if any of the seven axes of the robotic system are fixed. This means that, as soon as the interpolation of one of the axes is limited, the straightness error is considerably reduced, as shown in the results.

Regarding the results of measurement of topography in a flat area on the Necuron material, there is an independent variation in the process variables tested, feedrate and cutting speed. Possibly testing tools with greater rank angles improves the results of the Sa parameter. In the case of the Pz parameter it can be concluded that it is independent of the cutting conditions and it is influenced by the precision of the robotic arm. The surface morphology does not differ significantly between the selected cutting conditions.

Acknowledgements. The authors would like to thank the company Fundiciones Adrio S.L. for having allowed to use the machining robotized cell and particularly, Mr. Vicente Adrio, General Manager of the company. This research have been supported by the project IN-0337-2018 "Study of topography of polymeric pieces".

References

1. Pasca, N., Constantin Murariu, A., Marsavina. L.: Structure influence on the mechanical characteristics of polyurethane materials used in automotive industry. ModTech International Conference, Republic of Moldova (2011)
2. Amarandei, M., Virga, A., Berdich, K., Matteoli, S., Corvi, A., Marsavina, V.: The influence ODF defects on the mechanical properties of some polyurethane materials. J. Mater. Plast. **50**(2), 84–87 (2013)
3. Amarandei, M., Kun, K., Szigyarto, I., Kun, L., Marsavina, L.: Nondestructive evaluation of polyurethane materials using transient thermography. J. Key Eng. Mater. **21–24**, 525–526 (2012)
4. Kun, K., Bogdan, L., Nes, S., Faur, N., Andor, B.: Necuron 600 proposed as mechanical model for human cancellour bone from the femoral head. J. Weld. Mater. Test. **XXIV**(2), 13–15 (2015)
5. Patrascu, J.M., Amarandel, M., Kun, K., et al.: Compression and bending tests in order to evaluate the use of necuron for the manufacturing of transtibial prostheses. J. Mater. Plast. **51**(3), 263–266 (2014)
6. Zaragoza, V., Strano, M., Iorio, V., Monno, M.: Sheet metal bending with flexible tools. In: 18th International Conference on Sheet Metal, SHEMET 2019, Procedia Manufacturing, vol. 29, 232–239 (2019)
7. Barnfather, J.D., Goodfellow, M.J., Abram, T.: A performance evaluation methodology for robotic machine tools used in large volume manufacturing. Robot. Comput.-Integr. Manuf. **37**, 49–56 (2016). https://doi.org/10.1016/j.rcim.2015.06.002
8. Chen, Y.H., Hu, Y.N.: Implementation of a robot system for sculptured surface cutting. Part 1. Rough machining. Int. J. Adv. Manuf. Technol. **15**(9), 624–629 (1999)
9. Abellan-Nebot, V., Bruscas, G.M., Serrano, J., Vila, C.: Portable study of Surface roughness models in milling. Manufacturing Engineering Society International Conference, MESIC 2017. Procedia manufacturing, vol. 13, pp. 593–600 (2017)
10. Stephenson, D.A., Agapiou, J.S.: Metal Cutting Theory and Practice, 3rd edn. CRC Press (2016). ISBN: 9781466587533
11. Mohanty, S., Prakash, S., Daaseswara, V.: Investigation of influence of part inclination and rotation on Surface quality in robot assisted incremental sheet metal forming (RAISF). CIRP J. Manuf. Sci. Technol. **22**, 37–40 (2018)
12. Slamani, M. Gauthier, S., Chantelain, J.F.: Comparison of Surface roughness quality obtained by high speed CNC trimming and high speed robotic trimming for CFRP laminate. Robot. Comput.-Integr. Manuf. **42**, 63–72 (2016)
13. Prado, M.T., Pereira, A., Pérez, J.A., Mathia, T.G.: Methodology for tool wear analysis by a simple procedure during milling of AISI H13 and its impact on surface morphology. Procedia Manuf. **13**(Suppl C), 348–355 (2017). https://doi.org/10.1016/j.promfg.2017.09.090
14. Prado, M.T.: Análisis de desgaste de herramienta y optimización de proceso mecanizado mediante visión computarizada y consumo eléctrico (Analysis of tool wear and machining process optimization through computer vision and power consumption, PhD thesis under supervision of A Pereira), PhD, University of Vigo (2015)
15. Pereira, A., Hernandez, P., Martinez, J., Perez, J.A., Mathia, T.G.: Surface topographic characterization for polyamide composite injection molds made of aluminum and copper alloys. Scanning **36**(1), 39–52 (2014). https://doi.org/10.1002/sca.21083
16. Pereira, A., Martínez, J., Prado, M.T., Perez, J.A., Mathia, T.G.: Topographic wear monitoring of the interface tool/workpiece in milling AISI H13 steel. Adv. Mater. Res. 152–167 (2014)
17. Pereira, A., Riveiro, E., Pérez, J.A.: Machinability of hight -strength low alloy steel D38MSV5S forged crankshafts. Arch. Mech. Technol. Autom. **34**(4), 45–57 (2014)

18. Mathia, T.G., Pawlus, P., Wieczorowski, M.: Recent trends in surface metrology. Wear **271**, 494–508 (2011)
19. López, A., Pereira, A., Pérez, J.A.: Methodology to import cloud points from CMM to CADCAM software. Plizen - Czech Republic (2006)
20. EN-UNE-ISO 25178: Calidad Superficial: Areas. Parte2: Términos, definiciones y parámetros de calidad superficial, ed (2013)

Reverse Engineering as a Way to Save Environment with-in Patient-Tailored Production of Assistive Technology Devices – Based on Own Hand Exoskeleton Case Study

Izabela Rojek[1](✉), Marek Macko[2], Jakub Kopowski[1,3], and Dariusz Mikołajewski[1]

[1] Institute of Computer Science, Kazimierz Wielki University, Chodkiewicza 30, 85-064 Bydgoszcz, Poland
izarojek@ukw.edu.pl
[2] Faculty of Mechatronics, Kazimierz Wielki University, Chodkiewicza 30, 85-064 Bydgoszcz, Poland
[3] Faculty of Psychology, Kazimierz Wielki University, Chodkiewicza 30, 85-064 Bydgoszcz, Poland

Abstract. The paper shows own study on a hand exoskeleton described from environmental point of view: starting from constraints caused by patient-tailored therapy and healthy/disordered human biomechanics through possible problems associated with material engineering (mechanical properties, biocompatibility, etc.) and their compensation by exoskeleton' designers to material and technological limitations associated with recycling. The purpose of this article is to investigate how current opportunities in this area are being used, including reverse engineering as a part of the concept of the disabled person's hand exoskeleton. There is no doubt that more research is needed for a more complete understanding

3D printers emission, exposure, as far as effectiveness of indicators in real occupational conditions. But our solution can significantly improve situation concerning influence of the pollution and reuse of the materials. What more potential for harmful secondary changes from exposure to emission from certain additive manufacturing processes may influence further shape of the 3D printing within Industry 4.0 concept.

Keywords: Reverse engineering · 3D scan · 3D printing · Exoskeleton · Recycling

1 Introduction

Healthy environment becomes one of the most precious values worldwide. According to the World Health Organization (WHO) air pollution constitutes world's single biggest environmental health risk. By reducing air pollution, we could save a million lives a year worldwide by 2050 [1]. But there are possible much deeper solutions. Reasonable reverse

J. Machado et al. (Eds.): icieng 2021, LNME, pp. 82–91, 2022.
https://doi.org/10.1007/978-3-030-79165-0_8

engineering, proper material engineering, reuse of materials used in three dimensional (3D) printing processes may create another way to save environment within patient-tailored production of assistive technology devices. We should be aware that number of disabled people corresponds to fifteen percent of people living on Earth, constituting more than 1 billion persons in the world have some form of disability. This corresponds to huge number of assistive technology devices, because between 110–190 million people have very significant difficulties in functioning. No doubt novel technologies, such as 3D scanning, 3D printing, reverse engineering incorporated into Industry 4.0 paradigm may simultaneously meet apparently conflicting requirements:

- increased efficiency thanks to patient-tailored assistive devices and associated patient-tailored therapy, better adapted to needs of patients, stage of the therapy, etc.,
- low pollution thanks to at least partial recycling.

The purpose of this article is to investigate how current opportunities in this area are being used, taking into consideration reverse engineering - a part of the concept of the disabled person's hand exoskeleton.

2 Environmental Influence of 3D Printing

Unfortunately there is only a few studies concerning environmental influence of 3D printing, thus subject of our study, despite well defined, still requires many answers. There is many features which may be taken into consideration including i.a. concentration of particles, particle emission rates and characteristics (including their morphology and chemical composition), temperature conditions in different thermoplastic materials. Reliability and validity of research requires application of an exposure chamber. Moreover general recommendations for reduction of harmful particle emissions are following:

- use of low emission materials,
- use of low temperatures,
- using special cover with filter around the printer (even self-made is better than lack of it),
- regular control measures (with possible future obligatory certification - see exposure limit values recommended by the US National Institute for Occupational Safety and Health (NIOSH)), or even real-time measurement instruments.

There is need for further research concerning additive manufacturing process emission itself and, what more important, personal exposures in real-world environment (i.e. workplaces).

Study by Kwon et al. aimed to evaluate rates and characteristics of particles emission as far as used control methods and their effectiveness in lowering emission of particles during 3D printing. Thermoplastic materials (seven filaments), based on nanoparticle emissions, have been classified as:

- high emitters, i.e. $>10^{11}$ particles/min (i.e. #/min.),
- medium emitters, i.e. 10^9–10^{11} particles/min.,
- low emitters, i.e. $<10^9$ particles/min.

The highest removal efficiency of nanoparticles (99.95%) had a high efficiency particulate air (HEPA) filter [2], recommended to aforementioned applications.

Stefaniak et al. assessed atmosphere of 4 occupational conditions and observed huge variability in emission rate values. Main factors influencing aforementioned values were device construction and type, working environment, and used materials. Measured particle emission was following:

- from 9.4×10^9 to 4.4×10^{11} particles/min. for fused filament fabrication (FFF) 3D printers,
- from 1.9×109 to 3.8×109 particles/min. for sheer 3D printers.

What important even custom-built ventilated enclosure was capable to significantly reduce air pollution. Achieved values were following reduction of total organic chemical concentration by 53.2% accompanied by reduction of number of particles by 99.7%. Despite carbonyl compounds were observed in the air, none of them were directly attributed to the results of 3D printing process. NIOSH Recommended Exposure Limit of personal exposure to aluminium, iron and twelve different organic chemical agents have not been exceeded. Aforementioned results have not reflected all possible exposure scenarios thus deeper studies are needed [3, 4]. It may be significant that according to research by Potter et al. even printing 222 grams of filament can be enough to exceed the reference values of concentration for inhalation exposure 1 mg/m^3 according to the EPA's Integrated Risk Information System. Aforementioned values may be exceed even using amateur 3D printers, thus we should be aware existing severe threat caused by uncontrolled use of CNT nanocomposites in additive manufacturing [5].

Wojtyła et al. investigated thermal decomposition of popular commercial thermoplastic filaments: polylactic acid (PLA), acrylonitrile-butadiene-styrene (ABS), polyethylene terephthalate (PET), and nylon. This survey showed that PLA is significantly less toxic than ABS [6].

Research on 3D printer working in fused deposition modelling (FDM) technology using diverse materials and various extrusion temperatures showed, that all tested materials do not emit super-micron particles, while emission in the ultrafine range (10–30 nm) was detected. Emission increases as the extrusion temperature increases. What more even relatively short, 40-min-long 3D printing caused doses up to 200 mm^2 [7]. Observed ultrafine aerosol (UFA) emissions were much higher during printing standard objects using PLA than ABS (respectively 2.1×10^9 vs. 2.4×10^8 particles/min). Longer use of aforementioned 3D printer caused higher emission rates. What is important 3D print in spacious, well-ventilated room did not significantly increase concentrations of the UFA and VOC, whereas the same values were clearly detectable in a small, unventilated office [8]. This observation can be a guide when choosing a room for a 3D printing laboratory or a production plant.

Thermosetting photopolymers constitute almost 50% of the sold materials for 3D printing. Process of the two-step polymerization aiming at reprocessable thermosets (3DPRTs) for 3D printing purposes allow users to reuse materials reforming 3D printed objects into a new shape. This technology can be useful in repairing a broken parts (e.g. by reprinting material on damaged site) or recycle unwanted printed objects or their parts (material can be reused after this process) [9].

Fused particle fabrication (FPF) and fused granular fabrication (FGF) constitute particle material extrusion which can increase the use of recycled polymers in additive manufacturing. Recycled polycarbonate (PC) regrind may result in low-cost, heat-resistant and high-strength products [10]. Novel eco-polyol can base on PLA waste. Its use in manufacturing of inflexible polyurethane-polyisocyanurate (RPU/PIR) foams also matches with the current trends of sustainable development and green chemistry [11]. There is observed increased demand for new renewable printing materials and novel process technologies in 3D printing [12]. 3D printers such as Gigabot X and FPF/FGF can print from many recycled plastic (polymer) materials (PLA, ABS, PET, PP) with minimal post processing [13].

3 Own Concept of Exoskeleton

Exoskeleton constitutes external construction worn as suit (or its parts) supporting movements of the user based on her/his intent. Our own concept of hand exoskeleton is another breakthrough toward research on the family of 3D printed, both passive and active, hand exoskeleton for children and adults. Whole procedure contains:

- detailed description of the patient's hand anatomy and pathology,
- detailed description of patient's function,
- computational model of the hand with deficit,
- computational model of the hand exoskeleton adapted to the patient in the way decreasing deficit and improving function, based on both universal family of templates and features of the patient,
- 3D printing of the exoskeleton for specific patient,
- kinematic chain testing,
- testing the effectiveness and degree of improvement of health by the aforementioned exoskeleton in specific patient [14, 15].

It needs following procedures:

- functional assessment of the patient's hand,
- assessment of possible contraindications to application of hand exoskeleton in particular patient,
- scanning and imaging toward technical features of exoskeleton,
- adaptation of pre-programmed and well tested templates toward personalized hand exoskeleton design for particular patient,

- testing of the prototype and assessment of its efficacy everyday functioning, especially in so called activities of daily living (ADLs),
- long-term exploitation and helpdesk [14–16] (see Fig. 1).

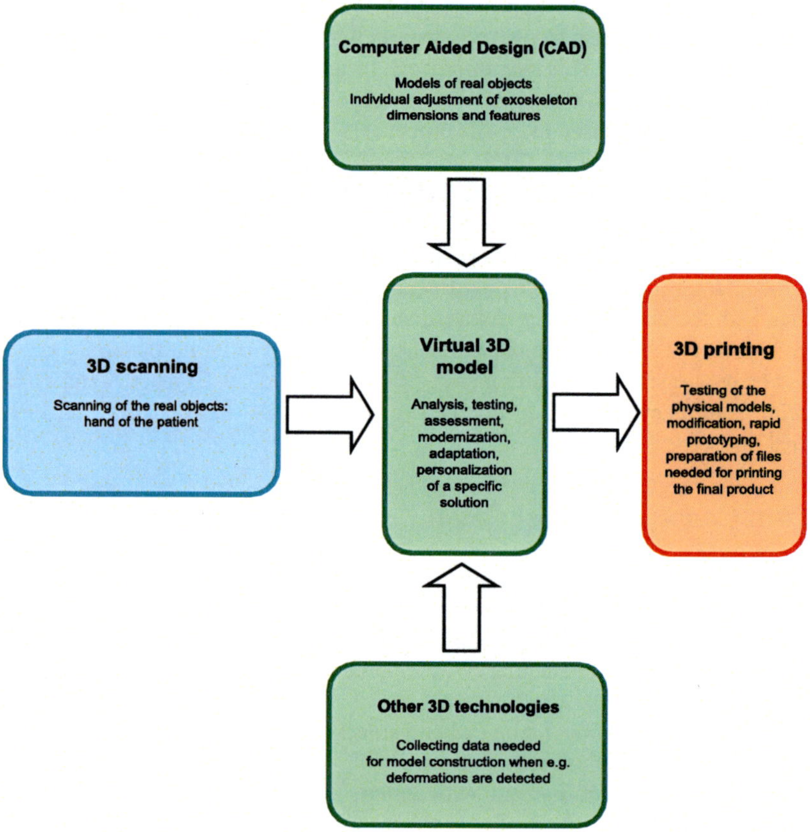

Fig. 1. The entire manufacturing process.

Aforementioned exoskeleton design is an iterative process taking into account the user perspective, especially by common goal setting, i.e. goals achieved by use of hand exoskeleton should be significant for the patient and her/his family acceleration recovery process [17, 18]. Patients' needs deficit(s) and are carefully assessed before the manufacturing and application/fitting, thus patient feels the appropriate device support [19]. Each period of hand exoskeleton use or every change of the patient's health status requires subsequent careful reassessment, in some cases demanding changes in exoskeleton support (e.g. matching of forces added). But current the offer is very diverse: 165 individual dynamic hand orthoses have been identified so far, and their function was categorized based on signal used, energy, mechanic features [20], training paradigm [21], intensity, duration of intervention [22], modification of pathological patterns [23], and degree of improvement of hand movement smoothness. We took into consideration current description of joint coordinate system for hand, wrist, elbow, and shoulder [24], including diverse factors such as global hand position, muscle activations, hand's parts velocities, angles in the joints of the hand, joint angular velocities, joint torques, as far as many variables: kinematic, kinetic, intrinsic, and extrinsic [25].

Our mechanical/computational human hand model was developed on the basis of high-resolution medical images similar to data sets obtained from the Visible Human Male [26] and Visible Human Viewer Project, for the thumb and all fingers [27], wrist movements [28], movement variability [29] and general assessment of the hand joint function [30, 31] (see Figs. 2, 3). Connection between the 3D hand scanning data and the 3D model of the exoskeleton is a part of our know-how.

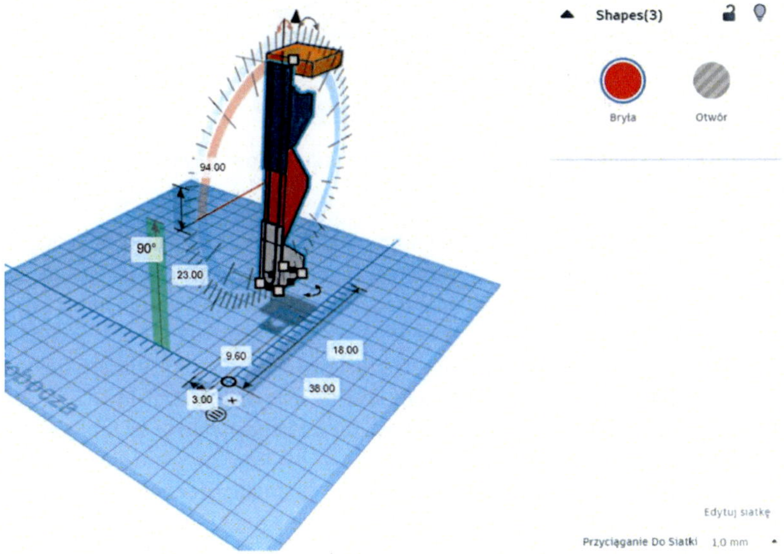

Fig. 2. Hand exoskeleton computational model - one finger: extension.

The main limitation of the current research remains initial research phase on flexible semi-automatic data gathering, computational modelling, adjustment and reassessment. Aforementioned procedures should take into account many various factors, including clinical, biomechanical, and material. They can be an effect of synergy on many processes, beginning from collecting clinical data (including functional testing of physiological and pathological function), hand clinical imaging, 3D hand scanning, data storage, processing and analysis, and use them for development and design templates of exoskeleton. The proposed concept combines physiological assessment of the entire patient reflected in biomechanical factors, advanced 3D computational modelling based on specific hand exoskeleton templates as far as advanced material engineering. There is need to determine the work necessary for the mechanical deformation of plastic parts (during further studies: plastic, metal, and even ceramic parts as far as magnetic joints) to

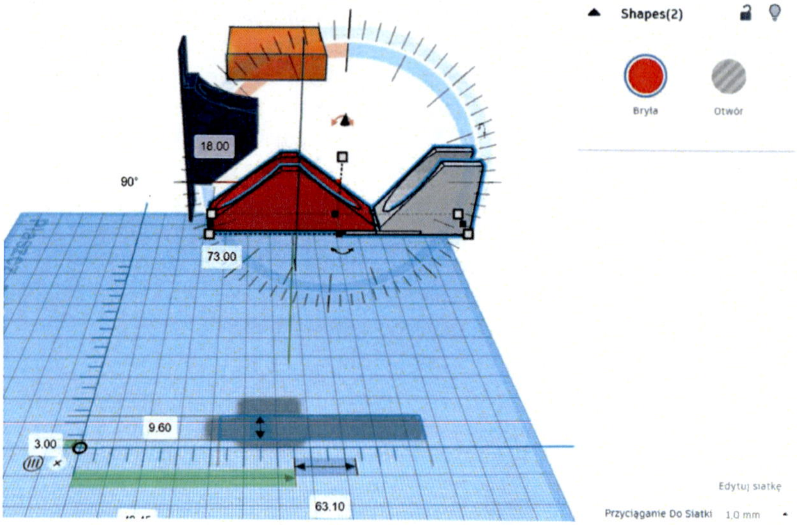

Fig. 3. Hand exoskeleton computational model - one finger: flexion.

specified dimensions (taking into account adjustment ranges where available) and deformation of various parts in diverse movements. It can provide an useful basement for a novel semi-automated method for user-centred hand exoskeleton design and assessment prior to patient-tailored manufacturing.

Our solution introduces new concept of crusher for visco-elastic materials, particularly plastics, including postconsumer, possible to use as a part of novel concept for recycling of materials for 3D printing of the patient-tailored assistive technology devices such as aforementioned hand exoskeleton [32–34].

Well known disadvantage of previous solutions of crusher for plastics was low productivity, high consumption of energy, high noise level, lack of possibility to produce precise grain distribution, danger of strong plastic heating into crusher chamber, as far as high risk of the damage of working parts, especially knives.

In addition, in the field of comminution, the work undertaken in terms of obtaining a product with a given geometric dimension, in a process with minimal energy demand, rational efficiency (minimum unit energy consumption), the proposed concept fully meets expectations. An experiment is needed to carry out the procedure related to modelling, which allows to further model testing and verification which requires check of comminution characteristics depending on design features, type of particles, movement features, energy grinding indicators, mechanical properties of the drive system, and many more process parameters [35–37].

4 Reverse Engineering as a Part of Industry 4.0 Concept

Industry 4.0 constitutes novel concept changing way of business/manufacturing. Resulting process of smart manufacturing is controlled by artificial intelligence (AI) systems which makes decentralized decisions. 3D printing processes can be aided with numerical computations. Using Finite Element Analysis and Genetic Algorithms, a proper set of printing parameters can be calculated, increasing practical applications and reducing material usage and manufacturing time, while maintaining required strength [37]. Future Internet of Things (IoT) systems based on also 5G mobile networks will communicate and coordinate both with each other, as a part of digital transformation. Aforementioned concept is strongly influenced by 3D printing, 3D scanning and reverse engineering, but it will depend on printing safety, quality and speed, as far as price, choice of printing materials and environmental impact. Thus 3D printing may:

– reduce waste,
– use Earth friendly, biodegradable materials,
– provide unique feature of making a single copy if it need be, and immediate reusing of the material to reduce unsold inventory.

Reverse Engineering and 3D Printing could be used in medicine for producing individualized orthoses, allowing reduction of both costs and time in comparison to traditional manufacturing process [38, 39]. Discontinued or unsold products may be reused to avoid unwanted pollution.

5 Conclusions

No doubt much more research based on advanced methodology within controlled environment is required to better explain 3D printers emission, exposure, and effectiveness of controls in various occupational and laboratory settings. But our solution can significantly improve situation concerning influence of the pollution and reuse of the materials. What more potential for harmful secondary changes from emission from certain uncontrolled additive manufacturing processes may influence further shape of the 3D printing within Industry 4.0 concept.

Acknowledgments. The work presented in the paper has been financed under grants to maintain research potential Kazimierz Wielki University.

References

1. WHO health an climate change survey report. https://www.who.int/phe/en. Accessed 25 Dec 2019
2. Kwon, O., et al.: Characterization and control of nanoparticle emission during 3D printing. Environ. Sci. Technol. **51**(18), 10357–10368 (2017)
3. Stefaniak, A.B., et al.: Evaluation of emissions and exposures at workplaces using desktop 3-dimensional printer. J. Chem. Health. Saf. **26**(2), 19–30 (2019)
4. Stefaniak, A.B., et al.: Insights into emissions and exposures from use of industrial-scale additive manufacturing machines. Saf. Health Work. **10**(2), 229–236 (2019)
5. Potter, P.M., Al-Abed, S.R., Lay, D., Lomnicki, S.M.: VOC emissions and formation mechanisms from carbon nanotube composites during 3D printing. Environ. Sci. Technol. **53**(8), 4364–4370 (2019)

6. Wojtyła, S., Klama, P., Baran, T.: Is 3D printing safe? Analysis of the thermal treatment of thermoplastics: ABS, PLA, PET, and nylon. J. Occup. Environ. Hyg. **14**(6), 80–85 (2017)
7. Stabile, L., Scungio, M., Buonanno, G., Arpino, F., Ficco, G.: Airborne particle emission of a commercial 3D printer: the effect of filament material and printing temperature. Indoor Air **27**(2), 398–408 (2017)
8. Steinle, P.: Characterization of emissions from a desktop 3D printer and indoor air measurements in office settings. J. Occup. Environ. Hyg. **13**(2), 121–132 (2016)
9. Zhang, B., Kowsari, K., Serjouei, A., Dunn, M.L., Ge, Q.: Reprocessable thermosets for sustainable three-dimensional printing. Nat. Commun. **9**(1), 1831 (2018)
10. Reich, M.J., Woern, A.L., Tanikella, N.G., Pearce, J.M.: Mechanical properties and applications of recycled polycarbonate particle material extrusion-based additive manufacturing. Materials (Basel) **12**(10), 1642 (2019)
11. Paciorek-Sadowska, J., Borowicz, M., Isbrandt, M.: New Poly(lactide-urethane- isocyanurate) foams based on bio-polylactide waste. Polymers (Basel) **11**(3), 481 (2019)
12. Gama, N.V., Ferreira, A., Barros-Timmons, A.: Polyurethane foams: past, present, and future. Materials (Basel) **11**(10), 1841 (2018)
13. Woern, A.L., Byard, D.J., Oakley, R.B., Fiedler, M.J., Snabes, S.L., Pearce, J.M.: Fused particle fabrication 3-d printing: recycled materials' optimization and mechanical properties. Materials (Basel) **11**(8), 1413 (2018)
14. Kopowski, J., Rojek, I., Mikołajewski, D., Macko, M.: 3D printed hand exoskeleton - own concept. In: Trojanowska, J., Ciszak, O., Machado, J.M., Pavlenko, I. (eds.) MANUFACTURING 2019. LNME, pp. 298–306. Springer, Cham (2019). https://doi.org/10.1007/978-3-030-18715-6_25
15. Kopowski, J., Mikołajewski, D., Macko, M., Rojek, I.: Bydgostian hand exoskeleton - own concept and the biomedical factors. Bio-Algor. Med-Syst. **15**(1), 1–7 (2019)
16. Kopowski, J., Rojek, I., Mikołajewski, D., Macko, M.: Bydgostian hand exoskeleton –concept and preliminary findings. Stud. Mater. Appl. Comput. Sci. **12**(1), 22–28 (2020)
17. Hill, D., Holloway, C.S., Morgado Ramirez, D.Z., Smitham, P., Pappas, Y.: What are user perspectives of exoskeleton technology? A literature review. Int. J. Technol. Assess. Health Care **33**, 160–167 (2017)
18. Wolff, J., Parker, C., Borisoff, J., Mortenson, W.B., Mattie, J.: A survey of stakeholder perspectives on exoskeleton technology. J. Neuroeng. Rehabil. **11**, 169 (2014)
19. Ngeo, J., Tamei, T., Shibata, T., Orlando, M.F., Behera, L., Saxena, A.: Control of an optimal finger exoskeleton based on continuous joint angle estimation from EMG signals. In: 35th Annual International Conference of the IEEE Engineering in Medicine and Biology Society Proceedings (EMBC), pp. 338–341. IEEE, Osaka (2013)
20. Bos, R.A., Haarman, C.J., Stortelder, T., Nizamis, K., Herder, J.L., Stienen, A.H.: A structured overview of trends and technologies used in dynamic hand orthoses. J. Neuroeng. Rehabil. **13**, 62 (2016)
21. Yue, Z., Zhang, X., Wang, J.: Hand rehabilitation robotics on poststroke motor recovery. Behav. Neurol. 3908135 (2017)
22. Rose, C.G., Kann, C. K., Deshpande, A.D., O'Malley, M.K.: Estimating anatomical wrist joint motion with a robotic exoskeleton. In: International Conference on Rehabilitation Robotics Proceedings (ICORR), pp. 1437–1442. IEEE, London (2017)
23. Proietti, T., Guigon, E., Roby-Brami, A., Jarrassé, N.: Modifying upper-limb inter-joint coordination in healthy subjects by training with a robotic exoskeleton. J. Neuroeng. Rehabil. **14**, 55 (2017)
24. Wu, G., van der Helm, F.C., Veeger, H.E., Makhsous, M., Van Roy, P., Anglin, C.: ISB recommendation on definitions of joint coordinate systems of various joints for the reporting of human joint motion – part II: shoulder, elbow, wrist and hand. J. Biomech. **38**, 981–992 (2005)

25. Chan, S.S., Moran, D.W.: Computational model of a primate arm: from hand position to joint angles, joint torques and muscle forces. J. Neural Eng. **3**, 327–337 (2006)
26. Garner, B.A., Pandy, M.G.: Musculoskeletal model of the upper limb based on the visible human male dataset. Comput. Meth. Biomech. Biomed. Eng. **4**, 93–126 (2001)
27. Li, J., Zheng, R., Zhang, Y., Yao, J.: iHandRehab: an interactive hand exoskeleton for active and passive rehabilitation. In: International Conference on Rehabilitation Robotics Proceedings, pp. 1–6. IEEE, Zurich (2011)
28. Wang, J., Li, J., Zhang, Y., Wang, S.: Design of an exoskeleton for index finger rehabilitation. In: Annual International Conference of the IEEE Engineering in Medicine and Biology Society Proceedings, pp. 5957–5960. IEEE, Minneapolis (2009)
29. Krüger, M., Eggert, T., Straube, A.: Joint angle variability in the time course of reaching movements. Clin. Neurophysiol. **122**, 759–766 (2011)
30. Kulig, K., Andrews, J.G., Hay, J.G.: Human strength curves. Exer. Sport Sci. Rev. **12**, 417–466 (1984)
31. Kapandji, A.I.: The clinical evaluation of the upper limb joints' function: back to Hippocrates. Hand Clin. **19**, 379–386 (2003)
32. Macko, M.: Economic-energetic analysis of multi-edge comminution of polymer recyclates. Przem. Chem. **92**(8), 1499–1502 (2013)
33. Macko, M., Flizikowski, J., Szczepanski, Z.: CAD/CAE Applications inv Mill's Design and Investigation. Book Series: Lecture Notes in Mechanical Engineering, pp. 343–351 (2017)
34. Flizikowski, J., Macko, M.: Method of estimation of efficiency of quasi-cutting of recycled opto-telecommunication pipes. Polimery. **1**, 53–59 (2001)
35. Flizikowski, J., Macko, M.: Competitive Design of Shredder for Plastic in Recycling Tools and Methods of Competitive Engineering, vol. 1–2, pp. 1147-1148. IOS Press, Amsterdam (2004)
36. Flizikowski, J., Macko, M., Czerniak, J.: Implementation of genetic algorithms into development of mechatronic multi-edge's grinder design. In: International Mechanical Engineering Congress and Exposition Proceedings (ASME), vol. 7, pp. 1227–1235, ASME, Denver (2012)
37. Macko, M., Mroziński, A., Flizikowski, J.: Design and utility of specialist comminution set-up for plastics and organic materials. In: International Mechanical Engineering Congress and Exposition Proceedings (ASME), vol. 3, pp. 397–402, ASME, Denver (2012)
38. Górski, F., et al.: Selection of fused deposition modelling process parameters using finite element analysis and genetic algorithms. J. Mult.-Val. Logic Soft Comput. **32**(3/4), 293–311 (2019)
39. Wierzbicka, N., Górski, F., Wichniarek, R., Kuczko, W.: Prototyping of individual ankle orthosis using additive manufacturing technologies. Adv. Sci. Technol. Res. J. **11**(3), 283–288 (2017)

Re-entry Vehicle Structural Optimization for Mass Minimization

Alexandru-Mihai Cismilianu[1,2](✉), Iulian Chirita[2], Alexandru Gabriel Persinaru[2], Alexandru Marin[2], Camelia Elena Munteanu[2], Ana-Maria Neculaescu[2], and Cornel Dragoman[2]

[1] Faculty of Industrial Engineering and Robotics (FIIR) Part of University Politehnica of Bucharest (UPB), Bucharest, Romania
[2] INCAS – National Institute for Aerospace Research "Elie Carafoli", Bucharest, Romania
{cismilianu.alexandru,chirita.iulian}@incas.ro

Abstract. The major efforts in the space industry have the aim of finding ways to reduce the overall costs of developing and launching a spacecraft (S/C). In the current paper, structural optimization proves to be a possible answer for respecting the most decisive elements like mass reduction and minimum lead time, which translates into cost saving. This paper aims to present an end to end process of a topological optimized structure of a reusable re-entry vehicle section, manufactured using conventional methods. The structural optimization is performed trough advanced commercial of the shelf (COTS) software tool. The results are presented as a mass comparison, ease of manufacturing, and other criteria's that are further presented in the paper.

Keywords: Aerospace design · Structural optimization · Topology optimization · Mass minimization · Mechanical structures · Stress analysis

1 Introduction

From the beginning of space exploration through launched and deployed spacecraft (S/C), into or after the Earth's orbit, the principal goal is to minimize the operational costs. The price/kg needed to take something into space represents the operational costs. As Harry W. Jones stated in his paper: "NASA's space shuttle had a cost of about $1.5 billion to launch 27,500 kg to LEO (Low Earth Orbit), $54,500/kg." [1]. Given the recent progress into transforming the launchers in reusable platforms, the overall cost was minimized greatly as represented in Fig. 1 below.

Even with the greatest shrinkage of the cost in launching a certain Kg (kilogram) into orbit, mass minimization is still a point of interest. For S/C and payloads that are not intended to be inserted into orbit with the latest reusable launchers, mass minimization is critical in the development of the spacecraft.

J. Machado et al. (Eds.): icieng 2021, LNME, pp. 92–100, 2022.
https://doi.org/10.1007/978-3-030-79165-0_9

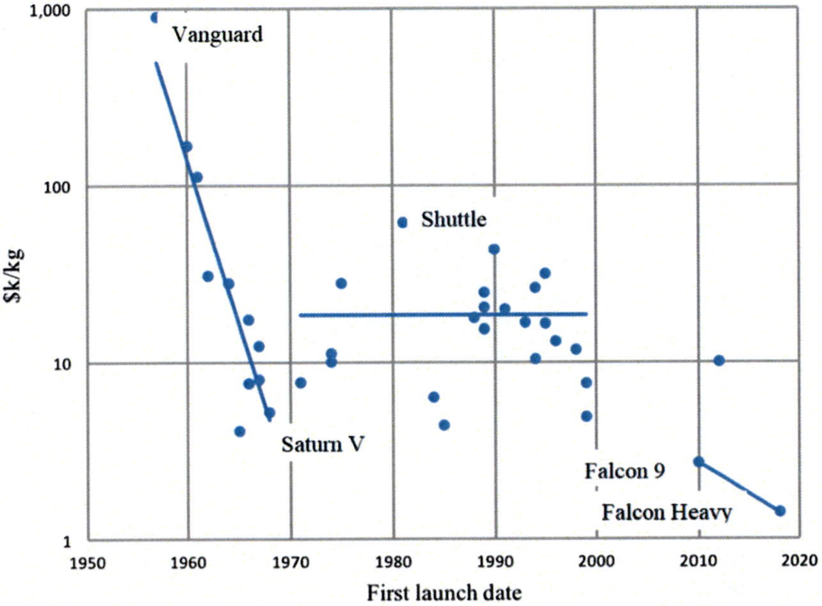

Fig. 1. Launch cost per kilogram to LEO versus first launch date [1]

In order to achieve minimization of mass, a classical method is to use the development team engineering expertise and feeling which is then validated through structural analysis and redesigned to obtain a part with the minimum mass. Another method is to minimize the mass trough structural optimization using in-house developed software [2, 3] or by using specialized COTS (components of the shelf) SW (software), in this case HyperMesh. Structural optimization has gained a lot of popularity in the latest years since it has a wide applicability range [4]. Another major improvement is that it allows to impose manufacturing constraints so the results obtained can also be produced using conventional manufacturing processes (like milling, casting, forging etc.). Even though this processes limits the fabrication of complex shapes, the progress made into additive manufacturing [5] (cost, tolerances, available materials, etc.) are still not enough to produce large parts as those needed for a S/C.

In the current paper, an initial structure is considered and a structural optimization is applied to understand how to re-design the structure in order to minimize the mass and to get a better loading of the structure. The new structure is validated through structural analysis. The added benefit is an end to end process where all the structural optimization set up cases are presented to ease up the process for further similar cases.

2 Optimization Process Definition

The section of a S/C is considered for the structural optimization. The initial thickness of the walls is of approximately 40 mm. The aim of the current optimization is to find the optimum shape and thicknesses of the given structure and implement it in the CAD

(Computer Aided Design) model which shall be then verified through means of structural analysis.

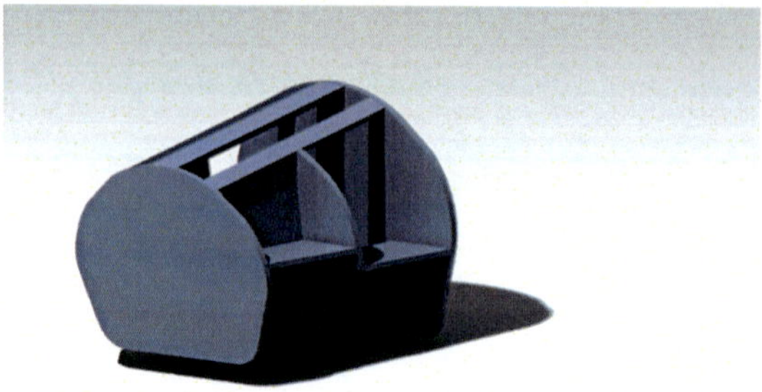

Fig. 2. Initial spacecraft section study

The model used as input it is taken from its initial state, Fig. 2, and simplified as connected surfaces. A mesh of shell elements with an overall element dimension of 10 mm is applied to all the surfaces using HyperMesh auto mesh. The FEM (Finite Element Modeling) is represented bellow in orange Fig. 3. The next step necessary as a pre-requirement to the structural optimization is splitting the FEM into design (burgundy) and non-design (blue) areas, Fig. 3.

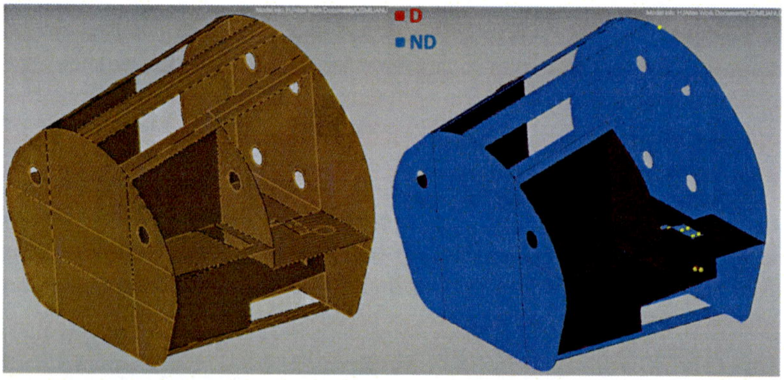

Fig. 3. FEM modeling; Design (D) and Non-Design (ND) separation

On the separated model into D (design) and ND (Non-Design), SPC (Single Point Constraint) are placed on the ND area on the outer margins of the closing panels. Also, on the ND area, forces are applied are into specific points which are represented in Fig. 4.

The forces applied to the model are presented Table 1. The point of application of the forces is shown only on one side but the points are symmetrically disposed with the XOZ plane. The second value of F2′ force has the vector intentionally driven in the opposite direction as F2 so an asymmetrical result should be expected. The reconstruction is made considering only a fully symmetrical structural section.

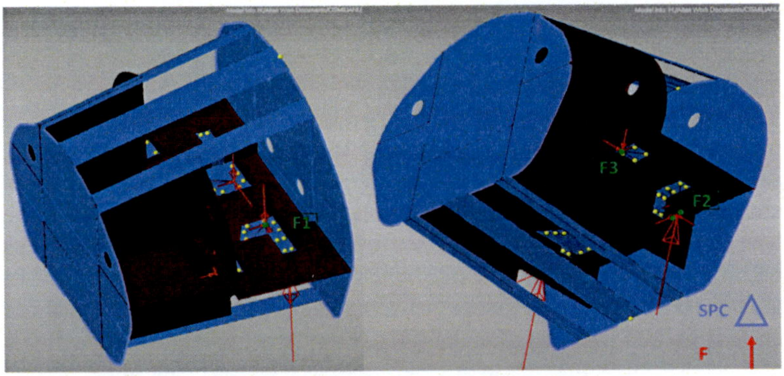

Fig. 4. Applied forces and constraints

Table 1. Forces applied onto the structure

	Fx [N]	Fy [N]	Fz [N]
F1	−60912	−33840	−76704
F2	53016	50760	197400
F2	−1353.6	47601.6	92496
F3	−43992	−103776	−133104
F1′	−60912	−33840	−76704
F2′	53016	50760	197400
F2′	−1353.6	−47601.6	−92496
F3′	−43992	−103776	−133104

The initial material considered in the model was EN AW 5083 as presented in Table 2, the same material is considered when writing the material card which will be used further on when defining the element properties. The material and element properties are assigned to all elements for the D and ND area.

Table 2. Material and element properties

EN AW 5083	
Material card	MAT1
Density [t/mm^3]	2,71e − 009
Young Modulus [MPa]	70000e + 003
Poisson ratio	0,33
Element thickness [mm]	40

A loadstep is defined where all the forces and the SPC's are assigned to a linear static case.

Previous experiences [6] showed that TO (topology optimization) is best to be used in the current case in order to observe the main directions of loading throughout the structure considering the imposed constraints and forces. The post processing of the results are implemented in the CAD model keeping in mind that the redesign is made for the targeted technology (i.e. design for milled, design for AM, etc.).

When setting the LC (load case) for the TO, since we are using shell elements, a design variable is imposed on them, which considers the mD (minimum member dimension), MD (maximum member dimension) and a stress constraint. Also, one plane symmetry is imposed with regard to the XOZ plane. The (mD) and (MD) will be varied along with the thickness in order to see their influence on the result of the optimization. The stress constraint will be kept constant as it represents the yield stress of the used material.

The chosen objective is the minimization of the response, which in the current case, it is the mass.

Table 3. Influence of the parameter variation on the results.

No	1.	2.	3.	4.	5.
Param. [mm]	T: 40 mD: 20 MD:100	T: 40 mD: 20 MD: 40	T: 30 mD: 15 MD: 30	T: 15 mD: 15 MD: 30	T: 20 mD: 52 MD: 40
Result	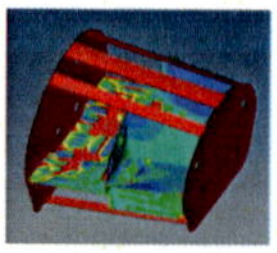	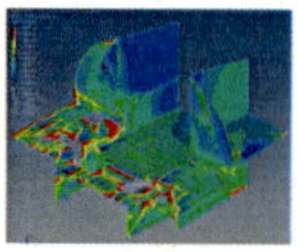	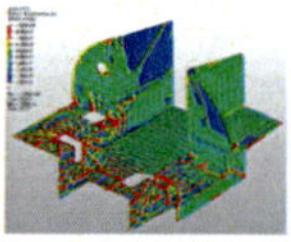	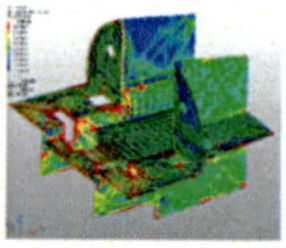	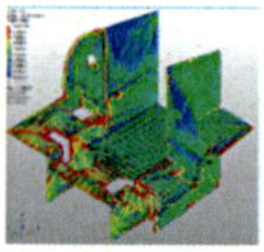

In Table 3, starting from the values of the parameters imposed in case 1, it can be observed that reducing the value of the MD to 40 mm increased the quality of the output results. Thus, the same proportion has been kept between the minimum member dimension and maximum member dimension, 1:2, throughout the rest of the cases. Continuing to cases 3 and 4, the mD and MD were reduced, with the 1:2 proportion kept between them, and the thickness was incrementally lowered in order to find the minimum admissible overall thickness. A case with the same mD and MD was performed with t = 15 mm thickness but no concluding results were obtained as the optimizer could not converge to such harsh requirements. Between cases 4 and 5, the same identified thickness was kept but in case 5 the same parameter values where used as in case 2.

Joggling with the thickness, mD and MD greatly improved the quality of the optimization result as it can be seen in Fig. 5.

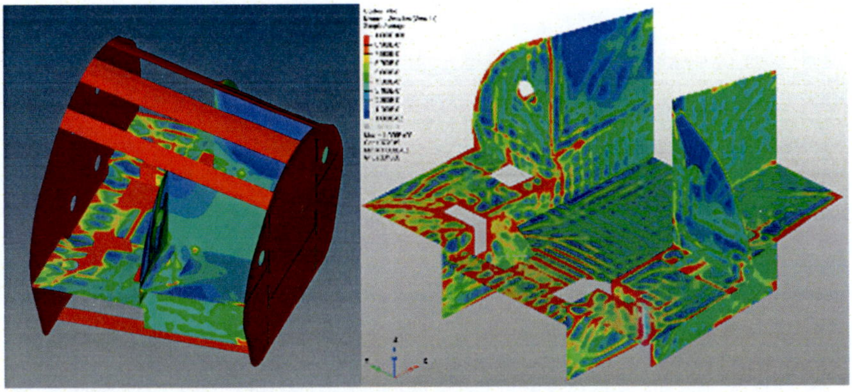

Fig. 5. Case 1 and case 5 comparison

3 Optimization Results Validation

3.1 Implementation of the Results

After the optimization process, the results were implemented as a redesign into a new CAD model. To ensure the structural performance of the new structure, in areas of minimum stress, pockets where introduced into the new CAD model, and in the critical area's ribs were added.

As a result from the TO, several pockets were identified. The manner in keeping a link between the results and the actual implemented changes in the CAD model was by considering a similar area of pockets restrained in one (sum of all the pockets in that area) disposed with an engineering feeling. The TO results are displayed graphically with a color scheme which starts from blue to red, represented through applied density to each element from 0 (blue) to 1 (red), based on the loads direction. The red color represents the loading path and the maximum stress that the structure needs to support. Therefore, ribs were implemented in the CAD model which ensure both the direction and the stress indicated by the red zone and the stress from the yellow areas. In the areas where the stress is minimum pockets are made, as represented in Fig. 6.

The initial thickness of the walls was approximately 40 mm and after the structural optimization their thickness is between 10–40 mm.

With these operations the mass was minimized and the new structure is verified also trough means of structural analysis in the next step.

3.2 Design for Manufacturing

In order to reduce the overall costs and to facilitate the manufacturing process it is necessary to compare various types of manufacturing processes. In this case, the most convenient manufacturing process is CNC (computer numerical control) machining. As mentioned above, in this paper the main objective is the reduction of mass and of the manufacturing time by keeping at the same time the lowest cost. Of course, the cost is also time dependent if we consider the entire process of delivering a product and shall include

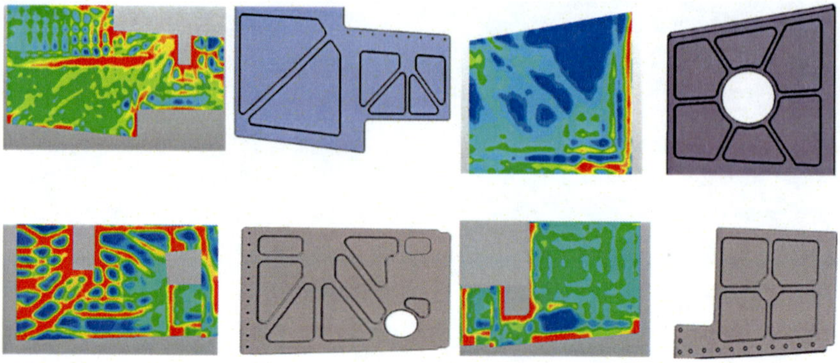

Fig. 6. Results from structural optimization implemented in the CAD model

all the steps from the design step up to the manufacturing step, activities like material selection, design constraints, machine CNC programming and other activities depending on the complexity and size of the component. The optimized structure requires more manufacturing time than the initial structure, but the time required for the manufacturing process is outweighed by the fact that the total mass decreased with 26,5%, Fig. 7. Given the fact that the price/Kg for a launch is quite high, a reduction of enough Kg's justifies the complexity of the parts.

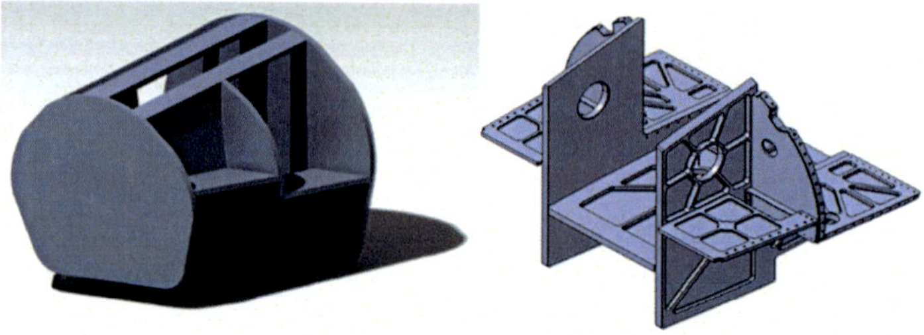

Fig. 7. Initial structure and optimized structure

As an example of mass reduction, for a component as represented in Fig. 8, by implementing the results based on the methodology mentioned above, the mass reduction is approximately 37%.

Also, besides the reduction of mass, time and cost it is absolutely necessary that the structure shall support the spacecraft subsystems with enough strength to avoid any failure [7].

3.3 Structural Validation

The S/C FEM and LC definition for the structural validation of the optimization result has been developed using the MSC PATRAN software and the processor used was MSC NASTRAN. The analysis performed is a linear static one, solved with sol 101.

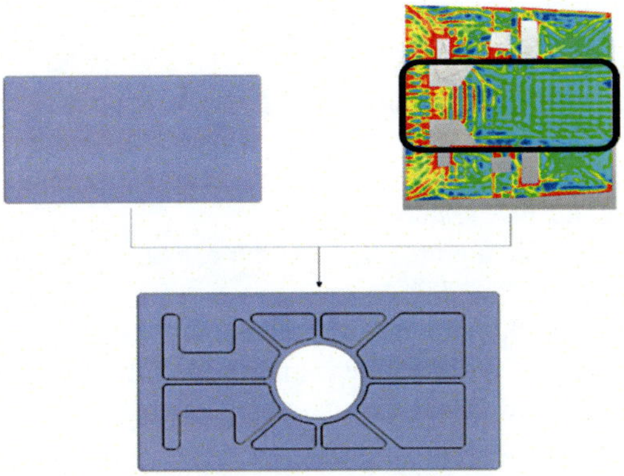

Fig. 8. Initial design and optimized design

The loads and boundary conditions are similar with the ones used in the optimization process presented in the previous chapter. The main purpose of this step is to identify if there are significant improvements regarding the structure's loading.

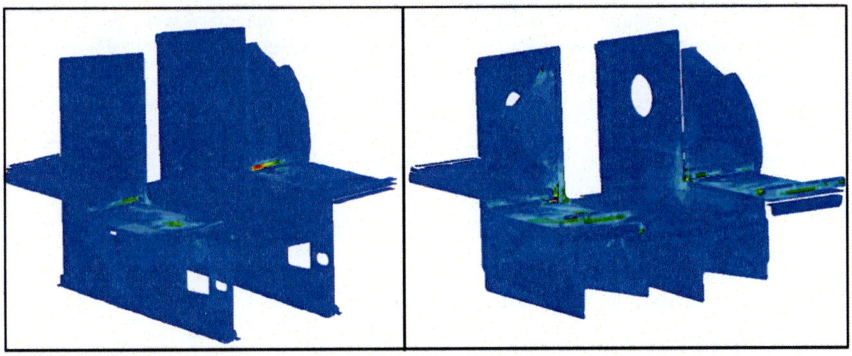

Fig. 9. Before and after optimization stress distribution

The analysis showed a significant improvement in the stress distribution by loading all the connected elements instead of concentrating the stress in a small area as it can be observed in Fig. 9. All the parts were loaded bellow the material yield allowable. Also, a significant improvement appeared on the fastening elements.

The optimization not only decreased the mass, but also improved the load distribution on all the linked elements.

4 Conclusion

Starting from the values of the parameters imposed in case 1, it can be observed that reducing the value of the maximum member dimension to 40 mm increased the quality of the output results. Thus, it was observed that keeping a ration between the mD and MD of 1:2 gives back the best results.

It is clearly observed that a more refined result can be obtained by joggling with the thickness, mD and MD.

The mass reduction for this structure has values between 21 and 37% with an average of 26,5% and the strength of the structure is improved. The use of topological optimization methods in the concept phase offers solutions for mass reduction that can reduce the number of design cycles.

Even though the complexity of the optimized structure requires more time than the initial structure, the time required for the manufacturing process is outweighed by the fact that the total mass decreased with 26,5%, since the price/Kg for a launch is quite high.

By associating a well-defined step of structural optimization into the typical engineering design approach it can add a slew of benefits for the final result.

Acknowledgment. This work has been funded by the European Social Fund from the Sectoral Operational Programme Human Capital 2014–2020, through the Financial Agreement with the title "Scholarships for entrepreneurial education among doctoral students and postdoctoral researchers (Be Antreprenor!)", Contract no. 51680/09.07.2019 - SMIS code: 124539.

This work has been funded by the National Institute for Aerospace Research "Elie Carafoli".

References

1. Harry, W.J.: The recent large reduction in space launch cost. In: 48th International Conference on Environmental Systems. ICES-2018–81, Albuquerque, New Mexico (2018)
2. Pricop, M.V.: Solid topology optimization using truss lattices. INCAS Bull. **4**(3) (2012). https://doi.org/10.13111/2066-8201.2012.4.3.4. ISSN 2066-8201
3. Pricop, M.V., Bunea, M.: Solid topology optimization using truss lattices for low volume fraction. MTA Rev. **XXIV**(3) (2014). ISSN 1843-3391
4. Pricop, M.V., Bunea, M., Nedelcu, R.: Truss systems and shape optimization, In: AIP Conference Proceedings of International Conference of Numerical Analysis and Applied Mathematics, ICNAAM 2016 (2016). https://doi.org/10.1063/1.4992592
5. Cismilianu, A.M., et al.: Spacecraft bracket design using additive manufacturing. In: 6th CEAS Air & Space Conference, CEAS Aeronautical Journals (Springer), CEAS 2017, Bucharest, Romania (2017)
6. Cismilianu, A.M., Petre, A.R., Liliceanu, C.D., Bibire, R.P.: Topology optimization of a payload adapter for a launcher. In: 5th International Workshop on Numerical Modelling in Aerospace Sciences, Book of Abstracts NMAS 2017 and DVD, NMAS 2017, Bucharest, Romania (2017)
7. Thomas, P.S., Wiley, J.L.: Spacecraft Structures and Mechanisms – From Concept to Launch. Springer, New York (1995).ISBN: 978-0-7923-3476-7

A Novel Integrated Framework Approach for TEBC Technologies in Distributed Manufacturing Systems: A Systematic Review and Opportunities

Veerababu Ramakurthi[1], Vijayakumar Manupati[1], M. L. R. Varela[2(✉)], and Goran Putnik[2]

[1] National Institute of Technology Warangal, Warangal 506004, India
[2] University of Minho, 4804-533 Guimarães, Portugal
leonilde@dps.uminho.pt

Abstract. Distributed manufacturing systems have become consensus, particularly networked manufacturing systems (NMS) owing to its flexible and adaptable nature in response to customized requirements. Highlighting of functions in NMS and their integration with recent key enabling technologies i.e., Artificial intelligence, Machine learning, Internet of Things, Block chain technology, and Agent-based techniques, etc. are of utmost importance to identify problems and increase its efficiency. Hence, besides the above-mentioned approaches, this paper surveyed and analysed various articles systematically related to networked manufacturing in the context of knowledge creation and information, security, interoperability, and reliability. To identify the most related papers, the search has been conducted with Web of Science and Scopus databases. Subsequently, after evaluation 30 most related papers were selected and analyzed that further extended by identifying the issues and gaps in the existing empirical knowledge. Finally, the paper presents a roadmap for future research directions and developments.

Keywords: Networked manufacturing system · Distributed manufacturing · Interoperability · Blockchain technology · Machine learning · Artificial Intelligence

1 Introduction

Network Manufacturing System (NMS) allows the transmission and incorporation of knowledge and data obtained from various processes starting from raw material to the finished product and it allows to share the resources between the firms which are established at various locations, where the manufacturing firms can respond to the immediate requirement in the market. However, the challenges in conventional manufacturing systems and its difficulties for enhancing the throughput and flexibility has been explained [1]. In this systematic literature review (SLR) paper, four important factors i.e., knowledge creation management, Interoperability, realiablity, and security which are responsible for NMS effectiveness in the context of text mining, evolutionary algorithms and

J. Machado et al. (Eds.): icieng 2021, LNME, pp. 101–112, 2022.
https://doi.org/10.1007/978-3-030-79165-0_10

block chain (TEBC) technologies is detailed. Inorder to respond to the frequent product changes, small product life cycle, customer customized requirements, etc. recent approach of enterprises shifted from intelligent manufacturing to knowledge creation and its management as one of the important aspect to achieve competitive gain between the manufacturing firms within the global market [2]. The second factor, Interoperability according to IS0161001 is the concept of effective transfer of information using a common platform between the enterprises to meet specific applications. Here, the works related to interoperability particularly in the context of distributed manufacturing is reviewed and presented. Thereafter, recent days' security of shared data between the enterprises becomes the most challenging as it is more vulnerable to cyber-attacks in distributed manufacturing systems and a brief review of security related aspects in the context of the distributed manufacturing system has been carried. Whether a system is able to meet the demand or not is measurable with a probability and that can be called system reliability [3]. Finally, a review on articles related to the reliability of the system in the context of the distributed manufacturing system has been conducted.

The main aim of this present article is to examine, synthesize the roles of the knowledge and Information management, Security, Interoperability, and Reliability in the network manufacturing system with help of SLR. The present literature review is typically illustrative and incipient in nature, the authors of this article is predetermined to identify the challenges in knowledge management in the context of networked manufacturing systems. The review also aimed at examining how aspects like security and reliability are affecting in maintaining networked manufacturing. Additionally, the importance of above mentioned manufacturing functions management related services such as identification of resources, Service management, and Managing Processes are discussed in the context of distributed manufacturing systems. Here, the aspects like response management and service reconfiguration in case of distributed manufacturing systems by: (i) reviewing in detail the effect of the mentioned referred areas in the NMS, with a special focus on knowledge and information management, Interoperability, security, and reliability; (ii) Finding some of the research gaps in the reviewed literature; (iii) suggesting directions for further research also discussed. The alignment of this present article is explained as follows. Section 2 explains the five-step based research methodology used in this article. Section 3 the locating studies of the SLR, and arrangement according to the conceptualization discussed in Fig. 1. Section 4 discussed the obtained results, presenting valid research suggestions and guidelines for future work. Section 5 illustrates the limitations and conclusions of the present SLR.

2 Research Methodology

This article adopted the systematic literature review (SLR) shows in Table 2 methodology [4] which is different from the traditional literature review. SLR methodologies allow constructing a framework for a thorough investigation of literature, adopting an organized and clear process [4]. As suggested by Denyer and [5], a scoping study of the area was performed before the SLR is to (i) Plan the existing foundation for the work that needs to be done; (ii) enumerate the fit of the mentioned SLR into the present body of knowledge; (iii) describe concepts; (iv) decide the research queries to be answered.

Therefore, this article systematically reviews related literature and their importance of knowledge and Information management, Security, Interoperability and Reliability in network manufacturing systems. The adopted SLR consist of a five-step procedure, as suggested by Denyer and Tranfield and Wong et al. [5]: Framing of the research question; Identifying the studies; Study selection and interpretation; Investigation and synthesis; Reporting and presenting the results.

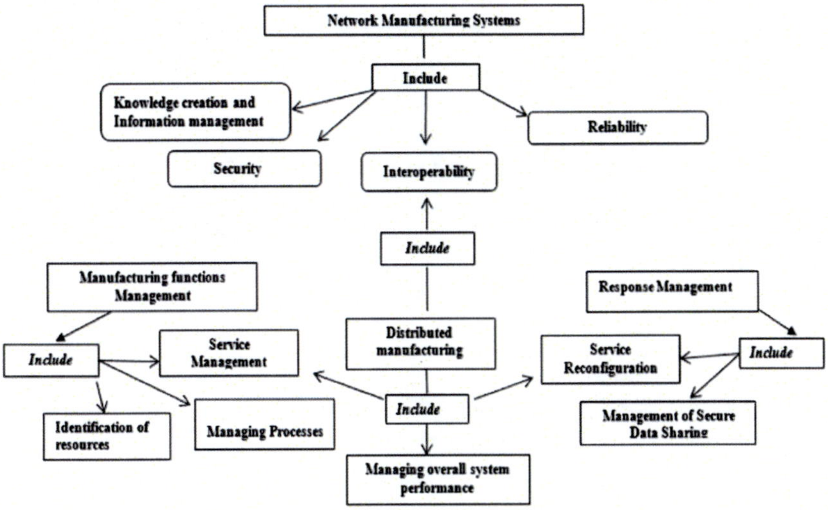

Fig. 1. Topics related to knowledge creation and information management, security, Interoperability, Reliability aspects considered in the literature addressing the Network manufacturing systems.

Step 1*: The mentioned research questions were formulated for this study*

RQ1**:** What is the role of knowledge and Information management, Security, Interoperability, and Reliability in network manufacturing systems.
RQ2: Which topics and functions are useful to describe the Manufacturing functions management, distributed manufacturing systems, and response management are considered the scope of the network manufacturing system.

Step 2*:Identifying the studies*

With the help of initial search strings which were mentioned in Table 1 are searched in the bibliographic database or search engine. The search was carried towards the research questions. This database is taken care of most of the research work that was published in Distributed Manufacturing, including almost all peer-reviewed articles on the area as shown in Tables 1 and 2.

Step 3*: Study selection and interpretation*

In the next step i.e. step 3 some filtering criteria are applied so that only the most related studies to be added in the review. To pay much more attention to the latest studies and latest procedures and applications, initially, a 10-year time span was taken into consideration. Only articles that fall into the category of peer-reviewed publications in english were taken into consideration. Colicchia and Strozzi [6] mentioned the filtering process of these peer-reviewed articles before they proceed for publication. Some more articles are filtered by restricting the search areas to Engineering, Manufacturing, Industrial, and Automation only.

This above exercise is limited to 363 papers, and then by using Mendeley and sEndNote duplicates were removed and further only 280 papers were considered. This procedure has been carried to find the authenticity of the work carried out in those articles. After, thorough review only papers satisfy the below mentioned three criteria are considered for further evaluation.

Table 1. Search strings and displayed corresponding results from Web of Sciences.

Search string	Search field	Date of search	No of articles
"Distributed Manufacturing" (or) "Distributed-Manufacturing" and "Integrated Process Planning and Scheduling"	Topic	08-08-2019	183
"Distributed Manufacturing" (or) "Distributed-Manufacturing" (or) "Agent-based systems"	Topic	08-08-2019	761
"Distributed Manufacturing" (or) "Distributed-Manufacturing" and "Integrated Process Planning and scheduling" and "Agent-based method"	Topic	08-08-2019	30
Distributed Manufacturing" (or) "Distributed-Manufacturing" and Blockchain	Topic	08-08-2019	21
Distributed Manufacturing" (or) "Distributed-Manufacturing" and Text- mining	Topic	08-08-2019	5

Table 2. Search strings and the displayed corresponding results from Scopus.

Search string	Search field	Date of search	No of articles
"Distributed Manufacturing" (or) "Distributed-Manufacturing" and "Integrated Process Planning and Scheduling"	Title-Abs-Key	08-08-2019	106
"Distributed Manufacturing" (or) "Distributed-Manufacturing"(or) "Agent-based systems"	Title-Abs-Key	08-08-2019	526
"Distributed Manufacturing" (or) "Distributed-Manufacturing" and "Integrated Process Planning and scheduling" and "Agent-based method"	Title-Abs-Key	08-08-2019	13
Distributed Manufacturing" (or) "Distributed-Manufacturing" and Blockchain	Title-Abs-Key	08-08-2019	55
Distributed Manufacturing" (or) "Distributed-Manufacturing" and Text- mining	Title-Abs-Key	08-08-2019	13

Lastly in the final stage a keener investigation of the 113 articles done by the reviewers with their expertise. To perform the review in the most effective way a recent articles i.e. last 10 years were taken into consideration. In this process, some of the articles did not align with the subject area were filtered out and finally, 30 papers were listed out in Table 3. The comprehesive investigation of journals has been carried out by the authors in step 4, using two reviewers when their investigations and findings are useful to reduce the errors, to sort out any differences. Detailed information from these 30 articles was classified based on methodologies and important contributions. The later parts of the paper describe the outcome of the current work in a systematic procedure.

Table 3. Summary of the systematic review articles selection and evaluation.

Bibliographic database	Search 1	Search 2	Search 3	Search 4	Search 5	Total
Web of sciences	185	761	30	21	5	1002
Scopus	106	526	13	55	13	713
Inclusive/Exclusive procedure for Web of Science						
Time period	57	306	30	24	5	422
Document type	52	302	10	21	5	390
Research area	51	287	10	18	5	371
Language	50	284	09	16	4	363
Inclusive/Exclusive procedure for Scopus						
Date range	35	272	08	55	10	380
Document type	35	173	08	53	09	278
Research area	19	92	02	28	03	144
Language	18	89	02	21	03	133
After checking duplicates (in each search)	48	291	26	22	4	391
After checking duplicates (in all search)	32	263	07	41	3	366
Title and abstract evaluation	113					
Finally detailed article analysis	30					

3 Findings

3.1 Knowledge and Information Management

Dinh et al., in 2014 developed a maintenance method that can map the attributes and associated theory while creating ontology evolution by finding the text to identify the most related attribute for the respective mapping of data [7]. Claire et al., in 2015, refered that the interoperability is the organized management and an application of an ontological reference model to decrease the risk occurs because of disruptions from various processes and capabilities to regulation and laws [8].

Park et al., in 2018 did built an activity-based cost ontology by using protégé software for stocking and distribute the knowledge of cost related to the family of products in case of distributed IT computer assembling and power unit systems [9]. Liang et al., in 2018, introduced a substitute ontology, called *AM-OntoProc*, with the help of Managing Engineering Knowledge (MOKA) to support the process planning in case of Additive Manufacturing (AM) [10]. To check the feasibility of the new ontology is applied in various AM types, Viz., Stereolithography (SLA), Selective Laser Sintering (SLS), and Fused deposition Modelling (FDM) [10]. Li et al., in 2019, constructed an ontology-based modeling framework for answering the problem of inconsistency during the distribution of manufacturing resources in the case of the cloud manufacturing environment [11]. Huang et al., in 2011 proposed an ontology framework that consists of the generation of data and evolution of data for allocating the manufacturing data in the mold production process in an intelligent manufacturing environment [12].

3.2 Security

In view of increasing trends in recent cyber-attacks on various systems, now manufacturing systems needs to concentrate more on their security to transfer the valuable information exchange between the firms. Blockchain Technology (BCT) is one such disruptive technology works on distributed ledger having the advantage parameters like security, reliability and transparency and immutability helps BCT to grow as a revolutionary technology throughout the world.

Risius et al., in 2017, proposed a research framework which explains blockchain technology beyond cryptocurrencies and to present research topics to allow meaningful scholarly collaborative in blockchain research [13]. Viriyasitavat et al., in 2018, developed a framework for service composition in industry 4.0. and explained the essential role of third party certificates in selecting services for a business project. They introduced a Block Chain based technique useful to transfer the authenticity of businesses into the text in the form of smart contracts [14]. Mondragon et al., in 2018, described the effect of incorporating blockchain technology in composite material supply chains. Blockchain is used to record the transactions that are taken during the supply of raw materials [15]. Chung et al., in 2019 proposed a Mining procedure to evaluate a variety of data related to a traceability system in cognitive manufacturing [16]. Li et al., in 2019, articulated a number of interoperable blockchains integrated with the smart contracts to manage energy and financial flows within the microgrids in a credible manner [17]. Lu et al., in 2019, reviewed opportunities and research challenges faced by blockchain technology while implementing it in Distributed manufacturing systems [18]. Wong et al., in 2019, explained the supply chain transformation with the incorporation of blockchain technology. A cognitive study carried out by considering the 14 expert's opinions in this area [19].

3.3 Interoperability

The way to achieve standardization in the distributed manufacturing system, knowing the aspects of interoperability is the key. Agostino and Goncalves, in 2016, explained that for exchanging information mutually in a geographically distributed environment, the important part is coordination and collaboration between the enterprises [20]. Panetto et al., in 2012, stated that for sharing the information during coordination, ontology is the mode for interoperating all the applications [21].

3.4 Reliability

Lin et al., in 2012, proposed a decomposition method that separates into two paths namely one for processing and other for reworking. Two algorithms are proposed to ensure the reliability of the system for various network models by creating a lower boundary vector of the machine [3]. Lin et al., in 2013, introduced a Capitalized Manufacturing Network (CMN) to create a vector that contains the least capacity to estimate the manufacturing system's reliability while producing touch panels. Li et al., in 2013, proposed a Stochastic-flow Manufacturing Network (SMN) framework by establishing a low capacity vector to estimate the reliability of the given manufacturing system and to satisfy the demand. SMN separated into two parts namely identification of production lines in parallel and the other one distinct parallel lines with the common station [22]. Ostadi et al., in 2018, developed a framework for optimizing the preventive maintenance plan in the context of robot paint sprayer and each component parameters like distribution fitting and Reliabilities' are calculated [23]. Permin, Lossie, and Schmilt, in 2018, proposed a functional reliability model based on fault tree analysis for enhancement in safety and security features in smart factories as a case study of high speed turning [24].

3.5 Manufacturing Functions Management

Barzanji et al., in 2019, proposed a Decomposition Algorithm or Integrated process planning and scheduling problem in a network manufacturing environment [25]. Li et al., in 2019, described Particle Swarm Hybridized with the Genetic Algorithm to solve uncertain Integrated Process Planning and Scheduling problem with interval processing time [11]. Kadir et al., in 2019, employed a mixed-integer linear programming model for the maintenance and repair operations in the airline industry to do production scheduling and workforce capacity planning, model [26]. Jens et al., in 2016, adopted a Genetic algorithm is applied to solve the Scheduling of Integrated Manufacturing and distribution systems [27].

3.5.1 Service Management

Cai et al., in 2011, explained the usage of semantic web technology to introduce man hub systems for manufacturing service modeling and acquisition and annotation at manufacturing service in the distributed manufacturing environment [28]. Li et al., in 2018, proposed a scheme based on the Shared Contract Net Protocol (SCNP) to support the sharing of unused resources data through the world wide web between the enterprises

[29]. They also proposed two heuristic algorithms named MAS-MSDT and MAS-LSS to solve the scheduling model. Wang et al., in 2019, adopted a multi user-oriented service distribution to study the kinetics of manufacturing service allocation an evolutionary game model developed to verify the manufacturing service global perspective into universal outlook [30]. Zhang et al., in 2019, proposed a memetic algorithm that can combine GA and simulated annealing to optimize the collaboration between manufacturing services. A blockchain-based platform also discussed to achieve the security, transparency aspects between users [31]. Hao et al., in 2005, proposed a manufacturing management framework to address the problems at various levels i.e. inter enterprise-level, intra enterprise, and shop floor level [32].

3.5.2 Identification of Resources

Identification of resources is the first and foremost task in any of the manufacturing processes. Conventional methods are time-consuming and not economical when large data needs to be classified. In this scenario, machine learning techniques like data mining or text mining are the best ones to classify such a large amount of data in manufacturing. Several authors Bianchi, et al., in 2018, Identified e-commerce in enterprises by text mining-based approach [33]. Manupati et al., in 2018, in his work classified the text data into two main categories of Manufacturing and Non- Manufacturing suppliers with the help of a support vector machine (SVM) algorithm [34]. Park et al., in 2017, explained that environmental sustainability indicators are classified by employing the Text Mining technique for manufacturing and service system, also established relationships between indicator utility levels and company characteristics [35]. Peyman and Ameri, in 2015, presented a text mining approach called the Naïve Bayes method is considered and executed for manufacturing supplier classification [36]. Choudhary et al., in 2009, described knowledge discovery techniques and text mining techniques for post project reviews [37].

3.6 Distributed Manufacturing System

Rauch et al., in 2017, discussed that plenty of literature is available to explain the several advantages of DM environment over a centralized manufacturing system. Some of them were, it has greater flexibility to satisfy local customer demand, improved sustainability, reduction in logistics costs and lesser delivery times [38]. Zanetti et al., in 2015, addressed how local manufacturing of green personalized products represents an opportunity for re-shoring the manufacturing industry through Distributed manufacturing [39]. Woern et al., in 2017, presented a DM platform for flexible products and discussed their technical feasibility and economic viability in the manufacturing of the household purpose commercial filaments [40].

3.7 Response Management for Service Reconfiguration

Andersen et al., in 2017, identified a generic design method for reconfigurable manufacturing systems (RMS) and also to realize the pros and cons of the specified design methodologies [41]. Xia et al., in 2017, described a reconfiguration opportunistic maintenance model to obtain agility in further RMS manufacturing. A degradation many attribute model is used to acquire preventive maintenance intervals constantly in machine-level scheduling, and to make dynamic scheduling at system level a Reconfigurable Maintenance Time Window (RMTW) is used at various stages [42]. Cheng et al., in 2000, described the RMS in five elements, information, equipment, product, process, and organization. They supported that reconfigurability is the backbone for the improvement of manufacturing systems [43]. Renna et al., in 2017, proposed a game theory algorithm (Gale -Shapely) to take the decision without any uncertainty thereby reducing the computational complexity of the system. A comparison of the proposed approach with the existing model has been done with the development of the simulation environment [44].

Rodrigues et al., in 2018, proposed a multi-agent-based approach to run reconfiguration in a distributed manner. They also designed a collaborative interaction protocol to avoid to reach nonoptimal configuration [45]. Chaube et al., in 2012, proposed a framework to generate real-time process plan for Reconfigurable Manufacturing System. The NSGA –II algorithm is adopted to solve the problem [46]. Abdi and Labib, in 2003, developed a method named Analytic Ranking Procedure, a useful tool to decompose to suit between the manufacturing unit and a batch of components in a sequential manner. The model contains the capacity to strengthen the product family basis in a whole variety of RMS [47]. Puik et al., in 2017, proposed an index methodology for classifying the process modules namely repeat', 'adapt', and 'expand', these modules are assessed for comparison of resources and lead-time in case of RMS [48].

4 Discussion and Research Agenda

Liang et al., in 2018, developed a new ontology called AM-onto proc. to help process planning in the context of the additive manufacturing industry [10]. Li et al., in 2019, used the ontology concept to sharing of resources in a distributed environment. Dinh et al., in 2014, utilized the knowledge management concept to identify the most influencing attributes in the available data. Houng et al., in 2019, have extended similar work done by Dinh, in 2014, thereby proposing ontology to manage the data in the mold production process. From this existing literature, it is observed that there is a lot of scope for proposing the ontologies in manufacturing industries like Automotive, steel and metal, paper, textile, coal, etc. in the case of a distributed environment. Alkohatani et al., in 2019, employed a data mining algorithm to detect faults in manufacturing and mapping them to design information [49]. Very few works of literature available in this area and there is a scope for research to apply the machine learning techniques to identify suitable manufacturing resources in a distributed environment. In Manufacturing function management, service management is an important aspect to survive any manufacturing system. Cai et al., in 2011, used semantic web technology. Further scope is available to do research in the service management of resources that are geographically distributed manufacturing systems [28].

5 Conclusions

Based on findings from the available literature and existing studies, a systematic review has been carried out in the related areas of network manufacturing systems. A framework that relates the relevant function of networked manufacturing systems has been proposed. Articles related to topics mentioned in the framework are reviewed in each section. SLR methodology one such methodology that can be a useful to keep away from conventional reviews of the literature, with including locating the important findings of the literature, the findings of research gaps, and the useful platform for further research. the results of this study may be useful to assist both academics and professionals to introduce or to build effective tools and methodologies. Additionally, aspects like manufacturing function management and response management of networked manufacturing systems are also discussed. Research gaps identified in this SLR might be useful for further researches in the area of networking manufacturing systems in specific to mentioned parameters. Some of the limitations of the SLR are presented in Sect. 5.

Acknowledgement. The project is funded by Department of Science and Technology, Science & Engineering Research Board (DST-SERB), Statutory Body Established through an Act of Parliament: SERB Act 2008, Government of India with Sanction Order No ECR/2016/001808 and FCT – Fundação para a Ciência e Tecnologia within the R&D Units Project Scope: UIDB/00319/2020.

References

1. Saygin, C., Kilic, S.E.: Integrating flexible process plans with scheduling in flexible manufacturing systems. Int. J. Adv. Manuf. Technol. **15**(4), 268–280 (1999). https://doi.org/10.1007/s001700050066
2. Omotayo, F.O.: Knowledge management as an important tool in organisational management: a review of literature. Libr. Philos. Pract. **1**(2015), 1–23 (2015)
3. Lin, Y.K., Chang, P.C.: System reliability of a manufacturing network with reworking action and different failure rates. Int. J. Prod. Res. **50**(23), 6930–6944 (2012)
4. Tranfield, D., Denyer, D., Smart, P.: Towards a methodology for developing evidence-informed management knowledge by means of systematic review. Br. J. Manag. **14**(3), 207–222 (2003)
5. Gangala, C., Modi, M., Manupati, V.K., Varela, M.L.R., Machado, J., Trojanowska, J.: Cycle time reduction in deck roller assembly production unit with value stream mapping analysis. In: Rocha, Á., Correia, A.M., Adeli, H., Reis, L.P., Costanzo, S. (eds.) WorldCIST 2017. AISC, vol. 571, pp. 509–518. Springer, Cham (2017). https://doi.org/10.1007/978-3-319-56541-5_52
6. Wilding, R., Wagner, B., Colicchia, C., Strozzi, F.: Supply chain risk management: a new methodology for a systematic literature review. Supply Chain Manag. **17**(4), 403–418 (2012)
7. Dinh, D., Dos Reis, J.C., Pruski, C., Da Silveira, M., Reynaud-Delaître, C.: Identifying relevant concept attributes to support mapping maintenance under ontology evolution. Web Semant. Sci. Serv. Agents World Wide Web **29**, 53–66 (2014)
8. Palmer, C., Usman, Z., Canciglieri Junior, O., Malucelli, A., Young, R.I.M.: Interoperable manufacturing knowledge systems. Int. J. Prod. Res. **56**(8), 2733–2752 (2015). https://doi.org/10.1080/00207543.2017.1391416
9. Park, J.: An ontology BOM for production costing. Int. J. Ind. Electron. Electr. Eng. **6**(2), 10–12 (2018)
10. Liang, J.S.: An ontology-oriented knowledge methodology for process planning in additive layer manufacturing. Robot. Comput.-Integr. Manuf. **53**, 28–44 (2018)
11. Li, X., Gao, L., Wang, W., Wang, C., Wen, L.: Particle swarm optimization hybridized with genetic algorithm for uncertain integrated process planning and scheduling with interval processing time. Comput. Ind. Eng. **135**, 1036–1046 (2019)

12. Huang, X., Xu, T., Gao, W., Jia, Z.: Ontology similarity measure and ontology mapping via fast ranking method. Int. J. Appl. Phys. Math. **1**(1), 54 (2011)
13. Risius, M., Spohrer, K.: A blockchain research framework. Bus. Inf. Syst. Eng. **59**(6), 385–409 (2017). https://doi.org/10.1007/s12599-017-0506-0
14. Viriyasitavat, W., Da Xu, L., Bi, Z., Sapsomboon, A.: Blockchain-based business process management (BPM) framework for service composition in industry 4.0. J. Intell. Manuf. **31**(7), 1737–1748 (2018). https://doi.org/10.1007/s10845-018-1422-y
15. Mondragon, A.E.C., Mondragon, C.E.C., Coronado, E.S.: Exploring the applicability of blockchain technology to enhance manufacturing supply chains in the composite materials industry. In: 2018 IEEE International Conference on Applied System Invention (ICASI), pp. 1300–1303. IEEE (2018)
16. Chung, K., Yoo, H., Choe, D., Jung, H.: Blockchain network based topic mining process for cognitive manufacturing. Wirel. Pers. Commun. **105**(2), 583–597 (2018). https://doi.org/10.1007/s11277-018-5979-8
17. Li, X., Zhuang, P., Yin, C.: A metadata based manufacturing resource ontology modeling in cloud manufacturing systems. J. Ambient Intell. Humaniz. Comput. **10**(3), 1039–1047 (2019). https://doi.org/10.1007/s12652-018-0964-3
18. Lu, H., Huang, K., Azimi, M., Guo, L.: Blockchain technology in the oil and gas industry: a review of applications, opportunities, challenges, and risks. IEEE Access **7**, 41426–41444 (2019)
19. Wong, L.W., Leong, L.Y., Hew, J.J., Tan, G.W.H., Ooi, K.B.: Time to seize the digital evolution: adoption of blockchain in operations and supply chain management among Malaysian SMEs. Int. J. Inf. Manag. **52**, 101997 (2019)
20. Agostinho, C., et al.: Towards a sustainable interoperability in networked enterprise information systems: trends of knowledge and model-driven technology. Comput. Ind. **79**, 64–76 (2016)
21. Panetto, H., Dassisti, M., Tursi, A.: ONTO-PDM: Product-driven ONTOlogy for Product Data Management interoperability within manufacturing process environment. Adv. Eng. Inform. **26**(2), 334–348 (2012)
22. Lin, Y.K., Chang, P.C.: A novel reliability evaluation technique for stochastic-flow manufacturing networks with multiple production lines. IEEE Trans. Reliab. **62**(1), 92–104 (2012)
23. Ostadi, B.: An optimal preventive maintenance model to enhance availability and reliability of flexible manufacturing systems. J. Ind. Syst. Eng. **11**(2), 47–61 (2018)
24. Permin, E., Lossie, K., Schmitt, R.: Functional reliability of cognitive control systems for manufacturing processes (2018)
25. Barzanji, R., Naderi, B., Begen, M.A.: Decomposition algorithms for the integrated process planning and scheduling problem. Omega **93**, 102025 (2019)
26. Ertogral, K., Öztürk, F.S.: An integrated production scheduling and workforce capacity planning model for the maintenance and repair operations in airline industry. Comput. Ind. Eng. **127**, 832–840 (2019)
27. Ehm, J., Freitag, M., Frazzon, E.M.: A heuristic optimisation approach for the scheduling of integrated manufacturing and distribution systems. Procedia CIRP **57**(2016), 357–361 (2016)
28. Cai, M., Zhang, W.Y., Zhang, K.: ManuHub: a semantic web system for ontology-based service management in distributed manufacturing environments. IEEE Trans. Syst. Man Cybern.-Part A: Syst. Hum. **41**(3), 574–582 (2011)
29. Li, K., Zhou, T., Liu, B.H., Li, H.: A multi-agent system for sharing distributed manufacturing resources. Expert Syst. Appl. **99**, 32–43 (2018)
30. Wang, T., Li, C., Yuan, Y., Liu, J., Adeleke, I.B.: An evolutionary game approach for manufacturing service allocation management in cloud manufacturing. Comput. Ind. Eng. **133**, 231–240 (2019)

31. Zhang, G., Zhang, Y., Zhong, R.Y., Wu, Y.: Extending augmented Lagrangian coordination for the optimal configuration of cloud-based smart manufacturing services with production capacity constraint. Robot. Comput.-Integr. Manuf. **58**, 21–32 (2019)
32. Hao, Q., Shen, W., Wang, L.: Towards a cooperative distributed manufacturing management framework. Comput. Ind. **56**(1), 71–84 (2005)
33. Bianchi, G., Bruni, R., Scalfati, F.: Identifying e-commerce in enterprises by means of text mining and classification algorithms. Math. Probl. Eng. **2018** (2018)
34. Manupati, V.K., Akhtar, M.D., Varela, M.L.R., Putnik, G.D., Trojanowska, J., Machado, J.: A text mining based supervised learning algorithm for classification of manufacturing suppliers. In: Rocha, Á., Adeli, H., Reis, L.P., Costanzo, S. (eds.) WorldCIST'18 2018. AISC, vol. 747, pp. 236–244. Springer, Cham (2018). https://doi.org/10.1007/978-3-319-77700-9_24
35. Park, K., Kremer, G.E.O.: Text mining-based categorization and user perspective analysis of environmental sustainability indicators for manufacturing and service systems. Ecol. Ind. **72**, 803–820 (2017)
36. Yazdizadeh, P., Ameri, F.: A text mining technique for manufacturing supplier classification. In: ASME 2015 International Design Engineering Technical Conferences and Computers and Information in Engineering Conference, pp. V01BT02A036–V01BT02A036. American Society of Mechanical Engineers (2015)
37. Choudhary, A.K., Oluikpe, P.I., Harding, J.A., Carrillo, P.M.: The needs and benefits of Text Mining applications on Post-Project Reviews. Comput. Ind. **60**(9), 728–740 (2009)
38. Rauch, E., Dallasega, P., Matt, D.T.: Distributed manufacturing network models of smart and agile mini-factories. Int. J. Agile Syst. Manag. **10**(3–4), 185–205 (2017)
39. Zaletelj, V., Hozdić, E., Butala, P.: A foundational ontology for the modelling of manufacturing systems. Adv. Eng. Inform. **38**(2018), 129–141 (2018)
40. Woern, A., Pearce, J.: Distributed manufacturing of flexible products: technical feasibility and economic viability. Technologies **5**(4), 71 (2017)
41. Andersen, A.L., Brunoe, T.D., Nielsen, K., Rösiö, C.: Towards a generic design method for reconfigurable manufacturing systems: analysis and synthesis of current design methods and evaluation of supportive tools. J. Manuf. Syst. **42**, 179–195 (2017)
42. Xia, T., Xi, L., Pan, E., Ni, J.: Reconfiguration-oriented opportunistic maintenance policy for reconfigurable manufacturing systems. Reliab. Eng. Syst. Saf. **166**, 87–98 (2017)
43. Wang, C.: Reconfigurable manufacturing: a response to turbulent changes. IFAC Proc. Vol. **33**(20), 83–87 (2000)
44. Renna, P.: Decision-making method of reconfigurable manufacturing systems' reconfiguration by a Gale-Shapley model. J. Manuf. Syst. **45**, 149–158 (2017)
45. Rodrigues, N., Oliveira, E., Leitão, P.: Decentralized and on-the-fly agent-based service reconfiguration in manufacturing systems. Comput. Ind. **101**, 81–90 (2018)
46. Chaube, A., Benyoucef, L., Tiwari, M.K.: An adapted NSGA-2 algorithm based dynamic process plan generation for a reconfigurable manufacturing system. J. Intell. Manuf. **23**(4), 1141–1155 (2012). https://doi.org/10.1007/s10845-010-0453-9
47. Abdi, M.R., Labib, A.W.: A design strategy for reconfigurable manufacturing systems (RMSs) using analytical hierarchical process (AHP): a case study. Int. J. Prod. Res. **41**(10), 2273–2299 (2003)
48. Puik, E., Telgen, D., van Moergestel, L., Ceglarek, D.: Assessment of reconfiguration schemes for Reconfigurable Manufacturing Systems based on resources and lead time. Robot. Comput.-Integr. Manuf. **43**, 30 (2017)
49. Alkahtani, M., Choudhary, A., De, A., Harding, J.A.: A decision support system based on ontology and data mining to improve design using warranty data. Comput. Ind. Eng. **128**, 1027–1039 (2019)

Risk Analysis and Risk Measures Applied to the Furniture Industry

Irene Brito[1(✉)], Celina P. Leão[2], and Matilde A. Rodrigues[2,3,4]

[1] Centre of Mathematics and Department of Mathematics, University of Minho, Guimarães, Portugal
ireneb@math.uminho.pt
[2] Algoritmi Centre, University of Minho, Guimarães, Portugal
cpl@dps.uminho.pt
[3] Research Centre of Environment and Health, School of Health of Polytechnic Institute of Porto, Porto, Portugal
[4] Center for Rehabilitation Research, School of Health of the Polytechnic Institute of Porto, Porto, Portugal
mar@estsp.ipp.pt

Abstract. In this work we show how concepts and methods from actuarial risk theory can be applied to risk analysis in industry. Risk analysis consists in identifying, quantifying and classifying or ordering risks. In the proposed methodology, the risks identified in industrial setting, are modeled by loss random variables. The loss random variables are used to calculate the expected loss, the loss variance and exceedance probabilities permitting risk quantification. In order to classify and order risks, besides measures of uncertainty, risk measures, such as Value-at Risk and Tail-Value-at-Risk (or Expected Shortfall) are determined and, together with risk quantification, the risk levels are analysed. To exemplify this methodology, a case study for risk analysis and classification of occupational accidents in the furniture industry was carried out.

Keywords: Risk analysis · Risk measures · Value-at-risk · Expected shortfall · Industry

1 Introduction

Risk analysis is important for many problems in industrial settings. In industry, risks are random losses which can for example represent monetary values or injuries as consequences of accidents. Risks can be modeled by loss random variables, which allow the description of exposure to risk, in particular the quantification and classification or ordering of risks. The level of exposure to risk can be described by risk measures, which inform the risk manager about the degree that the industry is liable to be confronted with specific details of risk. Risk measures have been developed in the context of actuarial risk theory and of risk management, however they are useful for the evaluation of any random variable.

J. Machado et al. (Eds.): icieng 2021, LNME, pp. 113–121, 2022.
https://doi.org/10.1007/978-3-030-79165-0_11

For example, risk managers often pay attention to the probability of an adverse outcome, which can be represented by the Value-at-Risk (VaR) at a specific level of probability. VaR refers to a loss level and it can represent an upper bound for the loss for a given confidence level and in a fixed time interval. This risk measure can be useful and relevant for the study of the risk of occupational accidents. So far, VaR was applied to the description of accident risks in various different sectors and activities, for example of accidents in the energy sector [1], in highway hazmat shipments [2] or in [3], where VaR permitted the identification of relevant accident scenarios in furniture industries. In [3] risks corresponding to occupational accidents were analyzed and classified taking into account the different injury categories. The data used in that study corresponds to accidents in the furniture industry in Portugal in the year 2010, this industry being one of the most relevant activity sectors in Portugal [4], which consists predominantly of small and medium-sized enterprises [5].

In this work, we present a more general analysis than in [3], showing how to apply the methodology to the analysis of risks in industry, and then extend the analysis of occupational accidents by generalizing certain concepts, such as the general prediction of lost work days due to accidents in industry, including a new risk measure, namely the Tail-Value-at-Risk (TVaR), also known as Expected Shortfall.

2 Risk Theory

In many problems a risk can be modeled by a loss random variable X defined as follows (see e.g. [6, 7]):

$$X = IB, \tag{1}$$

where B corresponds to a random amount of loss and I is an indicator random variable, with values $I = 1$ or $I = 0$; $I = 1$ indicates that some event, in particular a loss, has occurred.

The indicator random variable I can be described by a Bernoulli(q) distribution, $0 \leq q \leq 1$, where the probability of the event is $q = P(I = 1)$ and $1 - q = P(I = 0)$ corresponds to the probability of no risk occurrence. If $I = 1$ the loss X is drawn from the distribution of the random variable B and if $I = 0$, then $X = 0$, meaning that there is a zero loss.

The moments of X can be calculated with the iterative formula of conditional expectations. The expected loss is calculated in the following way

$$\mathrm{E}[X] = \mathrm{E}[\mathrm{E}[X|I]] = P(I = 1)\mathrm{E}[B] = q\mathrm{E}[B]. \tag{2}$$

Note that if $B = b$ is a fixed amount of loss, then the expected loss is simply given by

$$\mathrm{E}[X] = \mathrm{E}[Ib] = \mathrm{E}[I]b = qb. \tag{3}$$

The variance of $X = IB$ can also be determined using the conditional distribution of B, given I, from the following variance decomposition rule

$$\mathrm{Var}[X] = \mathrm{Var}[\mathrm{E}[X|I]] + \mathrm{E}[\mathrm{Var}[X|I]] = \mathrm{E}[B]^2 q(1-q) + \mathrm{Var}[B]q. \tag{4}$$

For a fixed amount of loss $B = b$ the variance of the loss is simply given by

$$\mathrm{Var}[X] = \mathrm{Var}[Ib] = b^2 q(1-q). \tag{5}$$

If X is the sum of a large number of independent random variables, i.e. when the general loss corresponds to the sum of independent random individual losses, then the probability for the random loss to exceed a certain value α can be calculated using the Central Limit Theorem:

$$P(X > \alpha) \approx 1 - \Phi\left(\frac{\alpha - \mathrm{E}[X]}{\sqrt{\mathrm{Var}[X]}}\right), \tag{6}$$

where $\Phi(\cdot)$ is the cumulative distribution function of a standard normal random variable; and $\mathrm{E}[X]$ and $\mathrm{Var}[X]$ are determined from (2) or (3) and (4) or (5), respectively.

A risk measure is a statistical tool used to assess risks by associating a real value to a random variable representing the loss. The real value is intended to quantify risk exposure. Examples of risk measures are the Value-at-Risk (VaR) or the Tail-Value-at-Risk (TVaR), also known as Expected Shortfall. The Value-at-Risk (VaR) is a standard risk measure, which is used in actuarial risk theory to assess the exposure to risk (see e.g. [6,7]). This measure is also called a quantile risk measure. The VaR of a loss random variable X at the 100p% level, represented by $\mathrm{VaR}_p(X)$, is the 100p percentile (or quantile) of the distribution of X. We use the notation π_p for the value $\mathrm{VaR}_p(X)$ satisfying

$$P(X \leq \pi_p) = p. \tag{7}$$

The VaR at the confidence level p, $0 < p < 1$, is the quantity π_p that will maximally be lost with probability p, so that there is a $1-p$ chance of exceeding π_p: $P(X > \pi_p) = 1 - p$.

The VaR has the disadvantage that it only informs about the probability of the shortfall of X over π_p being positive. However, the size of the shortfall should also be taken into account. Risk measures which consider the size of the shortfall $(X - \pi_p > 0)$ when the amount (e.g. capital) π_p is available, include the Tail-Value-at-Risk (TVaR). The TVaR, also known as Expected Shortfall or Mean Excess Loss (see e.g. [6,7]), of a random variable X at the 100p% level $\mathrm{TVaR}_p(X)$ is the expected loss, given that the loss exceeds the 100p percentile (or quantile) of the distribution of X, and is defined by

$$\mathrm{TVaR}_p(X) = \mathrm{E}[X|X > \pi_p] = \frac{\int_{\pi_p}^{\infty} x f(x) \mathrm{d}x}{1 - F(\pi_p)}, \tag{8}$$

where f and F denote, respectively, the density function and the cumulative distribution function of X. TVar gives therefore more information about the tail of the distribution than VaR does.

A simple way to judge risk is to consider the effect of the expected value of the risky outcome $\mathrm{E}[X]$ and also its variation or uncertainty given by the variance $\mathrm{Var}[X]$ or standard deviation $\sqrt{\mathrm{Var}[X]}$. In risk analysis it is therefore useful, first, to analyse and classify risks according to this criterion based on the mean and variance. Then one can take into account certain probabilities for the random loss to exceed certain values α using (6). And the risks can then be better classified and ordered using the risk measures $\mathrm{VaR}_p(X)$ and $\mathrm{TVaR}_p(X)$. Given two risks X and Y, we say that X is riskier than Y if

$$\mathrm{VaR}_p(X) > \mathrm{VaR}_p(Y) \tag{9}$$

and

$$\mathrm{TVaR}_p(X) > \mathrm{TVaR}_p(Y). \tag{10}$$

3 Application to Risk Analysis of Occupational Accidents in Industry

In this section we show how to generally apply the risk theory presented in the previous section to the risk analysis of occupational accidents in the furniture industry, generalizing the analysis done in [3], and extending that analysis with further results, including the risk measure TVaR, which will be a useful tool to decide the risk ordering of the injury categories with intermediate risk level. Workers from Portuguese furniture industries face several risks that can jeopardize their safety. These risks are related to common hazards in the sector, such as unsafe machinery, worker's unsafe behaviors and manual tasks (to saw, drill, cut, plane, polish or manual material handling) [4]. As a consequence, the accident frequency rate in furniture sector remains high [4]. Information about the most important accident mechanisms in the furniture sector, including the probability of the expected consequences, is critical to overall risk management process. This information will help enterprises to decide about the need of control measures and will contribute to authorities develop effective intervention programs. Official accident reports data provided by the Portuguese Office of Strategy and Planning (GEP) from the year 2010, which are aligned with European Statistics on Accidents at Work (ESAW III), described six categories of contact-modes of injuries, denoted by $i = 1, \ldots, 6$, which occurred in the furniture industry in Portugal in 2010 (see Table 1).

Table 1. Contact mode of injury categories.

i	Injury category
1	Contact with electrical voltage, temperatures, hazardous substances
2	Horizontal or vertical impact with or against a stationary object (victim in motion)
3	Struck by object in motion, collision with
4	Contact with sharp, pointed, rough, coarse Material Agent
5	Trapped, crushed, etc.
6	Physical or mental stress

Different injuries will lead to different numbers of lost work days; it is also possible to have zero lost working days. Therefore, the risk of accident is characterized by the probability of its occurrence and its severity, measured as lost work days.

The number of lost work days will be modeled using the loss random variable

$$X = IB, \tag{11}$$

which depends on the accidents' occurrence probability and on the estimated number of lost work days. The random variable B represents the number of lost work days. The random indicator variable I indicates if an accident leads to more than one lost work days, in which case $I = 1$, or if an accident has associated zero lost work days, in which case $I = 0$. In general, in the furniture industry one can identify a total number of $n = 4313$ accidents, from which $n0 = 1023$ had associated no lost work days. Therefore, the probability that an accident in the furniture industry will lead to at least one lost work day, $P(I = 1)$, is

$$q = \frac{n - n0}{n} = \frac{3290}{4313} = 0.76$$

and the probability that an accident in the furniture industry will lead to no lost work days, $P(I = 0)$, is

$$1 - q = \frac{n0}{n} = \frac{1023}{4313} = 0.24.$$

Table 2 contains the information about the number of accidents for each injury category, represented by n_i, $i = 1, \ldots, 6$, and about the number of accidents which had associated zero lost work days, denoted by $n0_i$, $i = 1, \ldots, 6$. Note that $n = \sum_{i=1}^{6} n_i = 4313$ and $n0 = \sum_{i=1}^{6} n0_i = 1023$. The probabilities $P(I = 1)$ for each injury category will be estimated by

$$q_i = \frac{n_i - n0_i}{n_i} \tag{12}$$

and $P(I = 0)$ by

$$1 - q_i = \frac{n0_i}{n_i}, \tag{13}$$

which represents the probability for injury category i that an accident will lead to zero lost work days.

Table 2 also contains the estimated number of lost days associated with an accident of category i given by

$$b_i = \frac{b_{Ti}}{n_i}, \tag{14}$$

where b_{T_i} stands for the total number of lost days associated with accidents of category i, and the occurrence probabilities of an accident of category i represented by

$$p_i = \frac{n_i}{n}. \tag{15}$$

Table 2. Results for the contact modes of injury categories.

i	n_i	$n0_i$	q_i	$1-q_i$	b_{Ti}	b_i	p_i
1	97	20	0.79	0.21	1135	11.70	0.02
2	523	162	0.69	0.31	17457	33.38	0.12
3	958	373	0.61	0.39	18082	18.87	0.22
4	1406	218	0.84	0.16	53661	38.17	0.33
5	331	61	0.82	0.18	13594	41.07	0.08
6	998	189	0.81	0.19	27062	27.12	0.23

In general, one can describe the risk X of lost days due to accidents in the furniture industry in the following way using the statistics and risk measures presented in Sect. 2.

In order to calculate the mean and variance of X one needs the mean and variance of B:

$$\mathrm{E}[B] = \sum_{i=1}^{6} p_i b_i = 30.51, \tag{16}$$

$$\mathrm{Var}[B] = \mathrm{E}[B^2] - \mathrm{E}[B]^2 = 68.82. \tag{17}$$

The expected number of lost work days due to accidents in the furniture industry is then given by (2)

$$\mathrm{E}[X] = q\mathrm{E}[B] = 23.19, \tag{18}$$

and the variance (4) by

$$\mathrm{Var}[X] = \mathrm{E}[B]^2 q(1-q) + \mathrm{Var}[B]q = 222.09. \tag{19}$$

The probabilities that the number of lost work days exceed one week and half a month are, respectively,

$$P(X > 7) = 0.86 \text{ and } P(X > 15) = 0.71. \tag{20}$$

The application of the Central Limit Theorem is justified by the fact that X corresponds in fact to a sum of a large number of independent random variables, since, assuming that the $n = 4313$ accidents are independent, one could write $X = X_1 + \cdots + X_{4313}$ or, taking into account the six injury categories with $n_1 = 97, \cdots, n_6 = 998$ (see Table 2), $X = X_{1,1} + \cdots + X_{1,97} + \cdots + X_{6,1} + \cdots + X_{6,998}$, where in (16) we used for each i the sum of n_i estimated number of lost days b_{T_i}, which is divided by n in order to have the expected number of lost work days due to an accident (cf. (14), (15)), and we used $P(I = 1) = q$ for each accident in the furniture industry.

The risk measures $\mathrm{VaR}_p(X)$ and $\mathrm{TVaR}_p(X)$ at the 95% level can be determined from (7) and (8) and one thus obtains

$$\mathrm{VaR}_{0.95}(X) = 47.7 \text{ and } \mathrm{TVaR}_{0.95}(X) = 53.95, \tag{21}$$

where we considered that X has approximately a Normal distribution, due to the reasons explained before.

These results indicate that given the occurrence of an accident in the furniture industry, there is a high probability, as indicated in (20), that the number of lost work days exceed one week or half a month. In fact the expected number of lost work days is 23.19 (approximately three weeks), with a high variance of 222.09, meaning that there is a considerable risk of occurring more extreme values of lost work days. Concerning the risk measures, the VaR indicates that the probability of lost work days being less than $\pi_{0.95} = 47.7$ days is 0.95, so that there is a 5% chance that the number of lost work days will exceed that number. The TVaR indicates that the expected number of lost work days being higher than $\pi_{0.95} = 47.7$ is 53.95, or, given that the number of lost work days exceed the threshold $\pi_{0.95} = 47.7$, the mean excess loss will be 53.95.

A detailed analysis about the risk of accident occurrence and number of lost work days due to each injury category can be found in [3]. Here we complement those results by calculating the risk measure TVaR for each injury category. For that purpose we use the particular loss random variable associated to each injury category $i = 1, \ldots, 6$,

$$X_i = I_i b_i, \qquad (22)$$

representing the lost work days due to an accident of category i, where I_i has a Bernoulli(p_i) distribution, with p_i and b_i given in Table 2. Note that in this case b_i is modeled as a fixed number of lost work days and therefore formulas (3) and (5) can be used to determine the mean and variance of X_i for each injury category. Calculating the risk measure TVaR at a 95% level for each injury category, one obtains the results listed in Table 3. This Table also contains the values of the risk measure VaR.

Table 3. Risk measures VaR and TVaR for each injury category.

i	1	2	3	4	5	6
$\text{VaR}_{0.95}$	3.12	21.97	17.10	41.87	21.14	25.09
$\text{TVaR}_{0.95}$	3.82	26.53	20.35	49.32	25.67	29.81

In [3], the injury categories were ordered in the following way according to their risk (see Table 1 for the identification of the injuries 1 to 6). Injury 4 has the highest risk level and injury 6 is the second problematic one for the industry, whereas injury 1 has the lowest risk level. Considering the injury categories with intermediate risk level, namely injuries 2, 3 and 5, their ordering was not straightforward taking into account the expected number of lost work days, the variance, the exceedance probabilities of 7 and 15 lost work days and the risk measure VaR. Additionally taking into account the new information about the risk measure TVaR for each injury category, the results reveal that for injuries 2, 3 and 5 with intermediate risk level the order would be: injury category 2 is riskier than injury category 5 and both are riskier than 3. Also with TVaR there is a strong evidence for classifying injury category 4 as the most problematic one, followed by category 6, and injury category 1 as the less risky one.

4 Conclusions

In this work we presented a methodology for risk analysis which can be adapted to the analysis of risks in the context of industry. The methodology was applied here to analyze in general the occurrence of accidents in the furniture industry, focussing on the risk of lost work days due to accidents. Based on the two risk measures, VaR and TVaR, both give similar evidence for the most problematic and for the less risky category in the furniture industry. These risk measures help to determine the ranking of the categories considered to be of intermediate risk, where the identification based on, for instance, the uncertainty measure or on the expected loss was not straightforward. The results indicate that there is a high probability of accidents leading to lost work days in this industry sector and the expected number of lost work days in the case of accident occurrence is around three weeks.

The general results presented in this work for the furniture industrial sector are relevant and useful if one wants to analyze in general how risk in the furniture industrial sector has evolved since 2010 and if one wants to compare the risk of accidents in the furniture industrial sector with other industrial sectors.

In the future we want to apply this methodology in order to compare accident risks, and in particular the risk of lost work days due to accidents, between different industry sectors and also continue the study in the furniture industry context using more actual data.

Acknowledgments. The authors acknowledge support from FCT, through the projects UIDB/00013/2020, UIDP/00013/2020 and UIDB/00319/2020 and UIDB/05210/2020.

References

1. Spada, M., Paraschiv, F., Burgherr, P.: A comparison of risk measures for accidents in the energy sector and their implications on decision-making strategies. Energy **154**, 277–288 (2018). https://doi.org/10.1016/j.energy.2018.04.110
2. Hosseini, S.D., Verma, M.: A value-at-Risk (VAR) approach to routing rail hazmat shipments. Transp. Res. Part D **54**, 191–211 (2017). https://doi.org/10.1016/j.trd.2017.05.007
3. Leão, C.P., Rodrigues M.A., Brito I. (2019) Analyzing and classifying risks: a case-study in the furniture industry. In: Arezes P., et al. (eds.) Occupational and Environmental Safety and Health. Studies in Systems, Decision and Control, vol. 202. Springer, Cham. https://doi.org/10.1007/978-3-030-14730-3_9

4. Rodrigues, M.A., Arezes, P., Leão, C.P.: Characterization of the portuguese furniture industry's safety performance and monitoring tools. In: Arezes, P.M.F.M., Carvalho, P.V.R. (eds.) Ergonomics and Human Factors in Safety Management. Chapter 5, pp. 93–109. CRC Press Taylors and Francis, London (2016). https://doi.org/10.1201/9781315370354-5
5. Eurostat: Homepage, Archive: Manufacture of furniture statistics - NACE Rev. 2 (2013). https://ec.europa.eu/eurostat/statistics
6. Klugman, S., Panjer, H., Willmot, G.: Loss Models, 4th edn. Wiley (2012)
7. Kaas, R., Goovaerts, M., Dhaene, J., Denuit, M.: Modern Actuarial Risk Theory, 2nd edn. Springer (2008). https://doi.org/10.1007/978-3-540-70998-5

Mathematical Model to Monitory Exposure of People to Occupational Risk in Manual Assembly Processes

Arminda Pata[1], José Carlos Sá[2,3](✉), Gilberto Santos[4], Francisco José Gomes da Silva[2], Luís Pinto Ferreira[2], and Luís Barreto[3]

[1] D. Dinis Higher Institute (ISDOM), Marinha Grande, Portugal
arminda.pata@isdom.pt
[2] Polytechnic of Porto, School of Engineering (ISEP), Porto, Portugal
cvs@isep.ipp.pt
[3] Instituto Politécnico de Viana do Castelo, Rua Escola Industrial e Comercial Nun Alvares, 4900-347 Viana do Castelo, Portugal
[4] Polytechnic Institute Cavado Ave., Design School, Barcelos, Portugal

Abstract. The manual assembly processes and the dynamic industrial environment usually requires fast adjustments between work teams and workstations to accomplish customers' needs. Sometimes, activities are scheduled without considering occupational health of work teams. Those occurrences can compromise work conditions and employee's health. Thus, the specific objective of this study was to develop a mathematical programming model, to permit monitoring the exposure to occupational risks of workers teams. The mathematical model proposed was validated through solutions generated via the CPLEX® optimization software and OPL language, which was applied in an assembly line of bicycle handlebar where 6 employees produce 800 units per day. First, the results for the original scenario were generated considering the people all the time in the same workstation. The solution was achieved in 12 ms and provided as solution OF = 26153. Then, to allow a deeper perception about the importance of monitoring people exposure at the shop floor, limits of subjection were established. The assignments between work teams and workstations were generated considering those limits. An admissible solution OF = 17888 with the potential to ensure the same output of bicycle handlebar was found in 14 ms. It is expected that the balanced employees' exposure to the work conditions may contribute to minimize occupational diseases, increase the active aging and ensure future healthy generations, in different manual assembly processes.

Keywords: Mathematical modeling · Scheduling optimization · Occupational health safety

1 Introduction

In the upcoming years, production paradigms will change frequently. New health problems can arise with those new paradigms. On one hand, people will continue to use their

J. Machado et al. (Eds.): icieng 2021, LNME, pp. 122–134, 2022.
https://doi.org/10.1007/978-3-030-79165-0_12

body to execute work (e.g., hands touching at monitors, eyes inspecting products, static work). On the other hand, those transformation processes (e.g., fourth industrial revolution) will not be immediate. Thus, many products are still and will be manufactured around manual assembly processes [1]. The production of highly customized systems is characterized by big product diversity, demanding the execution of tasks with different complexity levels, that may affect workers' performance [2, 3]. High-risk factors associated with the development of work-related musculoskeletal disorders (WMSDs) may result from lesser worker experience and cumulated exposure at manual and repetitive tasks. Older workers are more vulnerable to high physical demand occupations [4]. Faced with a society that is aging, with a longer life expectancy, and retirements that happen later, there is a need to keep people active for more years, without neglecting a good quality of life [5]. In this sense, the decisions making regarding the matching between people and workstations will continue to interfere in the quality of life of the people, for a long time. The issue of older workers at manual assembly processes has not yet received considerable critical attention from the scientific community. Researchers, in the last decades, have not studied Occupational Health and Safety (OHS) as well as older workers at the industry tasks in much detail, as they have done with the health area. Thus, production engineers, in the forthcoming decades, will have several challenges. A huge and distinct array of combinatorial optimization problems in the area of OHS will arise. The problem description is presented at Sect. 3. The purpose of the present study consists in the development of a work model to optimize affectations between people and workstations with manual assembly processes in order to minimize WMSDs. To respond to the literature gap about absence of limit values, was developed a mathematical model to monitor the real exposure of people at any industrial environment with manual assemble processes.

2 Literature Review

An assembly line is composed of workstations connected to a system that enables the manual or automatic transfer of work, which has to follow a certain sequence [6]. The balancing problems arise from the necessity of allocating the workload among all workstations. The importance of focusing on the process analysis rather than in the results is supported by some authors. There are several standard case studies whose main concern at the process's optimization are harmonizing machinery and equipment performance [7], as well as quality control [8]. Recently, some studies have been carried out about the integration of disabled people [9], ergonomic issues [10], rotation of workers [11] and satisfaction of customers [12]. According to the literature, assembly lines configurations may differ: (1) traditional assembly line, (2) two-sided line and (3) flexible "U" line. Regarding assembly line balancing problems, most of the case studies are conducted to reach specific purposes: (a) balancing and integrating people with disabilities (b) layout configurations and (c) ergonomic problems. Low-volume low-variety (LVLV) production systems are a hybrid form of low-volume high-variety (LVHV) and high-volume low-variety (HVLV) production systems, where products follow a pre-defined processing order through a series of unique tasks [13]. Thus, the workload distribution can cause a lot of diversity of assigning problems. Consequently, the decisions made by industrial managers can generate workers' injuries. Those damages (i.e., temporary or permanent) can affect the active workers' around the world [14]. They may emerge from the long and energetic exposure of workers at work performing conditions: (1) long professional careers [15]; (2) naturally aging of the active population [4, 16–20] (3) etiology; or (4) difficulties in transfer of knowledge between managers' and researchers due to virtual barriers between different areas of knowledge and skills profiles. The competitiveness of contemporary manufacturing systems is based on a high production rate and a high level of flexibility. Organizations are being installed in an increasingly unstable environment and rigorous in relation to supply and demand needs. Allocation problems play significant roles in the workforce scheduling domain [21]. Additionally, ergonomic evaluation of manual material handling (MMH) has mainly been based on task analysis approach where each job is broken down into simpler tasks and then studied [22, 23]. There is an absence of clarity in the use of terms defining various MMH activities. However, there exists a classification scheme that resolves some of the ambiguity about the terminology usage of MMH tasks. That classification scheme uses four significant work system dimensions: (1) material, (2) workplace, (3) task and (4) team [24]. On the other hand, decision making at the shopfloor can contribute to health people problems. Thus, the message of healthy workplaces campaign for the next years, is based on providing to workers of all ages a safe and healthy working environment. Such commitment during the working life is good for all workers, for the businesses and to the society. Nevertheless, the number of scientific articles that raise the question of how to incorporate OHS remains a minority [25, 26]. In the OHS literature, there are four major research domains: (1) safety atmosphere, (2) management systems integration, (3) voluntary OHS systems and (4) sustainable operations [27]. The ergonomic practices which aim at a reduction of exposures can be studied with regard to their impact on various performance measures [28]. The individual exposure of people to the conditions in which work is carried out accumulates over the years. The characteristics of individual effort depend on the characteristics of the performed work and the worker behaviors. Each person is an expert resource at certain workstations. To quantify the work condition that can be applied by each person (e.g., according to the worker that knows how to do, his/her historic of exposure) it is necessary a scale. The Katz scale [29] ranges from 1 (i.e., acceptable condition) to 7 (i.e., unacceptable condition) and classifies the condition of each worker. The risk of effort occurrence at each workstation are classified by the OCRA indices (i.e., ① acceptable, ② uncertain or very light, ③ light and ④ mean and ⑤ "high - exhaustion"). The Borg Scale (i.e., 0 to 10) was adapted to indicate the OCRA indexes [30]. Accordingly, up to 2,2 the risk is acceptable; from 2,3 to 3,5 uncertain or very light; mild, ranging from 3,5 to 4,5; from 4,5 to 9,0 medium, and ranging from 9,0 to 10 the risk is high. A lot of problems could be eliminated or minimized with combinatorial optimization. Therefore, it is essential to integrate different areas of knowledge (e.g., health, safety, engineering) to solve those problems. A recent study [31] about the adoption of ISO 45001:2018 has been performed, showing that 98% of the surveyed companies are aware about the benefits that OHSMS can provide, showing as well that 75% of those companies obey to the ISO 45001 requirements, namely concerning the establishment and monitoring of measurable annual targets for OHS, the implementation of the risk prevention program and internal audits to evaluate OHS concerns.

3 Problem Description

Products are assembled using manual processes and follow a pre-defined processing order, with specific working conditions and multiple resources. Work teams are responsible to complete a pre-defined statement of work, over the span of the imposed takt-time fitting customer's needs. This time is calculated through the division between daily net time available for production and the customers' daily demand. It is assumed that assembly line balancing problems are as follows: (1) determining the number of workstations and the time available at each one, (2) assigning tasks for each workstation, forming groups taking into account the cycle time necessary in each workstation and (3) evaluating the efficiency of the chosen cluster, were previously studied. According to a task group, each workstation has one or more work posts. Each work post is occupied by only one person. The work post is the physical location of work where people are subject to the work execution conditions. The tasks are all simple to complete with standardized work. The matching minimizes the idle time at work posts. A minor idle time means that a person is more occupied and more exposed to the work conditions. Thus, for each workstation is set the best sequence of tasks, that corresponds to all necessary movements to complete a pre-defined work. Nevertheless, the effort that people must have to complete the work, can change between work posts. Usually, this workload does not correspond to the best sequence of group tasks execution able to minimize the people exposure at the work conditions. Thus, after solving the aforementioned balancing problems, it is also important to analyze the hazard of exposing people at the work conditions, before advancing with the manufacturing order. When a production manager neglects the OHS in a workplace, this can result in a lot of losses (e.g., loss of costumers, slowdowns or stops in production). Therefore, in this study it is proposed a formulation for a mathematical programming model to extract information about the exposure time of each worker, for each work assignment.

4 Methodology

The mathematical programming model proposed in Sect. 4.4 employs a goal programming approach in the optimization of a dual-objective problem. The model aimed at minimizing the exposure time of people to risks when people conclude the maximum number of tasks without exceeding the exposure limited, previously established. When the model generates the result/solution, it does not exceed these established limits. It can be applied to solve capacity problems, at environments where exists a long people exposure at work conditions. The temporal horizon for completion of the assigned statement of work is minimized for each worker, demonstrating the work team maximum capacity. The objective function is solved iteratively. The solutions generated allow extract information based on the relative importance of exposure at work conditions. Finally, production manager analyses the results and make decisions aligned with the real work context.

4.1 The Scope of the Case Study

The dataset for the development of the case study was collected in an industrial environment, specifically in the context of a bicycle industry (i.e., identification of people and tasks, measurement of work, and so on). Based on the professional experience of managers, senior technicians of occupational safety, auditors and workers, were defined the assumptions, input data and decision variables of the problem. This is a study of a quantitative and exploratory nature, applied in a production system, more precisely, the manual assembly of a bicycles' handlebars. The approach to the objectives for the present case study is focused on a literature review covering operational research (i.e., affectation problems and methods of resolution) and occupational health and safety (i.e., OCRA indices). After the literature review, different lines of thought and knowledge were integrated. Some models of whole linear programming, suggested by researchers to solve problems of assignment were developed. Finally, the CPLEX® optimization software was chosen and it was used the OPL language to test the mathematical formulation that results from the involvement of all the research. The solutions are then generated in a Windows© environment and analyzed to test and validate the model.

4.2 Assumptions

To complete a set of tasks, within a time horizon, people are exposed to specific work conditions. According to assignments between workers and work posts, work team exposure at the work conditions can be changed.

Also, the own work conditions (e.g., intensity and frequency of movements of the upper limbs) and the people behavior (e.g., bad postures of the upper limbs) can affect individual work conditions. Table 1 provides an overview of the adopted assumptions.

Table 1. Assumptions of mathematical programming model.

Assumption	
• Production system is efficient with the work standardized	*Workstations*
• Physical effort is not the same at all work posts	
• Each work post is occupied by one person	
• Only people able to complete tasks at the workstation are candidates to occupy the work post	
• The number of people is equal to the number of work posts	*People*
• The number of able workers does not have to be the same of the number of work posts	
• People have restrictions of capacity by accumulated risk exposure	
• Workers have specific individual skill levels. They can work at only one or at various workstations	
• The body postures and intensities of movements are distinct between workers and even between workstations	
• People receive the risks of the station working conditions that they are assigned	*Scheduling (each period)*
• Workers at the same workstation are exposed to the same risks	
• The workstation rotation is permitted at the end of each period	
• The instantaneous worker production time is less than takt time to guarantee dysfunctions	
• The horizon plan is limited and pre-defined	

4.3 Restrictions and Decision Variables

P has been defined as representing the set of people and W as the set of workstations, respectively, as shown in Table 2. The presented values result from the individual history of the frequency of gestures and repetitive movements, in the accomplishment of the tasks and risk conditions, taking into account the values previously registered and recorded in a data base. All scores are assigned by the decision-maker to indicate the condition of each person to exert each type of effort and the risk of executed activities at each workstation.

Table 2. Sets, parameters and variables of the mathematical model

Component	Notation	Description	
Sets	P	$i \in P, P = \{1 \ldots m\}$	*Number of people*
	W	$j \in W, W = \{1 \ldots n\}$	*Workstation number*
Parameters	d_j	$d_j \geq 0$	*Duration of activities at workstation j [min]*
	c_i	$c_i \geq 0$	*Capacity of exposure of person i [min]*
	l_j	$l_j \leq C$	*Maximum exposure at workstation j – production time given in minutes*
	e_i	$1 \leq e_i \leq 7$	*Condition of person i to complete work – Scale of Katz (1 able to 7 unable)*
	r_j	$0 \leq r_j \leq 9$	*Risk of effort occurrence at workstation j – OCRA scale (1 acceptable risk to 9 high risks)*
Variables	x_{ij}	$x_{ij} \geq 0$	*Exposition of person i at workstation j [min]*

4.4 Formulation of the Mathematical Model

This formulation is proposed for scenarios where the statement of work is done completely and can be applied to solve people capacity problems. The individual exposure for completion of the assigned statement of work is minimized, demonstrating the capacity of all teamwork exposure. The objective function (OF) allows analyzing the affectation of the people to the workstations, in the scope of the people exposure at the conditions in which the work is carried out. The objective function can be defined as:

$$\min Z = \sum_{i}^{m} \sum_{j}^{n} e_i r_j x_{ij} \tag{1}$$

Subject to:

$$\sum_{i=1}^{m} x_{ij} = d_j \qquad \forall j \in W_j \tag{2}$$

$$\sum_{j=1}^{n} x_{ij} \leq c_i \qquad \forall i \in P_i \tag{3}$$

$$\sum_{i=1}^{m} x_{ij} \leq l_j \qquad \forall j \in W_j \tag{4}$$

$$x_{ij} \geq 0 \qquad \forall j \in W_j, \forall i \in P_i \tag{5}$$

Equation (1) is the OF that minimizes the penalty concerning people work conditions and risk of effort at the workstation, with monotonous and repetitive manual tasks. This objective function is subject to Eq. (2) which ensures that the manual work is entirely fulfilled by the work team; Eq. (3) certifies that people's capacity is not exceeded; Eqs. (4) confirms that the upper limits of the effort frequency are satisfied in all workstations. The suggested mathematical model can be applied to different contexts of manual effort (e.g., assembly or disassembly of physical goods, food preparation, fish handling, etc.). The model extracts information about permanence time of people in the workstations. This analysis allows monitoring the exposure of people to the work conditions.

5 Test Instances

The environmental and economic worries, as well as the choice of a healthy lifestyle, particularly regarding the practice of sports (e.g., biking, mountain, BMX, downhill or freeride), have stimulated the citizens to buy different bicycle models. The attention of the researchers in this area of knowledge also increased to: (a) analyze the impacts of using public bicycles [32, 33] (b) explore the use of cargo bicycles [34] and (c) study fat bikes [35]. However, there are no studies about OHS at the assembly lines of bicycles.

5.1 Presentation of the Environment of Tests Scenario

In this thinking line, it was considered, for the present study, a manual assembly process of bicycle handlebar. To complete a handlebar, it is necessary a set of seven components: (1) one handlebar, (2) one stem, (3) one pair of brake levers, (4) one pair of shift levers, (5) two handles, (6) one pair of brake control cables and (7) four gear cables. In the collection of the related data to the duration of each task, it was, in general, registered 10 observations with a 95% confidence level and a normal deviation of 1,96. Each worker occupies the work posts where is more skilled (i.e., positions where can complete the work). To do one handlebar, it is necessary to complete 7 simple tasks: (1) screw the stem at handlebar, (2) assign the brake levers, (3) assign shift levers and handles, (4) attach the brake levers and shift levers, (5) mount the brake control cable, (6) allocate the brake control cables at the brake levers and (7) put the gear cables at shift levers. The line configuration presents a traditional layout.

5.2 Characterization of the Manual Assembly Process

The solving of the problem (Table 3) begins with the data assembly (e.g. collection of times, assignment of tasks at each workstation, number of people at each workstation). Each task has specific duration and work conditions. The assembly sequence of the components respects all precedences until one unit is completed. It is intended to produce 800 handlebars per 8 h (one shift). The time needed (i.e., the sum of all tasks) to complete one handlebar is 3.33 min. However, the time allocated per cycle of time is 3.42 min, therefore exists an idle time of about 0.09 min. It is assumed at two workstations the existence of parallel work posts, so the number of people does not coincide with the number of workstations. At each workstation *j* the production time, the idle time and the number of task cycles are not the same. The total time the line is operating (i.e., given L in minutes) varies depending on the units to be produced. The configuration of the production system, line efficiency (E = 97,10%) and cycle time (C = 0.58 min) remain the same in all the test scenarios. The mathematical model enables standartize the people's positions to minimize the hazard of exposure at work conditions. To observe if the model retrieves a measurement for supervising people exposure, an optimization software was used. The process was repeated in the original scenario (i.e., scenario A) and in a new scenario (i.e., scenario B), thus allowing to make comparisons and to increase the reliability of the proposed solutions. At the original scenario, people stay at the same work post all the time (i.e., 444 min) until all 800 handlebars are completed. On scenario B some limits (lj) were established regarding people exposure at each workstation because there are no default values to indicate how long a person can do manual work before start collecting health problems. Each workstation has a risk of effort (r_j) measured with the scale 1 to 10. The limits of work capacity (c_i) (Table 4) permit control the exposure of people at the workstation. To punish the assignments established by the proposed model, it was defined a penalization (1 to 7) that indicates the people condition to complete the activities (e_i).

Table 3. Environment of work conditions.

W_j	N^er of able people	N^er of posts	Tasks	Duration tasks	Idle time	Risk	Scenario A and B D = 800 *handelbars*			
				min. per cycle	*min. per cycle*	r_j	d_j *min.*	*Idle time min.*	*N^er of cycles*	l_j *min.*
1	8	1	1|2	0.56	0.02	3,5	448.00	12.00	772	460
2	8	2	3|4	1.09	0.03	3,5	872.00	48.00	1503	920
3	8	2	5	1.13	0.01	2,3	904.00	16.00	1559	920
4	8	1	6|7	0.55	0.03	2,3	440.00	20.00	759	460
		6		3.33	0.09		2664.00	96.00		

Table 4. People capacity to complete work.

Person	1	2	3	4	5	6	7	8
Work capacity c_i (min.)	193.72	230.84	317.84	321.32	350.90	362.50	444	444
Number of cycles	334	398	548	554	605	625	765	765
Exposure condition e_i	6	5	4	3	2	2	1	1

6 Results

The efficiency of a line increases when the percentage of the idle time diminishes. With shorter idle times, longer is the time that a worker is submitted to the working conditions. The demand (D) of the handlebars increases synchronized with the cycle time (C). It is always the decision-maker that must analyze all the results regarding the advantages, disadvantages, and limitations of the performance of the assembly system and then choose the better admissible solution. Naturally, he/she also needs to create means to synchronize the supply of the components (e.g., supplying the workstations), avoiding certain failures ior excesses in their supply. All the adjustments, even the smaller ones, require high speed and efficiency, so that production runs efficiently. The results generated by the mathematical programming model are admissible solutions if there exists people that do not exceed the limit exposure time at each workstation. If the limits have already been reached, this means that people's health is compromised. In a real context, the objective is to monitor the subjection of people to the conditions of the work without exceeding these limits.

6.1 Results Generated by CPLEX®

The solutions generated by the software are now presented. When $p_{ij} = 0$ is used in the workstations, it means that the person p_i was not chosen to perform tasks at W_j. The costumer's needs are 800 handlebars per each working day or shift (8 h). Thus, the programming schedule was based on this quantity. Despite existing eight people that can do all activities to complete one handlebar, only six are working in this assembly line (i.e., original scenario – Table 5). In this way, people stay basically at the same workstation all the time. Thus, six people are active for 444 min. The solution OF = 26153 was obtained after 8 iterations in 0,12 ms. Consequently, the work condition of people is $\sum x_{ij} = 9768{,}00$ with a risk at workstations $\sum r_j = 7710{,}70$. The penalty for the people condition to exposure (e_i) and the risk about making work at the workstation (r_j) were analyzed separately to understand the individual results (i.e., x_{ij} and r_j), regarding people and workstations. Finally, it were considered limits to the work capacity of all people that can perform it (i.e., 8 people). Similarly, Table 6, the OF = 17888 solution was obtained in 0,14 ms, after 13 iterations. The work condition is defined by $\sum x_{ij} = 6862{,}92$. The risk at workstations have the same value, $\sum r_j = 7710{,}70$.

Table 5. Results of penalties at scenario A (Points)

W_j	1		2		3		4	
P_i	exposure x_{ij}	risk r_j	exposure x_{ij}	risk r_j	exposure x_{ij}	risk r_j	exposure x_{ij}	risk r_j
1	0	0	0	0	2664,00	1021,20	0	0
2	0	0	0	0	2220,00	1021,20	0	0
3	0	0	0	0	16,00	9,20	1760,00	1012,00
4	1296,00	1512,00	0	0	36,00	27,10	0	0
5	0	0	888,00	1554,00	0	0	0	0
6	32,00	56,00	856,00	1498,00	0	0	0	0
	1328,00	**1568,00**	**1744,00**	**3052,00**	**4936,00**	**2078,70**	**1760,00**	**1012,00**

Table 6. Results of penalties at scenario B (Points)

W_j	1		2		3		4	
P_i	exposure x_{ij}	risk r_j	exposure x_{ij}	risk r_j	exposure x_{ij}	risk r_j	exposure x_{ij}	risk r_j
1	0	0	0	0	1159,20	444,36	0	0
2	0	0	0	0	543,40	249,96	610,80	281,00
3	0	0	0	0	0	0	1271,36	731,00
4	0	0	0	0	963,96	739,04	0	0
5	701,80	1228,15	0	0	0	0	0	0
6	163,40	285,95	0	0	561,60	645,84	0	0
7	0	0	443,70	1552,95	0	0	0	0
8	15,40	53,90	428,30	1499,05	0	0	0	0
	880,60	**1568,00**	**872,00**	**3052,00**	**3228,16**	**2078,90**	**1882,16**	**1012,00**

6.2 Discussion of Results

The manual assembly processes enable to quickly respond to the market needs and demand, minimizing production costs, and maximizing profits. However, it is difficult for industrial managers to make efficient changes to keep the production process fitted to the market-driven demand. This is more highlighted when the decision-making process is dynamic, the demand is very unstable and the products extremely different from each other (e.g., a wide range of articles). Thus, in the case of assembly lines in which the product does not differ (i.e., a single article), or has small variations, the necessary adjustments and changes are in a small number and the decision making usually becomes simpler and more efficient. The fatigue compensation factor and performance factor are not the same at all tasks. After balancing the assembly line, the challenge is to solve assigning problems to supervise the hazard of exposing people to work conditions. This study considers two scenarios to better understand the results given by the mathematical programming model and to make comparison between each other. According to the original scenario, that will serve as the basis for the development of this work, the 800 handlebars are assembled with an efficiency of about 97%. Initially (i.e., scenario A) there is no rotation in the assignments established between people and workstations. Thus, the number of workstations matches the number of people who are busy. Consequently, people have OHS compromised if planning and programming are defined for several days (e.g., long exposure). Then (i.e., scenario B) it is verified that all the people that carry out the work are not always in the same position therefore it is important to have a rotation schedule in the workstations. The results generated by the mathematical model indicate some possible solutions. Thus, a person with less condition to work stays free (because OF is bigger than zero). A person with less exposure to the working conditions (OF is next to zero) and the person that is now available becomes more occupied and more exposed to these working conditions.

7 Conclusion

People exert a certain effort to carry out the work assigned to them. As their capacity is finite, there are exposure limits to the conditions in which the work is performed. The efforts of the workers are not always addressed in the assignments created between the people and the work to be realized. Thus, DSS should be created/used allowing to optimize the usage of resources without compromising workers' health and safety, and of course organizations' productivity. With this purpose it was used a dataset, together with an optimization software, allowing to generate results concerning the minimum exposure at the work conditions. Such mixing can contribute by minimizing a big diversity of problems. Scientific knowledge concerning real problems can increase the implementation of admissible solutions in a real context. The idea is indispensable and decisive, not only for the scientific development of different areas of knowledge (e.g., DSS, industrial simulation, operations research, management, supply chain management, ergonomics), but also to help production planners to know how to program and control the work, without compromising the health and safety of people over the span of the operations. The limitations of the case study are related to the difficulty on reorganizing the reference scenario in order to guarantee the proposed objective and to choose the most appropriate scenario with few acquisitions and without compromising workers' health and safety. The proposed mathematical model arises from the need to prevent the occurrence of WMSDs, as they cause slowdowns and frequent stops in production systems. The model ensures that the imposed limits are not exceeded, based on the OCRA indices. However, exposure time control may be a limitation because it is performed by imposing uncertain maximum limits (i.e., subjective indicative values) as there are no standardized exposure limit values to ensure the elimination of WMSDs. Finally, the lack of historical data from the application of the proposed model, in a real work context, clearly can limit the perception of the functioning and effectiveness of the practical results. Future scientific research should focus on people's performance, coupled with their quality of life. Monitoring of exposure to the risk of WMSDs should cover the entire human body.

In this sense, it should also be explored the assignment of tasks to workstations with different fatigue compensation values. An integrated information system with an attractive frontend graphical interface is urgent and fundamental so that the assessment of professional risks becomes really preventive, realistic and non-curative. The results can be presented in the format of working hours. Evaluations must be performed in real-time, before proceeding with the production order, based on the individual characteristics of the workers. The use of other layout line configurations, for example, supporting parallel workstations, should also be studied and considered.

References

1. Caputo, A.C., Di Salvo, G.: An economic decision model for selective assembly. Int. J. Prod. Econ. **207**, 56–69 (2019)
2. Anzanello, M.J., Fogliatto, F.S., Santos, L.: Learning dependent job scheduling in mass customized scenarios considering ergonomic factors. Int. J. Prod. Econ. **154**, 136–145 (2014)
3. Gonçalves, I., Sá, J.C., Santos, G., Gonçalves, M.: Safety stream mapping—a new tool applied to the textile company as a case study. Stud. Syst. Decis. Control **202**, 71–79 (2019)
4. Peng, L., Chan, A.H.S.: A meta-analysis of the relationship between ageing and occupational safety and health. Saf. Sci. **112**, 162–172 (2019)
5. Pata, A., Moura, A.: Applying metaheuristics to minimize work-related musculoskeletal disorders. Int. J. Technol. Hum. Interact. **14**, 17–34 (2018)
6. Stevenson, J.: Operations Management. Mc Graw Hill, New York (2002)
7. Kim, T., Glock, C.H.: Production planning for a two-stage production system with multiple parallel machines and variable production rates. Int. J. Prod. Econ. **196**, 284–292 (2018)
8. Xiao, X., Jiang, W., Luo, J.: Combining process and product information for quality improvement. Int. J. Prod. Econ. **207**, 130–143 (2019)
9. Mayron, M., Rafael, P., Alysson, C., Cristóbal, M.: The multi-objective assembly line worker integration and balancing problem of type-2. Comput. Oper. Res. **82**, 114–125 (2017)
10. Bian-bian, Q., Ji-ping, Z., Zai-xiang, Z., Hui, S.: Establishing a dynamic ergonomic evaluation index system for complex product designs based on the theory of product life cycle. Int. J. Ind. Ergon. **69**, 153–162 (2019)
11. Moreira, M.C.O., Costa, A.M.: Hybrid heuristics for planning job rotation schedules in assembly lines with heterogeneous workers. Int. J. Prod. Econ. **141**, 552–560 (2013)
12. Koos, S.E., Shaikh, N.I.: Dynamics of consumers' dissatisfaction due to stock-outs. Int. J. Prod. Econ. (2018)
13. Russell, A., Taghipour, S.: Multi-objective optimization of complex scheduling problems in low-volume low-variety production systems. Int. J. Prod. Econ. **208**, 1–16 (2019)
14. Occhipinti, E., Colombini, D.: A toolkit for the analysis of biomechanical overload and prevention of WMSDs: criteria, procedures and tool selection in a step-by-step approach. Int. J. Ind. Ergon. **52**, 18–28 (2016)
15. Ilmarinen, J.: Promoting active ageing in the workplace. In: European Year for Active Ageing and Solidarity Between Generations 2012: European Agency for Safety and Health at Work, pp. 1–7 (2012)
16. Streb, C.K., Voelpel, S.C.: Analyzing the effectiveness of contemporary aging workforce management: the case of daimler AG. Organ. Dyn. **38**, 305–311 (2009)
17. Fernandez-Ballesteros, R., Olmos, R., Santacreu, M., Bustillos, A., Molina, M.: The role of perceived discrimination on active aging. Arch. Gerontol. Geriatr. **71**, 14–20 (2017)

18. Bierwisch, A., Goluchowicz, K., Som, O.: Stocktaking of activities in active aging and work environment in policy, science and industry — The German case. Technol. Forecast. Soc. Chang. **89**, 343–357 (2014)
19. Voelpel, S., Streb, C.: A balanced scorecard for managing the aging workforce. Organ. Dyn. **39**, 84–90 (2010)
20. Aisa, R., Larramona, G., Pueyo, F.: Active aging, preventive health and dependency: Heterogeneous workers, differential behavior. J. Econ. Behav. Organ. **117**, 1–9 (2015)
21. Moussavi, S.E., Mahdjoub, M., Grunder, O.: Reducing production cycle time by ergonomic workforce scheduling (2016)
22. Correia, D., Silva, F.J.G., Gouveia, R.M., Pereira, T., Ferreira, L.P.: Improving manual assembly lines devoted to complex electronic devices by applying Lean tools. Procedia Manuf. **17**, 663–671 (2018)
23. Santos, J., Silva, F.J.G., Pinto, G., Baptista, A.: Lean and ergonomics: how to increase the productivity. In: Silva, F.J.G., Ferreira, L.P. (eds.) Lean Manufacturing - Implementation, Opportunities and Challenges, Nova Science Publishers, NY, U.S.A (2019)
24. Rajesh, R.: Manual material handling: a classification scheme. Procedia Technol. **24**, 568–575 (2016)
25. Bastos, A., Sá, J., Silva, O., Fernandes, M.C.: A study on the reality of Portuguese companies about work health and safety. Occup. Saf. Hygiene II. 687–691 (2014)
26. Badri, A., Boudreau-Trudel, B., Souissi, A.S.: Occupational health and safety in the industry 4.0 era: a cause for major concern? Saf. Sci. **109**, 403–411 (2018)
27. Li, H.-L., Huang, Y.-H., Fang, S.-C.: Linear reformulation of polynomial discrete programming for fast computation. Informs J. Comput. **29**, 108–122 (2017)
28. Thun, J.-H., Lehr, C.B., Bierwirth, M.: Feel free to feel comfortable—an empirical analysis of ergonomics in the German automotive industry. Int. J. Prod. Econ. **133**, 551–561 (2011)
29. Katz, S., Ford, A., Moskowitz, R., Jackson, B., Jaffe, M.: Studies of illness in the aged - the index of ADL: a standardized measure of biological and psychosocial function. JAMA (1963)
30. Nappi, F., Caselli, U., Rughi, D., Sarto, D., Todaro, N.: Conoscere il rischio movimenti retetuti degli arti superiori. Il protocollo OCRA. (2017)
31. Morgado, L., Silva, F.J.G., Fonseca, L.M.: Mapping occupational health and safety management systems in Portugal: outlook for ISO 45001:2018 adoption. Procedia Manuf. (2020, In press)
32. Zang, Y., Thomas, T., Brussel, M., Maarseveen, M.: Exploring the impact of built environment factors on the use of public bikes at bike stations: Case study in Zhongshan. China. J. Transp. Geogr. **58**, 59–70 (2017)
33. Ferreira, N., Santos, G., Silva, R.: Risk level reduction in construction sites: towards a computer aided methodology - a case study. Appl. Comput. Inf. **15**, 136–143 (2019)
34. William, R.: Cargo bikes as a growth area for bicycle vs. auto trips: exploring the potential for mode substitution behavior. Transp. Res. Part F **43**, 48–55 (2016)
35. Christopher, M., Andrew, K.: The emergence of "fat bikes" in the USA: Trends, potential consequences and management implications. J. Outdoor Recreat. Tour. **15**, 20–25 (2016)

Cycle Time Reduction in CNC Turning Process Using Six Sigma Methodology – A Manufacturing Case Study

Kakarla Manoj[1], Biswajit Kar[2], Rajeev Agrawal[3], Vijay Kumar Manupati[2(✉)], and José Machado[4]

[1] RR Donnelley, Chennai, India
[2] Mechanical Engineering Department, NIT Warangal, Warangal, Telangana, India
manupati.vijay@nitw.ac.in
[3] Mechanical Engineering Department, MNIT Jaipur, Jaipur, Rajasthan, India
[4] MEtRICs Research Centre, University of Minho, Guimaraes, Portugal

Abstract. Six-Sigma, a data-driven methodology, employed to improve the process in terms of Defect reduction or process optimization. In this paper, an experimental study is presented optimizing the cutting parameters while machining of shoulder bolt in a Computer Numerical Control (CNC) turning machine to reduce the cycle time. This study identifies, the effects of cutting speed, feed rate and dwell time on Thread rolling diameter (TRD) in CNC turning machine that was experimentally investigated. The experimentation plan is designed using six sigma D-M-A-I-C methodology, and the subsequent statistical analysis has been done using Minitab-16 software. Shainin based variable search tool has been used to investigate the design parameters that contribute to the reduction of the cycle time and factorial plots are employed to determine the contribution of important parameters. Later, the optimal values for the best cutting conditions are proposed for industrial production using the formulated mathematical model. Finally, this paper documents the analysis and tasks performed that reduced cycle time which resulted in increased productivity and also in annual savings.

Keywords: Thread rolling diameter · Turning · Machining · Six sigma · D-M-A-I-C · Variable search · Better vs Current · Factorial plot

1 Introduction

Today's manufacturing environment is experiencing a constant change with time due to variation in market demand as well as dynamic customer requirements. In order to support companies to overcome the difficulties and challenges imposed by their daily requirements in a dynamic and supported way, different kind of models, approaches and systems have been put forward, such as [1–7]. One of the major problems industries are facing today is to improve productivity without sacrificing quality [8]. The most widely used traditional processes of fabricating a workpiece to a useful product are metal forming and machining. Permanent plastic deformation is the key principle of

J. Machado et al. (Eds.): icieng 2021, LNME, pp. 135–146, 2022.
https://doi.org/10.1007/978-3-030-79165-0_13

the metal forming process, whereas machining involves the removal of material from a workpiece in the form of chip using a cutting tool [9]. The general forms of machining are turning, drilling and milling. Amongst all the machining processes, turning is the most widely used process, where the finished product from CNC turning machine is susceptible to dimensional accuracy, less tool force, minimal temperature, maximum material removal rate and maximum tool life [10]. To achieve the highest productivity, optimal settings for the design parameters can be achieved through various analyses available in the literature [11, 12]. In this paper, an attempt has been made to reduce cycle time and to optimize the design parameters like cutting speed, feed rate, and dwell time for the production of shoulder bolt.

2 Literature Review

Numerous practitioners studied [13] claim that "Lean Six Sigma has been viewed as a business change technique incorporating two philosophies: Lean and Six sigma integrating each other with a specific end goal to enhance the business performance" [14–16]. This coordination has been accomplished by blending their techniques and standards utilizing DMAIC (Define, Measure, Analyse, Improve, Control) cycle and DMADV (Define, Measure, Analyse, Design, Verify) as a foundation. DMAIC technique is most widely used for process improvements and in contrast, the DMADV technique is used for process re-designing [17, 18]. The optimal process improvements can be achieved when all the phases are analyzed in a sequential pattern. The motivation behind the Define stage is to decide the focus of the project, and to analyse the current process.

In the Measure stage, data is collected to evaluate the ability of the current process in meeting the design specifications [19]. The analysis stage yields the parameters of significance, the contribution of the significant parameters, optimum settings for the parameters of significance. In the Improve stage, the optimal settings for the significant parameters are validated to check for the consistency. Pyzdek and Keller, in 2009 [20] stated that the Control phase creates monitoring and recreation plans for the improved process. The paper considered the case study of a 'shoulder bolt' and the series of operations to be performed on it are rolling, forging, turning, threading, plating, heat treatment. The output part of the turning process is subjected to a diametric test and this diameter plays a vital role in the threading process. Therefore, the design parameters (Cutting speed, feed rate, and depth of cut) in the turning process [21] are to be optimized to improve productivity by reducing the cycle time. For the selection of the cutting parameters, several mathematical models with statistical regression or neural network as a base have been formulated [22] to analyse the relationship between the response and cutting parameters. [23] made use of Taguchi's design of experiments to optimize the design parameters by using the signal to noise ratio. This case study made an alternate approach of Shainin DOE for optimizing the cutting parameters [24–26].

3 Problem Description

Not many case studies in the literature adopted Dorian Shainin's design of experiments for their process optimization. Nonetheless, the common limitations of the aforesaid papers are: (i) They did not consider the optimization of design parameters with the diameter as a response factor. (ii) They did not consider dwell time as a design parameter in the optimization of the CNC turning process. Motivated by these factors, the Shainin Design of experiments (DOE) is modeled that considers the effect of design parameter levels in cycle time reduction for the CNC turning process. The objective of this paper is to reduce cycle time and to optimize the design parameters like cutting speed, feed rate, and dwell time for the production of 'shoulder bolt'. To accomplish the above-mentioned objective, as a first step, the cutting parameters of a CNC turning machine were identified. Next, with the help of factorial plots i.e., main effect plot and interaction plot, the influence of each factor on one another and with the response is interpreted. Later, a mathematical model along with constraints is developed to obtain the contribution of each design parameter on response. Moreover, the results obtained by the proposed method has been validated by using B Vs C technique. Finally, with the help of control charts, the process is proved to be within its control limits.

The paper has been designed as the following: Sect. 4 outlines the research methodology endorsed in this paper and the evolution of six sigma. The experimental procedure of Shainin DOE to fine-tune the design parameters is documented in Sect. 5. Section 6 outlines the essential conclusions are drawn depending on results from six-sigma methodology.

4 Methodology

Six-Sigma, as an approach guides firms to improve company efficiency and client fulfilment; reduces operational cost while increasing surplus. Many practitioner references demand Six-Sigma as the reason for the improvement of organisational performance. Empirical research in this area being limited, no thorough scrutiny was there on Six Sigma leading organizations to the improvement of performance (Sin et al. 2015) [19]. It is a methodology used to determine the root cause of any problem by assessing all the suspected sources of variation. Cost-effective solutions to optimize or eliminate the effects of the pinpointed causes are suggested by statistical analysis.

Defects can be reduced using tools like flowcharts, Pareto analysis, cause-and-effect diagram, Paired Comparison, Product/Process search, Component search, etc. Process optimization/Improvement can be achieved using Classical DOE, and Taguchi approach, the two most popular traditional techniques of Design of experiments [27–30]. The latest approach to the design of experiments, the Shainin DOE technique, introduces by Dr. D. Shainin is the best alternative compared to the above-mentioned techniques [26]. Some of the benefits of this technique are simple, efficient and cost-effective compared to the traditional DOE tools and techniques [31]. D-M-A-I-C is short for Define-Measure-Analyse-Improve-Control, one of the approaches to proceed with the problem [32]. Figure 1 depicts the tools used in each step of D-M-A-I-C to solve the problem.

The advent of the Shainin design of experiments (DOE) has replaced conventional approaches [9] of classical DOE and Taguchi DOE. [33] stated that Variable search, a key tool of Dr. Dorian Shainin provided a step by step analysis to optimize the cycle time. As stated above, the five-step by step analysis phases that form the backbone of six-sigma analysis are: define phase, measure and analyze phase, improve phase, control

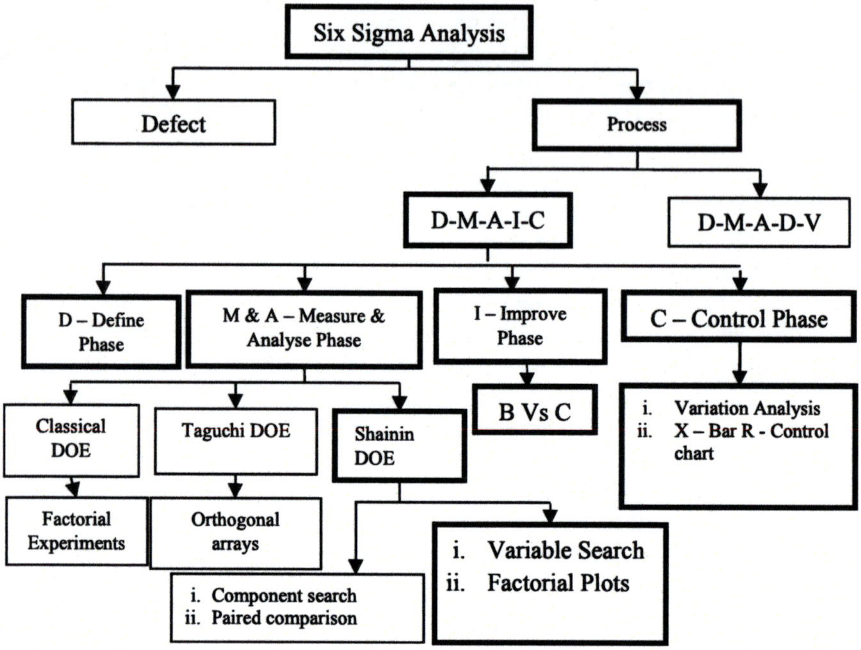

Fig. 1. Tools used in six sigma analysis

phase [24] to properly manage the underlying knowledge [34]. Define phase is further classified into four steps. 1) Selecting the process and machine for optimization based on the problem description. 2) Selecting the design parameters for optimization using tools like Delphi method, expertise knowledge, cause, and effect diagrams and etc. 3) Examining the current machining conditions by visually identifying the initial design setting levels. 4) Identifying the new setting levels for optimization based on the expertise knowledge. Measure and analyse phase can be accomplished by following a sequence of four stages: 1) using the value of the D/d ratio(D is the 'difference between median of existing setting and new setting' and d is 'the difference between the range of new setting and existing setting') as a basis in identifying the correct setting levels. 2) Significant parameters contributing to the response and cycle time have to be identified using a graph. 3) Finding the contribution of individual parameters using factorial analysis. 4) Identification of the optimal setting levels by framing a mathematical equation. Later in the next phase, Better Vs Current tool is used to validate the results and finally control charts monitors the process. This case study has been solved using the six sigma D-M-A-I-C cycle [35]. Furthermore, variable search, one of the Shainin methods [26] is used to analyze and optimize the design parameters. The results are then validated using Shainin B Vs C (Better Vs Current) Technique. The objectives and outputs for each phase of D-M-A-I-C are discussed in Table 1. Finally, the flow in which the problem is solved in each phase is shown in Fig. 2.

Table 1. Objectives and outputs of D-M-A-I-C

Description	Tasks	Output	Objective(s)
Define	Observe the process and collect the data	Identifying Suspected Sources of Variation (SSV's) for the problem	Understanding the Problem
Measure & Analyze	Pinpoint the real reasons which are creating the problem	Identifying the root factor of the problem	Knowing the cause of the problem
Improve	Validate the pinpointed causes	Identify solutions, evaluate and Implement	Implement the validated solution
Control	Monitor and control the pinpointed causes	Monitor and control the causes which are creating problem	Control the problem by implementing the results

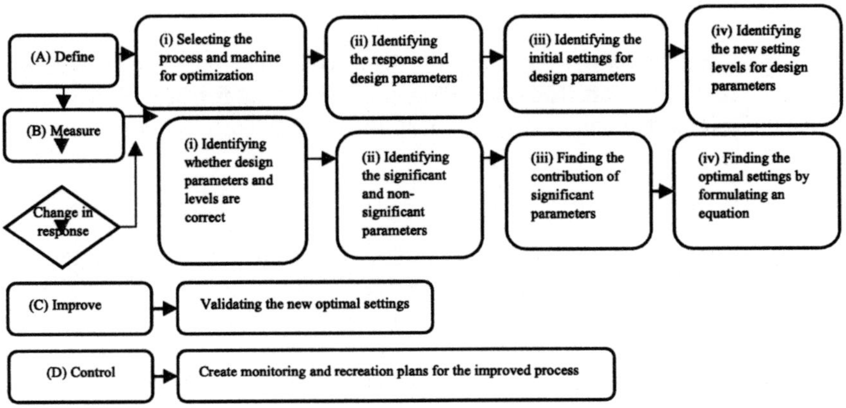

Fig. 2. Process flow diagram

5 Experimental Procedure

This research is aimed to reduce the cycle time for processing a part 'shoulder bolt' in a CNC turning machine by optimizing its design parameters. Henceforth, by reducing the cycle time, lead time can also be reduced leading to a decrease in production cost and an increase in customer satisfaction. This experimentation work is carried out in an Indian fasteners manufacturing company.

5.1 Define Phase

The process selected for optimization is turning, and the equipment employed for this purpose is a computer numerical control (CNC) machine. The response measured after processing the part is TRD (Thread Rolling Diameter). TRD is a variable feature of 'shoulder bolt' having specification as 6.24–6.28 mm. It is measured using a micrometre which is capable of measuring values up to three decimal units.

Design Parameters identified for Optimization are dwell time, feed rate and cutting speed. Dwell time is used to cause a delay in the process for the specified time (seconds or milliseconds). It stops the machine from moving for a set amount of time; this is to ensure that part is clamped in the spindle and all necessary material is removed from the component [36]. Feed rate and cutting speed are expressed in distance per spindle revolution and rpm (revolutions per minute) respectively. The specification and initial settings for the design parameters in which the equipment has been configured are given in Table 2.

Table 2. Design parameters and their existing settings, new settings

SN	Parameter	UOM	Specification	Existing (−) settings	New (+) settings
A	Dwell Time	Seconds	No specification	3	2
B	Feed Rate	mm/rev	0.08–0.15	0.1	0.11
C	Cutting Speed	Rpm	2500–3500	2800	2900

The cycle time for processing the part in the machine with the existing settings was found out to be 22 s per part. The new setting levels (+Setting) for the design parameters were identified based on the expert's opinions and tabulated in Table 2.

5.2 Measure and Analyze

Full Factorial analysis for optimizing the design parameters is performed as follows:

By configuring the machine to the (−) settings, a part is fabricated and the response is recorded in Table 4. Later, when the machine is configured to (+) settings, a part is fabricated, and the response is recorded in Table 4. This sequence is altered for all the three runs to distribute the variation due to common causes between the responses of (− and +) settings. Then D (difference between the medians of each trial) and d (the mean range for each trial is calculated) is determined to calculate the D/d ratio. All the setting levels for the four assumption trials are shown in Table 3.

Table 3. Trial settings for design parameters

S.No	Parameter	UOM	First trial settings		Second trial settings		Third trial settings		Fourth trial settings	
			−Setting	+Setting	−Setting	+Setting	−Setting	+Setting	−Setting	+Setting
A	Dwell Time	sec	3	2	2	2	2	2	2	1
B	Feed Rate	mm/rev	0.1	0.11	0.11	0.12	0.12	0.125	0.125	0.13
C	Cutting Speed	Rpm	2800	2900	2900	3000	3000	3100	3100	3200

Table 4. Optimal settings for design parameters

No	Parameter	UOM	Specification	Existing settings	Optimal settings
A	Dwell Time	sec	No spec.	3	1
B	Feed Rate	mm/rev	0.08–0.15	0.1	0.1325
C	Cutting Speed	Rpm	2500–3500	2800	3200

The existing settings and the optimal settings identified are tabulated in Table 4. The part 'Shoulder bolt' was then manufactured with better settings. Three runs were again carried out. The cycle time with the optimal setting was experimentally found to be 18 s.

5.3 Improve

The optimal settings obtained using the mathematical equation have to be validated before they are implemented in a real-time environment [19, 31, 37]. Hypothesis tests like 2t-test, ANOVA, Chi-square test, etc., are some techniques to statistically validate the improved settings. These tests help in identifying if the improvement is significant or marginal. Dr. Dorian Shainin proposed (Better Vs Current) tool, an effective substitute for the conventional tools used for the validation.

Since the D/d ratio for the initial and optimal settings in Table 5 is between 1.25 and 3, and there is no overlap, the optimal settings identified are better and they can be implemented in the real-time scenario to fetch the desired output. The improved settings are to be cross-validated using B(Better) Vs C(Current) tool to check for the consistency of improved setting levels. B(Better) Vs C(Current) is a Non- Parametric test where: 1) B = Better or new process and 2) C = Current or old process.

Table 5. D/d ratio for final trial (Optimal Settings)

Test	−Setting	+Setting (optimal)
1st Run	6.268	6.245
2nd Run	6.272	6.252
3rd Run	6.262	6.246
Median	6.268	6.246
Range	0.01	0.007
D	0.022	
d	0.0085	
D/d	2.59	

In consideration of the results from Table 6, there is no overlap in readings of current and better settings. Hence the optimal settings are better than the current settings. From Table 7, Since, $(X_b - X_c) >= K * \text{Sigma (b)}$, 19.5% improvement has taken place at 95% Confidence level. The results are now successfully validated.

Table 6. Data for B Vs C

Piece/Lot	Current (C)	Better (B)
1	6.268	6.248
2	6.270	6.252
3	6.278	6.248
4	6.265	6.255
5	6.270	6.250
6	6.274	6.258
7	6.270	6.244
8	6.278	6.255
9	6.268	6.249

Table 7. Analysis of B Vs C

1	The part number selected for validation	SPL000080
2	Average of B (X_b)	6.252
	Average of C (X_c)	6.271
3	D/d ratio	2.59
4	$X_b - X_c$ (Amount of Improvement)	0.0195
5	Sigma (B)	0.004
6	K	2.42 @ 95% CL for 9B, 9C
7	K * sigma(B)	= 2.42 * 0.004 = 0.00968
8	Is $X_b - X_c$ greater than k * Sigma (b)	Yes

5.4 Control

This is the most crucial phase of the process improvements and this phase ensures that the process continues to work well with the adjusted setting levels, producing parts within its specification limits of response. Variance analysis and control charts are mostly used to monitor the process.

The average part to part variation is obtained by finding the average of all the part to part variations in Table 8.

Table 8. Variance analysis

Time block	Response		Time block	Response		Time block	Response	
	6.268	6.255		6.26	6.258		6.254	6.272
1	6.254	6.258	7	6.25	6.266	13	6.255	6.279
	6.261	6.256		6.254	6.265		6.246	6.277
P–P	0.014	0.003	P–P	0.01	0.008	P–P	0.009	0.007
	6.276	6.256		6.264	6.285		6.268	6.259
2	6.261	6.255	8	6.263	6.288	14	6.276	6.264
	6.264	6.255		6.257	6.28		6.269	6.264
P–P	0.015	0.001	P–P	0.007	0.008	P–P	0.008	0.005
	6.261	6.269		6.276	6.265		6.271	6.251
3	6.256	6.264	9	6.269	6.254	15	6.267	6.259
	6.254	6.268		6.271	6.258		6.27	6.249
P–P	0.007	0.005	P–P	0.007	0.011	P–P	0.004	0.01
	6.299	6.263		6.243	6.251		6.264	6.276
4	6.304	6.259	10	6.255	6.259	16	6.259	6.271
	6.298	6.251		6.246	6.264		6.255	6.277
P–P	0.005	0.012	P–P	0.012	0.013	P–P	0.009	0.006
	6.28	6.278		6.289	6.257		6.265	6.255
5	6.277	6.265	11	6.284	6.246	17	6.262	6.258
	6.278	6.276		6.281	6.249		6.261	6.259
P–P	0.003	0.013	P–P	0.008	0.011	P–P	0.004	0.004
	6.262	6.287		6.259	6.244		6.242	6.268
6	6.269	6.279	12	6.272	6.245	18	6.254	6.267
	6.274	6.282		6.266	6.248		6.259	6.262
P–P	0.012	0.008	P–P	0.013	0.004	P–P	0.016	0.006

P–P = Part to part Variation

6 Conclusions

The conclusions that can be drawn based on the result obtained from experiment this case study are as follows:

- The impact of feed rate on the cycle time is rated highest followed by cutting speed and dwell time.
- Final affirmation result shows that at optimum cutting parameters, the feed rate is raised from 0.10 mm/rev to 0.13 mm/rev. Cutting speed is increased from 2800 rpm

to 3200 rpm and Dwell time is reduced from 3 s to 1 s. Henceforth Cycle time per part is reduced from 22 s to 18 s.

- In turn, the use of greater feed rate (0.13 mm/rev), moderate cutting speed (3200 rpm) and low dwell time (1 s) are recommended to obtain a better response with increased productivity.
- The improved settings resulted in an annual savings of INR 1,44,000 and the productivity increased by 800 components per day.
- Since the cycle time of turning process is declined, Lead time is then reduced and manufacturing wastes were eliminated.
- This article discusses an application of the Shainin method for optimization of cycle time, feed rate and cutting speed in turning operation. It also indicates that the Shainin DOE is a constructive method to figure out the optimal cutting conditions for calculating an optimal value of outputs with a comparatively fewer numbers of experiment runs.

Acknowledgement. The project is funded by MHRD Government of India as research seed money scheme National Institute of Technology Warangal, India with Order No NITW/DIR/2018/478/1138 and FCT – Fundação para a Ciência e Tecnologia who financially supported this work within the R&D Units Project Scope: UIDP/04077/2020 and UIDB/04077/2020.

References

1. Campanella. G., Ribeiro, R.A., Varela, M.L.R.: A model for B2B supplier selection. In: Melo Pinto, P., Couto, P., Serodio, C., et al. (eds.) EUROFUSE 2011: Workshop on Fuzzy Methods for Knowledge-Based Systems (2011). Adv. in Intell. and Soft Comp. Vol. 107, 221–228 (2011)
2. Varela, M.L.R., Silva, S.: An ontology for a model of manufacturing scheduling problems to be solved on the web. In: Azevedo, A. (ed.) BASYS 2008. ITIFIP, vol. 266, pp. 197–204. Springer, Boston, MA (2008). https://doi.org/10.1007/978-0-387-09492-2_21
3. Vieira, G., Varela, Maria Leonilde R., Putnik, Goran D.: Technologies integration for distributed manufacturing scheduling in a virtual enterprise. In: Putnik, Goran D., Cruz-Cunha, M.M. (eds.) ViNOrg 2011. CCIS, vol. 248, pp. 337–347. Springer, Heidelberg (2012). https://doi.org/10.1007/978-3-642-31800-9_34
4. Ribeiro, R.A., Varela, L.R.: Fuzzy optimization using simulated annealing: an Example set. In: Verdegay, J.L. (ed.) Fuzzy Sets Based Heuristics for Optimization, pp. 159–180. Springer, Berlin, Heidelberg (2003). https://doi.org/10.1007/978-3-540-36461-0_11
5. Manupati, V.K., Gangal, C., Varela, M.L.R., Mobin, M.: Application of value stream mapping for cycle time reduction in production of link and roller assembly. Int. J. Serv. Oper. Manage. **33**(2), 135–158 (2019)
6. Varela, M.L.R., Manupati, V.K., Manoj, K., Putnik, G.D., Araújo, A., Madureira, A.M.: Industrial plant layout analyzing based on SNA. In: Madureira, A.M., Abraham, A., Gamboa, D., Novais, P. (eds.) ISDA 2016. AISC, vol. 557, pp. 728–737. Springer, Cham (2017). https://doi.org/10.1007/978-3-319-53480-0_72
7. Reddy, M.S., Ratnam, C., Agrawal, R., Varela, M.L.R., Sharma, I., Manupati, V.K.: Investigation of reconfiguration effect on makespan with social network method for flexible job shop scheduling problem. Comput. Indus. Eng. **110**, 231–241 (2017)

8. Afrifa, J., et al.: Application of sigma metrics for the assessment of quality control in clinical chemistry laboratory in Ghana: a pilot study. Nigerian Med. J. **56**(1), 54 (2015)
9. Aksu, B., Baynal, K.: Shainin and Taguchi methods and their comparison on an application. In: International Symposium on Computing in Science & Engineering, pp. 801–809 (2010)
10. Biswas, R.K., Masud, M.S., Kabir, E.: Shewhart control chart for individual measurement: an application in a weaving mill. In: Proceedings of the 2015 Melbourne International Business and Social Science Research Conference. Australian Academy of Business Leadership (2015)
11. Cronemyr, P.: DMAIC and DMADV-differences, similarities and synergies. Int. J. Six Sigma Compet. Adv. **3**(3), 193–209 (2007)
12. Dixit, P.M., Dixit, U.S.: Metal forming and machining processes. In: Modeling of Metal Forming and Machining Processes: by Finite Element and Soft Computing Methods, pp. 1–32 (2008)
13. Hespanha, J.P., Morse, A.S.: Stability of switched systems with average dwell-time. In: Proceedings of the 38th IEEE Conference on Decision and Control, vol. 3, pp. 2655–2660 (1999)
14. Jones, M.B.: Organizational culture and knowledge management: an empirical investigation of US manufacturing firms, Nova Southeastern University (2009)
15. Kabra, A., Agarwal, A., Agarwal, V., Goyal, S., Bangar, A.: Parametric optimization & modeling for surface roughness, feed and radial force of EN-19/ANSI-4140 Steel in CNC turning using taguchi and regression analysis method. Int. J. Eng. Res. Appl. (IJERA) **3**(1), 1537–1544 (2013)
16. Kackar, R.N.: Off-line quality control, parameter design, and the Taguchi method. In: Dehnad, K. (ed.) Quality Control, Robust Design, and the Taguchi Method, pp. 51–76. Springer, New York (1989)
17. Kackar, R.N.: Taguchi's quality philosophy: analysis and commentary. In: Dehnad, K. (ed.) Quality Control, Robust Design, and the Taguchi Method, pp. 3–21. Springer, New York (1989)
18. Kurt, M., Bagci, E., Kaynak, Y.: Application of Taguchi methods in the optimization of cutting parameters for surface finish and hole diameter accuracy in dry drilling processes. Int. J. Adv. Manufact. Technol. **40**(5), 458–469 (2009)
19. Ledolter, J., Swersey, A.: Dorian Shainin's variables search procedure: a critical assessment. J. Qual. Technol. **29**(3), 237 (1997)
20. Logesh, B., Aezhisai, M.S., Gandhi, N.M., Velmurugan, C.: Determining the influence of various cutting parameters on MRR In Lm6/Sic composites. Int. J. Indus. Eng. Des. **1**(1), 11–17 (2016)
21. Nalbant, M., Gökkaya, H., Sur, G.: Application of Taguchi method in the optimization of cutting parameters for surface roughness in turning. Mater. Des. **28**(4), 1379–1385 (2007)
22. Pamfilie, R., Petcu, A.J., Draghici, M.: The importance of leadership in driving a strategic Lean Six Sigma management. Procedia Soc. Behav. Sci. **58**, 187–196 (2012)
23. Pande, P.S., Pepper, M.P.J., Spedding, T.A.: The evolution of lean six sigma. Int. J. Reliab. Manage. **27**(2), 138–155 (2010)
24. Pande, P.S., Neuman, R.P., Cavanagh, R.R.: The Six Sigma Way: How GE, Motorola, and Other Top Companies Are Honing Their Performance. McGraw-Hill, New York (2000)
25. Prashar, A.: Using shainin DOE for six sigma: an Indian case study. Prod. Plan. Control **27**(2), 83–101 (2016)
26. Pyzdek, T., Keller, P.A.: The Six Sigma Handbook, p. 25. McGraw-Hill Education (2010)
27. Ramesh, V., Kodali, R.: A decision framework for maximizing lean manufacturing performance. Int. J. Prod. Res. **50**(8), 2234–2251 (2012)
28. Shainin, D., Shainin, P.: Better than Taguchi orthogonal tables. Qual. Reliab. Eng. Int. **4**(2), 143–149 (1988)

29. Shanmugasundaram, P., Vikram, P.: total quality management, process analytical technology, five basic principles and pharmaceutical industry: an overview. Total Qual. Manage. **8**(6), 178–185 (2015)
30. Sharma, S., Chetiya, A.R.: Simplifying the Six Sigma toolbox through the application of Shainin DOE techniques. Vikalpa **34**(1), 13–30 (2009)
31. Sin, A.B., Zailani, S., Iranmanesh, M., Ramayah, T.: Structural equation modeling on knowledge creation in Six Sigma DMAIC project and its impact on organizational performance. Int. J. Prod. Econ. **168**, 105–117 (2015)
32. Sokovic, M., Pavletic, D., Pipan, K.K.: Quality improvement methodologies–PDCA cycle, RADAR matrix, DMAIC, and DFSS. J. Achiev. Mater. Manufact. Eng. **43**(1), 476–483 (2010)
33. Solanki, M., Desai, D.: Comparative study of TQM and six sigma. Int. J. Indus. Eng. Technol. (IJIET), 2277–4769 (2015)
34. Verma, A.K., Srividya, A., Mannikar, A.V., Pankhawala, V.A., Rathanraj, K.J.: Shainin method: edge over other DOE techniques. In: Proceedings. 2004 IEEE International on Engineering Management Conference, vol. 3, pp. 1110–1113 (2004)
35. Yang, W.P., Tarng, Y.S.: Design optimization of cutting parameters for turning operations based on the Taguchi method. J. Mater. Process. Technol. **84**(1), 122–129 (1998)
36. Yusuf, M., Ariffin, M.K.A., Ismail, N., Sulaiman, S.: Surface quality and chip formation in turning of LM6 aluminum alloy and particulate reinforced metal matrix composite. Mater. Sci. Forum **773**, 894–901 (2014)
37. Zerti, O., Yallese, M.A., Belhadi, S., Bouzid, L.: Taguchi design of experiments for optimization and modeling of surface roughness when dry turning X210Cr12 steel. Appl. Mech. Behav. Mater. Eng. Syst. 275–288 (2017)

Research of Quasi-static Tests and Static Loading on Hybrid Adhesive Bonds

Martin Tichý(✉), Miroslav Müller, Vladimír Šleger, and Petr Valášek

CULS, Department of Material Science and Manufacturing Technology, Kamýcká 129, 165 00 Praha 6, Suchdol, Czech Republic
martintichy@tf.czu.cz

Abstract. The research was focused on quasi-static tests of adhesive bonds based on modified epoxy resin by adding filler in microparticle form reached by treatment from seed pressing process Jatropha Curcas L. The service life and mechanical properties (the tensile strength and elongation at break) of adhesive bond were evaluated for quasi-static test by exposure to 1000 cycles with loading 5–30% (165–989 N) and 5–70% (165–2307 N) of adhesive bond static tensile strength without modification. The results of hybrid adhesive bonds (with filler) did not prove significant increase in adhesive bond tensile strength but proved an increase in the fatigue strength of adhesive bonds. Research results proved that the service life of adhesive bond decreased at loading 70% of adhesive bond static tensile strength, i.e. 100 to 400 cycles according to filler concentration.

Keywords: Epoxy · Biological reinforcement · Jatropha Curcas L. · Adhesive bond service life · Single lapped adhesive bond

1 Introduction

Adhesive bonding technology is perspective method of bonding various materials. Adhesives can achieve good mechanical properties in comparison with other bonding technology. When adhesives are exposed to degradation aspects or cyclic stress, which is typical for practical application, there is a significant decrease in mechanical properties and often adhesive bonds do not perform a function [1, 2]. The cyclic fatigue stress significantly influences the integrity of adhesive bonds strength at low periodically loads and causes deformations of adhesive and cohesive interactions [1, 2]. This adhesive layer degradation decreases the final static strength or service life of an adhesive bond [3–5].

Adhesives can be modified by adding a filler [6, 7]. It is necessary to include e.g. type of filler and its shape and dimension, chemical properties etc. Utilization of biologically based filler is a trend in the field of material engineering. The research results confirmed significant influence on mechanical properties of composite material by adding biological filler [8–11]. The global topic is biological waste production and possibilities of material utilization. Series of biological wastes are burned and use as fertilizer or otherwise disposed of without the possibility of alternative use. The material utilization of biologically based wastes is positive due to environmental and new possibilities in the

J. Machado et al. (Eds.): icieng 2021, LNME, pp. 147–154, 2022.
https://doi.org/10.1007/978-3-030-79165-0_14

field of material engineering [12–15]. Fillers such as post-harvest line residues, bamboo powder, powders from fruit stones, fibres of sisal, flax, hemp, banana tree, jute etc. are used in microparticle, short and long fibre base [7, 14, 16–18]. It is well known that biological fillers can optimise mechanical properties [6–12, 19].

In the research was used microparticle filler reached by treatment of press cake from oil pressing of seed Jatropha Curcas L. This plant produce oil seeds. The waste from pressing process cannot be used for animal feeding. The example of effective utilization of this waste is to improve tribological properties [8, 20, 21].

The aim of this paper is to investigate the mechanical properties of an adhesive bond based on a modified epoxy resin with filler from seed pressing process of Jatropha Curcas L. Quasi-static tests at various loads were performed in order to establish the fatigue strength in context of the adhesive bond service life.

2 Materials and Methods

In this work, adhesive bonds with epoxy adhesive Lepox 1200 and composite based on epoxy adhesive Lepox 1200 with filler in microparticle form from oil pressing of seed Jatropha Curcas L. (whole seed – kernel, shell) were used. The filler was crushed in Retsch MM 44 crushing machine. Then the microparticles were dried in temperature 105 ± 5 °C for 24 h. The microparticle size was: Arithmetic mean 53.74 μm, Median 47 μm, Mode 35 μm. The microparticles were measured by Gwyddion programme from SEM images. Figure 1 presents microparticle filler which was used in the composite material. The composite adhesive layer was prepared by mixing 10 g of two-components epoxy adhesive Lepox 1200 and 1 g of filler (marked C1), 2 g of filler (marked C2) and 3 g of filler (marked C3). Adhesive bonds were prepared according to ČSN EN 1465 standard. Bonded material was structural carbon steel S235J0. Adhesive bonds with epoxy Lepox 1200 are marked ML1200. Adhesive bonds with composite adhesive are marked Cx-ML1200-JC (x – filler mass in grams, JC - Jatropha Curcas L.). The adhesive bonds loaded by the static tensile test are marked ST, by the quasi-static test are marked Q and the number indicates loading value (%, N). The bonded material was mechanically blasted by Garnet MESH80 and chemically cleaned in Acetone. The surface roughness of structural carbon steel S235J0 was Ra = 1.52 ± 0.15 μm, Rz 10.62 ± 0.72 μm. The adhesive bonds were load with 750 g and cured for 72 h in room temperature. The mechanical properties were measured on universal strength testing machine LABTest 5.50ST (sensing unit AST type KAF 50 kN, evaluating software Test & Motion) in room temperature.

The quasi-static test measured adhesive bonds on shear tensile strength and elongation at break. A quasi-static parameter is determined by the static tensile test according to ČSN EN 1465 standard from 6 adhesive bonds by the test speed 0.6 mm.min^{-1} (the reference value was determined from test samples without added filler JC). The quasi-static test was loaded with various percentage of the static tensile strength with 1000 cycles and 6 mm.min^{-1} test speed between. The quasi-static loading interval was between 5%, 30% (165–989 N) and 70% (165–2307 N) of the static tensile strength. A stability on the lower and higher limit was determined at 0.5 s. The strength reference value was determined from static tensile tests results of adhesive bond without filler (ML 1200). The

Fig. 1. SEM images of microparticle filler based on press cake from oil pressing of seed Jatropha Curcas L.

fatigue strength was determined after finishing 1000 cycles by the speed 0.6 $mm.min^{-1}$ until a total failure of the adhesive bond [5].

3 Results and Discussion

The cyclic stress negatively influences adhesive bonds at low loading values already and causes layer delamination of adhesive bonds which caused strength decrease of the adhesive bond [1, 2, 5, 22]. The strength and deformation results of adhesive bonds at static and quasi-static test are presented in Fig. 2, 3 and Table 1. The static test proved that the static tensile strength increases up to 4% and elongation up to 27% by adding 1 g of the filler in the adhesive. The static tensile strength decreases up to 8–10% by increasing filler concentration, i.e. 2 and 3 g of the filler. The elongation increases up to 6–8%. A similar increase in strength with low concentration of microparticle filler was also reported in literature [23].

The quasi-static test proved that the filler JC increase the resistance of adhesive bonds against the long-term cyclic loading. The best results of quasi-static test were achieved with the adhesive bonds C1-ML1200-JC (1 g of filler JC). The adhesive bonds could resist the quasi-static test at 5–30%, i.e. the loading in interval 165 to 989 N. The adhesive bonds resisted this cyclic loading interval with 1000 cycles which is evident from Table 1. On the other hands, the adhesive bonds couldn´t resist the quasi-static test with 1000 cycles at 5–70%, i.e. the loading in interval 165 to 2307 N. The cycle number that adhesive bonds could absorb is evident from Table 1. Again, the best results were achieved with the adhesive bonds C1-ML1200-JC (1 g of filler JC), which resisted until 406 cycles in interval 165 to 2307 N. The number of finished cycles was different according to adhesive layer.

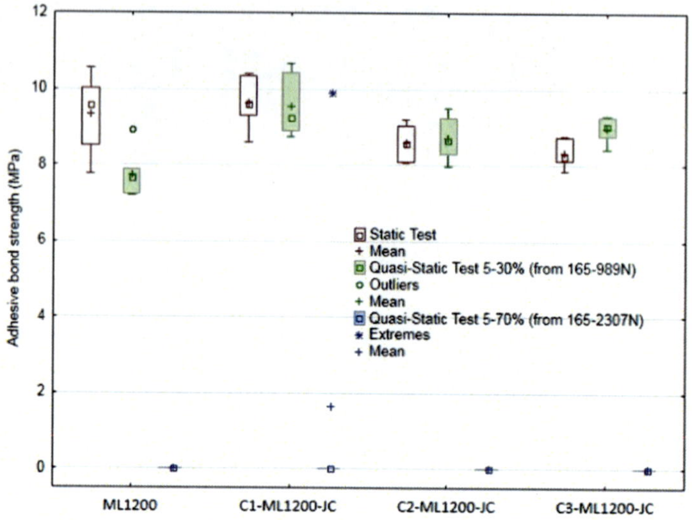

Fig. 2. Research results – adhesive bond strength

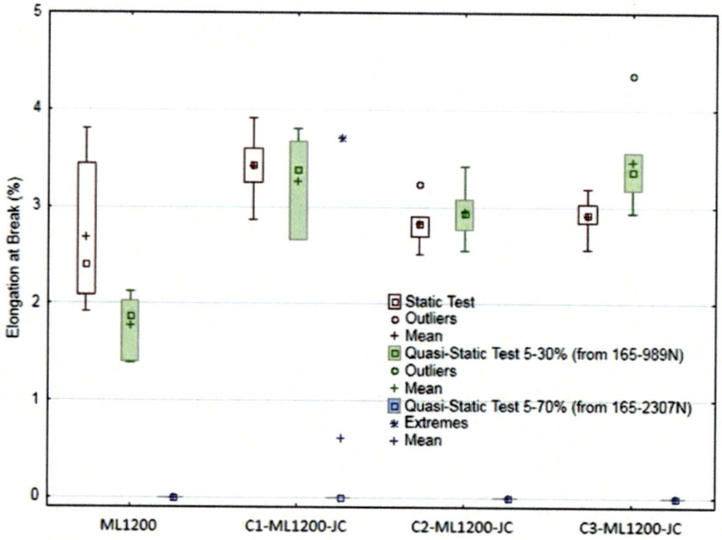

Fig. 3. Research results – elongation at break of adhesive bond

The loading of the adhesive layer with filler JC (composite hybrid layer) by the quasi-static test at 5–70%, i.e. the loading in interval 165 to 2307 N, caused a viscoelasticity behaviour, i.e. the creep. Between the first and last cycle before destruction of adhesive, deformation occurs inside the adhesive layer. The viscoelasticity behaviour of adhesive layer is evident from Table 1, where is an influence of increasing values of cyclic loading force. The relative deformation after finishing the first cycle increased up to 1.26% and

Table 1. Results of quasi-static test of adhesive bond

Characteristics of adhesive bond/Quasi-static test	Number of cycles	Number of test samples (number of finished cycles/total number of tests)	Relative deformation after finishing 1st cycle (%)	Relative deformation after last cycle (%)
ML1200/QS5-30% (165–989 N)	1000 ± 0	6/6	0.14 ± 0.06	0.15 ± 0.06
ML1200/QS5-70% (165–2307 N)	266 ± 169	0/6	0.35 ± 0.07	0.50 ± 0.09
C1-ML1200-JC/QS5-30% (165–989 N)	1000 ± 0	6/6	0.81 ± 0.03	0.83 ± 0.05
C1-ML1200-JC/QS5-70% (165–2307 N)	406 ± 286	1/6	1.21 ± 0.06	1.41 ± 0.11
C2-ML1200-JC/QS5-30% (165–989 N)	1000 ± 0	6/6	0.90 ± 0.05	0.92 ± 0.05
C2-ML1200-JC/QS5-70% (165–2307 N)	159 ± 92	0/6	1.26 ± 0.07	1.42 ± 0.08
C3-ML1200-JC/QS5-30% (165–989 N)	1000 ± 0	6/6	0.93 ± 0.10	0.97 ± 0.10
C3-ML1200-JC/QS5-70% (165–2307 N)	99 ± 137	0/6	1.39 ± 0.09	1.68 ± 0.22

relative deformation after last cycle increased up to 1.42%. The adhesive bonds without filler modification do not have this type of behaviour. Similar behaviour was reported for adhesive bonds modified by microfibres and microparticles of cotton [5].

The research results confirmed hypothesis, that cyclic loading of adhesive with higher value of loading force up to 2307 N (the maximal average force was 3296 ± 313 N) caused adhesive deformation at relatively low number of cycles. This hypothesis states results of authors with different adhesives and fillers type [3, 5].

Results of the statistical testing with ANOVA F-test are evident from Table 2. Hypothesis H0 presents a statistically insignificant difference among measured data ($p > 0.05$) and a hypothesis H1 presents a refusal of the hypothesis H0, i.e. there is the statistically significant difference among measured data ($p < 0.05$).

A progress of quasi-static test is evident on Fig. 4. Figure 4 A presents quasi-static test with 1000 cycles at 5–30% (165–989 N). After that, adhesive bond was loaded until destruction (without taking a sample from the machine between the last cycle and the test until destruction). Figure 4B presents Quasi-static test with unfinished 1000 cycles at 5–70% (165–2307 N). The adhesive bond resisted 471 from 1000 cycles. After 471 cycles, the destruction has occurred.

Based on results of the research it is possible to agree with statement that the filler in adhesive can decrease negative influence of cyclic loading on adhesive bonds [4].

Table 2. Statistical evaluation of static tensile test according ANOVA F-test with stated parameter p in significance level α 0.05

Tensile test	Static Test	Quasi-Static Test 5–30% (165–989 N)	Quasi-Static Test 5–70% (165–2307 N)
Tensile strength (MPa)	0.0116	0.0007	0.4133
Elongation at break (%)	0.0555	0.0000	0.4123

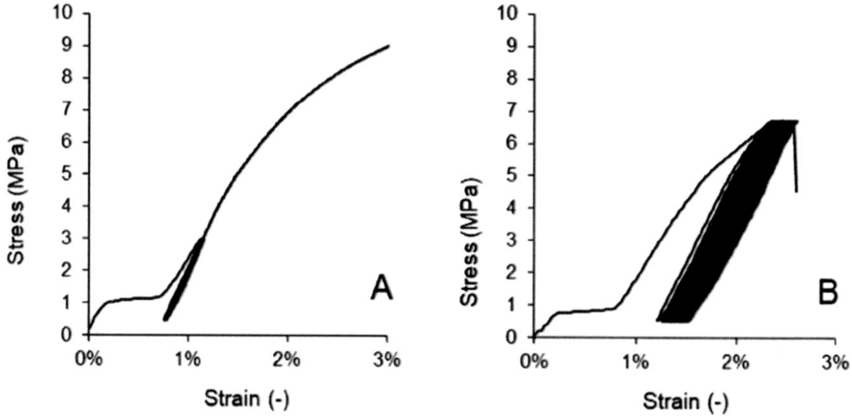

Fig. 4. Quasi-static test ML1200: A The test with 1000 cycles at 5 to 30% (165–989 N) and destruction test after the period, B: The test with unfinished 1000 cycles at 5 to 70% (165–989 N) – 471st cycle finished.

4 Conclusions

Results of the research proved significant influence of low cyclic fatigue on the service life and mechanical properties of the adhesive bond. It is not possible to eliminate cyclic stress at adhesive technology application. One option for fatigue elimination is creating a hybrid adhesive layer, i.e. adding a filler into the adhesive.

A significant increase at the static tensile strength of adhesive bonds with the microparticles of press cake from oil pressing of seed Jatropha Curcas L. (JC) was not proved. A slight increase of the static tensile strength occurred at lower concertation of the filler, i.e. C1-ML1200-JC 10 g of adhesive and 1 g of filler). However, the static test results proved an increase at the elongation at break.

From the quasi-static test results is evident that the microparticles of press cake from oil pressing of seed Jatropha Curcas L. (JC) in the adhesive increased resistance against a long-time cyclic stress. The viscoelasticity properties (creep) of the adhesive were significantly demonstrated at cyclic stress value in interval from 5% to 70%, i.e. the loading force in interval 165 N to 2307 N. During this test an adhesive destruction occurred, i.e. the 1000 cycles was unfinished. The viscoelasticity properties were not demonstrated at cyclic stress value in interval from 5% to 30%, i.e. the loading force in interval 165 N to 989 N.

Acknowledgments. Supported by Internal grant agency of Faculty of Engineering, Czech University of Life Sciences Prague.

References

1. Hafiz, T.A., Wahab, M.A., Crocombe, A.D., Smith, P.A.: Mixed-mode fracture of adhesively bonded metallic joints under quasi-static loading. Eng. Fract. Mech. **77**(17), 3434–3445 (2010)
2. Kelly, G.: Quasi-static strength and fatigue life of hybrid (bonded/bolted) composite single-lap joints. Composite Struct. **72**(1), 119–129 (2006)
3. Broughton, W.R., Mera, R.D., Hinopoulos, G.: Project PAJ3 - combined cyclic loading and hostile environments 1996–1999. In: Cyclic Fatigue Testing of Adhesive Joints Test Method Assessment. Environments (1999)
4. Messler, R.W.: Joining of materials and structures: from pragmatic process to enabling technology. Elsevier (2004)
5. Müller, M., et al.: Material utilization of cotton post-harvest line residues in polymeric composites. Polymers **11**(7), 1106 (2019)
6. Valášek, P., Ruggiero, A., Müller, M.: Experimental description of strength and tribological characteristic of EFB oil palm fibres/epoxy composites with technologically undemanding preparation. Compos. Part B Eng. **122**, 79–88 (2017)
7. Müller, M., Valášek, P., Rudawska, A.: Mechanical properties of adhesive bonds reinforced with biological fabric. J. Adhes. Sci. Technol. **31**(17), 1859–1871 (2017)
8. Ruggiero, A., D'Amato, R., Merola, M., Valašek, P., Müller, M.: Tribological characterization of vegetal lubricants: Comparative experimental investigation on Jatropha curcas L. oil, Rapeseed Methyl Ester oil, Hydrotreated Rapeseed oil. Tribology Int. **109**, 529–540 (2017)
9. Zieleniewska, M., Leszczyński, M.K., Szczepkowski, L., Bryśkiewicz, A., Krzyżowska, M., Bień, K., Ryszkowska, J.: Development and applicational evaluation of the rigid polyurethane foam composites with eggshell waste. Polymer Degra. Stabil. **132**, 78–86 (2016)
10. Cheung, H.Y., Ho, M.P., Lau, K.T., Cardona, F., Hui, D.: Natural fibre-reinforced composites for bioengineering and environmental engineering applications. Compos. Part B Eng. **40**(7), 655–663 (2009)
11. Satheesh Kumar, M.N., Yaakob, Z., Mohan, N., Kumaresh Babu, S.P.: Mechanical and abrasive wear studies on biobased jatropha oil cake incorporated glass–epoxy composites. J. Am. Oil Chemists' Soc. **87**(8), 929–936 (2010)
12. Yan, L., Kasal, B., Huang, L.: A review of recent research on the use of cellulosic fibres, their fibre fabric reinforced cementitious, geo-polymer and polymer composites in civil engineering. Compos. Part B Eng. **92**, 94–132 (2016)
13. Yan, L., Chouw, N., Jayaraman, K.: Flax fibre and its composites–A review. Compos. Part B Eng. **56**, 296–317 (2014)
14. Müller, M., Valášek, P., Ruggiero, A.: Strength characteristics of untreated short-fibre composites from the plant ensete ventricosum. BioResources **12**(1), 255–269 (2017)
15. Bajpai, S.K., Mary, G., Chand, N.: The use of cotton fibers as reinforcements in composites. In: Biofiber Reinforcements in Composite Materials, pp. 320–341, Woodhead Publishing (2015)

16. Aziz, S.H., Ansell, M.P.: The effect of alkalization and fibre alignment on the mechanical and thermal properties of kenaf and hemp bast fibre composites: Part 1–polyester resin matrix. Compos. Sci. Technol. **64**(9), 1219–1230 (2004)
17. Lu, X., Zhang, M.Q., Rong, M.Z., Shi, G., Yang, G.C.: Self-reinforced melt processable composites of sisal. Compos. Sci. Technol. **63**(2), 177–186 (2003)
18. Rao, K.M.M., Rao, K.M.: Extraction and tensile properties of natural fibers: Vakka, date and bamboo. Compos. Struct. **77**(3), 288–295 (2007)
19. Yan, L., Kasal, B., Huang, L.: A review of recent research on the use of cellulosic fibres, their fibre fabric reinforced cementitious, geo-polymer and polymer composites in civil engineering. Compos. Part B Eng. **92**, 94–132 (2016)
20. Shivamurthy, B., Murthy, K., Joseph, P.C., Rishi, K., Bhat, K.U., Anandhan, S.: Mechanical properties and sliding wear behavior of jatropha seed cake waste/epoxy composites. J. Mater. Cycles Waste Manage. **17**(1), 144–156 (2015)
21. Mohan, N., Natarajan, S., KumareshBabu, S.P.: Investigation on sliding wear behaviour and mechanical properties of jatropha oil cake-filled glass-epoxy composites. J. Am. Oil Chem. Soc. **88**(1), 111–117 (2011)
22. Šleger, V., Müller, M.: Quasi static tests of adhesive bonds of alloy AlCu4Mg. Manuf. Technol. **15**, 694–698 (2015)
23. Cho, J., Joshi, M.S., Sun, C.T.: Effect of inclusion size on mechanical properties of polymeric composites with micro and nano particles. Compos. Sci. Technol. **66**(13), 1941–1952 (2006)

Technology Foresight to Enable New R&D Collaboration Partnerships: The Case of a Forestry Company

José Coelho Rodrigues[1](✉) and Vinicius Delfim[2]

[1] Faculty of Engineering of the University of Porto and INESC TEC, Porto, Portugal
jpcr@fe.up.pt
[2] School of Economics of the University of Porto, Porto, Portugal

Abstract. The increasing need to innovate to keep or create competitive advantage and the difficulty in retaining resources that enable innovation, force companies to look for partnerships. Partnering with the appropriate and leading R&D (Research and Development) institutions is critical to be able to innovate. One of the first challenges that emerges when building a collaboration is the selection of organizations to form the consortium. Technology foresight might be an interesting process to start building such partnership, as it helps identify technology opportunities and the entities that lead technology development.

This paper uses case research to study the creation and use of a technology foresight process to build new R&D collaborations, while identifying of technology development opportunities. Findings from the case study led to the identification of how the technology foresight process was used to create a new R&D collaboration in the company under study, while identifying what motivated such collaboration and how it was managed. Furthermore, important characteristics of the team responsible for that process and of the process management are highlighted.

Keywords: Technology foresight · Technology collaboration · R&D collaboration partnership

1 Introduction

Companies increasingly need to innovate in order to create competitive advantage. Furthermore, there is an increasingly scarce availability of resources that are able to contribute to such innovation and, therefore, companies face severe difficulties in retaining those resources. Conducting appropriate technology foresight efforts and partnering with other players in the innovation ecosystem, namely R&D (research and development) institutions, is critical to overcome such difficulties and be able to innovate and develop the desired competitive advantage.

According to Nieto and Rodrigues [1], the exploration of knowledge external to the company, despite the barriers and complexity of its transfer, is an interesting source of innovation for companies. In the 1990s, the Triple Helix concept was developed by Henry Etzkowitz and Loet Leydesdorff [2], pointing to three innovation actors and their

J. Machado et al. (Eds.): icieng 2021, LNME, pp. 155–163, 2022.
https://doi.org/10.1007/978-3-030-79165-0_15

respective interactions, namely: university-industry-government. This concept has been evolving in recent decades [3]. Over time, according to Augustinho and Garcia [4], the interaction model of the main actors of innovation has changed and demonstrates, regardless of criticism, that the relationship between its elements is necessary for technological knowledge generated in universities to be transferred to companies and reach the whole society.

R&D collaboration projects to develop technology can include the participation of different institutions, such as companies, associations, research institutes, universities, and governments. The R&D collaboration process involves understanding the actors, their respective strategies and the convergence of interests in the scope of innovation.

This paper focuses on collaborations between companies and R&D institutions built based on technology foresight efforts and the identification of technology development opportunities. It starts by recognizing the process of identifying technology development opportunities, based on the existing literature and a case study of an organization that is organizing itself to use technology foresight to build technological collaborations and foster technological innovation. Afterwards, based on the case study, the emergence of new R&D collaborations is identified during that process, highlighting when and how they emerge, what motivates such collaborations and how they are managed.

2 Theoretical Background

R&D collaborations consists of a consortium of organizations formed to innovate together. Such collaborations, and particularly the R&D projects emerging from them face numerous challenges due to the complexity of having several and different types of organizations involved. One of the first challenges that emerges when building such collaboration projects is the matchmaking process among organizations, to select the organizations to form an initial consortium. Individuals play a particularly important role in this matchmaking process, as that is mostly a relational process where built-up trust is crucial [5]. Therefore, learning dynamics and stabilizing teams of collaboration among organizations from one project to subsequent projects is important to form trust, bridge time periods of latency, and completely integrate new ideas into established series of projects [6], which supports the idea of creating R&D collaboration partnerships. These, and other challenges that emerge during collaborative R&D projects, are usually managed using appropriate project management tools [7]. However, some adaptations might have to be made in comparison to the standard formal project management approach proposed by the new product development literature, as that approach seems not to be the most appropriate for science-based partners [8]. Particularly concerning building collaborative R&D partnerships, considering the need to find appropriate partners to develop technological innovations, technology foresight might be an interesting process to support building such partnership.

According to Martin [9], technology foresight might be defined as a process for looking into long-term technology and science evolution to identify areas of promising strategic research and technological development, which are likely to bring economic and social benefits. The technological foresight activity, according to Andrade, Chimendes, Rosa, Silva and Chagas Jr. [10], can be defined as a systematic means of mapping scientific and technological developments. These developments are capable of significantly influencing an organization, an industrial sector, a specific product or process, or the economy or the society.

Following Bahruth, Antunes, and Bomtempo [11], technology foresight can be divided into four stages:

1. Preparatory stage: including the definition of goals for the technology foresight, as well as its scope, approach, and methodology
2. Pre-prospective stage: including surveying data sources and defining the methodology to be used
3. Prospective stage: including conducting the prospection, i.e. the collection, treatment and subsequent analysis of the data
4. And, post-prospection stage: communicating the results to internal and external stakeholders and implementing actions that result from the previous efforts.

Popper [12] proposed Foresight Diamond, a tool that classifies methods of technological foresight into two groups according to their basic attributes, namely their nature (quantitative, qualitative, or semi-quantitative) and the capabilities they enhance (creativity, expertise, interaction, and evidence). It is expected that when applying methods that enhance the interaction capability new collaboration projects may emerge from such interaction with other institutions.

Tomioka, Lourenço and Facó [13] argue that technology foresight activities can be used in academic or business contexts to: anticipate technological changes, understand technological evolution, provide support for decision making in R&D, support the technology protection process, and assist in the commercialization of technologies.

Because of the different prospecting approaches, Caruso and Tigre [14] propose a classification of prospecting approached, according to its goals:

- Assessment: action carried out systematically, consisting of the act of following the evolution of facts and the identification of events with changes
- Forecasting: making projections based on historical information and modeling trends
- Foresighting: the anticipation of future possibilities based on unstructured interaction between specialists, each one supported exclusively by their knowledge and subjectivities.

A fundamental aspect of technology foresight is the establishment of technological collaboration with identified partners. Collaboration is of fundamental importance for innovation, since companies alone may, often, have difficulties in gathering all the necessary skills to implement new products or processes [15, 16].

3 Case Study

The case study focuses on the Technological Center (TC) of a Brazilian forestry company, the global leader in its industry. The TC organizational structure reports directly to the CEO (Chief Executive Officer) of the company and consists of a director and seven managers, six of them technical, distributed according to the different parts of the company's value chain, and a cross-sectional manager focused on activities including innovation projects, financial management, intellectual property, funding, and technology cooperation. The TC has facilities in four countries, with laboratories, pilot plants, a portfolio of several patents, and highly qualified staff.

Technologies developed in TC are mostly in the biotech segment, biorefinery, and consumer goods. The TC already collaborates with Brazilian research institutes and universities in several technological development projects. However, the organization has been recently exploring methodologies to adapt the TC for a better and broader technological foresight process, and, as a result of that adaptation, started approaching other R&D institutions (from different countries) to start collaborating with. It was, therefore, possible to conduct a longitudinal study of these adaptations and the emergence of a new collaboration partnership, which led to particularly interesting results.

4 Methodology

The methodology found to be most appropriate for this exploratory study was case research [17, 18] because it provides a systematic and in-depth approach to study real-life settings [18]. A single case study was carried out, where the researchers conducted interviews with the director and seven managers involved in the process under study (the definition of a new technology foresight process focused on the creation of a new R&D collaboration), and observed the changes as they occurs *in situ*, taking careful notes about what was observed. A semi-structured questionnaire was prepared for the interviews, and extensive notes were taken during the interviews.

Data was carefully organized in the specific topics under study: the process being created, the timing, reason and motivation of collaborations that emerged, how those collaborations emerged (as part of the process being created), and how was the process managed.

Data was then confirmed be using reports that were supporting the restructuring of the company under study, and informal conversations with people involved in and affected by the process. Furthermore, some of the people of contact in the R&D institution that started to collaborate with the company were also interviewed.

The development of this work was structured in four stages (Fig. 1). The first stage was dedicated to alignment of the gaps in the literature with the context of the company under study, and consisted in a literature review and collection of information concerning the company under study and its context of operations. The second stage was dedicated to planning the development of the study. In the third stage, activities were concentrated on data collection and discussion among the research team of learning assimilated up to that point. The fourth stage consisted on the analysis of results and recommendations about using technology foresight to establish new collaboration partnerships.

The literature review was carried out using reliable scientific databases, namely SCOPUS and Web of Science, and focusing on finding papers about technological exploration/foresight, technological innovation, and technological cooperation. Moreover, additional information was sought in publications pointed out in the papers found as relevant to the topic and relevant to the object of study.

The analysis of the company's documents was focused on the organization's innovation management system and on how the technological cooperation team operates

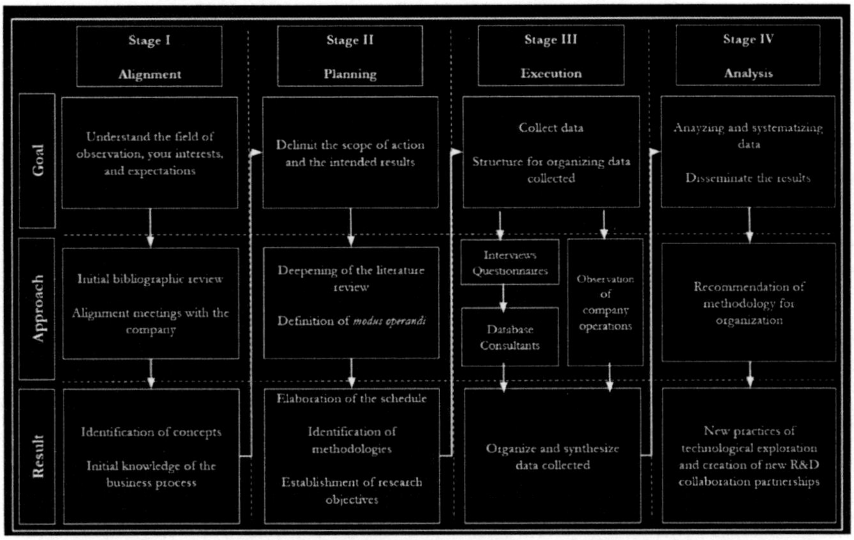

Fig. 1. Research stages.

within the management of the technological innovation board. From this analysis, it was possible to obtain an overview of the process and raise questions that were addressed to the TC director in a semi-structured interview carried out later, within the scope of the following steps.

This research was carried out very close to the TC director, who is responsible to make implementation and development decisions regarding the tool to support the technology foresight and cooperation activities. A formal interview was performed with this director and complemented with frequent informal contacts due to the presence of a researcher in the company. Furthermore, interviews were conducted with the other TC managers involved in the technology foresight and cooperation activities.

In addition to the data collected within the company (through interviews and analysis of documentation), the results were continuously being confronted with the previous literature review and with continuous updates to that review, particularly focusing on identifying methodologies relevant to the technological foresight for establishing the organization's technological cooperation.

5 Results

Findings from the case study led to the identification of the technology foresight process used in the company under study. That process, which is later presented in this section based on the technology foresight process proposed by Bahruth et al. [11], was focused on the creation of a new R&D collaboration together with the identification of technology development opportunities. Finally, it was possible to identify when and how collaborations emerge in the process and assess what motivates such collaborations and how they are managed. The most relevant findings are presented in the following paragraphs, initiated by a contextualization of the emergence of the particular technology foresight process under study.

The technology foresight process studied in this work emerged in a company that is restructuring its processes and creating a new technology foresight division/team. Such restructuring is being led by the top management of the company, who provide the broader view of the business and its positioning in the sector (Fig. 2). The strategic orientation defined by the top management is used by the R&D team and the Intellectual Property team, using literature review combined with patent analysis, to assess the state of the art of the technologies of interest for the company. Then, a panel of experts was consulted to contribute for that assessment. This initial restructuring effort is closed by the elaboration of a technological roadmap that then feeds a set of goals to be addressed by TC managers, who later start new prospecting cycles more focused on those defined objectives. One of such cycles was the focus of this analysis and is described next.

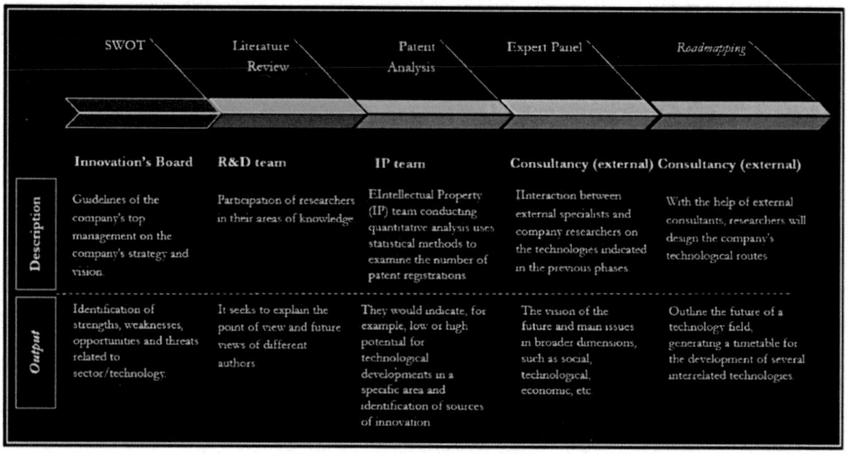

Fig. 2. Recommendation of methods

The technology foresight process studied started with the identification of R&D institutions that developed technologies of interest to the company. Firstly, the goals for the prospection process were defined (step 1 – preparatory stage of Bahruth et al. [11]), which were already focused on finding partners to develop new technology. Furthermore, in this first step methodologies to be used during the process were defined. A combination of benchmark with networking was used (step 2 – pre-prospective stage [11]) to identify an R&D institute of interest to partner with (in this case it was a university), and a person in charge to establish a first contact. Next, a prospection for the possibility of the collaboration was conducted (step 3 – prospective stage [11]), by contacting and meeting with that person, and visiting on-campus facilities. Step four (post-prospection stage [11]) consisted in the use of surveying techniques to identify research topics of interest for both partners (company and university) that would then be used to define the collaboration strategy. Afterwards, the company's prospection team interacted with the person responsible for formalizing new collaborations, and a Non-Disclosure Agreement (NDA) was elaborated and signed between the company and the university to start creating R&D projects. The company and the university started then to align their interests and designed a collaboration project that met both needs and contributed to the motivations of each partner to collaborate: innovation for the company and scientific advance with empirical evidence for the university.

To create a successful technology foresight method the company created a new team dedicated to that process (the prospection team). Such team had to be able to communicate with several departments of the organization, particularly R&D, financial department, and legal department, which are the most important internal partners to carry out collaboration projects with external institutions focused on technology development. The team had also to be able to easily interact with important external stakeholders for the innovation process, including government, universities, and other companies. It was important that the team members were comfortable in considering combining methods for technology foresight, according to the time horizon for prospection and the expected output of the process. For instance, if the goal is to develop collaborative R&D projects, as was the case of the prospection undertook while conducting this case study, patent analysis, benchmarking, and/or technological networking might be the most appropriate methods to be used. Furthermore, the team members should also feel comfortable with adopting standards, such as the TRL (technology readiness level) for measuring the maturity of the technology, and with conforming with industry certifications in general, which can facilitate the establishment of a new collaborations by exhibiting compliance with standards and evidencing quality of operations.

The new collaboration was the main objective of the technology foresight process, i.e., to find partners to develop new technologies that could bring a competitive advantage for the company. This purpose for the technology foresight process shaped that process towards becoming more aligned with developing a collaboration with an R&D institute. However, the collaboration itself, i.e., the agreement between both parties to work together emerged in step 3 and was only consolidated at the end of step 4, with the non-disclosure agreement signing. The possibilities of entities to collaborate with were carefully analyzed before step 3, considering the motivation for that collaboration, i.e., find a partner recognized as a leader in technology development in the sector where the company operates that could be interested in and benefit from partnering with the company. Then, after identifying the desired partner and the right person to talk with within that partner, efforts to confirm the conditions from the partner in collaborating and its interest in collaborating were carefully conducted. These efforts already involved both parties, which is why they are considered the beginning of the partnership. Afterwards, diligences were taken to make sure that the partnership was formalized, and strategy and objectives were defined for such partnership. And, from that point on, the collaboration started with a first definition of a collaborative project.

6 Discussion and Conclusion

This paper analyses the contribution from a case of formulating a collaboration between a company and the academy focused on the creation of a new R&D collaboration together with the identification of technology development opportunities. The case occurred in a company that is restructuring its departments towards becoming more innovative. In this context, the company is looking to establish new partnerships with R&D institutions and decided to use the technology foresight process for the prospection of institutions to collaborate with.

From the case it is possible to systematize the practices that led to a successful integration of the technology foresight process with the prospection of new R&D collaboration partners. This case shows that using the technology foresight process (as the one proposed by Bahruth et al. [11]) to establish a first contact for technology cooperation with R&D institutions, may increase its probability of success. The process was used to prospect for the most appropriate institution to partner with, and to prepare and manage the first contact with that institution in detail, helping to focus on the benefits for both partners. It seems to be very important for this process to be led by a specific team within the company to ensure its continuity, while coordinating the different roles of each external and internal agent involved in the process. Considering adhering to certifications (standards), such as a TRL scale to measure the maturity of technology and other certifications of the sector in which it operates, is particularly important for companies and R&D institutions to be seen as reliable by potential partners, and provide access to highly qualified networking. Management profile used throughout the process should be adapted to the preferential management approach used in the company: top-down or bottom-up. The process should therefore be designed to be started by some guidance from top-management to start prospection (top-down) or to be started by prospecting technology to then deliver a recommendation to the organization's top management (bottom-up). Motivations of both partners should also be carefully considered and addressed while establishing the collaboration partnership. Finally, the team managing this whole process, must be particularly comfortable with using technology foresight methods, as the members of that team should be able to select the most appropriate method or set of methods to be used in the context of the cooperation being established, particularly for the prospection stage of the process. All these practical recommendations are not particularly disruptive for the scientific knowledge we have so far, and are very much aligned with common-sense practices. However, the use of technology foresight process to guide the process of establishing a new technology cooperation is shown to be very useful, which provides evidence for a new setting of use and purpose for such kind of processes, which might be interesting to address in future research.

This study had some limitations related to the nature of case research. Particularly regarding availability of professionals from the company to be interviewed and changes in the hierarchy of the company while the study was conducted, implying further data collection efforts. Furthermore, confidentiality concerns involving not mentioning departments' names, professionals' names and positions, strategic technologies, and other details about management were extremely restrictive for data analysis.

Future research should now focus on the implementation of the methodologies indicated in this work in other contexts of use for a better understanding of what may define the development of an appropriate technological exploration methodology for the company that uses it.

Acknowledgments. This work is financed by National Funds through the Portuguese funding agency, FCT - Fundação para a Ciência e a Tecnologia, within project UIDB/50014/2020.

References

1. Nieto, M.J., Rodrigues, A.: Offshoring of R&D: looking abroad to improve innovation performance. J. Int. Bus. Stud. **42**(03), 345–361 (2011)
2. Leydesdorff, L., Etzkowitz, H.: The Triple Helix as a model for innovation studies. Sci. Public Policy **25**(3), 195–203 (1998)
3. Stanford University – Triple Helix Research Group. https://triplehelix.stanford.edu/3helix_concept. Accessed 20 Jan 2020
4. Augustinho, E.O., Garcia, E.N.: Inovação, transferência de tecnologia e cooperação. Revista Direito Desenvolvimento **09**(01), 223–239 (2018)
5. Holzmann, T., Sailer, K., Katzy, B.R.: Matchmaking as multi-sided market for open innovation. Technol. Anal. Strat. Manage. **26**(6), 601–615 (2014)
6. Manning, S.: The rise of project network organizations: Building core teams and flexible partner pools for interorganizational projects. Res. Policy **46**(8), 1399–1415 (2017)
7. Munkongsujarit, S., Srivannaboon, S.: Managing open innovation: a case study of the National Science and Technology Development Agency (NSTDA) in Thailand. In: The 2017 Portland International Conference on Management of Engineering and Technology (PICMET), pp. 1–9. IEE – Institute of Electrial and Electronics Engineers, Portland (2017)
8. Du, J., Leten, B., Vanhaverbeke, W.: Managing open innovation projects with science-based and market-based partners. Res. Policy **43**(5), 828–840 (2014)
9. Martin, B.R.: Foresight in science and technology. Technol. Anal. Strat. Manage. **7**(02), 139–168 (1995)
10. Andrade, H.S., Chimendes, V.C.G., Rosa, A.C.M., Silva, M.B., Chagas Jr., M.F.: Prospecting and technological readness level to support R&D activities. ESPACIOS Rev. **39**(08), 12–26 (2018)
11. Bahruth, E.B., Antunes, M.A.S., Bomtempo, J.V.: Prospecção Tecnológica na Priorização de Atividades de C&T: caso Q-Trop_Tp. In: Gestão em Biotecnologia vol. 1(18), pp. 300–324 (2006)
12. Popper, R.: Foresight methodology. In: Georghiou, L., Harper, J.C., Keenan, M., Miles, I. (eds.) The Handbook Of Technology Foresight, Concepts and Practices, 1st edn. Edward Elgar Publishing Ltd., Cheltenham (2008)
13. Tomioka, J., Lourenço, S., Facó, J.F.: Patentes em nanotecnologia: prospecção tecnológia para tomada de decisão. INGEPRO Rev. **02**(10), 01–12 (2010)
14. Caruso, L.A., Tigre, P.B.: Modelo SENAI de prospecção: documento metodológico. In: 1st Oficina Internacional del Trabajo, pp. 9–14. Cintefor/OIT, Montevideo (2004)
15. IBGE Report. https://biblioteca.ibge.gov.br/visualizacao/livros/liv99007.pdf. Accessed 20 Jan 2020
16. Guimarães, S.M.K., Pecqueur, B.: Inovação, territórios e arranjos cooperativos – experiências de geração de inovação no Brasil e na França, 1st edn. Open Edition Press, Marselha (2015)
17. Yin, R.: Case Study Research: Design and Methods (Applied Social Research Methods, vol. 5), 3rd edn. SAGE Publications Inc, California (2002)
18. Voss, C., Tsikriktsis, N., Frohlich, M.: Case research in operations management. Int. J. Oper. Prod. Manage. **22**(2), 195–219 (2002)

Simulation of Crashworthiness Performance of Thin-Walled Structures with Adapted Trigger Design

Nuno Peixinho(✉) and Pedro Resende

Universidade do Minho, Campus de Azurém, 4800-058 Guimarães, Portugal
peixinho@dem.uminho.pt

Abstract. This paper presents numerical results for crashworthiness parameters of thin-walled structures with different section and trigger geometries. The improvement potential of the introduction of configurable crush triggers is analyzed. The main objective is to absorb impact energy in a progressive and controlled manner with higher efficiency and moderate peak loads. In the studied implementation different alternatives for trigger geometry and section design are proposed. The application is studied recurring to numerical simulation of representative octagonal geometries. The introduction of a complex section design suitable for extrusion manufacturing process allowed for significant improvements in specific absorbed energy. The ability to tailor peak loads was validated for the original octagonal geometry allowing for a peak load reduction of approximately 20%.

Keywords: Crashworthiness · Trigger design · Numerical simulation

1 Introduction

The objectives of reduction of fuel usage and improved crashworthiness have led to new developments for lighter and more efficient crash related structures. These have resulted in the extensive use of high-strength steels and aluminum alloys. However, crashworthiness is also largely dependent on geometry design of thin-walled structures and of particular relevance is the study and improvement of deformation modes.

One of the design elements worthy of further development is the trigger geometry. The trigger is a geometrical feature introduced to lower the peak load and induce a favorable deformation mode in the crushing process. In this manner several trigger designs have been proposed. Zhang and co-authors [1] presented a buckling initiator design comprised of a pre-hit column and pulling strips located near the impact end. With this approach it was possible to significantly reduce the peak load (approx. 30%) yet still retaining energy absorption and comparable deformation modes. Marsolek [2] introduced non-axisymmetric folding patterns for improved management of crash energy absorption. Their trigger design was suitable for different load levels and provided for a reduction of maximum forces [2]. The more common geometrical trigger implementation

J. Machado et al. (Eds.): icieng 2021, LNME, pp. 164–172, 2022.
https://doi.org/10.1007/978-3-030-79165-0_16

was further developed by Rai and co-authors [3] favoring cut-out holes in comparison with other trigger design features.

A different concept has explored the use of localized thermal modification of aluminum alloys therefore forming a "thermal trigger" zone with lower yield stress [4]. In relevant applications the deformation mode of crash boxes may be controlled by such localized soft zones or thermal induced triggers [5, 6]. In order to implement such strategy appropriate design tools are required. Also practical techniques to achieve the thermal treatment must be developed. Regarding the design process numerical simulation tools are required to include appropriate description of geometry, loading and contact conditions and material properties [7–9].

This paper presents results for numerical simulation of crashworthiness in octagonal thin-walled structures with different overall section and trigger geometry designs. A numerical simulation procedure is detailed and results are presented and discussed for representative geometry section design and materials.

2 Numerical Simulation Procedure

The reference structure analyzed in the numerical study is a prismatic column with octagonal cross-section on aluminum 6060T5. Figure 1 presents such geometry and dimensions that include 41 mm overall width, 1.2 mm wall thickness for the section, and 100 mm overall length of the part

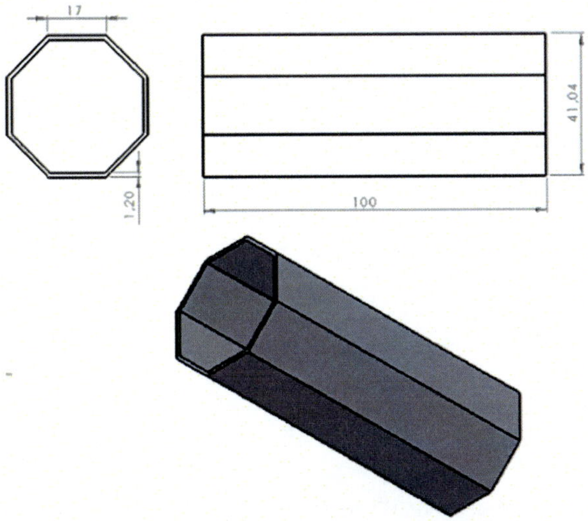

Fig. 1. Reference crashbox design.

A different section geometry targeted for crashworthiness improvement is presented in Fig. 2. This multicellular thin-walled structure evolves from the original octagonal geometry seeking to optimize the possibilities offered by the extrusion process for more complex and efficient designs [10].

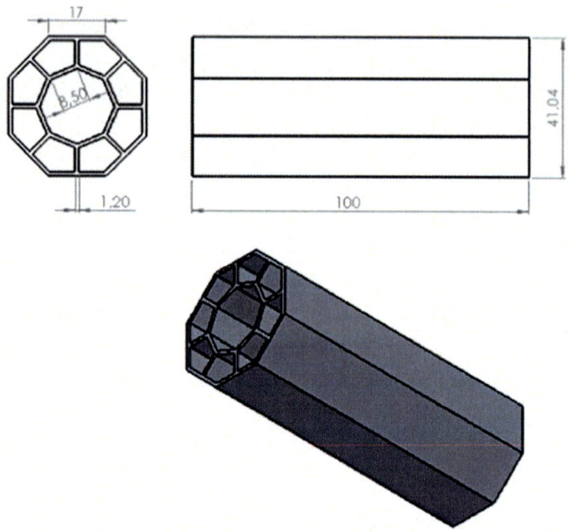

Fig. 2. Improved crashbox design.

The mechanical properties of the aluminum alloy 6060T5 were obtained by static tensile tests [5]. The true stress–strain curve is presented in Fig. 3. Since this particular aluminum alloy is insensitive to the strain rate effect such effect was not considered in the numerical models.

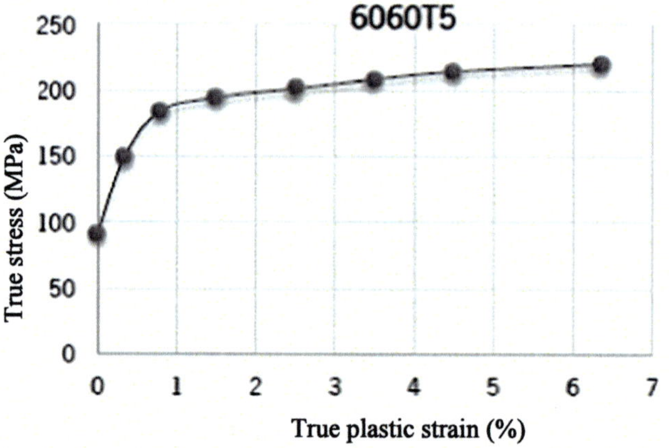

Fig. 3. 6060T5 true stress vs true plastic strain

The numerical simulations were performed recurring to software ANSYS Workbench, in particular the module "Explicit Dynamics" [11] This module is suitable for simulation of non-linear behavior and large deformation in impact events [12, 13]. The loading condition considers the impact of a rigid mass of 200 kg at an initial speed of 7 m/s on the top face of the prismatic columns while the lower part is considered as clamped. The finite element suitable for this type of modelling needs to have good bending and membrane behavior for large in plane deformations while allowing for axial loads. Therefore the Belytschko-Lin-Tsay shell element was adopted and four integration points were used in the thickness direction. The contact definition between the rigid wall and the model was defined as surface-surface interaction having a friction coefficient of 0.2 [14]. The same coefficient value is used for the self-contact defined on the model surfaces.

The numerical simulation program included analysis of structures without trigger and with the inclusion of different trigger geometries, as indicated in Table 1 and Figs. 4 and 5. The selection of trigger geometry is based on a more extensive study that evaluated trigger design [14]. The presented results include a geometry based on a localized thickness reduction of 0.6 mm (Type I – Fig. 4) and rectangular holes at prescribed locations (Type II – Fig. 5). These trigger geometries were used in both reference design (Figs. 4 and 5) and multicell octagonal structure (Fig. 6).

Table 1. Numerical simulation program.

Section geometry	Trigger description	Simulation Ref.
Baseline octagonal (Fig. 1)	Without trigger	OCT_NT
	Trigger Type I	OCT_TI
	Trigger Type II	OCT_TII
Multicell octagonal (Fig. 2)	Without trigger	MTC_NT
	Trigger Type I	MTC_TI
	Trigger Type II	MTC_TII

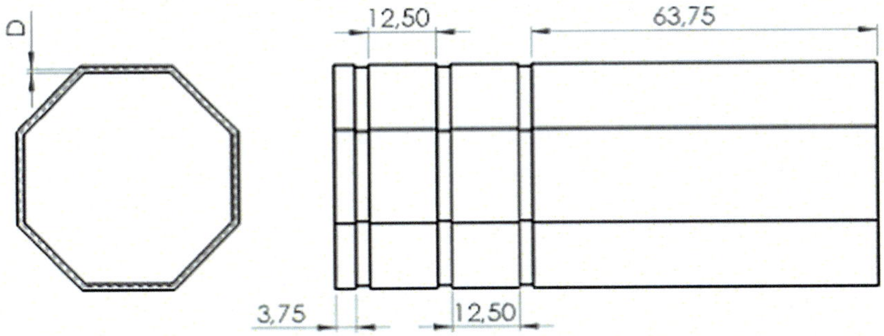

Fig. 4. Trigger design Type I.

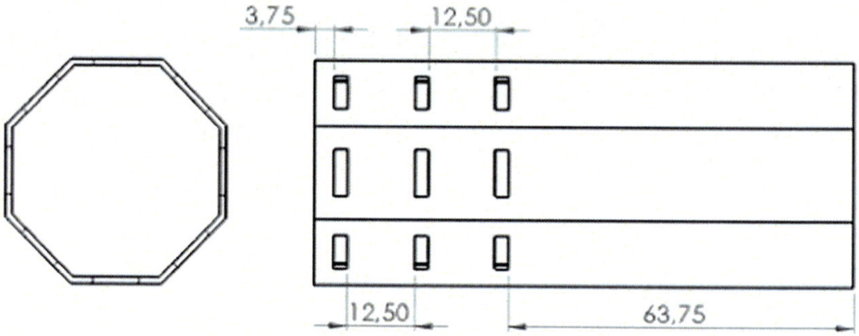

Fig. 5. Trigger design Type II.

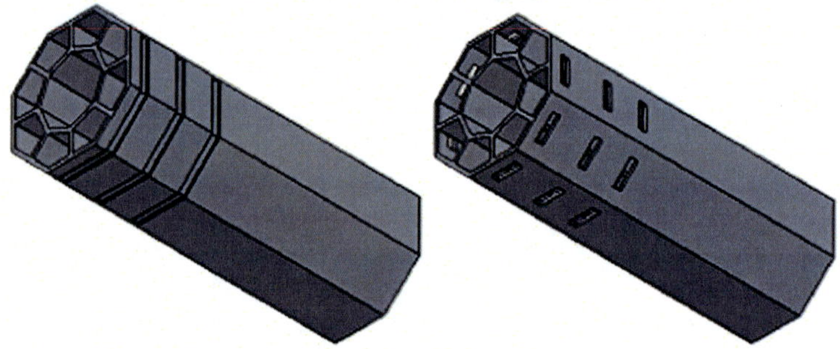

Fig. 6. Trigger geometries as implemented in multicell octagonal design.

3 Results and Discussion

The numerical simulation results demonstrated the formation of plastic folds for both reference design and multicell geometry. These folds form initially in the upper part of the columns and continue to develop gradually down into the distal end. Plastic folds are an important mechanism for energy absorption in the compressive deformation of the column and their development can be observed in the load-time curves (Fig. 7 and 8). Their formation contributes to impact energy absorption and as such an efficient initiation and development are critical for an improved crashworthiness design. The final deformed shapes that resulted from the crushing process are presented in Fig. 9.

From the analysis of results it transpires that both crashbox designs are suitable for crashworthiness applications. However, the multicellular section design supports much higher loads which must be further analyzed for efficiency comparisons.

Additional results are therefore presented in Table 2 that includes relevant crashworthiness parameters: absorbed energy (E_a); peak load (F_{max}); Crushing Force Efficiency (CFE); mass and specific absorbed energy (S_e). The presented CFE parameter is indicative of the uniformity of collapse loads, being defined as the ratio of mean crushing load (F_{avg}) to initial peak load (F_{max}). Ideally a highly efficient impact energy absorber would have a crush force efficiency of 100%, which is difficult to achieve in practical applications.

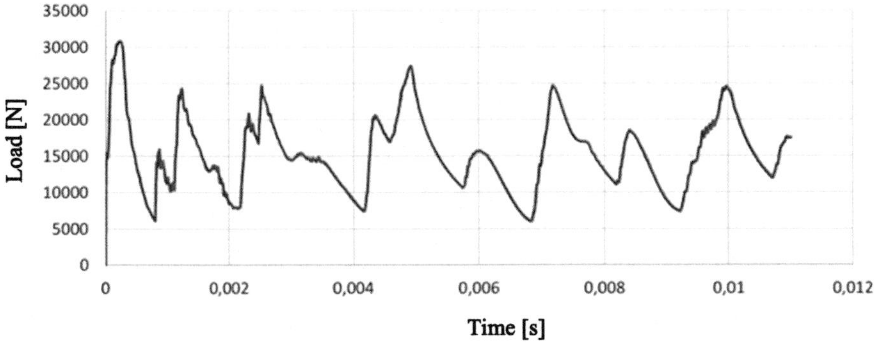

Fig. 7. Load-time curve of base reference geometry (OCT_NT).

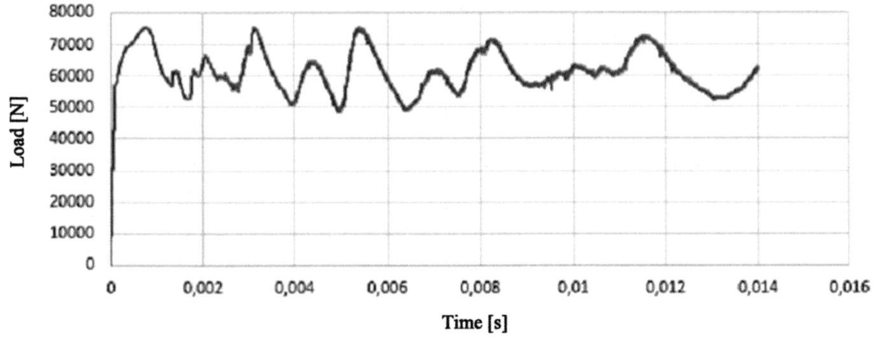

Fig. 8. Load-time curve of multicell geometry (MTC_NT).

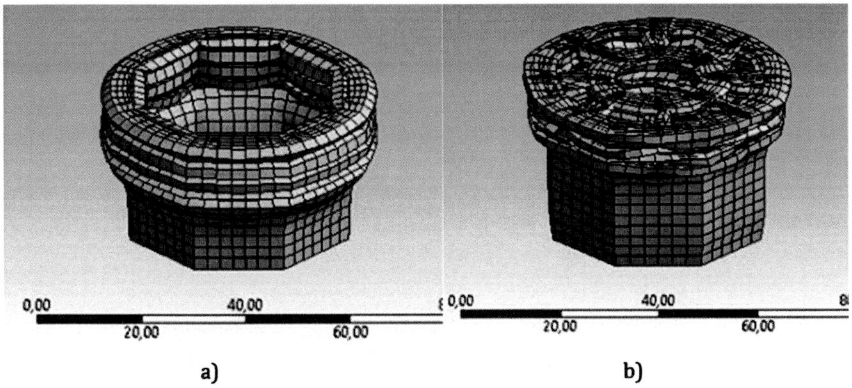

Fig. 9. Final deformed shapes: a) OCT_NT; b) MTC_NT

The comparison of such different geometries highlights the load carrying capacity of the multicellular geometry, as expected from its higher mass and extensive section

Table 2. Crashworthiness parameters for the studied crashbox designs.

Simulation Ref.	Mass [kg]	E_a [J]	$F_{máx}$ [N]	S_e [kJ/kg]	CFE
OCT_NT	0.0452	1109	30905	24.56	0.497
MTC_NT	0.0928	4125	75527	44.47	0.807

inertia. Furthermore, regarding efficiency, the results are significantly promising: the multicellular geometry allows for an improvement in specific absorbed energy (81%) and CFE parameter (62%). Such results are a testimony of the efficiency that is possible to obtain with complex extruded sections.

The crashworthiness parameters and response of the geometries with triggers are presented in Tables 3 and 4 as well as a comparison with the geometries without triggers. Such comparison is presented as percentage variation in relation to the original geometry without trigger.

Table 3. Crashworthiness parameters for the base design and trigger variations.

Simulation Ref.	Mass [kg]	E_a [J]	$F_{máx}$ [N]	S_e [kJ/kg]	CFE
OCT_NT	0.0452	1109	30905	24.56	0.497
OCT_TI	0.0435	1081 (−2.5%)	25281 (−18.2%)	24.87 (+1.3%)	0.590 (+18.7%)
OCT_TII	0.0435	1146 (+3.3%)	24445 (−20.9%)	26.36 (+7.3%)	0.646 (+30.0%)

Table 4. Crashworthiness parameters for the base design and trigger variations.

Simulation Ref.	Mass [kg]	E_a [J]	$F_{máx}$ [N]	S_e [kJ/kg]	CFE
MTC_NT	0.0928	4125	75527	44.47	0.807
MTC_TI	0.0911	4284 (+3.9%)	71286 (−5.6%)	47.04 (+5.8%)	0,889 (+10.2%)
MTC_TII	0.0911	4244 (+2.9%)	71744 (−5.0%)	46,60 (+4.8%)	0.857 (+6.2%)

The numerical results presented in Tables 3 and 4 indicate a possibility for improvements in specific energy absorption and Crush Force Efficiency (CFE). These improvements are more significant for the base geometry that also presents a higher reduction in peak loads with the introduction of triggers. This ability to tailor the peak load (a reduction of 20% was possible) is critical to the design of crashworthiness parts that need to absorb energy while limiting loads transmitted to the ocupants of the vehicle [15]. Regarding the multicell geometry (Table 4) the triggers are only able to provide a peak load reduction of approximately 5% although improvements in efficiency were observed and the overall crashworthiness of the geometry presents important advantages.

4 Conclusions

This paper presented a geometry approach for crashworthiness improvement of parts that are representative of automobile structures. A base octagonal geometry was compared with a candidate geometry suitable for extrusion process that uses multicell section. These geometry designs were analysed with the introduction of geometric initiators that are developed to obtain a controlled deformation process and obtain a reduction of peak loads. The study was performed recurring to numerical simulation of the impact event in Ansys Explicit Dynamics.

The analysed geometries are suitable for crash energy absorption through plastic deformation. The complex section presents significant improvements in efficiency, namely in specific absorbed energy (81%) and CFE parameter (62%). These results are indicative of the structural gains possible with the use of aluminum extruded sections. The introduction of different trigger designs allowed for a stable crushing behaviour and improvements in relevant crashworthiness parameters, namely reduction of peak loads and increase of crush force and energy absorption efficiency. The ability to tailor peak loads was particularly significant for the trigger application in the original octagonal geometry allowing for a peak load reduction of approximately 20%.

Acknowledgments. This work has been supported by FCT – Fundação para a Ciência e Tecnologia within the R&D Units Project Scope: UIDP/04077/2020.

References

1. Zhang, X.W., Su, H., Yu, T.X.: Energy absorption of an axially crushed square tube with a buckling initiator. Int. J. Impact Eng **36**(3), 402–417 (2009)
2. Marsolek, J., Reimerdes, H.: Energy absorption of metallic cylindrical shells with induced non-axisymmetric folding patterns. Int. J. Impact Eng **30**, 1209–1223 (2004)
3. Rai, V., Ghasemnejad, H., Watson, J., Gonzalez-Domingo, J., Webb, P.: Developed trigger mechanisms to improve crush force efficiency of aluminum tubes. Eng. Struct. **199**, 109620 (2019)
4. Bjørneklett, B., Myhr, O.: Materials design and thermally induced triggers in crash management. In: Proceedings IBEC Conference (2003)
5. Peixinho, N., Soares, D., Vilarinho, C., Pereira, P., Dimas, D.: Experimental study of impact energy absorption in aluminum square tubes with thermal triggers. Mater. Res. **15**(2) (2012)
6. Fjær, H., Bjørneklett, B., Myhr, O.: Microstructure based modelling of Al-Mg-Si alloys in development of local heating processes for automotive structures. In: Proceedings 2005 TMS Annual Meeting (2005)
7. Peixinho, N., Doellinger C.: Characterization of dynamic material properties of light alloys for crashworthiness applications. Mater. Res. **13**(4), 471–474 (2010). ISSN: 1516-1439
8. Peixinho, N., Jones, N., Pinho, A.: Application of dual-phase and TRIP steels on the improvement of crashworthy structures. Mater. Sci. Forum **502**(5), 181–186 (2005)

9. Peixinho, N., Jones, N., Pinho, A.: Determination of crash-relevant material properties for high-strength steels and constitutive equations. SAE Trans. J. Mater. Manufact. **111**, 1019–1025 (2002). Section 5
10. Chen, Y., Bai, Z., Zhang, L., Wang, Y., Sung, G., Cao, L.: Crashworthiness analysis of octagonal multi-cell tube with functionally graded thickness under multiple loading angles. Thin-Walled Struct. **110**, 133–139 (2017)
11. Chen, X., Liu Y.: Finite Element Modeling and Simulation with ANSYS Workbench, 2nd edn. CRC Press (2018)
12. Peixinho, N., Pinho, A.: Dent resistance of aluminum and magnesium alloys. Proc. ImechE Part D J. Automobile Eng. **220**, 1191–1198 (2006)
13. Costa, S., Mendonça, J., Peixinho, N.: Study on the impact behavior of a new safety toe cap model made of ultra-high-strength steels. Mater. Des. **91**, 143–154 (2016)
14. Resende, P.: Simulação numérica do comportamento ao impacto de estruturas smart com iniciadores de deformação configuráveis. MSc thesis, Universidade do Minho (2018)
15. Marshall, N., Nurick, G.: The effect of induced imperfections on the formation of the first lobe of symmetric progressive buckling of Thin-Walled Square Tubes. In: Jones, N., Talaslidis, D.G., Brebbia, C.A., Manolis, G.D. (eds.) Structures Under Shock and Impact. WIT Press, UK (1998)

Study of Heat Transfer Conditions in the Cutting Zone When Grinding

Mykhaylo Stepanov[1], Maryna Ivanova[1](✉), Petro Litovchenko[2], Larysa Ivanova[2], and Yurii Havryliuk[1]

[1] National Technical University «Kharkiv Polytechnic Institute», 2 Kyrpychova St., Kharkiv 61002, Ukraine
Maryna.Ivanova@khpi.edu.ua

[2] National Academy of the National Guard of Ukraine, 3 Maidan Zahysnykiv Ukrainy, Kharkiv 61001, Ukraine

Abstract. A large influence on the accuracy and quality of the surface being machined is shown by the temperature processes occurring in the cutting zone at finishing operations, especially such as grinding. Despite a liquid coolant is used to reduce the temperature in a cutting zone, it could become an additional heat source and influence on heating of the grinding machine. To avoid the ingress of heated cutting fluids on the machine parts, devices are used that create an air curtain. The simultaneous effect of flows of cutting fluids and air jets cause a change in heat transfer processes occurring in the cutting zone. The changes of the heat transfer coefficient along the contact line of the grinding wheel and the workpiece are considered in the paper. The phenomenon of overlapping sections of the contact zone of the grinding wheel and the workpiece, that occurs because of the recesses in the coolant under the action of an air jet from both sides is uncovered. It's determined that the overlap pattern largely depends on the input air pressure and the size of the annular gap between the workpiece and the inner surface of the end elements of the coolant supply device. Construction measures that make it possible to minimize the length of the overlapping sections by air currents are proposed.

Keywords: Grinding wheel · Liquid coolant · Air curtain · Air flow rate · Heat transfer coefficient

1 Introduction

Modern manufacturing is characterized by a high level of fierce competition [1]. The competitiveness of the enterprise largely depends on the quality and reliability of the product, which, in turn, is greatly influenced by the accuracy and quality of processing of each individual detail that is part of the final product. Achievement of sufficient accuracy of machining operations with high its productivity is possible in a stable condition and proper setup of machine tools [2]. This dependence is especially noticeable in finishing operations, such as grinding [3, 4], because grinding is one of the most important

J. Machado et al. (Eds.): icieng 2021, LNME, pp. 173–181, 2022.
https://doi.org/10.1007/978-3-030-79165-0_17

manufacturing processes, especially when high surface qualities have to be realized or if hard or brittle materials have to be machined [5].

It should be take into account that all the parameters of the technological system, namely: wheel parameters (size, shape, binder, structure, grade, grain size, abrasive materials); blanks (chemical composition, mechanical properties, fracture mode, etc.); statistical and dynamic characteristics of the machine tool; and cutting conditions, affect the occurrence and behavior of thermal processes in the system, which in turn determine the quality indicators of the machined part [6]. Often the grinding performance is limited by workpiece failure due to a high thermal load in the contact zone [5]. Therefore, the development of methods and devices to reduce thermal loads during grinding is an urgent scientific task.

2 Literature Review

The grinding force and power play an important role in the grinding process as they have a direct influence on the wheel wear, grinding accuracy, grinding temperature and surface integrity [7]. In addition, increased cutting forces can cause surface burns, which is one of the most important grinding problems [8].

Errors caused by elastic deformations of the technological grinding system can occur, as well, as a result of the action of the cutting force on a not sufficiently rigid workpiece [9]. As a result, the workpiece has become deformed and acquires a barrel-shaped (curvilinear) form. In addition, the temperature deformations of cylindrical grinding machines cause errors in the size and shape of the machined parts.

In the grinding process, coolant lubricant is used to lubricate and mainly to transmit the heat generated in the contact zone [10]. Reducing the heat load in the contact zone can be achieved with an improved coolant flow directly through the cutting zone [5]. At the same time, low jet velocity and flow rate tend to reduce the efficiency of the cutting fluid, however, excessive flow or jet velocity during grinding is useless and does not lead to an additional reduction in the temperature of the grinding zone [11].

One of the sources of heat that determine the temperature field and the temperature deformations of the units and parts of the machine tool is a liquid coolant heated in the grinding zone [12]. It is necessary to highlight the elimination of the contact of the liquid coolant with parts and units of grinding machine among the measures to reduce the temperature deformations and their influence on the accuracy of the workpieces. To this purpose the device has been developed for supplying liquid coolant to the cutting zone using airflows for shielding (insulation) liquid coolant [13–15]. The device [15] makes it possible to solve a number of problems (Fig. 1).

One of the problems solved by the developed device is the reduction of the elastic deformation of the workpiece due to the formation of the force created by the air, to the opposite direction of the cutting force. This solution helps to reduce joint deformations by 21%.

When air and liquid coolant contact in the proposed device, the bubbles of air can dissolve in the liquid. At this time, they turbulize and mix the liquid, intensifying heat transfer from the ground workpiece.

Studies [16] have confirmed the hypothesis of a decrease in temperature in the contact zone in the case of the use of gas-saturated liquids. Moreover, the use of gas-saturated

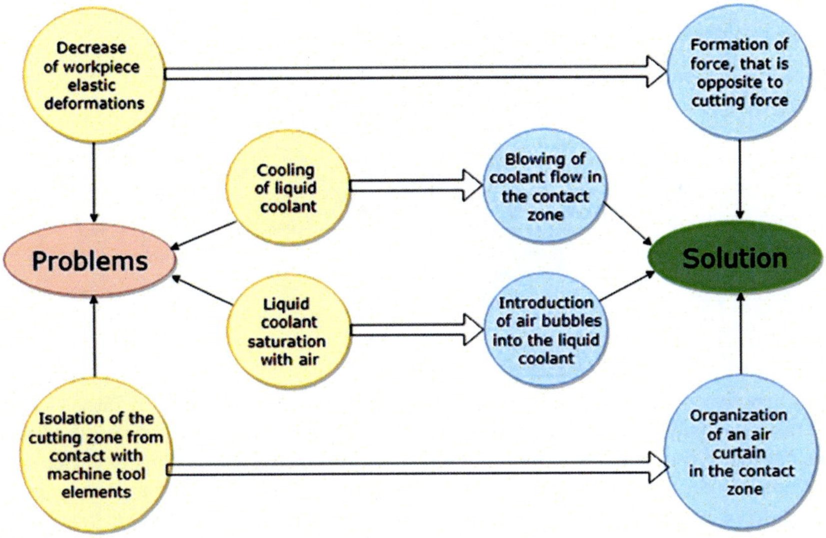

Fig. 1. Tasks that are solved by using the developed device for supplying liquid coolant when grinding.

coolant liquids in comparison with conventional liquids can reduce the temperature on the contact surface by 10…50%, and cutting forces - by 10…40%. The reason for this may be the effervescing of bubbles from the coolant, which changes the heat transfer mode from convective to nucleate boiling, which is determined by a sharp increase in the heat-transfer coefficient.

3 Research Methodology

The mechanical effect of an air jet (air barrier) on the liquid coolant flow in the device [15] channel leads to a decrease in the size (width) of the contact of liquid coolant with the grinding wheel and workpiece. As a result, the heat transfer conditions change, because two more zones are formed: the heat exchange zone between the coolant and the surface of the workpiece (circle); the heat exchange zone between the surface of the workpiece (circle) and air. The foregoing can be represented as a conventional diagram of the interaction of a turbulent air jet with a liquid coolant surface in the channel (Fig. 2). The gap between the workpiece and the element of the coolant supply device can be considered as a nozzle formed by the above elements.

The heat transfer coefficient is different for the contact zones washed by the coolant flow and blown by air jets (Fig. 3).

When the contact line is blocked by air, the heat transfer coefficient α over the entire contact line is the same. When blowing air in sections *I* and *V* the heat transfer coefficient depends on the air flow rate and is in the range from 10 to 100 W/(m^2 ·°K). In sections *II*, *IV* heat transfer is characterized by nucleate boiling with a heat transfer coefficient $\alpha = 18 \cdot 10^4$ W/(m^2 · °C). In section *III*, for which the convective heat transfer regime is characteristic, α equals from $3.1 \cdot 10^4$ to $5 \cdot 10^4$ W/(m^2 · °C).

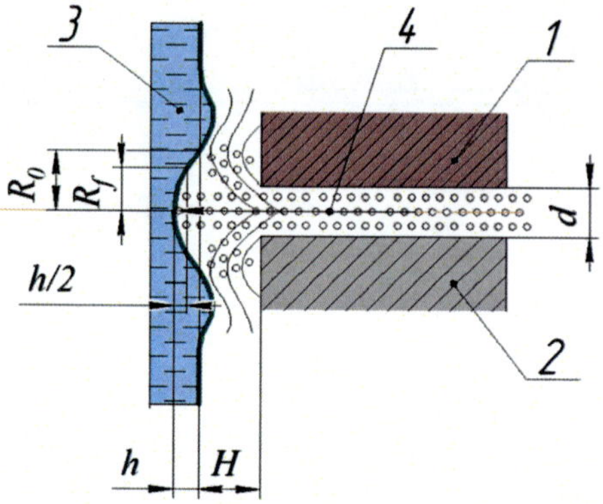

Fig. 2. Scheme of the interaction of a turbulent air jet with a liquid cooling surface: 1 - element of the coolant supply device; 2 - workpiece; 3 - liquid coolant; 4 - air jet; *d* - the width (diameter) of the channel formed by the surface of the end element of the coolant supply device and the workpiece; *H* - the distance from the surface of the end element of the coolant supply device to the liquid coolant flow; *h* - the depth of the cavity formed in the liquid coolant flow under the influence of an air jet; $2R_0$ - the width of the cavity formed in the liquid coolant flow under the influence of an air jet.

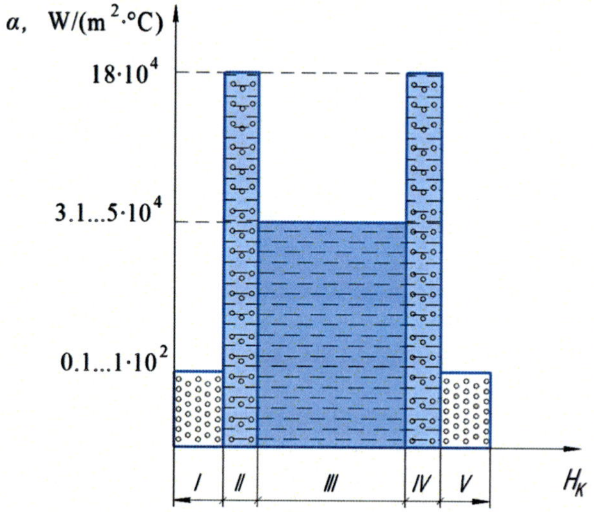

Fig. 3. A qualitative hypothetical illustration of the change of the heat transfer coefficient along the contact line of the grinding wheel and the workpiece: *I, V* - sections of air blowing; *II, IV* - sections washed by gas-saturated liquid coolant; *III* - section washed by liquid coolant.

If rational parameters of the liquid coolant jet supplied to the cutting zone and air jets flowing from the end elements of the coolant supply device are not provided, next conditions may be created:

- air jets will cool the liquid coolant flow from both sides;
- air jets will prevent the liquid coolant from entering the contact zone of the wheel and the workpiece (Fig. 4).

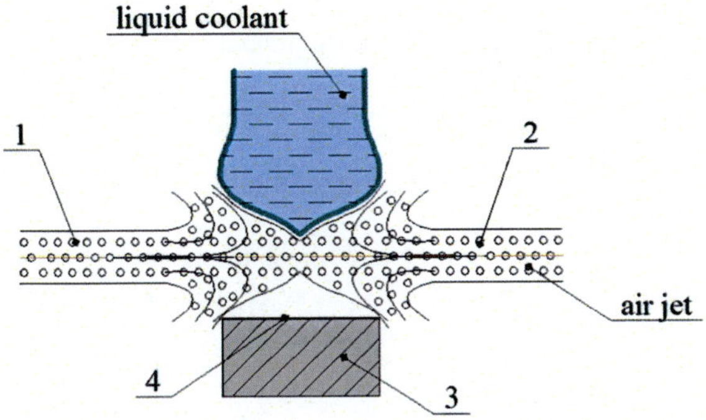

Fig. 4. Scheme of overlapping of the liquid coolant flow with air jets in the contact zone of the wheel and the workpiece: *1, 2* - air supply channels; *3* - workpiece; *4* - the line of contact of the wheel with the workpiece.

When solving the problem of the interaction of an air jet with a fluid flow, it is necessary to determine the dependence of the size of the formed cavity on the intensity of the jet and the distance between the source of the jet and the liquid surface.

In many cases, to determine the size of the zone of influence of the air jet and the mathematical description of the connection of the parameters, the equation of the balance of forces on the interface of the two phases is used, which is based on momentum conservation law. Moreover, since surface tension has a significant effect on the shape of the recess only with small dimensions of the recess, it can be ignored.

To determine the size of the recess in the coolant under the action of an air jet, the equation can be used [17].

$$\frac{F_j}{\rho \cdot g \cdot h \cdot d^2} = \frac{\pi}{2K^2} \cdot \left(\frac{H+h}{d}\right), \tag{1}$$

where F_j - pressure force of the air jet on the liquid coolant surface; ρ - liquid coolant density; g - acceleration of gravity; K - dimensionless coefficient characterizing the rate of decrease of the axial velocity of air in the stream as it moves away from the outflow hole, $K = 6.4$ [18].

The pressure force of the air jet on the liquid coolant surface is determined on the basis of the assumption that all the kinetic energy of the jet central part is converted into potential one.

The value F_j can be determined with a sufficient degree of accuracy based on the fact that it is approximately equal to the force acting from the side of the jet on the plane perpendicular to the direction of air flow [19].

$$F_j = \rho_a \cdot f_a \cdot v_a^2, \tag{2}$$

where ρ_a - air density; f_a - cross sectional area of a jet of air; v_a - the average air flow rate.

In determining the cross-sectional area of the air jet, it is necessary to take into account the fact that when it expires from the hole, its diameter decreases. This is due to the compression of the jet, as a result of which the diameter of the jet becomes smaller than the diameter of the hole. The compression diameter of the jet can be determined by the formula [20]

$$d_c = \sqrt{\frac{\mu}{\varphi}} d, \tag{3}$$

where μ - flow rate coefficient, $\mu = 0.67$; φ - speed field coefficient, $\varphi \approx 0.97$.

In practice, to calculate the force F_j, it is more convenient to use a formula that takes into account the pressure in front of the outflow hole

$$F_j = k_0 \frac{\pi \cdot d^2 \cdot \mu}{2} P, \tag{4}$$

where P - overpressure in front of the outflow hole, k_0 - shape factor, the value of which depends on the change in direction and modulus of the vector of the amount of air movement in the jet during interaction with the liquid surface, $k_0 = 1.06 \ldots 1.47$.

The air flow rate at the outflowing can be determined by the well-known formula [21]

$$v_a = \sqrt{\frac{2k}{k-1} \frac{P_1}{\rho_a} \left[1 - \left(\frac{P_2}{P_1} \right)^{\frac{k-1}{k}} \right]}, \tag{5}$$

where k - adiabatic exponent for air, $k = 1.4$; P_1, P_2 - pressure in front and behind the nozzle, respectively.

Taking into account the foregoing, the condition under which the coolant access to the contact line will be blocked by air is as follows

$$H_k = 2\text{h}. \tag{5}$$

The reason for the occurrence of sections *I* and *V* blown by air is a high level of air flow speeds, issued from the gap between the workpiece and the inner surface of the end elements of the study device for supplying liquid coolant. And here it should be said about the technical contradiction arising in the process of work. On the one hand, to maintain the load-bearing capacity of the device by reducing the elastic deformations of the workpiece, a high pressure level P_1 must be provided. On the other hand, a high level of pressure P_1 leads to an increase in the speed of air leaving the gap, which determines the extent of the formed aforementioned sections *I* and *V*.

4 Results

The speed at the exit from the gap can be suppressed by increasing the area of air exit from the gap (in case of maintaining the gap at the width of the main part of the inner hole of the end element of the device). Structurally, this can be done by making groove 4 at the exit of the gap (Fig. 5).

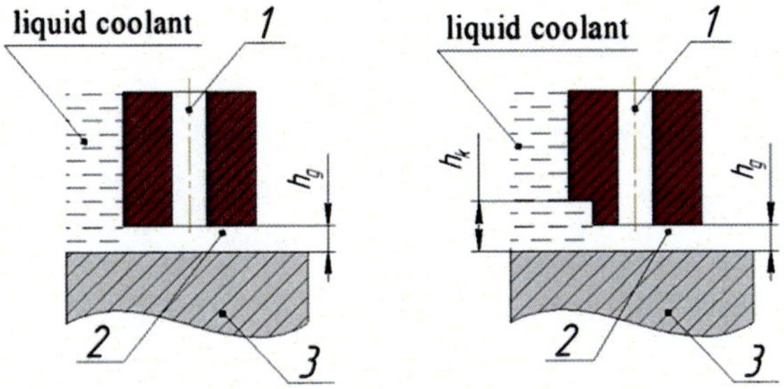

Fig. 5. Schematic representation of the gap: a - without the groove, b - with the groove, *1* - nozzle for air supply; *2* - annular gap; *3* - workpiece; *4* - annular groove.

The magnitude of the decrease in speed depending on the inlet air pressure and the size of the annular groove can be estimated by analyzing Fig. 6. By varying the parameter h_k, it is possible to reduce the air velocity at the exit from the gap by more than 4 times.

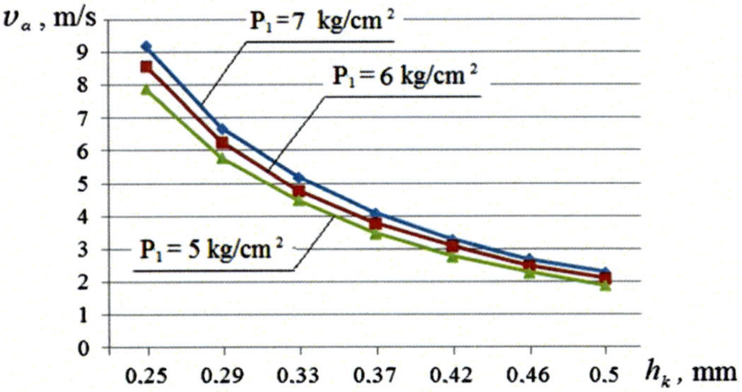

Fig. 6. The influence of the annular groove parameter and the air supply pressure on the value of the air flow rate at the outlet of the annular gap $h_g = 0.17$ mm.

As can be seen from the graph (Fig. 7), it is possible to almost completely minimize the size of the overlap areas that affect the formation of the thermal pattern in the grinding zone by varying the parameter h_k.

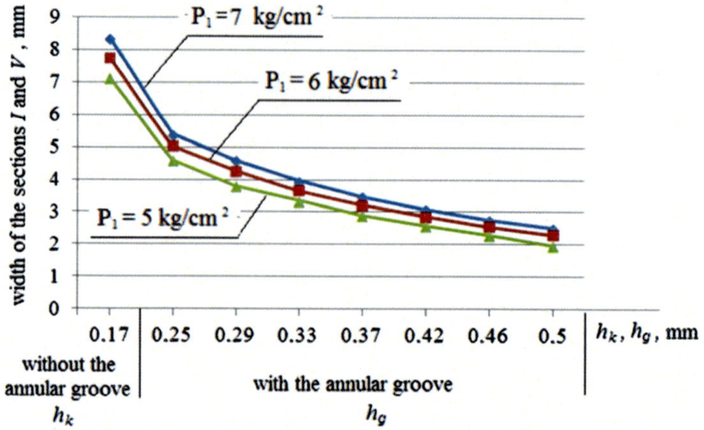

Fig. 7. The effect of parameters h_k and h_g on the width of the overlap of the contact zone (Sects. 1 and 5) by air jets flowing from the gap.

5 Conclusions

The phenomenon of overlapping sections of the contact zone of the grinding wheel and the workpiece with air jets that occurs when air barriers are formed for preventing the liquid coolant from contacting the grinding machine is revealed. Overlapping the contact zone by air flow prevents the liquid coolant from ingress to the cutting zone, which affects the heat transfer coefficient.

The parameters forming the pattern of overlap are determined, the most important of which are the input air pressure and the size of the annular gap between the workpiece and the inner surface of the end elements of the coolant supply device.

Construction measures are proposed that make it possible to minimize (decrease more than 4 times) the length of the overlapping sections by air currents and, thus, ensure unimpeded admission of liquid coolant to the contact zone of the grinding wheel and the workpiece.

References

1. Ali, A.Y.S., Awdini, A.O., Adan, H.D.: The effect of globalization on local industries: a case of Mogadishu manufacturers. Int. J. Bus. Manage. Tomorow **2**(11), 1–15 (2012)
2. Otaghvar, M.H., Hahn, B., Werner, H., Omiditabrizi, H., Bähre, D.: Optimization of centerless through-feed grinding using 3D kinematic simulation. In: 12th CIRP Conference on Intelligent Computation in Manufacturing Engineering, vol. 79, pp. 308–312. Gulf of Naples, Italy (2019). https://doi.org/10.1016/j.procir.2019.02.072
3. Wu, M.F., Chen, H.Y., Chang, T.C., Wu, C.F.: Quality evaluation of internal cylindrical grinding process with multiple quality characteristics for gear products. Int. J. Prod. Res. **57**(21), 6687–6701 (2019). https://doi.org/10.1080/00207543.2019.1567951
4. Kumar, S., Bhatia, O.S.: Review of analysis & optimization of cylindrical grinding process parameters on material removal rate of En15AM steel. J. Mech. Civil Eng. **12**(4), 35–43 (2015). https://doi.org/10.9790/1684-12423543

5. Denkena, B., Grove, T., Göttsching, T.: Grinding with patterned grinding wheels. CIRP J. Manuf. Sci. Technol. **8**, 12–21. https://doi.org/10.1016/j.cirpj.2014.10.005
6. Ozay, C., Ballikaya, H., Savas, V.: Application of the Taguchi method to select the optimum cutting parameters for tangential cylindrical grinding of AISI D3 tool steel. Materiali in Tehnologije **50**(1), 81–87 (2016). https://doi.org/10.17222/mit.2014.293
7. Agarwal, S., Venkateswara Rao, P.: Predictive modeling of force and power based on a new analytical undeformed chip thickness model in ceramic grinding. Int. J. Mach. Tools Manuf. **65**, 68–78 (2013). https://doi.org/10.1016/j.ijmachtools.2012.10.006
8. Aslan, D., Budak, E.: Semi-analytical force model for grinding operations. Procedia CIRP **14**, 7–12 (2014). https://doi.org/10.1016/j.procir.2014.03.073
9. Ropyak, L., Shatskyi, I., Makoviichuk, M.: Analysis of interaction of thin coating with an abrasive using one-dimensional model. Metallofiz. Noveishie Tekhnol. **41**(5), 647–654 (2019). https://doi.org/10.15407/mfint.41.05.0647
10. Vesali, A., Tawakoli, T.: Study on hydrodynamic pressure in grinding contact zone. Considering grinding parameters and grinding wheel specifications. In: 6th CIRP International Conference on High Performance Cutting, pp. 13–18, HPC (2014). https://doi.org/10.1016/j.procir.2014.03.053
11. Lavisse, B., et al.: The effects of the flow rate and speed of lubricoolant jets on heat transfer in the contact zone when grinding a nitrided steel. J. Manuf. Process. **35**, 233–243 (2018). https://doi.org/10.1016/j.jmapro.2018.07.029
12. Stepanov, M., Ivanova, L., Litovchenko, P., Ivanova, M., Basova, Y.: Model of thermal state of the system of application of coolant in grinding machine. In: Ivanov, V. et al. (eds.) Advances in Design, Simulation and Manuf. DSMIE 2018. Lecture Notes in Mechanical Engineering, pp. 156–165. Springer, Cham (2019). https://doi.org/10.1007/978-3-319-93587-4_17
13. Strongin, A.S., Nikulin, M.V.: On the issue of calculating air-thermal curtains. In: ABOK, vol. 1 (2004). (In Russian). https://www.abok.ru/for_spec/articles.php?nid=2324
14. Titov, V.P.: Features of a jet of air curtains. Thermal regime of heating, ventilation, air conditioning and heat and gas supply systems. In: Collection of Papers of MGSU, Moscow, vol. 177, pp. 3–15 (1980). (In Russian)
15. Stepanov, M., Ivanova, M., Litovchenko, P., Ivanova, L., Kotliar, A.: Improvement of the accuracy of grinding by means of coolant supply. In: Ivanov, V., Trojanowska, J., Pavlenko, I., Zajac, J., Peraković, D. (eds.) Advances in Design, Simulation and Manufacturing III. DSMIE 2020. Lecture Notes in Mechanical Engineering, pp. 325–335. Springer, Cham (2020) https://doi.org/10.1007/978-3-030-50794-7_32
16. Ryabov, G.K.: Development of technology for the use of gas-saturated cutting fluids in the grinding operations of steel billets. Synopsis of Ph.D. thesis, Kuibyshev (1987). (In Russian)
17. Banks, R.B., Chandrasekhara, D.V.: Experimental investigation of the penetration of a high velocity gas jet through a liquid surface. J. Fluid Mech. **15**(1), 13–34 (1963). https://doi.org/10.1017/s0022112063000021
18. Cheslak, F.R., Nicholls, J.A., Sichel, M.: Cavities formed on liquid surfaces by impinging gaseous jets. J. Fluid Mech. **36**(1), 55–63 (1969). https://doi.org/10.1017/S0022112069001509
19. Bashta, T.M.: Engineering Hydraulics. Reference Manual. Mashinostroenie, Moscow (1971). (In Russian)
20. Chugaev, R.R.: Hydraulics (Technical Fluid Mechanics), 4th edn. Energoizdat, Leningrad (1982). (In Russian)
21. Mordasov, M.M., Savenkov, A.P., Chechetov, K.E.: Method for analyzing the gas jet impinging on a liquid surface. Technical Physics. Russ. J. Appl. Phys. **61**(5), 659–668 (2016). https://doi.org/10.1134/s1063784216050170

Material Selection Guidelines for the Product Designer

Pedro Ferreira[1], Maria João Félix[1], Ricardo Simoes[1,3], Olga Silva[2], and Gilberto Santos[1](✉)

[1] Design School, Polytechnic Institute Cávado Ave, Barcelos, Portugal
gsantos@ipca.pt
[2] ESCE, Polytechnic Institute Viana do Castelo, Valença, Portugal
[3] Institute for Polymers and Composites (IPC), University of Minho, Guimarães, Portugal

Abstract. The aim of this work is to analyze the importance of the selection of materials in the development of products. As such, this work can provide an important contribution to the field of development of product design.

A questionnaire was sent to professional designers. 29 responses were validated, this being the sample of our study. An application model of the guidelines to support the product designer in the selection of materials was developed. The relationship of parameters related to the model are presented. Thus, the selection of materials is considered of extreme importance for the designers. Several aspects influence it, including, among others, the conditions of service, the nature of the mechanical efforts, the availability of the materials, the cost of the materials, the safety, the ease of manufacture and assembly, the previous experience, the cycle product life. All this, looking for a better use of the products, capable of providing an improvement in the standard of living of the people. Furthermore, they can also create emotional connections with future users. For this, it is necessary the materials classification in simple and definitive categories. But it is a complex work, due to the amount and also, the diversity of options of materials available in the market able to materialize the ideas of the designers. Hence, the survey results are very important for professionals who need to select materials. The authors intend to extend this study to more professionals, to improve the results, through a more significant sample.

Keywords: Materials selection · Material properties databases · Product design

1 Introduction

In view of an empirical and subjective review of the selection of materials with a view to Product Design, it is necessary to create a library containing methods and instruments that allow the development of projects to be monitored. For that, it is necessary to prioritize a reflection on the use of the best materials and the respective transformation processes, so that they are sustainable [1]. It is very important to provide as much information as necessary so that the design engineer can make correct decisions in the selection of materials, particularly in the early stages of product development.

J. Machado et al. (Eds.): icieng 2021, LNME, pp. 182–190, 2022.
https://doi.org/10.1007/978-3-030-79165-0_18

When using different materials, such as metals, plastics, wood, ceramics or even composites [2], the design engineer has at his disposal a wide variety of technical manuals applied to the different environments and the respective conditions of use. However, as can be seen in this study regarding the selection of materials in Design, there are some gaps in the methods of selecting materials, as well as in the way they are made available. Therefore, it is necessary to provide more information, namely, to all those who have little experience. Most of the time, the support that designers have on this topic is developed by specialist engineers, in which the technical language used is not well interpreted, in the same way, by the language of a simple designer. Therefore, the time has come to break this current paradigm that exists, both in companies and universities.

The designer, that is, the professional who works more in the aesthetic and sensory aspects, must understand the main concepts related to the materials to be applied to the products. But they, often coming from secondary education based on the arts, do not need to state all the physical and mechanical properties, analyze and interpret all the graphic data, as for example, they are presented to us through the excellent CES program [3], selection of materials, developed by Professor Mike Asbhy [4] and respective collaborators from the University of Cambridge.

Following a review of the literature on this subject, interviews were scheduled with former students who studied product design [5], who are working in the market. It was planned to make a survey about the difficulties encountered on the day to day regarding the selection of materials. A critical analysis of the existing teaching materials will be presented, as well as a proposal for their improvement.

2 Literature Review

There are, surely, over 100,000 materials in our world[1]. It is known that materials play a very critical role throughout the product design, as well as, the manufacturing process. Hence, suitable material selection is a very important factor, and it is a very challenging and multifaceted task for diverse engineering applications. Thus, existence of numerous material selection approaches is an apparent sign for significance of this issue in all materials context [6] and, in the evaluation of alternative materials, various factors should be taken into account. To specify the performance requirements, to relate them to the main material properties and processing requirements is the first step in material selection process. Normaly, the most important properties for the materials are its mechanical properties, namely its ability to withstand certain efforts without breaking, such as, strength, stiffness, hardness and others. For sustainable development of products, the selection of the most appropriate materials is a very important process and sometimes also difficult. Hence, materials play an important role during the entire product design and manufacturing phase. It is known that a wrongly selected material may often lead to premature product failure, often with disastrous consequences, causing sometimes disasters, other times loss of repute and revenue of the concerned manufacturing organization. While selecting the most suitable material for a specific application, the designers and the engineers, often need a systematic and sound methodology to deal with this problem having various alternative choices and, oftentimes, conflicting objectives [7] difficult to solve. It is known that in engineering design selection of material for a particular product plays an very important role. Nowadays, manufacturing environment, where a large number of manufacturing processes and engineering materials are available to select, many difficult decisions need to be taken while selecting a material for a determined application. A large number of factors, such as, physical, mechanical and electrical properties, as well as, namely economic considerations should be taken into account by the engineers and designers depending upon the application of the final product. The complicated interrelationships that exist among various selection criteria of different type of materials make the material selection process more challenging and, simultaneously, time consuming to the engineers and designers [8]. Related on the fulfillment of the desired properties, some materials may be rejected and some of them may be selected for the works. Seeking to achieve a sustainable society, design practitioners must eco-innovate through material selection. Material selection resources evolve every year, where are included databases, physical material libraries, software and tools that connect materials with engineers and designers. In spite of that, the material-properties focus of these resources risks over-rationalising material selection within the context of material properties. Hence, material selection needs consideration of a broader material system, reflecting legislation, sectoral behaviour, stakeholders, access to knowledge and networks [9]. Thus, the designers and engineers must have the knowledge, namely, of material properties, design concept, cost, and their interrelationship. The inappropriate choice of a material may often lead to an early failure of the product in the field of application and, consequently, an overall increase in production cost, thereby affecting the prestige of the manufacturing organization. Therefore, the engineers and the designers must identify and select the most appropriate material for a product taking into account the minimum possible cost and good specific performance considerations [10]. Mostly, an inadequate selection of materials may result in failure and consequently damage of an assembly and decreases the performance [11]. In the process of selection of a suitable material for a determined product, so many influencing factors, such as cost, operating environment, desired properties, availability of supplying sources, production process, etc., need to be contemplate, which make the materials selection process a multi-criteria decision-making (MCDM) problem [11]. Selection of suitable materials for a determined work is one of the hardest tasks in the project and product improvements in many industrial and other applications. Thus, materials play an important function during the entire project and manufacturing process of the product. The incorrect selection of materials often leads to price increase and can results in product breakdown. Hence, the engineers and designers need to identify and select the most appropriate materials with very good functionalities in order to attain the correct output with the minimum cost concern for a specific applicability [12].

There has never been such a time in which the knowledge of materials properties, characteristics and the diversity of materials were so great. Engineers and designers are at risk of seeing their knowledge on materials easily go out of date, because the permanent changes. As a consequence of this situation, it exist an ever greater number of designers, architects and engineers far from the real potentialities that the diversity of materials can offers. We know there are many ways of innovating and one of them is, certainly, through materials. The engineers and the designers have to meet this way of innovation. This is why they must consider the widest number possible of materials in the beginning of the design process. After that, and as the concept evolves, the options regarding form and function take place, decreasing the number of selected materials for the product. In the final stage of the project, the possibilities are reduced to few materials and the important information becomes increasingly more detailed, as the process is developing and finishing [13]. Selection guidelines are very important criterion for optimization of materials selection for certain specific applications in order to meet simultaneous, and often conflicting requirements [14], at a time when customers are very demanding [15] and when it is necessary the improvement the education system [16], as well as the quality management system [17–19], to improve and to add value to products [20–22].

The main objective of this work is to study and to analyze the importance of material selection, with the purpose to provide professionals, who need to select materials for the development of a new products, some improvement of current methodologies, according to responses of professionals, that work in the sector.

3 Methodology

The survey was basically done through a questionnaire, that was divided into three parts. The first part, composed of five questions, sought to survey the relationship that exists between material selection and product design; the second part, composed of seven questions intended to question what are the selection methodologies and programs that professionals can adopt in their work and, finally, the third part composed of five questions, one of them open question, for a reflection on the best tools used for material selection, looking to know what they miss. The questionnaire was conducted through an online survey and had a total of 29 participants aged between 18–42 years, mostly from the northern region of Portugal - Braga, Porto and Viana do Castelo - and also Viseu and Lisbon already in center zone, what constitutes the sample of this survey. Target population were professionals that work in product design, or in industrial design [23]. It was intended to survey the main problems that designers encounter when they intend to select materials for the products they design, as well as, what are the best tools/programs they know that contribute most to solving the problems related to material selection.

In the questionnaire, the main difficulties that designers encounter in their professional activity when they have to select materials for their products were surveyed. As main issues we can highlight: as a designer, do you feel the need to have these methods and tools in your workplace?; In your opinion, do you think a study is needed that demonstrates the importance of having material selection tools to support the product designer? why?; In your opinion, do you find it interesting to develop a new material selection methodology/tool that meets the needs of the current product designer? do you have any suggestions to make?

Of the 29 participants, 96% feel the need to have material selection methods and tools in their work because: Makes work more analytical and the process faster and more efficient; It is important to keep up to date in order to help with the work, serving as a guide; The selection of material for the product is fundamental and indispensable in the product development process; These tools help to understand the limits of the material, the manufacturing process and the capacity of the material; They are important in the development of new equipment or in the improvement of existing equipment; It is important to know very well the difference between materials to create designs for the next applications; It makes work more methodical and effective.

4 Results

All design engineers understand the importance of material selection in the development of new products. But they need something more, in order to validate their choices. The problems and theories can be analyzed, seeking to find solutions to this complex problem. For those just starting out, or for those with little experience, it is known that the methods of the materials selection aren't easily available and there are gaps in the way, where the methods are used by the product designer.

Sometimes the tools available for selecting materials are not intuitive for professionals who intend to use them, namely, if they have little experience. In interpreting the data, difficulties are often encountered. The existing tools use a specific language, too deep, that some professionals have difficulty understanding. It is therefore necessary to clarify the language, making the concepts simple and accessible, where sometimes only some physical and mechanical properties are needed.

In addition to the physical and mechanical characteristics of the materials, it is very important to understand the emotional attributes, such as curiosity, reaction, pleasure, emotion, desire, among others, in order to incorporate them into the object, putting the designer in place of the possible future user. And, also understand that the selection of materials is an important component of the creative process. A methodology to be used in a specific phase of the project, or in the progression of the proposal, in cycles, can be adopted. According to Table 1, most respondents of the survey were industrial designers (54%) male (55%).

Table 1. Occupation and gender of the respondents

Occupation	Product designer	54%
	Industrial designer	46%
Gender	Male	55%
	Female	45%

In view of the result of the analysis of the inquiries made to product designers and Industrial designers, who are working in the market and students who are graduating, a new approach to material selection was developed.

This consists of creating a modular methodology that achieves a whole set of core requirements for the selection of materials. Inspired by the "Canvas" model, a business model that is based on helping people to make decisions both in the management of projects within their companies, or in the segmentation of the market, defining the target audience where they intend to be inserted. In this case, the model presented (Fig. 1) will be a tool to assist the product designer in the selection of materials.

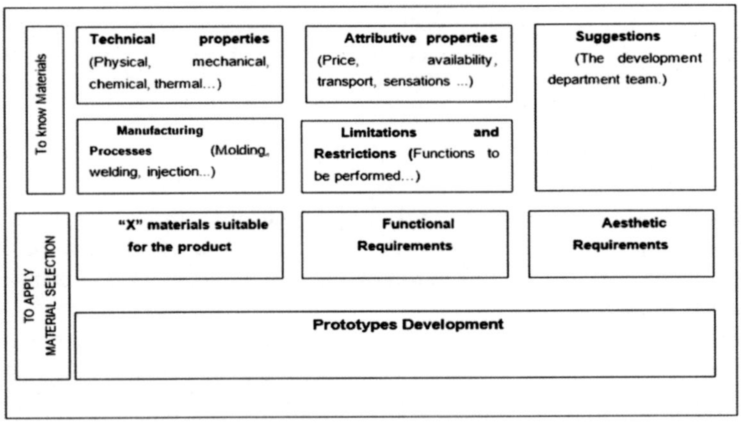

Fig. 1. Application model of the guidelines to support the product designer in the selection of materials [23].

The model is based on the essential guidelines to be taken into account in a project to be developed in the industry or in the development of new products in collaboration with universities. The application model of the guidelines to support the product designer in the selection of materials is presented in the Fig. 1.

In the Table 2 are presented a relationship of parameters related to the model presented in the Fig. 1. Of them we can highlight aspects related with technical properties to manufacturing processes; the availability of materials; the functional requirements; materials suitable for the product.

Table 2. Relationship of parameters related to the model presented [23].

Relate technical properties to manufacturing processes;
Relate the availability of materials with the price and transport;
Identify suitable manufacturing processes for different materials;
Observe samples and identify sensory and aesthetic aspects, such as, touch, texture and colours;
Identify the functional requirements such as, "check what the product's function is";
Identify "x" materials suitable for the product taking into account the function it will serve, the environment where it will act and the user who will buy.

4.1 Technical Properties Common to the Main Family of Materials

This first guideline was developed through research on the main properties of materials and corresponds to the common characteristics that the main materials have among themselves. Thus, the main materials were divided by families and the main common properties were characterized and presented, as an innovative support for the model presented, since of all the supports found, none of them expresses this importance -

the common analysis of the main properties of the materials. Thus, the designer has at his disposal a range of intrinsic characteristics that can help him in the best choice of material for his projects.

4.2 Attributive Properties

Attributive properties are already more difficult to categorize than technical properties, as there is no logical approach, and it will always depend on the price of the material, where it comes from and how it will be transported. Each company has its own databases with which it works.

However, it is important at this stage to emphasize that the selection of materials is the interface with the product (aesthetic, visual, technical, emotional and environmental aspects; viability; risks; costs; performance; production) and that the choice of material for the product has its influences and vice versa. (quality/price factor; product useful life; productive processes; environment - environmental impact, ecological footprint and waste; material accessibility, availability and durability; production feasibility; market - where you are operating; valuing product). The selection of materials can be designated as "an interface with the product".

4.3 Manufacturing Processes Common to Different Materials

As with the properties of materials, it is important to emphasize in manufacturing processes, the common that exists between materials. Thus, in a clear and objective way it is possible to draw conclusions. It can be emphasized in this field that the majority of new materials - nanomaterials, biomaterials, semiconductors and intelligent materials - are developed in laboratories, so the manufacturing processes can be different. Each designer can list the concepts that are essential to his project.

5 Conclusion

The survey results are very important for professionals who need to select materials. Hence, the contribution of this work intends to stimulate future developments/improvements of material selection activities. It is known that the selection of materials plays a very important role in the design and development of the product design, as well as, in their respective production. Several methodologies have already been proposed and others have been addressed, to assist design engineers in the selection of the most suitable materials, for the most diverse engineering applications. These methods may employ some criteria weighting values in their calculations that are based on the subjective needs of the design engineers. The guidelines to support the product designer, in the selection of materials, taking into account several aspects addressed in the study regarding the selection of materials in design were proposed. They take into account, namely: a- the contribution of the selection of materials, for the development of useful objects; b- an improvement in the product's useful life; c- an increase in the product's life cycle; d- the valorization of environmental culture and materials capable of being recycled; e- the replacement of materials that may be harmful to human health, as well as to the environment; f- greater use of materials from natural resources, which are biodegradable.

This is a first step work of a more complex study to be continued. Thus, the authors intend to extend this survey to more professionals, to obtain a more significant sample, seeking to help all professionals to find the best answers when they need to select materials for their products.

References

1. Ljungberg, L.Y.: Materials selection and design for development of sustainable products. Mater. Des. **28**(2), 466–479 (2007)
2. Lefteri, C.: Making It: Manufacturing Techniques for Product Design. Continuum (2007)
3. Ashby, M.F., Cebon, D.: Teaching Engineering Materials : The CES EduPack. Cambridge University (2007)
4. Ashby, M.F., Johnson, K.: Materials and Design: The Art and Science of Material Selection in Product Design, 3rd edn. (2014)
5. van Kesteren, I.E.H.: Product designers' information needs in materials selection. Mater. Des. **29**(1), 133–145 (2008)
6. Mousavi-Nasab, S.H., Alireza Sotoudeh-Anvari, A.: A new multi-criteria decision making approach for sustainable material selection problem: a critical study on rank reversal problem. J. Clean. Prod. **182**(2018), 466–484 (2018)
7. Prasad, K., Chakraborty, S.: A quality function deployment-based model for materials selection. Mater. Des. **49**, 525–535 (2013)
8. Edwards, K.L.: Selecting materials for optimum use in engineering components. Mater. Des. **26**, 469–472 (2005)
9. Prendeville, S., O'Connor, F., Palmer, L.: Material selection for eco-innovation: SPICE model. J. Cleaner Prod. **85**, 31–40 (2014)
10. Edwards, K.L.: Materials influence on design: a decade of development. Mater. Des. **32**, 1073–1080 (2011)
11. Jahan, A., Mustapha, F., Ismail, Y., Sapuan, S.M., Marjan Bahraminasab, M.: A comprehensive VIKOR method for material selection. Mater. Des. **32**, 1215–1221 (2011)
12. Ipek, M., Selvi, I.H., Findik, F., Torkul, O., Cedimoglu, I.H.: An expert system based material selection approach to manufacturing. Mater. Des. **47**, 331–340 (2013)
13. Ramalhete, P.S., Senos, A.M.R., Aguiar, C.: Digital tools for material selection in product design. Mater. Des. **31**, 2275–2287 (2010)
14. Vitorino, N., Abrantes, J.C.C., Frade, J.R.: Quality criteria for phase change materials selection. Energy Convers. Manag. **124**, 598–606 (2016)
15. Bravi, L., Murmura, F., Santos, G.: Attitudes and behaviours of Italian 3D prosumer in the era of additive manufacturing. Procédia Manuf. **13**, 980–986 (2017)
16. Santos, G., Doiro, M., Mandado, E., Silva, R.: Engineering learning objectives and computer assisted tools. Eur. J. Eng. Educ. **44**(4), 616–628 (2019)
17. Sá, J.C., Amaral, A., Barreto, L., Carvalho, F., Santos, G.: Perception of the importance to implement ISO 9001 in organizations related to people linked to quality-an empirical study. Int. J. Qual. Res. **13**(4), 1055–1070 (2019)
18. Santos, G., Afonseca, J., Murmura, F., Félix, M.J., Lopes, N.: Critical success factors in the management of ideas as an essential component of innovation and business excellence. Int. J. Qual. Serv. Sci. **3**(3), 214–232 (2018)
19. Santos, G., Gomes, S., Braga, V., Braga, A., Lima, V., Teixeira, P., Sá, J.C.: Value creation through quality and innovation – a case study on Portugal. TQM J. **31**(6), 928–947 (2019)

20. Bravi, L., Murmura, F., Santos, G.: Manufacturing Labs - where digital technologies help improve life quality. Int. J. Qual. Res. **12**(4), 957–974 (2018)
21. Zgodavova, K., Bober, P., Majstorovic, V., Monkova, K., Santos, G., Juhaszova, D.: Innovative methods for small mixed batches production system improvement: the case of a bakery machine manufacturer. Sustainability **12**(15), 1–20 (2020)
22. Félix, M.J., Silva, S., Santos, G., Doiro, M., Sá, J.C.: Integrated product and processes development in design: a case study. Procedia Manuf. **41**, 296–303 (2019)
23. Ferreira, P.M.A.M.: Conceção de modelo de aplicação das diretrizes de apoio à seleção de materiais. Elaboração de diretrizes para a seleção de materiais no apoio ao designer de produto. Master thesis, Inst. Polit. Cávado Ave, Portugal (2019)

Hemodynamic Studies in Coronary Artery Models Manufactured by 3D Printing

Violeta Carvalho[1(✉)], Paulo Sousa[2], Vânia Pinto[2], Ricardo Ribeiro[3], Pedro Costa[3], Senhorinha Teixeira[4], and Rui Lima[1,5]

[1] MEtRICs, Minho University, Guimarães, Portugal
violeta.carvalho@dem.uminho.pt
[2] CMEMS, Minho University, Guimarães, Portugal
[3] BIOFABICS, Rua Alfredo Allen 455, 4200-135 Porto, Portugal
[4] Algoritmi, Minho University, Guimarães, Portugal
[5] CEFT, Faculty of Engineering of the University of Porto, Porto, Portugal

Abstract. Atherosclerosis is one of the leading causes of death worldwide. It is a chronic inflammatory disease of the arterial wall that progressively reduces the lumen size because of plaque formation. To understand this pathological process, several hemodynamic studies have been carried out, either experimentally or numerically. However, experimental studies have played an important role to validate numerical results. Recent advances in computer-aided design (CAD), medical imaging, and 3D printing technologies have provided a rapid and cost-efficient method to produce physical biomodels for flow visualization. As a manufacturing process, 3D printing techniques have attracted significant attention due to the low cost and potential to rapidly fabricate biomodels to perform flow hemodynamic studies.

In the present work, a study was performed using biomodels manufactured by 3D printing that mimic both healthy and stenotic coronary arteries with different degrees of stenosis (0%, 50% and 70%). Firstly, it was evaluated the influence of the printing resolution on flow visualization, and the results showed that, when comparing to 150 μm, the 100 μm resolution biomodel was the most adequate for performing the proposed experimental studies, presenting an arithmetic average roughness of 7.24 μm. Secondly, the effect of stenosis severity on velocity and flow behavior was studied. It was concluded that as the severity of stenosis increases, the velocity at the stenosis throat also increases. In addition to this, it was also observed a recirculation zone downstream the stenosis, when the diameter was reduced to 70%.

Keywords: Atherosclerosis · Hemodynamic · 3D printing · *In vitro* · Biomodels · Print resolution

1 Introduction

Cardiovascular diseases (CVDs) remain the biggest cause of death worldwide, responsible for over 17.3 million deaths per year and the underlying disease process is known as

P. Costa, S. Teixeira and R. Lima—Shared senior authorship

J. Machado et al. (Eds.): icieng 2021, LNME, pp. 191–200, 2022.
https://doi.org/10.1007/978-3-030-79165-0_19

atherosclerosis [1]. This is a complex disease characterized by fatty material deposition in the inner most layer of the arterial wall. These deposits cause the lumen to become narrow, affecting the blood flow behavior [1]. Given its significant negative socioeconomic impact, atherosclerosis has been intensively studied in the last decades. *In vivo* studies are of capital importance as they allow a direct observation of the physiological phenomena. However, these are expensive, it is hard to control and obtain detailed measurements of the flow behavior and sometimes the selected animal model does not completely represent the disease condition [2–4].

With recent advancements on microscopy, computational power and software, image analysis techniques, and additive manufacturing techniques, experimental flow studies have gained widespread attention [3, 5–9]. In particular, 3D printing technologies have become an effective tool for rapidly producing artery phantoms with high-fidelity at low cost [8, 10]. In the past few years, many researchers used these technologies to manufacture arterial biomodels, namely by means of inkjet-based printing [11], Fused Deposition Modelling (FDM) [3, 12], Poly Jet [13], Binder jetting [12, 14], Inkjet printing [14] and stereolithography (SLA) [2, 15]. For instance, Geoghegan et al. [14], used two different 3D printing technologies (Binder Jetting and InkJet) to manufacture rigid and flexible stenotic phantoms. In a study conducted by Doutel et al. [15], secondary flows were studied using 3D printed coronary artery biomodels by SLA technology. By using the same additive manufacturing technique, Costa et al. [2], studied arterial thrombosis in stenosed and healthy coronary arteries, employing biomodels fully printed with inlets, outlets and a box-like container. In another perspective, Lai et al. [16] combined SLA with a compliant photopolymer to produce vessel-like structures. Recently, Stepniak et al. [17] produced coronary biomodels with three different technologies (FDM, SLA and PolyJet) and concluded that SLA and Polyjet techniques are appropriate to manufacture arterial phantoms. As observed, SLA is a suitable technique for phantom production, particularly, due to its high resolution, which allows to obtain extremely smooth surfaces. In this technology, the models are manufactured layer-by-layer from a photocurable liquid resin and cured by exposure to light [18].

Despite extensive research work in this field, there is still a lack of knowledge regarding the influence of printing parameters in flow visualization and measurement. In this study, it was performed an experimental study that shows how the printing resolution can affect flow visualization. Moreover, the hemodynamic effects of atherosclerosis were studied, using both healthy and stenotic idealized coronary arteries.

2 Materials and Methods

2.1 Coronary Artery 3D Biomodels Design and Fabrication

The geometry employed in this study is a simplification of a 3D model of a coronary artery, although based on real data [19]. The online platform BIOFABICS TOOLBOX (www.biofabics-toolbox.com) was used to automatically generate the studied geometries, by employing its design tool number 3. The biomodels were custom-designed with a 24-well plate layout, 59 mm total channel length, 3 mm channel 1 diameter, 3 mm channel 2 diameter, 6 mm stenosis length, 0.5 mm base thickness and a total device height of 4.5 mm (Fig. 1). To capture the changes in flow behavior, the degrees of stenosis selected were 0%, 50%, and 70%. The first one represents the healthy case. On the other hand, the 50% model represents a moderate stenotic condition, while 70% represents the most severe case. After defining all parameters, the 3D biomodels were custom-manufactured (Biofabics, Porto, Portugal) at 100 μm and 150 μm resolutions.

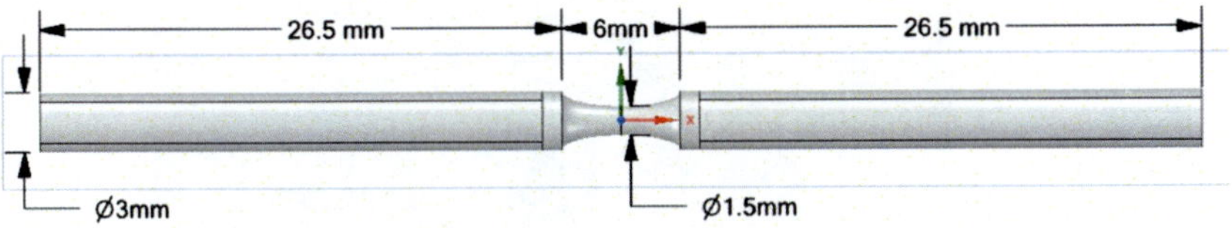

Fig. 1. Dimensions of 50% stenosis model.

2.2 Experimental Setup for Roughness Measurements and Flow Characterization

Roughness measurements were performed in order to understand the influence of printing resolution in the flow visualizations. To evaluate the roughness of the biomodels, a surface profiler (Dektak® 150, Veeco) was used which is represented in Fig. 2. Due to the difficulty of performing these measurements on circular surfaces, it was decided to print 3D cubes with the various resolutions employed.

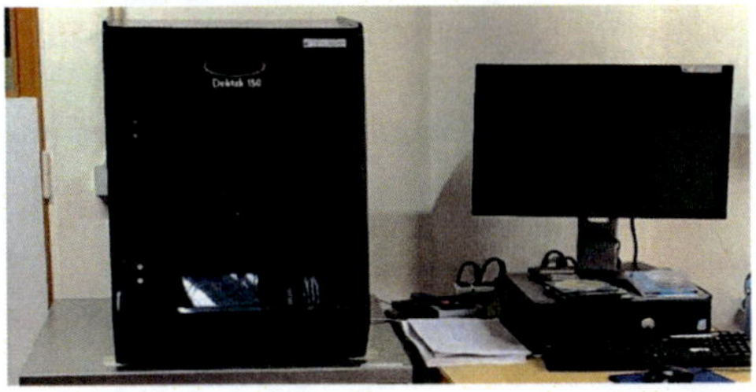

Fig. 2. Experimental setup for roughness measurements.

For the flow visualization, all experiments were conducted by using a high-speed video microscopy system (Fig. 3). The main equipment of this system consists of an inverted microscope (IX71, OLYMPUS, Japan) and a high-speed camera (Fastcam SA3, Photron, USA). To impose a constant inlet flow rate was used a syringe pump (LEGATO® 100, USA), which was kept constant and equal to 5 ml · min^{-1}.

The fluid selected as blood analogue was a dimethyl sulfoxide (DMSO)/water mixture (52%/48%-w/w), which is similar to blood in its viscosity and density. In suspension of the blood analogue fluid were used 40 μm monosized spherical microparticles, (CA 40, Spheromers®) at a concentration of 0.2%.

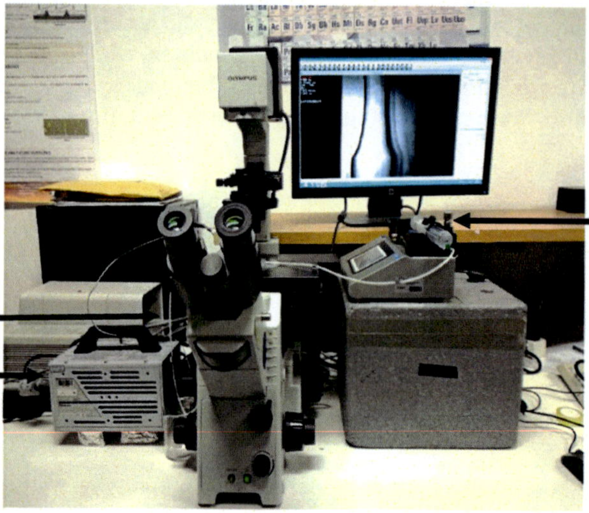

Fig. 3. Experimental setup used to control and visualize the flow.

2.3 Image Processing

To understand the diameter reduction effect on flow behavior, the video sequences captured by the high-speed camera were digitally processed using an image handling software ImageJ [20]. This is a tool frequently used to process and analyze images. In this study, it was used to perform the pre-treatment of the acquired frames, aiming to remove the noise and image artifacts. Firstly, it was selected the median projection type by means of the Z Project function in ImageJ. Secondly, it was used the image calculator tool to subtract the median from the original image. Finally, the maximum intensity projection was applied in order to obtain the images presented in this work.

To track the particles, the process previously described were applied together with a manual plugin (MTrackJ) of the image analysis software ImageJ. This plugin allows, not only to track moving particles at several image sequences, but also to measure the particle velocities.

Notice that, the regions of interest in this study are the stenotic and post stenotic region of the biomodel, however, for each case the frame rate selected was different. At the post-stenotic section, a sequence of 1000 frames per second were used and the same for the healthy model. Contrarily, to track the particles at the stenosis throat, the frame rate selected was higher (6000 frames per second). In this location, the velocity increases and so, to reliably track the particles it is necessary increase the frames number.

3 Results and Discussion

3.1 Print Resolution Effect on Flow Visualization

Looking at Fig. 4A and B, it is possible to observe the differences between the two printing resolutions, verifying that the best visualization is obtained in model B (100 μm). Smoother lines are observed in the second case, making it easier to visualize the flow.

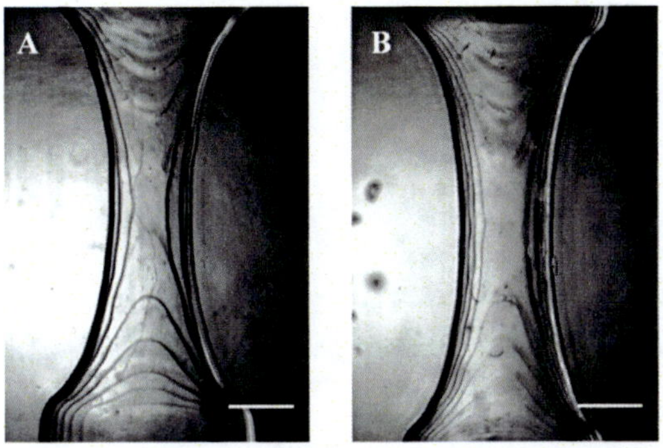

Fig. 4. Images obtained with a printing resolution of (A) 150 μm and (B) 100 μm. Scale bar, 1 mm.

To better support this observation, roughness measurements were performed in both models. One of the most widely used parameters is the arithmetic average roughness (Ra). This parameter represents the arithmetic mean of the absolute ordinate Z(x) within the sampling length [21]. The Ra parameter for the 150 μm and 100 μm model was 19.23 μm and 7.24 μm, respectively. These results demonstrate that the best biomodel was obtained with a printing resolution of 100 μm.

For these reasons, the measurements were performed in the 100 μm model.

In summary, the choice of the printing resolution is a valuable step, as it can compromise the reliability of the experimental results. Without quality images, it is difficult to analyze the flow behavior and to accurately perform the particle tracking.

3.2 Velocity Measurements

The presence of plaques in the coronary artery is responsible for obstructing blood flow to the myocardium, consequently affecting the flow velocity. Because of this, the velocities were evaluated quantitatively in the healthy model and in the stenotic section of the diseased models. For each case, several particles were tracked, in order to ensure the accuracy of the measurements (Fig. 5). The results effectively show the influence of diameter reduction on velocity. As expected, comparing the measured velocities in the healthy model with the remaining, it can be observed the increasing velocity as the severity of stenosis increases, due to the reduction in the arterial area.

One aspect that should be highlighted is the irregularities of the different particles trajectories performed for each case. For the healthy model, the measured velocities were practically the same. However, for the 50% and 70% model it was detected some oscillations at the measurements of each particle. Especially for the severe model (70%), those variations were found to be the highest. This might be due to the extremely high velocities of the particles flowing at this region and some possible measurements errors due to limitations of our high-speed video camera. The tracking of the particles is extremely difficult to perform at high flow rates and consequently the velocity error may increase [22].

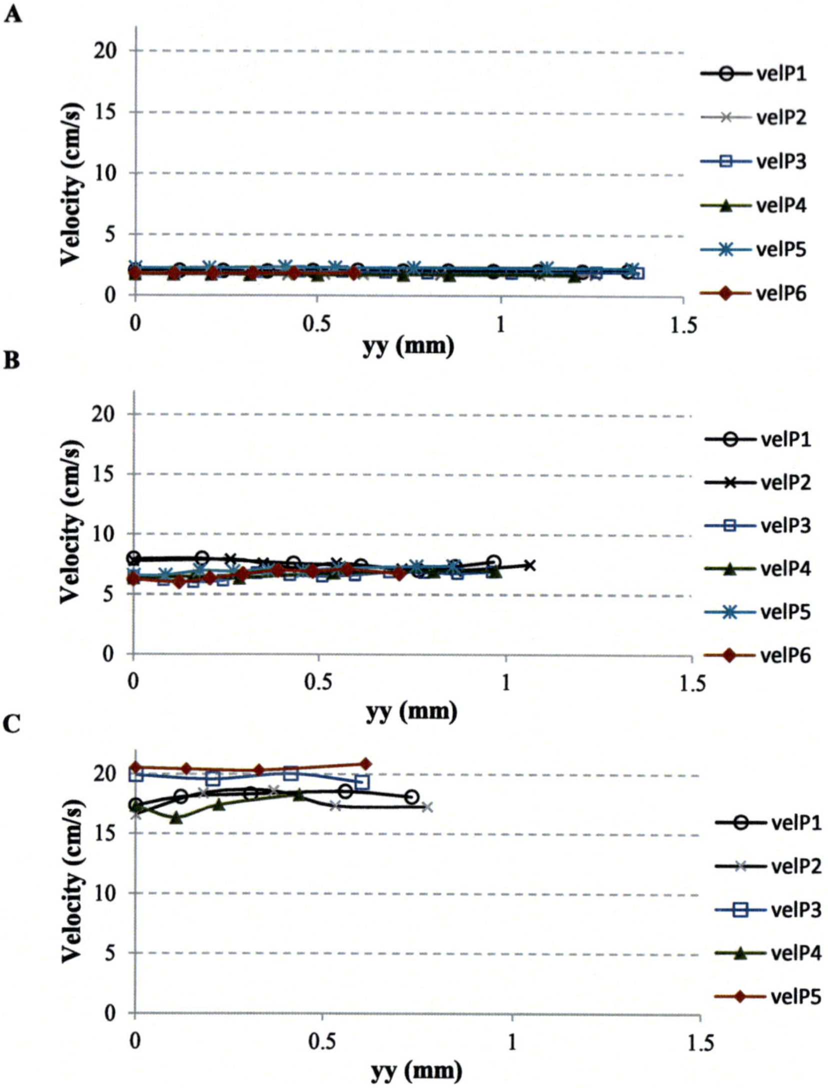

Fig. 5. Particle velocity measurements for (A) healthy model; (B) 50% stenosis and (C) 70% of stenosis.

Through the previous results, the average velocity was calculated for each particle and the results are shown in Fig. 6. The lowest values were obtained for the healthy model (1.92 cm/s). In contrast, when the diameter reduces 50% and 70%, the velocities are significatively higher, and the average value was approximately 6.98 cm/s and 18.72 cm/s, respectively.

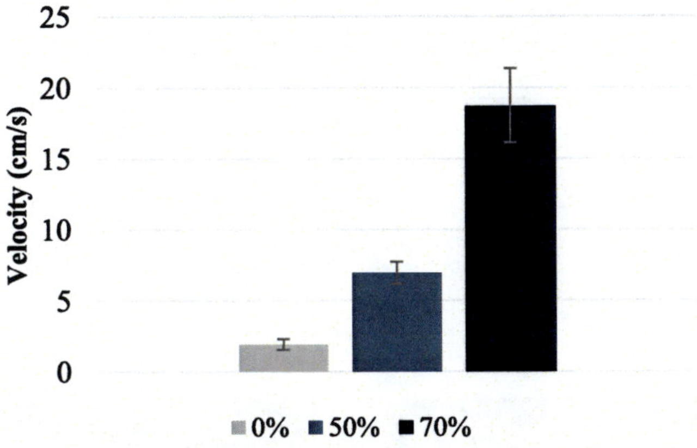

Fig. 6. Average velocity measured for each stenosis degree. The measurements are expressed as the mean ± standard deviation according to a t-test analysis at 95% confidence interval.

3.3 Effect of Stenosis Degree on Flow Behavior

Figures 7 and 8 show the images obtained by applying Z Project maximum intensity to all images of a movie. Figure 7 represents the healthy model and the alignment of the particles within the flow was observed. On the other hand, the effect of stenosis on hemodynamics is shown in Fig. 8(A) and (B). Similarly, to the healthy model, in the 50% stenosis, there is no recirculation and the fluid flows in a laminar way (Fig. 8(A)). When the degree of stenosis reaches 70% (Fig. 8(B)), there is a significative recirculation zone at the post-stenotic section. This clearly shows the effect of stenosis severity on changing the flow behavior. This phenomenon may be explained by the significative increase in velocity observed at the stenotic section of the 70% model. In contrast, when the diameter reduces 50%, the velocities at the throat are not high enough to cause these changes in flow.

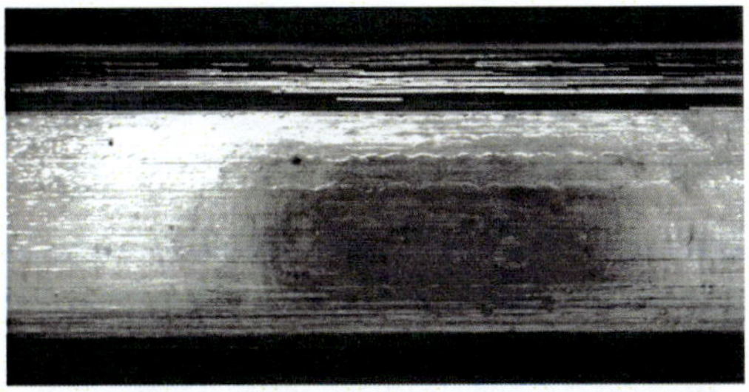

Fig. 7. Z Project maximum intensity image of the healthy model.

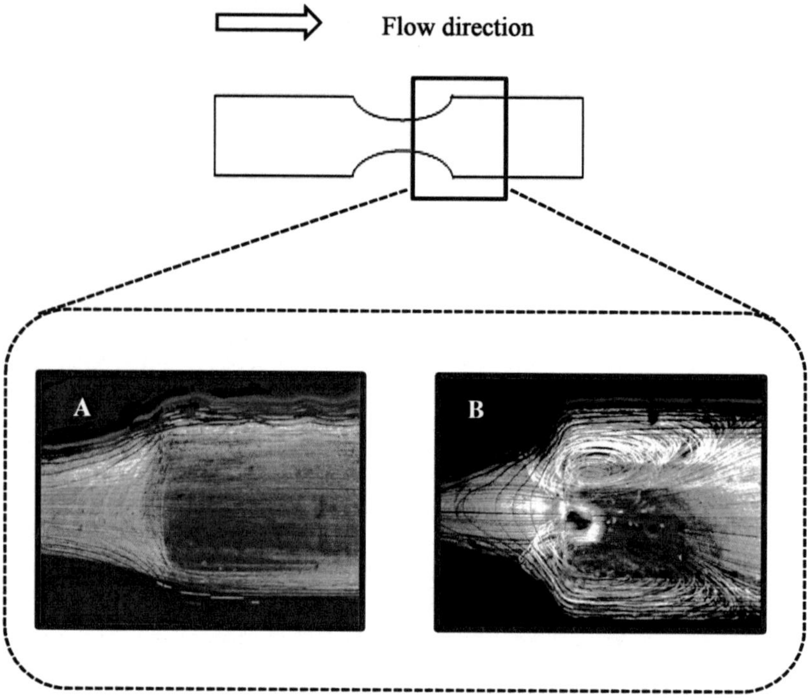

Fig. 8. Z Project maximum intensity image: (A) 50% stenosis (B) 70% stenosis.

4 Conclusions

The present study shows flow measurements and visualizations of idealized coronary arteries with atherosclerotic plaques. The results acquired in this study shows effectively that the printing resolution affects the flow visualization and reveals the importance of this parameter when 3D printing technologies are used to obtain biomodels for experimental flow studies. In addition, some changes in the flow behavior were clearly observed. For example, it was observed the formation of recirculations immediately at downstream of the stenosis when a diameter reduction of 70% was reached. Comparing all the obtained results, it is clear that diameter reduction has a significant impact on the velocities and flow behaviors. The study of these phenomena is extremely important, helping not only to better understand what happens in physical models but also to validate and improve numerical models.

Acknowledgements. The authors acknowledge the financial support provided by Fundação para a Ciência e a Tecnologia (FCT), through the projects UIDB/04077/2020, UIDB/00319/2020, and UIDB/04436/2020, NORTE-01-0145-FEDER-029394 and NORTE-01-0145-FEDER-030171, funded by COMPETE2020, NORTE 2020, PORTUGAL 2020 and FEDER. This project has received funding from the European Union's Horizon 2020 research and innovation programme under the Marie Sklodowska-Curie grant agreement No 798014. This project has received funding from the European Union's Horizon 2020 research and innovation programme under grant agreement No 828835.

References

1. Thomas, H., et al.: Global atlas of cardiovascular disease prevention and control. World Health Organ. (2011). https://doi.org/10.1016/j.gheart.2018.09.511
2. Costa, P.F., et al.: Mimicking arterial thrombosis in a 3D-printed microfluidic: in vitro vascular model based on computed tomography angiography data. Lab Chip **17**, 2785–2792 (2017). https://doi.org/10.1039/c7lc00202e
3. Queijo, L., Lima, R.: PDMS anatomical realistic models for hemodynamic studies using rapid prototyping technology. In: Lim, C.T., Goh, J.C.H. (eds.) 6th World Congress of Biomechanics (WCB 2010). August 1–6, 2010 Singapore. IFMBE Proceedings, vol. 31. Springer, Berlin, Heidelberg (2010). https://doi.org/10.1007/978-3-642-14515-5
4. Geoghegan, P.H., Jermy, M.C., Nobes, D.S.: A PIV comparison of the flow field and wall shear stress in rigid and compliant models of healthy carotid arteries. J. Mech. Med. Biol. **17**, 1–16 (2017). https://doi.org/10.1142/S0219519417500415
5. Kabinejadian, F., et al.: In vitro measurements of velocity and wall shear stress in a novel sequential anastomotic graft design model under pulsatile flow conditions. Med. Eng. Phys. **36**, 1233–1245 (2014). https://doi.org/10.1016/j.medengphy.2014.06.024
6. Banerjee, R.K., Ashtekar, K.D., Helmy, T.A., Effat, M.A., Back, L.H., Khoury, S.F.: Hemodynamic diagnostics of epicardial coronary stenoses: in-vitro experimental and computational study. Biomed. Eng. Online **7** (2008). https://doi.org/10.1186/1475-925X-7-24
7. Friedman, M.H., Giddens, D.P.: Blood flow in major blood vessels - modeling and experiments. Ann. Biomed. Eng. **33**, 1710–1713 (2005). https://doi.org/10.1007/s10439-005-8773-1
8. Catarino, S.O., Rodrigues, R.O., Pinho, D., Miranda, M., Minas, G., Lima, R.: Blood cells separation and sorting techniques of passive microfluidic devices: from fabrication to applications. Micromachines **10**, 593 (2019)
9. Bento, D., et al.: deformation of red blood cells, air bubbles, and droplets in microfluidic devices: flow visualizations and measurements. Micromachines **9** (2018). https://doi.org/10.3390/mi9040151
10. Wang, K., Ho, C.C., Zhang, C., Wang, B.: A review on the 3D printing of functional structures for medical phantoms and regenerated tissue and organ applications. Engineering **3**, 653–662 (2017). https://doi.org/10.1016/J.ENG.2017.05.013
11. Aycock, K.I., Hariharan, P., Craven, B.A.: Particle image velocimetry measurements in an anatomical vascular model fabricated using inkjet 3D printing. Exp. Fluids **58**, 1–8 (2017). https://doi.org/10.1007/s00348-017-2403-1
12. Faria, C.L., Pinho, D., Santos, J., Gonçalves, L.M.: Low cost 3D printed biomodels for biofluid mechanics applications. J. Mech. Eng. Biomech. **3**, 1–7 (2018)
13. Jewkes, R., Burton, H.E., Espino, D.M.: Towards additive manufacture of functional, spline-based morphometric models of healthy and diseased coronary arteries: in vitro proof-of-concept using a porcine template. J. Funct. Biomater. **9** (2018). https://doi.org/10.3390/jfb9010015
14. Geoghegan, P.H., Buchmann, N.A., Spence, C.J.T., Moore, S., Jermy, M.: Fabrication of rigid and flexible refractive-index-matched flow phantoms for flow visualisation and optical flow measurements. Exp. Fluids **52**, 1331–1347 (2012). https://doi.org/10.1007/s00348-011-1258-0

15. Doutel, E., Carneiro, J., Campos, J.B.L.M., Miranda, J.M.: Experimental and numerical methodology to analyze flows in a coronary bifurcation. Eur. J. Mech. B/Fluids **67**, 341–356 (2018). https://doi.org/10.1016/j.euromechflu.2017.09.009
16. Lai, S.S.M., Yiu, B.Y.S., Poon, A.K.K., Yu, A.C.H.: Design of anthropomorphic flow phantoms based on rapid prototyping of compliant vessel geometries. Ultrasound Med. Biol. **39**, 1654–1664 (2013). https://doi.org/10.1016/j.ultrasmedbio.2013.03.015
17. Stepniak, K., Ursani, A., Paul, N., Naguib, H.: Novel 3D printing technology for CT phantom coronary arteries with high geometrical accuracy for biomedical imaging applications. Bioprinting **18**, (2020). https://doi.org/10.1016/j.bprint.2020.e00074
18. Kalaskar, D.M.: 3D Printing in Medicine. Elsevier (2017)
19. Zhou, F.-F.: Coronary artery diameter is inversely associated with the severity of coronary lesions in patients undergoing coronary angiography. Cell. Physiol. Biochem. **43**, 1247–1257 (2017). https://doi.org/10.1159/000481765
20. Abràmoff, M.D., Magalhães, P.J., Ram, S.J.: Image processing with image. J. Biophotonics Int. **11**, 36–41 (2004). https://doi.org/10.1017/CBO9781107415324.004
21. Srivani, A., Xavior, M.A.: Investigation of surface texture using image processing techniques. Procedia Eng. **97**, 1943–1947 (2014). https://doi.org/10.1016/j.proeng.2014.12.348
22. Feng, Y., Goree, J., Liu, B.: Errors in particle tracking velocimetry with high-speed cameras. Rev. Sci. Instrum. **82** (2011). https://doi.org/10.1063/1.3589267

Optimization of the Flowing Part of the Turbine K-310-240 Based on the Object-Oriented Approach

Olena Avdieieva(✉), Oleksandr Usatyi, and Iryna Mykhailova

National Technical University "Kharkiv Polytechnic Institute", 2 Kyrpychova Street, Kharkiv 61002, Ukraine

Abstract. This article describes the application of the optimization methodology of a complex technical system using the K-310-240 turbine as an example based on a block hierarchical approach. The methodology for optimizing the flow part of powerful steam turbines has been developed taking into account operating conditions. The complex hierarchical structure of the optimization task is implemented in CAD "Turboagregat", which is based on the principles of a single integrated information space by adding new optimization objects. To organize effective information exchange, the process of optimal design is implemented using recursive bypass of optimization levels. Application of the methodology for solving a two-level multi-parameter and two-criteria optimization problem allowed us to find the optimal combination of 55 design parameters of the K-310-240 turbine, while increasing the absolute efficiency by 0.83% and the turbine power by 6.179 MW (~1.87%) regarding the prototype. By calculation, the mutual influence of turbine objects on its optimal characteristics was identified and evaluated.

Keywords: Object-oriented approach · Cylinder · Steam turbine · Blade and nozzle cascades

1 Introduction

In the world, the demand for electricity is growing every year. At the moment, there is a huge amount of work in the field of renewable energy, but all this cannot replace the already existing turbine park, which generates the bulk of all electricity in the world. Therefore, the design of new and modernization of existing steam turbines is still an urgent task in the energy sector.

At the moment, vast experience has already been accumulated in the field of optimal design of multi-cylinder turbine units, samples of flow parts (FP) of each cylinder having rather high technical and economic indicators have been created. Further search for reserves to increase the efficiency of steam turbines is possible only if a powerful computing techniques along with new methods and approaches implemented in the framework of modern computer aided design systems (CAD).

J. Machado et al. (Eds.): icieng 2021, LNME, pp. 201–213, 2022.
https://doi.org/10.1007/978-3-030-79165-0_20

2 Literature Review

There are many works devoted to optimization of steam and gas turbines [1–9], in which various methods and algorithms are used: genetic algorithm [10–12], surrogate modeling [13, 14], bee colony algorithm [15], DOE methods [16, 17].

At the moment there are many methods, algorithms for finding the optimal solution and also a large number of software complexes [1, 10, 13, 14]. No exception is the software package CAD "Turboagregat", aimed at finding the optimal solution for complex technical systems (CTS). The problem of optimal design of such a system, taking into account constraints and inequalities in general form, can be represented as follows:

$$\vec{Y}_{opt}\left(\vec{x}_k^{opt}\right) = \max\vec{Y}(\vec{x}_k), \vec{x}_k \in X, \vec{v}(\vec{x}_k) \in V,$$
$$\vec{Y}\left(\vec{Y}_1(\vec{x}_k), \vec{Y}_2(\vec{x}_k), \ldots \vec{Y}_n(\vec{x}_k)\right),$$
$$N_{X\min} \le |X| \le N_{X\min} < \infty, N_{V\min} \le |V| N_{V\min} < \infty, \quad (1)$$

where $\vec{Y}$ is the vector of objective functions; $\vec{x}_k$ is a vector of constructive parameters; $\vec{v}$ is the vector of functional constraints; V, X-regions of existence of functional and constructive constraints; $N_{V(\min,\max)}$, $N_{X(\min,\max)}$ - the boundaries of the regions of existence of the corresponding constraints.

The solution (1) is the extremum of the objective function that satisfies the constraints. A well-known fact is that CTS are basically either hierarchically structured constructions or various schematic solutions in which the elements of the circuit can also have their own structure. That is why for solving the problem (1) of the CTS, appropriate methodologies, methods and algorithms are required.

One of the methodologies for the search for the optimal design of the flow section of a turbo-aggregate was proposed in [18] (Fig. 1).

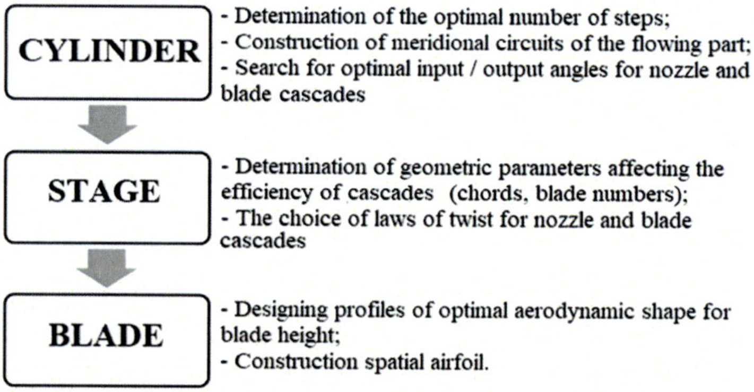

Fig. 1. Distribution of tasks by optimization levels

The paper [18] describes a three-level block-hierarchical approach for optimizing a cylinder of a turbine. The disadvantage of this approach is the lack of accounting for the operation of the regulatory system in conjunction with the rest of the flowing part. This became the basis for developing a methodology for the integrated optimization of the flowing part of powerful steam turbines using an object-oriented approach [19, 20]. This methodology is universal for CTS and was used in obtaining the results of optimization of the turbine K-310-240, given in this article.

3 Research Methodology

As an object of research, take the flowing part of the turbine K-310-240 produced by PJSC "Turboatom". In Fig. 2, consider the structure of the object in question from the point of view of the object-oriented approach.

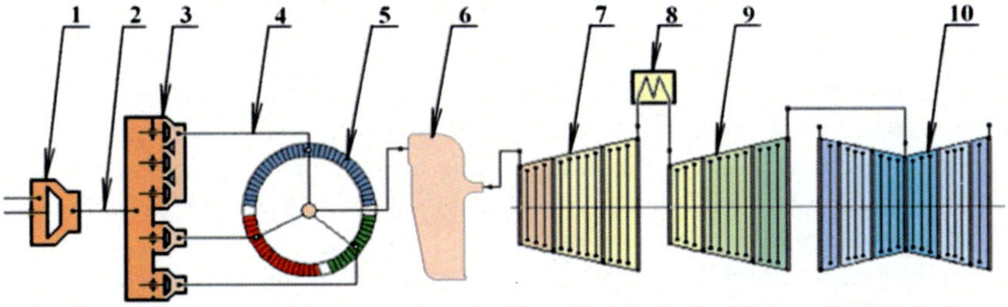

Fig. 2. Diagram of the flowing part of the turbine K-310-240: 1 - shut-off valve (NVD); 2 - stop valve line (NVD); 3 - box with control valves (SPR); 4 - segment pipelines (NVD); 5 - segments of the control stage (NVD); 6 - equalization chamber (EC); 7 - the cylinder of a high pressure (the Cylinder); 8 - high pressure cylinder (Cylinder); 9 - medium pressure cylinder (Cylinder); 10 - Low-pressure cylinder (Cylinder)

Elements that make up the structure can be divided into three objects according to their purpose. The first object is a "nozzle vapor distribution" (NVD), which includes a stop valve, a check valve line, a box with control valves, segment pipelines and a regulating stage (segments of the control stage). To the second object can be attributed, the link between the NVD and the rest of the flowing part - equalization chamber (EC), which is designed to equalize the flow at the entrance to the first stage of the high-pressure cylinder. The third object - "Cylinder" - included cylinders of high, medium and low pressure (HPC, MPC, LPC). Each of the objects, except the second one, can be divided into objects subordinate to it. Division into sublevels can be carried out until the simplest optimization object is determined.

Figure 3 shows the hierarchical structure of the information model of the structural diagram of the flowing part of a steam turbine. From the block diagram it is seen that the highest (zero) level is the turbine itself. At the first level there are previously described heterogeneous objects with their subordinate hierarchy. Each of the optimization objects has its own mathematical model and a quality assessment system.

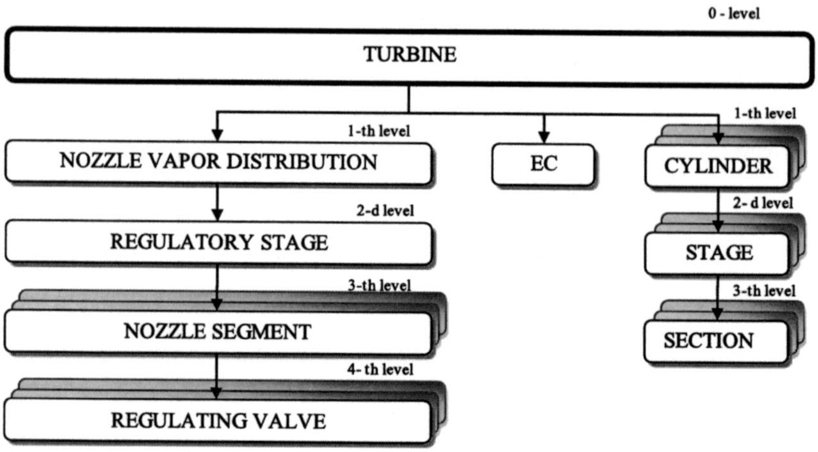

Fig. 3. Block diagram of the information model of the flowing part of the steam turbine

Mathematical models of objects of different levels used in the program complex are given in [19]. The proposed structure of the solutions to the optimization problem is implemented in such a way that it is possible to solve the optimization problem of the whole object (a powerful steam turbine) and its individual parts (NVD, HPC, MPC, LPC, separate stage, separate blade, etc.).

As a method of search optimization in the optimization subsystem, pseudo-random sequences of LPτ numbers are used. At optimizing levels, when searching for optimal solutions for each point from the set of points of the LPτ sequence, the FMM of the functional constraints is calculated. Therefore, for the points that satisfy these constraints, the FMM of the quality criteria is calculated. This optimization algorithm allows solving multicriterial problems using the convolution of the vector quality criterion. Applying the convolution of the criteria for the proposed method in solving optimization problems for various combinations of weight coefficients, we find the points farthest from the origin, thus obtaining a set of unmodified solutions corresponding to the Pareto front.

The algorithm is constructed in such a way that when choosing the optimal solution, both the solutions obtained in the computation process, the calculation of the experimental mathematical model, and the 5 best solutions using the LPτ search are involved. The software complex CAD "Turboagregat" is implemented on the principles of a single integrated information space and implies a hierarchically structured format for describing information models of optimal design objects. According to the proposed methodology of optimal design of the flowing part of powerful steam turbines in the CAD "Turboagregat" created the highest level of "Turbine" with its opera. In Fig. 4 from the window for forming the optimization task for the highest level of the "Turbine". The left part of the figure shows the structure of the project. When selecting the level of interest in the rest of the window, it becomes possible to perform optimized parameters, functional constraints, quality criteria, parameter settings, design type and optimization method. It is also seen from the figure that you can select a suitable condition that determines the status of the optimized parameter.

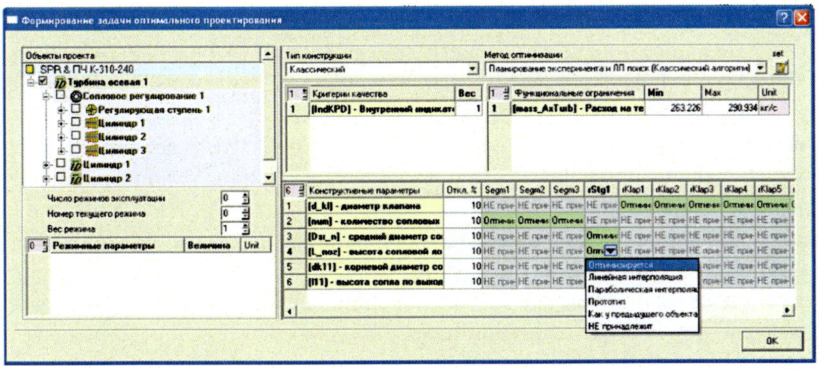

Fig. 4. The dialog for forming the optimization task of a steam turbine

Optimization of the turbine K-310-240 was carried out according to the scheme in Fig. 5, where as the object of optimization of the first level is a turbine, and at the second level, objects such as NVD, HPC, MPC and LPC are optimized. At each level, the problem is solved in accordance with the above algorithm.

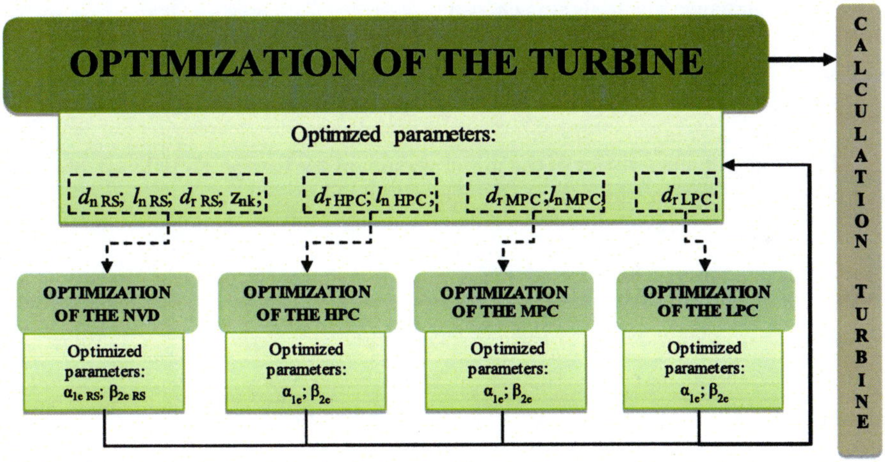

Fig. 5. Turbine optimization diagram

At the first level of the "Turbine" 16 parameters are optimized. These include the main parameters taken from the lower level: diameters of control valves ($d_{r\,RS}$); number of nozzle channels in each segment (z_{nk}); average diameter of the regulating stage ($d_{n\,RS}$); length of the nozzle blade of the regulating stage ($l_{n\,RS}$); the root diameter of the directing device of the first pressure stage of the HPC, the MPC and the LPC (d_{rCHP}, d_{rMPC}, d_{rLPC}); the height of the guide vane blade of the first pressure stage of the HPC and the MPC ($l_{n\,HPC}$, $l_{n\,MPC}$).

At the second level of “NVD” and “Cylinder” 39 parameters are optimized: the effective yield angles from all nozzle and blades grids except for the effective exit angles from the nozzle grilles of the first stages of the cylinders that ensure the throughput of the cylinders.

For the first level of optimization, the following objective functions are selected: absolute efficiency, turbine power and in equal weight fractions absolute efficiency and turbine power. At the second level, the Optimization Cylinder for HPC, MPC and LPC, the search for optimal solutions is carried out using the same goal functions as in the first approach. Optimization at the level of “NVD” was carried out by three separate objective functions: efficiency of the control stage; the power of the regulating stage; efficiency and power of the regulating stage.

When evaluating the efficiency of the initial design and solving the optimization problem, the following methods for estimating energy losses were used:

- to estimate the profile energy losses in the gratings - the Craig and Cox methods with the KhPI corrections;
- for estimation of secondary energy losses in lattices - the method of G.Yu. Stepanova;
- to estimate the energy losses from periodic nonstationarity - the technique of S.Z. Kopelev;
- To calculate the losses associated with radical leaks, the methodology given in the technical guidance materials was chosen;
- to assess the moisture losses of steam - the GE methodology;
- to determine the amount of moisture to be removed as a result of separation, the algorithms described in the book of G.A. Filippova, O.A. Povarov and V.V. Pryakhin.

The integral characteristics of the initial version of the turbine K-310-240, obtained as a result of the design studies and are given in Table 1.

Table 1. Integral characteristics of the original version of the turbine K-310-240

Parameter	Value	Parameter	Value
Absolute efficiency of the cycle	0,4441	Turbine power, MW	330,577
Efficiency NVD	0,5817	Power RS, MW	10,5111
Efficiency of RS	0,7367	Power of HPC, MW	88,7690
Efficiency of HPC	0,8098	Power of the MPC, MW	136,434
Efficiency of MPC	0,8587	Power of the LPC, MW	47,4315
Efficiency of LPC	0,7819	Theoretical work of the cycle, kJ	1380,11

- Thus, for each optimized object four tasks were solved:
- “prototype” - calculation of the prototype;
- “η” - optimization by the quality criterion of the efficiency of the optimized object;
- “N” - optimization by the criterion of quality of the power of the optimized object;

- "$\eta + N$" - optimization by the objective function, which includes the quality criterion of efficiency and the quality criterion of the power of the optimized object in equal weight fractions.

4 Results

In accordance with the methodology and algorithm described previously, a comprehensive optimization of the steam turbine K-310-240 for various target functions has been carried out, the integral characteristics are given in Table 2.

Table 2. Integral characteristics of the turbine

Parameter	Results of calculations			
	Prototype	η	N	$\eta + N$
Absolute efficiency of the cycle, ηa	0,4441	0,4525	0,4521	0,4524
Increase in the absolute efficiency of the cycle, $\Delta\eta a$,%	0	0,84	0,8	0,83
Turbine power N, MW	330,58	336,43	336,96	336,76
Capacity increase ΔN, MW	0	5,853	6,379	6,179

Such integral parameters of the turbine as the absolute efficiency of the cycle and the power of the turbine are contradictory, therefore, in the presence of selections in the flowing part of the turbine, the use of the two-criterion objective function ensures finding the best constructive solutions in terms of power quality and turbine efficiency (Table 2).

The distribution of the power gain and the level of absolute efficiency values for the turbine objects are shown in Figs. 6 and 7, respectively. The least influence on the increase in turbine power is provided by the MPC, and the largest is by HPC (Fig. 7).

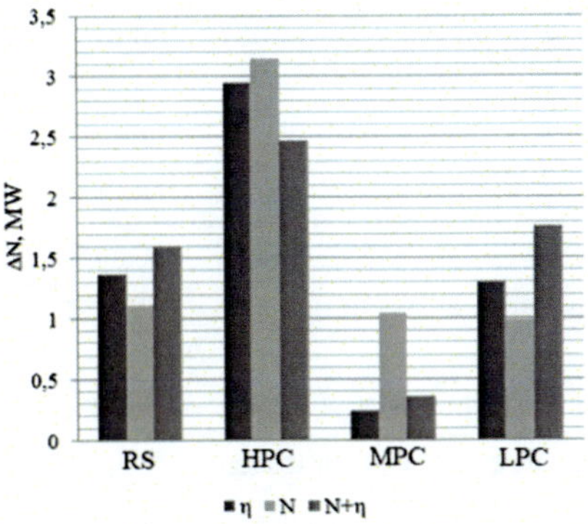

Fig. 6. Increase in the capacity of turbine objects relative to the prototype for various objective functions

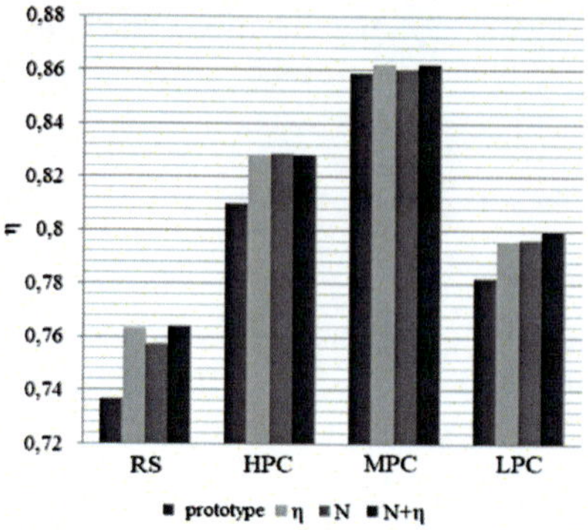

Fig. 7. Power efficiency of turbine objects depending on the objective function

The values of the optimized parameters as a result of solving optimization problems are presented in Tables 3, 4, 5 and 6. The change in the diameters of the control valves is associated with an increase in the heights of the valve lifts that result from minimizing throttle losses and the need to skip the required flow. Redistribution of the number of nozzles by segments is associated with a change in the flow area of the nozzle channels of the segments, caused by a change in the diameter and height of the nozzle grid of the regulating stage.

Table 3. The values of the optimized parameters at the level of "Turbine" (the first level)

Parameter	Results of calculations			
	Prototype	η	N	$\eta + N$
Number of channels in segment I	40	48	49	49
Number of channels in segment II	23	19	18	18
Number of channels in segment III	15	11	11	11
Diameter 1-th of the valve, m	0,0750	0,0713	0,0745	0,0741
Diameter 2-d of the valve, m	0,0750	0,0713	0,0751	0,0747
Diameter 3-th of the valve, m	0,1120	0,1065	0,1131	0,11191
Diameter 4-th of the valves, m	0,1120	0,1065	0,1104	0,1117
Diameter 5-th of the valve, m	0,1250	0,1188	0,1216	0,1207
Diameter 6-th of the valve, m	0,1250	0,1188	0,0126	0,1230
Average diameter of the nozzle cascades of the regulating stage, m	1,1750	1,1610	1,170	1,1615
Height of nozzle blade, m	0,0230	0,0225	0,023	0,0225
Root diameter of the nozzle cascade of the first stage of the HPC, m	0,982	0,944	0,950	0,9585

(continued)

Table 3. (*continued*)

Parameter	Results of calculations			
	Prototype	η	N	η + N
Root diameter of the nozzle cascade of the first stage of the MPC, m	1,202	1,202	1,202	1,202
Root diameter of the nozzle cascade of the first stage of the LPC, m	1,620	1,703	1,732	1,749
Height of nozzle blade of the first stage of the HPC, m	0,022	0,0225	0,023	0,023
Height of nozzle blade of the first stage of the MPC, m	0,081	0,081	0,081	0,081

Table 4. Optimized parameters of HPC (the second level)

Parameter	Type task	Stage number							
		1	2	3	4	5	6	7	8
Angle α_{1e}, degree	Prototype	13,83	13,85	13,90	13,95	14,03	14,08	14,15	14,28
	η	14.14	13,84	13,93	14,04	13,66	13,93	13,54	14,24
	N	13.16	13,18	13,67	13,69	14,01	14,11	13,86	14,33
	η + N	13.65	13,45	13,32	13,45	13,93	13,51	14,10	13,49
Angle β_{2e}, degree	Prototype	21,17	21,22	21,27	21,33	21,45	21,53	21,63	21,77
	η	22,65	21,85	22,35	22,49	22,3	22,61	21,92	22,52
	N	20,98	20,96	22,32	22,07	22,54	23,04	22,65	22,55
	η + N	22,34	21,93	21,35	21,61	21,84	21,84	22,13	21,47

Table 5. Optimized parameters of MPC (the second level)

Parameter	Type task	Stage number										
		1	2	3	4	5	6	7	8	9	10	11
Angle α_{1e}, degree	Prototype	13,3	13,4	14,15	13,7	15,27	15,53	14,9	17,02	15,65	16,07	17,32
	η	13.34	13,26	13,94	13,57	15,09	15,32	14,72	16,91	15,82	15,85	17,19
	N	13.13	13,36	14,03	13,62	15,13	15,68	15,11	17,23	15,80	16,17	17,17
	η + N	13.27	13,27	14,01	13,56	15,11	15,38	14,75	16,85	15,49	15,91	17,14
Angle β_{2e}, degree	Prototype	20,58	20,57	20,5	21,28	21,22	21,08	20,8	24	21,07	21	20,57
	η	21,29	20,85	20,85	21,06	20,90	20,79	20,51	23,65	20,75	20,52	20,28
	N	20,82	20,97	21,13	21,87	21,39	21,22	20,84	24,21	21,35	20,92	20,50
	η + N	21,31	20,87	20,91	21,28	21,00	20,87	20,59	23,76	21,17	20,82	20,46

Table 6. Optimized parameters of LPC (the second level)

Parameter	Type task	Stage number			
		1	2	3	4
Angle α_{1e}, degree	Prototype	14,1	15,33	17,65	17,583
	η	13.55	14,486	17,65	17,583
	N	13.24	14,618	17,65	17,583
	$\eta + N$	13.18	14,442	17,65	17,583
Angle β_{2e}, degree	Prototype	20,283	18,9	18,283	26,183
	η	19,151	19,165	18,283	26,183
	N	19,528	19,934	18,283	26,183
	$\eta + N$	19,13	19,429	18,283	26,183

With regard to the optimized geometric parameters of the MPC, as a result of optimization, the root diameter of the nozzle cascade of the first stage of the HPC decreased, and the root diameter of the nozzle cascade of the first stage of the LPC increased. The height of the nozzle cascade of the first stage of the HPC after optimization has increased slightly.

The increase in power, as noted earlier, is achieved due to the redistribution of the heat transfer along the pressure steps of the flowing part of the turbine, therefore, consider the distributions of heat differences in the cylinders shown in Figs. 8, 9 and 10.

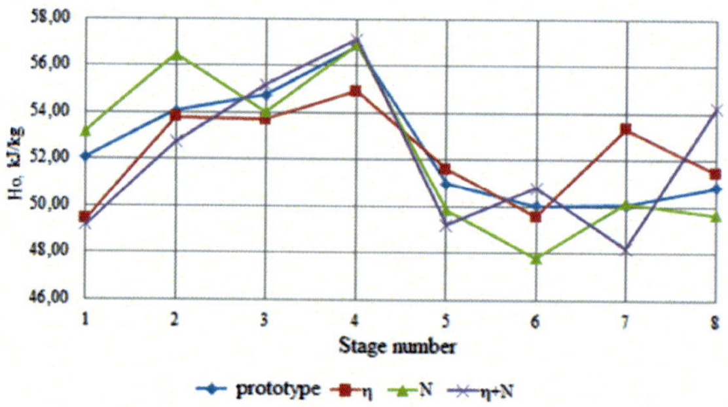

Fig. 8. Distribution of the available heat transfer along the HPC stages

Depending on the objective function, the curves for the variation of the heat transfer along the HPC and LPC steps have a similar character, except for the first stage of the LPC, where the changes are caused by abrupt changes in the effective exit angle from the nozzle array.

Proceeding from the fact that the best design option is adopted as a result of optimization of the turbine based on the two-criterion objective function, we compare the distribution of the heat drop of the HPC in this variant with the prototype. The heat

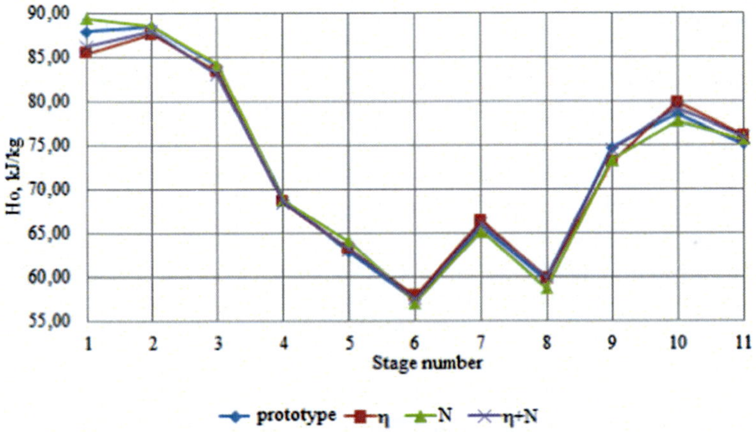

Fig. 9. Distribution of the available heat transfer along the MPC stages

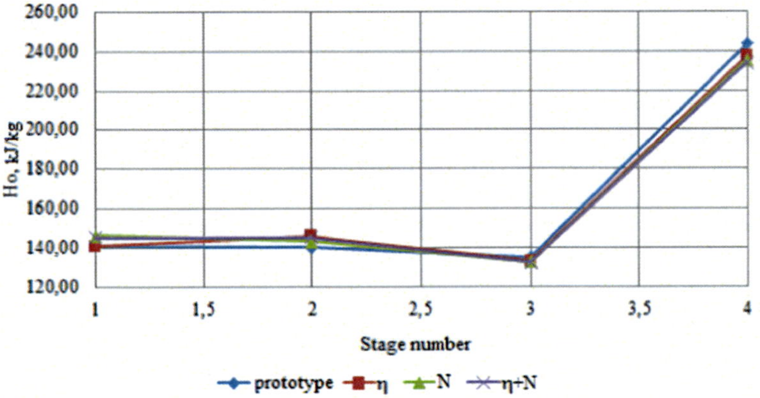

Fig. 10. Distribution of the available heat transfer along the LPC stages

transfer at the fourth stage is almost the same, in contrast to the other stages. At the 1st, 2nd, 5th and 7th stages the heat transfer of the received structure is lower than that of the prototype. Despite this, the power gain was ~2.5 MW due to the increase in heat dissipation of the 3.6 and 8 stages, as well as the power efficiency and flow rate in these stages.

In addition, the increase in power is also affected by the increase in power efficiency due to a significant decrease in radical and radial leakage, caused by a decrease in the degree of reactivity. A slight increase in the efficiency of the nozzle cascades also affects the efficiency of the turbine.

Based on the results obtained and the comparative analysis, depending on the objective function, the distribution of the integral characteristics of the turbine according to its objects is very diverse.

5 Conclusions

The proposed object-oriented approach for complex optimization of the flowing part of a powerful turbine implemented in the CAD "Turboagregat" has shown its effectiveness in the example of the turbine K-310-240.

For the first time, the mutual influence of the turbine objects on its optimal characteristics was identified and assessed.

Application of the universal methodology for CTS has shown its effectiveness in multi-level and multi-criteria optimization of the turbine K-310-240 in the composition, which includes different types of objects. As a result of complex optimization of this turbine, its power was increased by 6,179 MW (~1.87%), and the increase in the absolute efficiency of the cycle was 0.83% relative to the prototype.

To obtain a simultaneous increase in two quality indicators, such as efficiency and power as an objective function, it is necessary to take them in equal parts.

Analysis of the obtained results shows that the optimization carried out not only leads to a change in the geometric parameters, but also to the redistribution of heat drops between the stages of the pressure cylinder, which in turn contributes to an increase in the efficiency of the drive, cylinder in the stages, which have higher values of efficiency and flow.

References

1. Xu, C., Amano, R.S.: A turbomachinery blade design and optimization procedure. In: Proceedings of ASME Turbo Expo 2002, pp. 927–935. Amsterdam, The Netherlands (2002)
2. Smith, R.W.: Steam turbine cycles and cycle design optimization: combined cycle power plants. Advances in Steam Turbines for Modern Power Plants, pp. 57–92. Woodhead Publishing (2017)
3. Cao, L., Si, H., Lin, A., Li, P., Li, Y.: Multi-factor optimization study on aerodynamic performance of low-pressure exhaust passage in steam turbines. Appl. Therm. Eng. **124**, 224–231 (2017)
4. Shao, S.: Aerodynamic optimization of the radial inflow turbine for a 100kw-classmicro gas turbine based on metamodel-semi-assisted method. In: Proceedings of ASME Turbo Expo 2013: Turbine Technical Conference and Exposition. Volume 6B: Turbomachinery, V06BT37A032. San Antonio, Texas, USA (2013)
5. Nowak, G.: Pareto multicriteria optimization of airfoil cooling system. In: Proceedings of the 8th European Turbomachinery Conference, Graz (2009)
6. Turner, M.G., Park, K. [at alias]: Framework for multidisciplinary optimization of turbomachinery. In Proceedings of ASME Turbo Expo: Power for Land, Sea, and Air. Volume 7: Turbomachinery, Parts A, B, and C, pp. 623–631. Glasgow, UK (2010)
7. Goulos, I., Hempert, F., Sethi, V., et al.: Rotorcraft engine cycle optimization at mission level. Proc. ASME Turbo Expo: J. Eng. Gas Turbines Power. **135**(9), 091202 (2013) (GT2013-95678)
8. Wu, Y., Li, B., Teng, J., et al.: Automated design optimization and experimental validation for intermediate casing duct of aeroengine. In: Proceedings of ASME Turbo Expo: Turbine Technical Conference and Exposition. Volume 6B: Turbomachinery. San Antonio, Texas, USA. V06BT43A004, GT2013-94137 (2013)

9. Lytvynenko, O., Tarasov, O., Mykhailova, I., Avdieieva, O.: Possibility of using liquid-metals for gas turbine cooling system. In: Ivanov, V., Pavlenko, I., Liaposhchenko, O., Machado, J., Edl, M. (eds.) Advances in Design, Simulation and Manufacturing III. DSMIE 2020. Lecture Notes in Mechanical Engineering. Springer, Cham (2020). https://doi.org/10.1007/978-3-030-50491-5_30
10. Horn, J., Nafpliotis, N., Goldberg, D.E.: A niched pareto genetic algorithm for multiobjective optimization. In: Proceedings of the First IEEE Conference on Evolutionary Computation, IEEE World Congress on Computational Intelligence, IEEE Service Center, vol. 1, pp. 82–87. New Jersey, Piscataway (1994)
11. Qin, X., Chen, L., Sun, F., Wu, C.: Optimization for a steam turbine stage efficiency using a genetic algorithm. Appl. Therm. Eng. **23**(18), 2307–2316 (2003)
12. Safari, A., Lemu, H. G., Assadi, M.: A novel combination of adaptive tools for turbomachinery airfoil shape optimization using a real-coded genetic algorithm. In: Proceedings of ASME Turbo Expo: Turbine Technical Conference and Exposition. Volume 6B: Turbomachinery. San Antonio, Texas, USA. V06BT43A003 (2013)
13. Ogaday, W., Moore, W., Mala-Jetmarova, H., Gebreslassie, M., Tabora, G.R., Belmont, M.R., Savic, D.A.: Comparison of multiple surrogates for 3D CFD model in tidal farm optimization. Procedia Eng. **154**, 1132–1139 (2016)
14. Mehmani, A.: Uncertainty-integrated surrogate modeling for complex system optimization. Ph.D. thesis, Syracuse University (2015)
15. Yang, X., Liu, BO., Cao, Z.: Opposition-based artificial bee colony algorithm application in optimization of axial compressor blade. In: Proceedings of ASME Turbo Expo, GT2013-95177 (2013)
16. Usatyi, O., Avdieieva, O., Maksiuta, D., Tuan, P.: Experience in applying DOE methods to create formal macromodels of characteristics of elements of the flowing part of steam turbines. In: AIP Conference Proceedings, vol. 2047, no. 1, p. 020025, (2018)
17. Kelin, A., Larin, O., Naryzhna, R., ... Vodka, O., Shapovalova, M.: Mathematical modelling of residual lifetime of pumping units of electric power stations. Advances in intelligent systems and computing, 1113 AISC, pp. 271–288 (2020)
18. Boiko, A., Govorushchenko, Y.: Optimization of the Axial Turbines Flow Paths. Science Publishing Group, New York, NY 10018, U.S.A. (2016)
19. Boiko, A.V., Usaty, A.P., Avdieieva, O.P.: Methodology of the object-oriented complex optimization of the flow passes of powerful steam turbines taking into consideration the variable operation mode. NTU "KhPI" Bulletin: Series "Power and Heat Engineering Processes and Equipment", vol. 13, pp. 5–10 (2014)
20. Avdieieva, O., Usatyi, O., Vodka, O.: Development of the typical design of the high-pressure stage of a steam turbine. In: Ivanov, V., Pavlenko, I., Liaposhchenko, O., Machado, J., Edl, M. (eds.) Advances in Design, Simulation and Manufacturing III. DSMIE 2020. Lecture Notes in Mechanical Engineering. Springer, Cham (2020). https://doi.org/10.1007/978-3-030-50491-5_26

Collaborative Mass Customization of Footwear: Conceptualization of a Three-Stage Holistic Model

Nelson Oliveira(✉), Helder Carvalho, and Joana Cunha

Department of Textile Engineering, 2C2T – Centre for Textile Science and Technology, University of Minho, Guimarães, Portugal
id6963@alunos.uminho.pt

Abstract. Despite Mass Customization being already well known and quite studied in the literature, the interest in this productive approach has not been exhausted. This happens due to the continuous evolution of the paradigm based on technological evolution and on the increasing heterogeneity of the consumer profile. In fact, in the fashion industry and specifically in the footwear sector, there is a continuous interest in approaches that answer to commercial pretensions of the brands/industry, to consumers differentiation needs and to requirements of sustainability. Thus, Collaborative Mass Customization of Footwear (CMCF) is a solution to be considered by business strategy evaluators, such as managers, marketers and designers. The objective of the present study is to define the principles for a more holistic methodology to assist the development of CMCF, based on a literature review. As results, a proposal of a CMCF model and framework are presented. It is aimed to contribute to the theoretical reflection of the Co-design applied to the Mass Customization of footwear.

Keywords: CMCF model · Shoes · Co-design · Framework · Workflow

1 Introduction

1.1 Theoretical Framework

Fashion markets are particularly sensible to the consumers' incessant search for novelty. In fact, this has been one of the main key drivers of the industry, over the last decades [1]. This mindset reveals itself very dangerous to society, economy and Nature, resulting in intensive approaches, such as mass production and fast-fashion systems that seek profit at all cost in growing demand markets for low-cost products [2]. However, with the growth of the current millennium, contemporary society has been debating the harmful impact that the excesses of production and consumption have inflicted on communities and Nature. Therefore, a gradually collective adoption of sustainability approaches has been taken. In the fashion universe, new paradigms have emerged, such as Ethical Fashion, Green Fashion, Upcycling, Slow Fashion, Minimum Waste, among others, and combined with other practices, such as Fair Trade – all having in common to counteract side effects

J. Machado et al. (Eds.): icieng 2021, LNME, pp. 214–225, 2022.
https://doi.org/10.1007/978-3-030-79165-0_21

of capitalist ambition [3]. Predictably, in this conjecture, the design assumes a preeminent role of mediation between the commercial pretensions of brands/industry, the interests of the consumer and the requirements of sustainability [4–6]. Nevertheless, despite the "disposable" being "out of fashion" and increasingly rejected in favor of the precepts of preserving resources and the environment, the hedonistic needs of consumers remain, and it is up to social responsibility of brands/industries to contribute to educating the consumer offering conscious and balanced proposals [6–10]. For this purpose, managers, marketers and designers must make use of methodologies and approaches capable of assisting them in pursuing that ambitious goal. The knowledge, application and mastery of different ways of thinking, designing and creating products/services are the greatest assets that a professional in this area should desire to achieve [11, 12]. Over the last years, the footwear sector has been witnessing a transition from Mass Production to Mass Customization (MC), generating new opportunities and business models [13, 14]. The present paper explores these notions and proposes a conceptual model with the key elements to the operationalization of Collaborative Mass Customization of Footwear (CMCF). In this way, the paper starts with an overview of the key concepts, a brief review of consumers' profiles, and an overview of footwear customization. In the following section is presented de theoretical definition of the CMCF process. To synthesize all the research, a framework of the developed conceptual model is presented.

1.2 Research Objective and Methodology

The main objective is to define the bases for a methodology to assist the development of CMCF process, in a holistic way. The selected methodology is divided into four phases: theorization – a literature review; operationalization – exploratory nature with the development of the analysis model (based on literature review); results – construction and discussion of a synthesis framework; review of the contents – inferences and conclusions of the study.

2 State of the Art

2.1 Mass Customization and Co-design

The literature presents several definitions of MC [15]. The present study considered MC as the production approach based on offering customized goods on a large scale and satisfying more adequately customers' individual needs at an affordable price [16]. Fogliatto et al. [15] consider MC a response of the industry to consumers' demand for affordable customized products/services, connoted as "customer-centric". Concerning Collaborative Design (Co-design), this creative method especially favors any act of cooperative creativity shared at least by two people, and it could be practiced across the entire design process [17]. Thus, MC and Co-design have in common being user-centered, promoting cooperation, interaction, and articulation between different players, in order to satisfy interdependent needs. Therefore, these approaches are intrinsically related. The main differentiation factor of MC is customer collaboration in the design process. Thus, the concept of "Collaborative Mass Customization" (CMC) becomes clear [18, 19]. This methodology is grounded on the cooperation between the brand (endorsing corporative image and values) and the customer in the creative process, promoting customer's emotional connection with the product and brand. This phenomenon is strongly advocated by Emotional Design approaches, which should be considered in the CMC process because through the co-creation of products and immersive experiences it is possible to evoke or avoid specific emotions in the consumer [20]. CMC consequently tends to increase customer loyalty, when trust is achieved between the players, and the customer is satisfied with the created product [21].

2.2 Changes in Consumers Profiles

Consumers profiles are continuously changing as a result of society's evolution and the growing heterogeneity [2, 4]. However, two main common characteristics define the present consumption paradigm – individualism and sustainability. Although they may seem to be mutually exclusive, these concepts can exist in harmony. In the course of individualization, consumers aim for products with a high level of differentiation, strongly associated with slow-fashion and sustainability approaches as CMC [22, 23]. Compared to previous generations, today´s consumers reveal more interest in decision-making management [24], finding in CMC a way to state their uniqueness by customizing products that best match their own personal convictions, distinguishing or approaching them from the others [25]. In addition, this active role improves symbolic benefits (intrinsic and social) required by consumers, namely pride-of-authorship effect [26]. This results in customers' higher willingness to pay a premium price for customized products, revealing a greater awareness of value incrementation conceded to products that better match consumer's needs and expectations [27]. However, these expectations tend to be higher than for standard or regular products, coming from massification [24, 28, 29]. The need for satisfying the individual customers' specificities is now stronger than ever. Customers expect that products fulfill their personal requirements in an individualized manner [30].

Individualism and hedonism are not exclusive to younger generations. In fact, the increase of the elderly population generates a consumer group highly interested in premium quality, tailor-made products and little willingness to compromise [24]. Thus, a new profile of hybrid customer is growing fast in worldwide markets, strongly characterized by switching between different products, basing their behavior and decisions on alternating key criteria – fluctuating between quality, price and brand. In general, the pure desire for change and novelty leads consumers to take risks and demand higher product/service differentiation [10, 21, 31].

2.3 Footwear Customization Overview

The industrial footwear sector is traditionally focused on mass markets, but in the present these markets are becoming more and more heterogeneous. Consumer trends indicate that segments are becoming niches and micro-niches [24, 32]. Thus, the footwear market is undergoing restructuring. The days when mass production was absorbed by the market are gone, opening the opportunity to new business strategies as CMC. The consumer of the present is more informed and knowledgeable, becoming more and more an active player [23]. In fact, consumers play an increasingly important role in the design process, interacting with creatives to tailor the product to their real needs. This is reflected in the growing demand not for large amounts of product, but rather for a wide range of products. To match this reality, footwear brands and producers are adapting, following market segmentation policies, focusing on very specific niches. This leads to the need for companies to adjust and develop appropriate technologies and processes across the value chain, from creation, production, distribution, sales and after-sales service. The demand for highly differentiated products and consequent high added value has been a strong trend for the footwear sector across Europe [32–34].

Based on the study of the literature, was verified that CMC models lack a holistic view and do not attend to the specificities of the current footwear industry. For instance, some models are focused on the product manufacturing process and others on the communication process. Head and Porter [35] developed a footwear personalization tool for use in physical stores and not in online context, focusing only on the product. Simpson et al. [36], Kwon, Ha and Kowal [37], and Kang [38] carried out studies related to the development and analysis of customizable products and online customization platforms, but not specifically for footwear. The EuroShoe [24] study focused on the product and manufacturing dimensions. However, some analyses, methodologies and results of these studies were considered.

3 CMCF Model

The proposed CMCF model is divided in three main stages: Understand (Discovery and Evaluation), Explore (Conception and Customization), and Materialize (Production and Distribution). Although this sequential three-stage approach is based on the IDEO´s design thinking process [39], the subsequent six phases result from other backgrounds, namely EuroShoe [24] for manufacturing and PURAM [40] for communication.

3.1 Discovery

The first phase is conducted by the industry; however, the consumer should participate actively. Consists of three main steps – Strategy, Target and Requisites:

- **Strategy:** the cycle starts with the necessary business-to-consumer (B2C) strategy plan with internal and external analysis. CMCF is a hybrid competitive strategy combining differentiation and cost control [24, 33, 41].
- **Target:** definition of a specific consumer niche (market segmentation), ensuring an effective marketing mix alignment, leading to factual benefits, which includes a clearer understanding of consumer differences and specificities, able to improve the match of organizational assets and consumer requirements [42].
- **Requisites:** objective definition of the consumers' needs, desires, experiences and expectations specificities. The center of the CMC process has to be this information gathering concerning the brand, product and communication dimensions [24, 42].

3.2 Conception Stage

The second phase is conducted by the industry; however, the consumer should participate actively. Consists of three main steps – Approach, Solutions and Interface:

- **Approach:** definition of the footwear customizable dimensions, namely: style/aesthetics, fit/comfort and functionality/performance [5, 43]. The style/aesthetic is the easiest, prompt, popular, and flexible dimension to intervene, where the consumer designs the product choosing between pre-defined options. The fit/comfort is more complex to carry on since it requires specific tailoring of the footwear size width and shape (interfering with the footwear last), along with components and materials. The functionality/performance dimension is related to technical requirements, concerning the selection of design configuration, components and manufacturing procedures to best match the user's intentions of usage. Analyzing cost impact, the style customization is the more immediate, feasible and appealing dimension both to customers and producers. Nevertheless, concerning the value creation there are no empirical facts that confirm if this dimension is the most significant [26, 41, 43].
- **Solutions:** transfer the consumers' expectations and needs into product specifications (critical step) [24]. CMC solutions are mostly based on modular structural design of the elements, allowing the customer to tailor the product based on different modular combinations [15, 44]. In this way, customers perform Co-design activities within a list of options and pre-defined components [19, 26]. CMCF is strongly supported by technology. Nevertheless, taking into account that this type of product presents some constraints associated with the manufacturing process and the performance of materials and components [45], it is possible to change several attributes in the modular architecture of footwear, namely upper (surface module), accessories (components module) and sole (bottom module) [24, 45]. The intervention based on materials is generally easy, attending to the large range of usable options in the upper module, from traditional materials (several types of leather) to innovative leather tanning techniques, specific textiles structures, new polymeric materials and advanced smart textiles. However, the intervention in the sole module tends to be more difficult, particularly in injected soles from traditional production, being more feasible to customize only colors [43, 46, 47]. In fact, the intervention at the color level (keeping the same material) is the simplest and most viable method to customize all the modules [43]. Other possibilities are laser cut and engraving [48, 49], embroidery [49] and seamless production [50]. 3D-Printing process, based on additive manufacturing, is revealing the potential to become a prominent technique to increase footwear customization level, in the near future. The production of components, such as embellishment elements or soles, is expected, but the technology community is working towards the disruptive possibility of developing full-printed footwear. This potential has been already partially evidenced through fully implemented processes that print 3D layers on a textile substrate surface, then used in the upper module. This technology has the potential to allow creating new designs and choosing textures, colors and gloss levels in an easy way [51, 52].
- **Interface:** the distinctive factor of CMCF is based on the imperative inclusion of the customer into the value chain. To allow this, a multi-channel strategy combining

innovative online platforms offers further possibilities. From the marketing strategy perspective, the interface between producers and consumers presents many differentiation possibilities. Companies have to understand the importance of transforming the configuration process into a truly engaging customer experience [24]. This particular step is deeply studied in the PURAM model [40], developed within the scope of the same investigation of this article. The model defines the main requirements to develop an online interface of footwear customization, through the conceptualization of five key dimensions – Entertainment, Organization, Informativeness, Trust and Convenience (that become operable through their division in categories of analysis, subcategories and elements).

3.3 Customization Stage

In the third phase, the consumer and the interface play important roles; however, the industry participates actively. Consists of three main steps (more deeply presented in the PURAM model [40]) – Experience, Design and Purchase:

- **Experience:** apply the principles of experience marketing and Emotional Design. Being consumption primarily motivated by hedonism [24], it is important to trigger appropriate emotions, aligned with the brand, product and consumer background. Entertainment emerges as a catalyst for interactive and engaging experiences, with the ability to evoke positive emotional and cognitive states in the user, resulting in the willingness to spend more time and money on the experience [40, 53].
- **Design**: critical step where the user proceeds to the customization design, carrying out a logical and sequential set of tasks that guide him in the process. Users require skilled advising and consulting, particularly when customization options are in large number [24]. Interface functionality and usability are extremely important to assist users in the identification of solutions and minimizing process complexity [20, 40, 53].
- **Purchase**: order accomplishment; CMCF is a make-to-forecast or build-to-order production approach (the production is preceded by the order), usually based on the online sales channel – most convenient for customers and producers, saving time and minimizing the effort required to make a purchase and obtain the good. Being a transaction, it is important to promote trust and convenience [24, 29, 30, 40, 53].

3.4 Production Stage

The fourth phase is conducted by the industry; however, the consumer and the interface also participate actively. Consists of three main steps – Supply, Manufacture and Control:

- **Supply:** all the information collected from the customization phase is deconstructed to determine all the necessary components and production requisites, to fulfill each specific part of the product and to interact with the warehouse and suppliers. This is a very critical step to ensure the appropriate workflow process [24, 34, 35]. Through the Industry 4.0 principles is possible to reduce the complexity [54].

- **Manufacture:** up to this moment, the CMCF process was entirely on the information level. This step requires planning and manufacturing the footwear after the order has been placed and based on the design configuration made by the user. A flexible manufacturing system is essential to accomplish the specific tasks of each small batch, or even a single product. The main conventional subphases are cutting, stitching, lasting and finishing. [24, 34, 35, 41, 43, 55].
- **Control**: quality control of the customized product and order correspondence verification [34, 53].

3.5 Distribution Stage

The fifth phase is conducted by the industry; however, the consumer and the interface also participate actively. Consists of three main steps – Packing, Shipping and Delivery:

- **Packing:** starts with the verification of the order correspondence, packaging the product in an appropriate container and getting it fully prepared for shipment. The package also can be part of the customization features [24, 52, 56].
- **Shipping:** making the product available for the consumer. This can be done directly by the producer or service provider. Shipping information should be provided to the customer. This service can be customized attending to the customer convenience and processual information should be available, indicating the expected time to manufacturing, tracking and expected date of delivery [40].
- **Delivery:** customer physical access to the product. This service also can be customized attending to the customer convenience [40, 53].

3.6 Evaluation Stage

The sixth phase is conducted by the industry; however, the consumer and the interface also participate actively. Consists of three main steps – After sales, Feedback and Improvement:

- **Customer service:** help and information service provided to customers before, during and after the customization. Directly related to customer satisfaction and retention, providing information and convenience, e.g. ease of exchange/refund/return. This service is also responsible for collecting data on consumer behavior, essential to generate feedback [14, 24, 40].
- **Feedback:** measurement and report of the data collected by interacting with customers and consumers, in order to deepen and re-direct market-driven and technological competencies [12, 24, 30, 40].
- **Improvement:** consequences of the analysis and practical application of all the generated and collected knowledge of each and all step of the CMC process, in order to improve efficiency and quality, generating constant improvements and innovation opportunities of processes, products and services [5, 24, 32, 53].

4 CMCF Framework

The CMCF framework (Fig. 1) is the synthesis of all the entire developed model. The image in the center illustrates the key subject (CMCF by digital interface), surrounded by the three stages of the model (Understand, Explore and Materialize), respective phases and corresponding steps. The circular configuration seeks to demonstrate the sequence and continuity of the process, reinforced by the dashed line and arrows. Next to each phase, different icons represent the industry, consumers and the interface – its size differs according to the impact that each one assumes in the respective phase (big/small). In the lower right corner, appears the considered type of business-model (B2C). Through a simple and appealing workflow design, all the abstract concepts are presented, in order to an easier understanding of the advocated model.

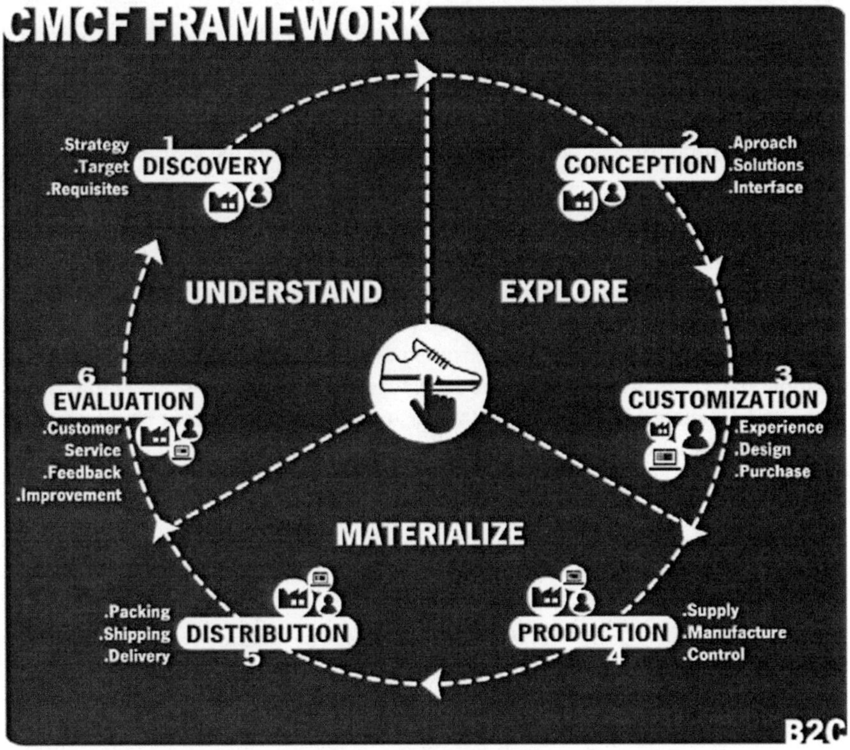

Fig. 1. CMCF framework.

5 Conclusions

In the present social and economic conjecture, the design assumes a preeminent role of mediation between the commercial pretensions of the brands/industry, the interests of the consumer and the requirements of sustainability. The footwear sector has been witnessing a transition from Mass Production to Mass Customization (MC), generating new opportunities and business models based on collaborative approaches. Thus, the Collaborative Mass Customization of Footwear (CMCF) reveals to be a win-win shared experience for the industry and specific market niches, revealing a B2C solution to be considered by business strategy evaluators, such as managers, marketers and designers. Thus, is essential to understand the operationalization process, to assist professionals in the implementation of this productive, creative and relational approach.

The proposed objective was fulfilled by the development of the CMCF conceptual model. Thus, through the conceptualization of three key stages – Understand, Explore and Materialize (that become operable through their division into respective phases and corresponding steps) – it was possible to congregate a holistic approach of continuous improvement which includes industry, consumer and interface, emphasizing the product and communication. Results were the conceptual model and the corresponding framework, in order to an easier understanding of the advocated abstraction.

6 Limitations and Final Considerations

The proposed model has no ambition to cover all the specificities of every single firm, rather it was designed to be specific enough to provide a base plan for action and implementation of CMCF. Moreover, it is assumed that in this stage the model needs further validation and verification in the field. However, the results reveal new dimensions of discussion and analysis. Thus, it is intended to continue the present study, in order to give it a greater foundation and scientific legitimacy, namely through the integration of professionals and consumers.

Acknowledgment. This work is supported by Project UID/CTM/00264/2019 of 2C2T – Centre for Textile Science and Technology, funded by National Funds through FCT/MCTES and by FAMEST Project (RTD Project in mobilizer co-promotion nr. 24529, 2017-2020).

References

1. Zeba, F., Ganguli, S.: Novelty seeking as driving factor for fashion apparel innovators: a qualitative investigation. J. Glob. Fash. Mark. **10**, 246–266 (2019)
2. Hernández, A.: Fast and Slow Fashion as Seen Through the Millennial Mindset (2018). http://rave.ohiolink.edu/etdc/view?acc_num=osu1524092309937056
3. Gardetti, M.A., Torres, A.L.: Sustainability in Fashion and Textiles: Values, Design, Production and Consumption. Routledge, New York (2017)
4. Cyrklaff, D.: Sustainability as a Trend in Fashion Industry. An Empirical Study about Changing Consumer Behaviour (2017)

5. Homburg, C., Schwemmle, M., Kuehnl, C.: New product design: concept, measurement, and consequences. J. Mark. **79**, 41–56 (2015)
6. Vezzoli, C., Kohtala, C., Srinivasan, A., Xin, L., Fusakul, M., Sateesh, D., Diehl, J.C.: Product-service System Design for Sustainability. Routledge, New York (2017)
7. Calderon Huerta, M.R.: Corporate Social Responsibility (CSR) in times of Global Financial Crisis. A study of CSR Reporting practices (2013)
8. Emmitt, S.: Design Management for Sustainability. Routledge, New York (2017)
9. Goworek, H., Fisher, T., Cooper, T., Woodward, S., Hiller, A.: The sustainable clothing market: an evaluation of potential strategies for UK retailers. Int. J. Retail Distrib. Manag. **40**, 935–955 (2012)
10. Noh, M.: Hedonic/utilitarian cool attitudes: a comparison of young female and male consumers. Curr. Trends Fash. Technol. Text. Eng. **1**, 322–340 (2017)
11. Fernanda, G.-L.: Business process architecture design methodologies – a literature review. Bus. Process Manag. J. **25**, 1317–1334 (2019)
12. Qu, M., Yu, S., Chen, D., Chu, J., Tian, B.: State-of-the-art of design, evaluation, and operation methodologies in product service systems. Comput. Ind. **77**, 1–14 (2016)
13. World Footwear: How design is passing on to consumers. https://www.worldfootwear.com/news/how-design-is-passing-on-to-consumers/2494.html. Accessed on 01 June 2017
14. Oliveira, N., Cunha, J., Carvalho, H.: Online footwear customization: experts' overview. In: Textiles, Identity and Innovation: In Touch: Abstract book of the 2st International Textile Design Conference (D_TEX 2019), p. 16. June 19–21, 2019, University of Lisbon, Lisbon, Portugal (2019)
15. Fogliatto, F.S., da Silveira, G.J.C., Borenstein, D.: The mass customization decade: an updated review of the literature. Int. J. Prod. Econ. **138**, 14–25 (2012)
16. Pine, B.J., Victor, B., Boynton, A.C.: Making mass customization work. Harv. Bus. Rev. **71**, 108–111 (1993)
17. Sanders, E.B.N., Stappers, P.J.: Co-creation and the new landscapes of design. CoDesign **4**, 5–18 (2008)
18. Wu, W., Jo Anderson-Connell, L., Ulrich, P.V: Consumer co-design of apparel for mass customization. J. Fash. Mark. Manag. Int. J. **7**, 398–412 (2003)
19. Gilmore, J.H., Pine, B.J.: The four faces of mass customization. Harv. Bus. Rev. **75**, 91–101 (1997)
20. Espejo, R., Dominici, G.: Cybernetics of value cocreation for product development. Syst. Res. Behav. Sci. **34**, 24–40 (2017)
21. Dominici, G., Roblek, V., Abbate, T., Tani, M.: Click and drive: consumer attitude to product development: towards future transformations of the driving experience. Bus. Process Manag. J. **22**, 420–434 (2016)
22. Gupta, S., Gwozdz, W., Gentry, J.: The role of style versus fashion orientation on sustainable apparel consumption. J. Macromark. **39**, 188–207 (2019)
23. Oliveira, N., Cunha, J.: Footwear customization: a win-win shared experience. In: Textiles, Identity and Innovation: Design the Future: Proceedings of the 1st International Textile Design Conference (D_TEX 2017), pp. 273–280. November 2–4, 2017, CRC Press, Lisbon, Portugal (2018)
24. Consortium EuroShoe: The market for customized footwear in Europe: market demand and consumer's preferences, Munich (2002)
25. Zhang, M., Tseng, M.M.: A Product and Process Modeling Based Approach to Study Cost Implications of Product Variety in Mass Customization (2007)
26. Piller, F.T.: Mass customization: reflections on the state of the concept. Int. J. Flex. Manuf. Syst. **16**, 313–334 (2004)
27. Franke, N., Piller, F.: Value creation by toolkits for user innovation and design: the case of the watch market. J. Prod. Innov. Manag. **21**, 401–415 (2004)

28. Wind, J., Rangaswamy, A.: Customerization: the next revolution in mass customization. J. Interact. Mark. **15**, 13–32 (2001)
29. Sandrin, E., Trentin, A., Grosso, C., Forza, C.: Enhancing the consumer-perceived benefits of a mass-customized product through its online sales configurator: an empirical examination. Ind. Manag. Data Syst. Data Syst. **117**, 1295–1315 (2017)
30. Mourtzis, D., Doukas, M., Psarommatis, F., Giannoulis, C., Michalos, G.: A web-based platform for mass customisation and personalisation. CIRP J. Manuf. Sci. Technol. **7**, 112–128 (2014)
31. Joshanloo, M., Jarden, A.: Individualism as the moderator of the relationship between hedonism and happiness: a study in 19 nations. Pers. Individ. Dif. **94**, 149–152 (2016)
32. Marques, A.M.D.R., da Guedes, M.G., Ferreira, F.: How innovation creates value in the Portuguese fashion value chain. In: 8th International Textile, Clothing and Design Conference-ITC&DC 2016, pp. 388–393. University of Zagreb, Faculty of Textile Technology, Croatia (2016)
33. Georgoulias, K., Papakostas, N., Chryssolouris, G.: Novel business models for manufacturing firms (2009)
34. Pandremenos, J., Georgoulias, K., Chryssolouris, G., Jufer, N., Bathelt, J.: A shoe design support module towards mass customization (2010)
35. Head, M., Porter, C.S.: Developing a collaborative design toolkit for the personalisation of running shoes. Des. Princ. Pract. **5**, 303–325 (2011)
36. Simpson, T.W., Umapathy, K., Nanda, J., Halbe, S., Hodge, B.: Development of a framework for web-based product platform customization. J. Comput. Inf. Sci. Eng. **3**, 119–129 (2003)
37. Kwon, S., Ha, S., Kowal, C.: How online self-customization creates identification: antecedents and consequences of consumer-customized product identification and the role of product involvement. Comput. Human Behav. **75**, 1–13 (2017)
38. Kang, J.-Y.M.: Customer interface design for customer co-creation in the social era. Comput. Human Behav. **73**, 554–567 (2017)
39. Goller, I., Bessant, J.: Design thinking. In: Creativity for innovation management, pp. 281–291. Routledge, New York (2017)
40. Oliveira, N., Cunha, J.: Co-design and footwear: breaking boundaries with online customization interfaces. Int. J. Vis. Des. **13**, 1–26 (2019)
41. Oliveira, N., Cunha, J., Carvalho, H.: Co-design and mass customization in the portuguese footwear cluster: an exploratory study. Procedia CIRP **84**, 923–929 (2019)
42. Dolnicar, S., Grün, B., Leisch, F.: Market Segmentation Analysis: Understanding It, Doing It, and Making It Useful. Springer Singapore, Singapore (2018)
43. Boër, C.R., Dulio, S.: Mass Customization and Footwear: Myth, Salvation or Reality?. Springer, London (2007). https://doi.org/10.1007/978-1-84628-865-4
44. Pandremenos, J., Chryssolouris, G.: Modular product design and customization. In: Proceedings of the 19th CIRP Design Conference–Competitive Design, pp. 94–98. Cranfield University Press (2009)
45. Luh, Y.-P., Wang, J.-B., Chang, J.-W., Chang, S.-Y., Chu, C.-H.: Augmented reality-based design customization of footwear for children. J. Intell. Manuf. **24**, 905–917 (2013)
46. APICCAPS: World Footwear Yearbook 2018. Porto (2018)
47. Deselnicu, V., Crudu, M., Zainescu, G., Albu, M.G., Deselnicu, D.C., Guþã, S.A., Ioannidis, I., Gurău, D., Alexandrescu, L., Constantinescu, R.R.: Innovative materials and technologies for sustainable production in leather and footwear sector. Rev. Piel. Incaltaminte (Leather Footwear Journal) **14**, 147–158 (2014)
48. Roma, A. di: Footwear design. The paradox of "tailored shoe" in the contemporary digital manufacturing systems AU - Di Roma, Annalisa. Des. J. **20**, S2689–S2699 (2017)
49. Waskito, M.A., Mulyana, A., Wahyuning, C.S., Jamaludin, J.: Digital modification techniques application on footwear upper decorative design process. J. Arts Humanit. **7**, 30–37 (2018)

50. Power, E.J.: Advanced knitting technologies for high-performance apparel. In: Woodhead Publishing Series in Textiles, pp. 113–127. Woodhead Publishing (2018)
51. Schmidt, T.W., Fang, V.Z.: Haptic-a new additive manufacturing technology for footwear, apparel and accessories applications. In: Proceedings of the 2nd International Conference on Progress in Additive Manufacturing (Pro-AM 2016), pp. 97–102. NTU Library, Singapore (2016)
52. Arribas, V., Alfaro, J.A.: 3D technology in fashion: from concept to consumer. J. Fash. Mark. Manag. An Int. J. **22**, 240–251 (2018)
53. Marconi, M., Papetti, A., Scafà, M., Rossi, M., Germani, M.: An innovative framework for managing the customization of tailor-made shoes. Proc. Des. Soc. Int. Conf. Eng. Des. **1**, 3821–3830 (2019)
54. Hermann, M., Pentek, T., Otto, B.: Design principles for Industrie 4.0 scenarios. In: 2016 49th Hawaii International Conference on System Sciences (HICSS), pp. 3928–3937 (2016)
55. Aqlan, F., Lam, S.S., Ramakrishnan, S.: An integrated simulation–optimization study for consolidating production lines in a configure-to-order production environment. Int. J. Prod. Econ. **148**, 51–61 (2014)
56. Anselmsson, J., Bondesson, N.V., Johansson, U.: Brand image and customers' willingness to pay a price premium for food brands. J. Prod. Brand Manag. **23**, 90–102 (2014)

Check for updates

A Case Study on Scheduling of Repairs in an Automobile Shop

M. Fátima Pilar, Eliana Costa e Silva(✉), and Ana Borges

CIICESI, ESTG, Politécnico do Porto, Porto, Portugal
eos@estg.ipp.pt
http://estg.ipp.pt

Abstract. The classical combinatorial problem of scheduling is extensively studied and arises in several economic domains. However, there are few studies in the automobile sector, particularly in scheduling vehicle repair tasks and using real instances. This paper intends to contribute to fill this gap, focusing on the scheduling of the repairs of the mechanical section of a Portuguese firm in the automobile sector. A mathematical model is presented that will assist the shop manager on the scheduling of the repairs, taking into account: the mechanics and other resources that are available, the mechanical interventions to be performed on each vehicle and its expected processing time. The aim is to reduce the time of inactivity of the vehicles between interventions, as well as, the downtime of mechanics, and therefore improve productivity. The results, from real instances extracted from the data provided by the firm, show that the interventions are scheduled in a suitable form, and there is a reduction of the downtimes.

Keywords: Scheduling · Mixed integer linear programming · Automobile sector · Real-world application

1 Introduction

Most of the works concerning scheduling found in literature use simulated data instead of using real data. Some of the reasons for it are the hindrance to data transfer by firms, the difficulty in collecting data or even the existence of data, that require preparation and aggregation, which are time consuming [10]. Furthermore, due to the complexity of most real-world problems researchers and practitioners reduce its complexity either by simplifying it or by making further assumptions [12]. Additionally, until 2014, only 9% of the works were on the Automotive industry, and 2% on the Transportation industry [1]. For Flexible Job Shop Problems (FJSP), which will be the focus of this paper, only 7% use Integer/Linear programming (IP/LP) and mathematical programming [4]. Thus, problems related to the automotive industry are not very frequent in literature. The present work propounds a contribution for filling that gap, by addressing a real problem proposed by a Portuguese firm in the automobile sector.

J. Machado et al. (Eds.): icieng 2021, LNME, pp. 226–236, 2022.
https://doi.org/10.1007/978-3-030-79165-0_22

The FJSP is strongly NP-hard and consists of: **(i)** a routing subproblem - assigning each operation to a machine out of a set of capable machines; and **(ii)** the scheduling subproblem - sequencing the assigned operations on all the machines to obtain a viable schedule minimizing an objective function. Brucker and Schlie [2] were the first to address and solve such problems, by developing a polynomial algorithm to solve the FJSP with two jobs [13]. According to [11], the problem of FJSP can be defined as follows: considering a set of n independent jobs j, each with its own processing order across a set of available k machines, a number of l_j ordered operations $(O_{j1}, \cdots, O_{jl_{(j)}})$ must be performed to complete job i. Operation i of job j, O_{ji}, can be processed by any machine in a given set M_i for a given processing time p_{ik} [4].

The present work focus on the mechanical section of a Portuguese firm in the automobile sector, and uses data on the mechanical repairs carried out over a time horizon period of one year. The main objective of the firm is to reduce the time that the vehicles are at the repair shop without any intervention being performed. Likewise, the firm also intends to reduce the downtime of its workers, specifically its mechanics. For this, a mathematical model was developed to schedule the necessary interventions to be carried so that the vehicles are ready to be delivered to their customers as soon as possible and with the lowest possible downtime.

The rest of this paper is organized as follows: Sect. 2 presents the developed model. Section 3 presents the computational results on real instances extracted from the database provided by the firm. Finally, Sect. 4 presents the conclusions and future research.

2 Methodology

The firm's specificities taken into account in the development of the model were: **(1)** the sequence of mechanical interventions to be executed in each vehicle is established by the repair order and it is assumed to be known; **(2)** the time each mechanic takes to perform a mechanical intervention is the same for all mechanics and it is specified by the *barème* time defined by each car brand; **(3)** the interventions may be interrupted and returned to, at any moment; **(4)** only one mechanic can be working at the same time on each vehicle; **(5)** only one intervention can be executed at a time in each vehicle; **(6)** the firm considers two levels of the mechanics qualification senior mechanics who can perform any intervention, and senior specialized mechanics, who can perform any intervention and diagnosis; **(7)** a delivery date is negotiated between the receptionist and the customer when the vehicle arrives, for almost all cases; **(8)** a working period of 4 h in the morning, 4 h in the afternoon, a lunch break of 1 h, and two flexible 10 min breaks (one in the morning and one in the afternoon), is practiced.

In the dataset provided by the firm the repairs are composed of a set of lines that may correspond to the same intervention. Therefore, an intervention actually performed in repair shop is different from the interventions considered in the development of the proposed model. In fact, for the developed model the interventions are defined according to the intervention for which a *barème* time is specified. Although the focus is on the mechanical section of the repair shop, the times associated with the interventions that are not executed in the mechanics are also consider, since they take time.

The mathematical model developed in the present work is a Mixed Integer Linear Programming (MILP) model and combines linear ordering variables to sequence the interventions performed in each vehicle, with assignment variables and linear variables for the starting times. It was inspired by the model proposed in [3] and by the models presented in [5], and was initially addressed in [9].

The sets and indexes that are used in the model are: $o \in O$ – vehicles to be repaired or the repair orders (ROs); $m \in M$ – mechanics working at the repair shop; $i \in I_{all}$ – mechanical interventions that the repair shop can perform; I_o – pre-established ordered sequence of interventions to be performed in $o \in O$; $(i, o) \in I_o \subseteq (I_{all} \times O)$ – ordered pair of interventions to be performed on repair order $o \in O$; $M_{(i,o)}$ – mechanics that can perform the intervention $(i, o) \in I_o$.

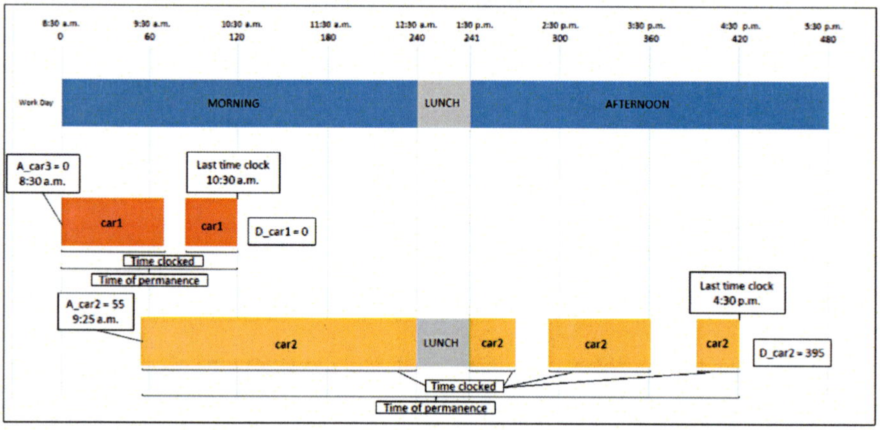

Fig. 1. Illustration of the parameters of the model.

The parameters are: $t_{(i,o)}$ – processing time for each intervention $(i, o) \in I_o$, in minutes, and it is defined for the combination of the intervention performed and the vehicle; D_o – time between the moment that vehicle $o \in O$ arrives at the repair shop until the expected delivery date, in minutes, which was computed considering the working minutes of the repair shop from the moment the vehicles enter the repair shop until the expected delivery time agreed with the customer (see Fig. 1); A_o – time of arrival of vehicle $o \in O$ into the repair shop, in minutes, and was computed as the minutes from the reference day and hour (01/08/2016, 8:30 a.m.) until the moment the vehicle arrives at the repair shop[1]; c_o – is 1 if

[1] E.g., $A_o = 0$ for a vehicle that has entered the repair shop at 8:30 a.m. on 01/08/2016, i.e. has as entry minute 0, while $A_o = 480$ for a vehicle that has entered the repair shop at 8:30 a.m. on 02/08/2016, i.e. has as entry minute 480 (see Fig. 1).

vehicle $o \in O$ has an expected delivery date, and 0 otherwise, i.e. it indicates whether an expected delivery date of the vehicle was been established with the customer or not; L – a sufficiently large number.

The following decision variables are consider: $x_{(i,o),(j,o')}$ – defines the precedence relations between interventions (i, o) and (j, o'), with $M_{(i,o)} \cap M_{(j,o')} \neq \emptyset$ and $i \neq j$ or $o \neq o'$, it is 1 if intervention (i, o) is performed before intervention (j, o'), and 0 otherwise; $a^m_{(i,o)}$ – defines the assigning each mechanical intervention i of vehicle o to a mechanic m, it is 1 if intervention (i, o) is assigned to mechanic m, and 0 otherwise; $S_{(i,o)}$ – starting time of intervention i on vehicle o, in minutes; T_o – tardiness of vehicle o.

The developed MILP model is:

$$\min \sum_{o \in O} T_o \, c_o + \sum_{o \in O, i \in I_o} S_{(i,o)} \, (1 - c_o) \tag{1}$$

s.t.

$$\sum_{m \in M_{(i,o)}} a^m_{(i,o)} = 1, \quad \forall (i,o) \in I_o \tag{2}$$

$$S_{(j,o)} \geq S_{(i,o)} + t_{(i,o)}, \quad \forall o \in O, \quad \forall i, j \in I_o : j = \text{next}(i) \tag{3}$$

$$S_{(j,o')} \geq S_{(i,o)} + t_{(i,o)} - L \left(1 - x_{(i,o),(j,o')}\right), \quad \forall (i,o), (j,o') \in I : M_{(i,o)} \cap M_{(j,o')} \neq \emptyset \wedge (i,o) \neq (j,o') \tag{4}$$

$$T_o \geq S_{(i,o)} + t_{(i,o)} - (A_o + D_o), \quad \forall o \in O, \forall i \in I_o \tag{5}$$

$$x_{(i,o),(j,o')} + x_{(j,o'),(i,o)} \leq 1, \quad \forall (i,o), (j,o') \in I : M_{(i,o)} \cap M_{(j,o')} \neq \emptyset \wedge (i,o) \neq (j,o') \tag{6}$$

$$x_{(i,o),(j,o')} + x_{(j,o'),(i,o)} \geq a^m_{(i,o)} + a^m_{(j,o')} - 1, \quad \forall (i,o), (j,o') \in I, (i,o) \neq (j,o'), m \in M_{(i,o)} \cap M_{(j,o')} \tag{7}$$

$$x_{(i,o),(j,o)} = 1, \quad \forall o \in O, \forall i \in I_o, i \neq \text{last}(I_o), j = \text{next}(i) \tag{8}$$

$$x_{(j,o),(i,o)} = 0, \quad \forall o \in O, \forall i \in I_o, i \neq \text{last}(I_o), j = \text{next}(i) \tag{9}$$

$$S_{(i,o)} \geq A_o + 15, \quad \forall o \in O, \forall i \in I_o \tag{10}$$

$$T_o \geq 0, \quad \forall o \in O \tag{11}$$

$$S_{(i,o)} \geq 0, \quad \forall o \in O, \forall i \in I_o \tag{12}$$

The objective function (1) is to minimize the sum of the total tardiness of the ROs for which an expected delivery date was established (i.e. $c_o = 1$) with the sum of the start times of all interventions for the ROs for which an expected delivery date was not established (i.e. $c_o = 0$). Constraints (2) guarantee that each intervention (i, o) is only assigned to exactly one mechanic $m \in M_{(i,o)}$. Constraints (3) ensure that there is no overlap of interventions, since it requires that the starting time of intervention j is larger than the starting time plus the processing time of intervention i. The disjunctive constraints (4) are only activated if intervention i precedes intervention j in the same mechanic, i.e. it guarantees that there is no overlapping of interventions for the same mechanic.

On the other hand, constraints (5) compute lower bounds for the tardiness of each RO. Constraints (6) and (7) are for linear ordering of the interventions. More specifically, (6) ensure that, either intervention (i, o) precedes intervention (j, o'), or intervention (j, o') precedes intervention (i, o), or interventions (i, o) and (j, o') are not carried out by the same mechanic. Constraints (7) are only active when (i, o) and (j, o') are processed by the same mechanic, guaranteeing that either (i, o) precedes (j, o') or (j, o') precedes (i, o). Constraints (8) and (9) are related to the precedence between interventions performed on the same vehicle. These constraints establish that if i precedes j then the corresponding variables is 1 and zero otherwise. Finally, constraints (10) ensure that vehicles can only be repaired at least 15 min after arriving at the repair shop, since this is the average duration of filling up the RO form.

3 Computational Results

The model in (1)–(10) was implemented using AMPL ([6,7]) and the real instances, extracted from the real data provided by the firm, were solved using Gurobi, with AMPL interface [8]. All numerical tests were performed using an Intel Core i7-4510U@2.0 GHz 4 GB RAM running Windows 10 64-bits. The real data provided by the firm concerned the vehicles that arrived at the repair shop over a period of one year, with a total of 251 working days and 5,848 ROs.

The instances selected to test the scheduling correspond to the most challenging situations of the everyday work at the repair shop. The results obtained in these instances will allow to verify if the proposed model guarantees that the vehicles are in the repair shop without being intervened for as little time as possible (either at the beginning, end or between interventions), i.e., with the lowest downtime. This will contribute to a more efficient use of the resources available in the repair shop.

Table 1. Number of opened and time clocked ROs, and total downtime, per month.

Year	Month	# opened ROs	# time clocked ROs	Total downtime (in min.)
2016	August	441	540	387,709.2
	September	455	646	354,881.2
	October	498	687	392,858.0
	November	473	664	306,310.0
	December	454	620	253,616.0
2017	January	516	707	278,735.8
	February	430	592	368,473.4
	March	519	748	369,077.0
	April	450	596	192,531.4
	May	547	754	255,557.8
	June	532	737	257,939.0
	July	533	713	264,060.6

The five months with the highest number of ROs entries[2] are May, July, June, March and January; the months with the highest number of time clocked ROs are May, March, June, July and January; and the months with the lowest downtime are April, December, May, June and July (see Table 1). From these top 5 lists the three months that belong simultaneously to all of them, i.e. May, June and July, were selected. These months correspond to the most challenging months in terms of volume of work. Furthermore, for these months the days with the lowest downtimes were May 19th and 26th, June 2nd and 30th, July 14th and 27th (see Table 2). For all months the number of time clocked ROs is larger than the number of ROs entries (Table 1). This suggests that there are ROs opened in previous months that are only intervened in latter months.

Table 2. Characterization of the real instances with the lowest downtimes in the firm.

Month	Day	Downtime (min.)	# of opened ROs	Total time clocked	Total *barème* time (min.)
May	19th	3,040.2	18	1,549.8	2,032.0
	26th	5,081.8	21	1,895.4	4,441.9
June	2nd	2,111.6	13	1,311.6	1,752.4
	30th	2,857.6	20	1,622.4	2,349.3
July	14th	4,187.6	19	1,727.4	2,102.3
	27th	4,239.2	26	2,296.8	1,881.6

Table 3 presents the results for the two most challenging instances: May 19th and June 2nd. June 2nd is the instance with the lowest firm's downtime (2,111.6 min.). Furthermore, this is the instance with the lowest number of opened ROs (13) and mechanical interventions (31), and as expected, it presents the smallest number of decision variables (1,186) and constraints (9,056).

Table 3. Number of ROs, interventions, variables and constraints, objective function value, total downtime of the firm and the model, and percentage, for each instance.

Month	Day	# O	# I_o	# var	# constr.	Objective function	Total downtime		
							Firm	Model	%
May	19th	18	53	3,181	27,071	26.7	3,040.2	271.6	8.9%
June	2nd	13	31	1,186	9,056	51.9	2,111.6	103.9	4.9%

The downtimes of the scheduling of all vehicles obtained using the developed model is smaller than the downtimes of the schedule that was performed by

[2] Note that for some of the ROs the interventions are not completed on the same day.

the firm. In particular, for June 2nd the total downtime obtained using the developed model represents approximately 4.9% of the downtime observed in the firm. However, the firm also performed on June 2nd repairs on other vehicles that are not considered on this instance. Also there were cases of vehicles that were not repaired because some mechanical part that needed to be substituted was not available. The instance that presents the second best reduction in the downtimes is May 19th (8.9%).

The 13 vehicles that arrived on June 2nd[3] were repaired by the firm on three different days, whereas in the result of the developed model the repairs were scheduled so that almost all vehicles were completed on the day of arrival, with the exception of car4812 that was completed after June 2nd (Fig. 2, top), where the repairs end after 5:30 p.m.

Using the developed model, 11 of the 13 vehicles that arrived at the repair shop on June 2nd started to be intervened immediately after completing the RO form (see Figs. 2 and 3, top). Vehicles car4807 and car4812 had to wait so that some mechanic was available to start their repairs. With the exception of car4812, all vehicles have completed the repairs on this day, and were available to be delivered to the customers on time. Furthermore, the interventions on each vehicle were performed almost without any breaks. Mechanics mec02 and mec06 started to work as soon as the repair shop opened (Fig. 3, top). Mechanics mec01, mec03, mec04, mec05, mec07 and mec08 started working 14, 29, 60, 44, 40 and 7 min, respectively, after the beginning of the working day. Note that June 2nd starts at minute 101,280 and ends at minute 101,760.

The 18 vehicles that arrived on May 19th (minute 96,480) were repaired by the firm on three different days, whereas in the result of the scheduling, the repairs were scheduled so that all vehicles were completed on the day of arrival, with the exception of car4586, car4587, car4588 and car4589 that were completed after day May 19th (Fig. 2, bottom), where the repairs end in the end of day (Fig. 2, bottom). For the solution of the developed model, 15 of the 18 vehicles that arrived at the repair shop on May 19th started to be intervened immediately after completing RO form (see Figs. 2 and 3, bottom). Furthermore, the interventions on each vehicle are performed almost without any breaks. Mechanics mec01 and mec05 started to work as soon as the repair shop opened (Fig. 3, bottom), while mec02, mec03, mec04, mec06, mec07 and mec08 started 68, 38, 57, 49, 28 and 26 min, respectively, after the beginning of the working day. Note that May 19th starts at minute 96,480 and ends at 96,960.

For the firm, on June 2nd and May 19th, respectively, the lowest downtime registered in a vehicle was 0 and 2.8 min. In contrast, the highest downtime registered was 938.6 and 1,011 min (Table 4). Regarding the values obtained with the model, on both days, there were vehicles that obtained a downtime equal to 0, i.e. vehicles that started to be repaired as soon as they arrived the repair shop, without any break. The vehicles with highest downtime, recorded an

[3] These vehicles arrived at minute 101,280. However, only after completing the RO form, i.e., 15 min later, at the minute 101,295 the mechanics may start working on them.

Table 4. Downtimes of the firm *vs.* the developed model.

		Min	Mean	Median	Max	Total
June 2nd	Model	0.0	8.0	0.0	60.0	103.9
	Firm	0.0	162.4	72.4	938.6	2,111.6
May 19th	Model	0.0	15.1	2.8	92.9	271.6
	Firm	2.8	168.9	107.3	1,011.0	3,040.2

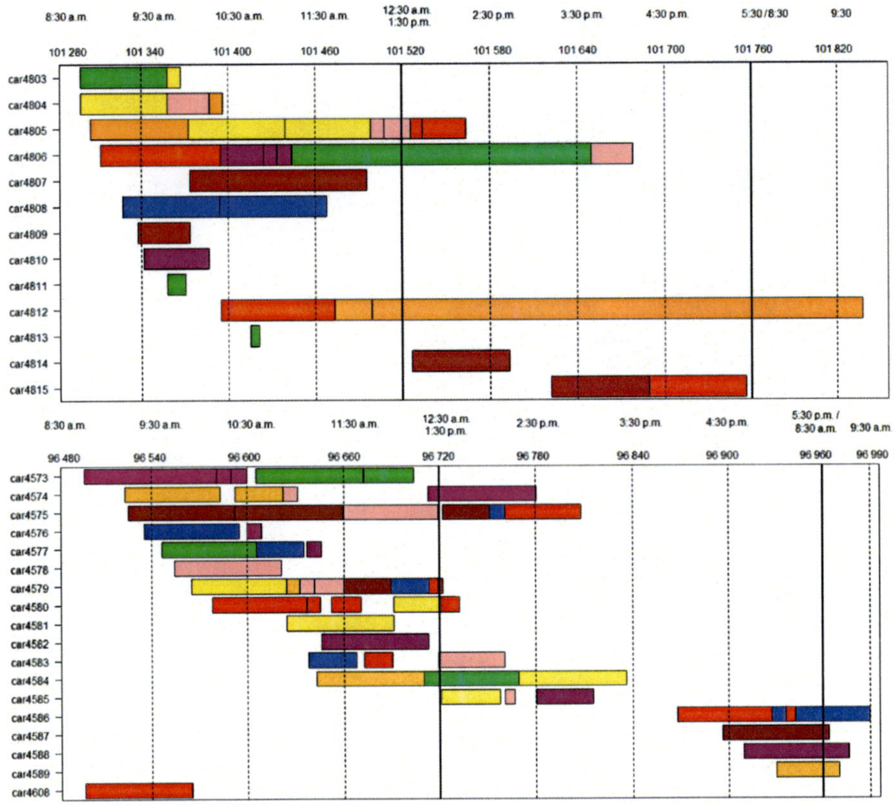

Fig. 2. Solution of the model, per vehicle, for the interventions of the June 2nd (top) and May 19th (bottom) instances.

average downtime of 60 and 92.9 min, respectively, on June 2nd and May 19th. On June 2nd, on average, the firm recorded a downtime per vehicle of 162.4 min, whereas through the developed model registered a downtime of 8 min. On May 19th, on average, the firm recorded a downtime of 168.9 min, and 15.1 min using the developed model (Table 4).

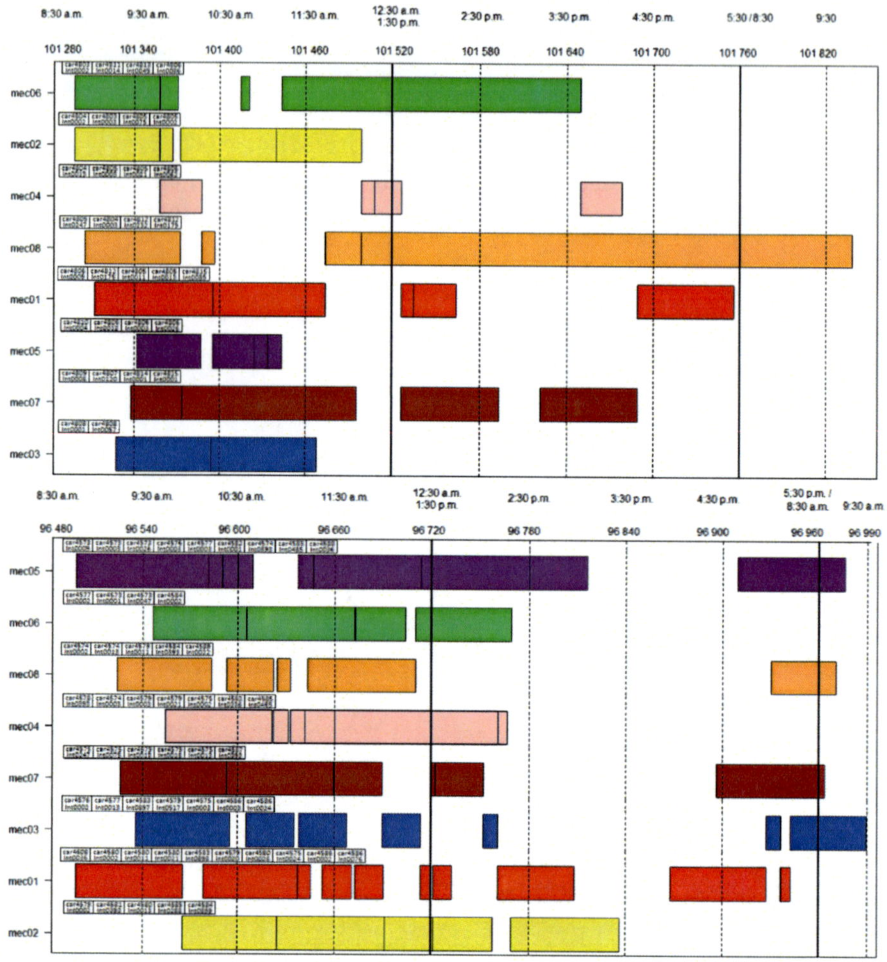

Fig. 3. Solution of the model, per mechanic, for the interventions of the June 2nd (top) and May 19th (bottom) instances.

4 Conclusions

In today's competitive marketplace, the high level of delivery performance has become increasingly necessary to satisfy customers, as failure to do so can lead to customer dissatisfaction or even customer loss. Thus, in the management of the repair shop, the existence of a well-designed scheduling plan is important as it can lead to better performance.

The real problem addressed in this paper fits into the FJSP context where operations can be processed on any machine within a set of alternative machines. This type of problem is quite complex because it consists of a routing subproblem with the assignment of machines to process each operation, and a scheduling subproblem where all assigned operations are sequenced [4].

In this paper a MILP model was developed to address a real problem proposed by a Portuguese firms in the automobile repair sector. The objective was to reduce the downtime between interventions performed in the vehicles, by minimizing the time that the vehicles stay without any intervention being performed, and simultaneously, reducing the inactive time of the mechanics. Thus optimizing both the resources of the repair shop and the mechanics' performance. To this end, real instances, taken from the database provided by the firm regarding repairs carried out during a one year time horizon, were used.

Comparing the results obtained with the scheduling of the developed model, with what actually happened in the firm, it is possible to observe that the downtimes between the repairs in each of the vehicles have decreased. For all the days under analysis, the idle time of the mechanics is longer than the time that mechanics actually would be idle in the shop. This is due to the fact that the number of mechanics actually working in the repair shop on each of the days considered was smaller than the number of available mechanics indicated by the firm (most probably due to vacation or absences of the employees).

Further test on other instances were performed but not discuss in the present paper. These results will be presented in future work. Also, the developed model will be improved so that downtimes by mechanics decrease, maximizing the mechanics work. Another factor not yet considered in the model is the incorporation of mechanics' breaks and the lunch break. Furthermore, the interventions of the vehicles that had already arrived the repair shop in previous days will also be incorporated in the model.

Acknowledgement. This work has been supported by national funds through FCT - Fundação para a Ciência e Tecnologia through project UIDB/04728/2020.

References

1. Brandenburg, M., Govindan, K., Sarkis, J., Seuring, S.: Quantitative models for sustainable supply chain management: developments and directions. Eur. J. Oper. Res. **233**(2), 299–312 (2014)
2. Brucker, P., Schlie, R.: Job-shop scheduling with multi-purpose machines. Computing **45**(4), 369–375 (1990)
3. Cerdeira, J.O., Lopes, I.C., Costa e Silva, E.: Scheduling the repairment of aircrafts. In: 2017 ICCAIRO, pp. 259–267. IEEE (2017)
4. Chaudhry, I.A., Khan, A.A.: A research survey: review of flexible job shop scheduling techniques. Int. Trans. Oper. Res. **23**(3), 551–591 (2016)
5. Demir, Y., İşleyen, S.K.: Evaluation of mathematical models for flexible job-shop scheduling problems. Appl. Math. Model. **37**(3), 977–988 (2013)
6. Fourer, R., Gay, D.M., Kernighan, B.W.: A modeling language for mathematical programming. Manage. Sci. **36**(5), 519–554 (1990)
7. Fourer, R., Gay, D., Kernighan, B.: AMPL: A Modeling Language for Mathematical Programming. Brooks/Cole-Thomson Learning, Pacific Grove (2003)

8. Gurobi Optimization, LLC. Gurobi Optimizer Reference Manual (2018). http://www.gurobi.com
9. Pilar, M.M.F., Costa e Silva, E., Borges, A.: Scheduling in an automobile repair shop. In: Machado, J., Soares, F., Veiga, G. (eds.) Innovation, Engineering and Entrepreneurship. HELIX: Lecture Notes in Electrical Engineering, vol. 505. Springer, Cham (2018). ISBN 13: 978-85-352-5193-7
10. Melo, M.T., Nickel, S., Saldanha-Da-Gama, F.: Facility location and supply chain management-a review. Eur. J. Oper. Res. **196**(2), 401–412 (2009)
11. Özgüven, C., Özbakır, L., Yavuz, Y.: Mathematical models for job-shop scheduling problems with routing and process plan flexibility. Appl. Math. Model. **34**(6), 1539–1548 (2010)
12. Sarker, R.A., Newton, C.S.: Introduction to Operations Research. CRC Press (2007)
13. Shen, L., Dauzère-Pérès, S., Neufeld, J.S.: Solving the flexible job shop scheduling problem with sequence-dependent setup times. Eur. J. Oper. Res. **265**(2), 503–516 (2018)

Design Methodology for the Research and Development of Polygonal Artefacts

Bernardo Providência[1](✉) and Daniel Vieira[1,2]

[1] Landscapes, Heritage and Territory Laboratory, University of Minho, Guimarães, Portugal
providencia@arquitectura.uminho.pt

[2] Centre for Textile Science and Technology, University of Minho, Guimarães, Portugal

Abstract. This work is part of the project "2^3 Design Methodology", which describes the design process applied to creating two-dimensional polygonal artefacts for the construction of three-dimensional forms to support scenographic spaces and presents a design method applied to the educational activity. If, on the one hand, the design process and the design of the artefact are related. On the other hand, to justify its normative position regarding the design process, a design methodology was used that considered the versatility of the projected artefact. Several polygonal forms were explored as an enhancement of the research process. After obtaining the artefact's final design, a workshop was prepared for students of the product design degree at the University of Minho. In this sense, the students, based on the presented work methodology, idealized and prepared a scenographic space for the show "Piano Caos", organized by the Educational Service – Casa da Música. The project developed by the students approaches the design process suggested by the authors, based on (i) selected artefact; (ii) articulation between usage and language; finally, (iii) the combination of artefacts in the scenic space.

Keywords: Design approach · Design methodology · Design process · Design studies · Polygonal artefacts · Workshop

1 Introduction

According to Watkins, it is asked for a designer to accept a problem to find reasons, devote time and effort to its resolution. Analyze it, finding out as much as possible about the problem. Define it, defining which are the most critical aspects of the problem. Idealize it, developing as many ways as possible to solve the problem, selecting the best ideas. Implement it, acting and experimenting with the best idea. Finally, evaluate it, analyzing what happens when an action was taken [1].

The design methodology adopts a normative position about design, mainly oriented towards the process, while the research methodology is descriptive and strongly oriented towards the product. In contrast to a deductive methodology that is strongly oriented towards a normative approach, the design methodology, closer to an inductive interpretation, aims to improve the design process [2]. Based on this premise of design

J. Machado et al. (Eds.): icieng 2021, LNME, pp. 237–245, 2022.
https://doi.org/10.1007/978-3-030-79165-0_23

methodology versus research methodology, we move on to the research and survey phase of existing projects, some more interesting than others. Still, all framed in the theme related to research. This transposition of design methodologies for research can be of interesting applicability, even when a model is adapted [3]. The applied model (see Fig. 1) is divided into three stages: analysis, conception/design, and, finally, materialization. Each of these stages is divided into three semiotic dimensions: semantic dimension (designates the shape of the artefact), syntactic dimension (implies the definition of the artefact) and the pragmatic dimension (expresses the use of the artefact).

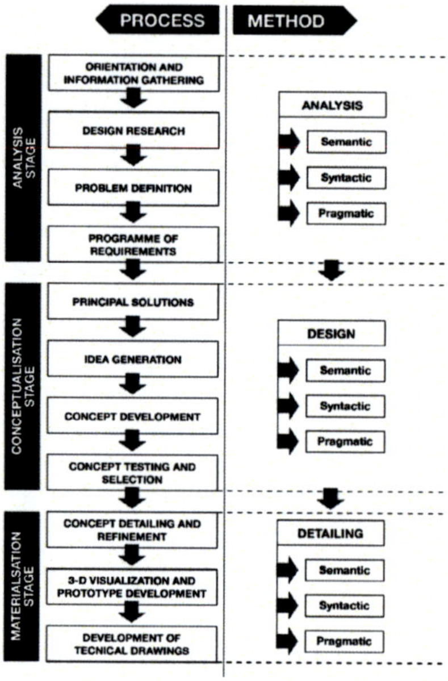

Fig. 1. The model used for the design and development process of the artefact [3].

This article aims to report what happens at each stage of the design process adopted for designing the artefact. It reports the students' work in designing the scenographic space in the workshop and the benefits of this educational initiative.

2 Research, Shape Mapping and Materialization of the Artefact

A mind map was created through the Coggle platform (see Fig. 2), where auxiliary information and relevant research were inserted. The creation of this mind map allows a total visualization of the investigated projects, identifying projects worldwide, such as Switzerland, China, passing through the United States to Chile. Plastic artists, designers or architects were interested in the creative potential of three-dimensional geometric design.

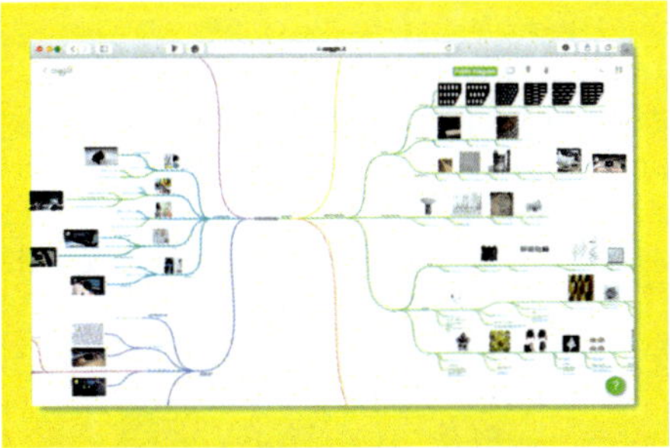

Fig. 2. Part of the mind map developed for the research process.

Of the various projects included in the platform, three should be highlighted. First, the facade of the "La Suisse Vigilante" pavilion for the 1964 Swiss National Exhibition in Lausanne (see Fig. 3). Designed by Swiss architect Carl Fingerhuth as an architect in Jan Both's atelier. It is an expressive, bold and controversial project, with a brutalist architecture that does not reflect that time's language. On the other hand, it had some clear and indisputable responses to a grammar of form—formed by the junction of "shells" in a hexagonal grid/structure, allowing quick execution based on weight reduction and the lightening of the entire structure. This project reinforced the artefact's hexagonal base matrix for the workshop, a grid of points to gain three-dimensionality, in a semantics similar to the "concrete hedgehog". In this case, a fragile character given by the material in enhancing the scenic space.

Fig. 3. "Shells" facade of the "La Suisse Vigilante". Source: Image provided by the author.

Second, the pilot project "Hyposurface" from 2000 assumes itself as an innovative product for communication (see Fig. 4). Designed by a multidisciplinary team led by Mark Goulthorpe, it shows us the value of three-dimensional surfaces in the symbolic load as a means of forming meaning in creating a context. The prototype was installed at the entrance to the IMTS (International Manufacturing Technology Show), with a structure composed of actuators that responds with articulated movement directly to the sound and movement of visitors, which in turn causes waves up to 96 km/h, deforming a surface of metal and rubber with 6 meters of height. Despite being a prototype, the emergence of this type of products allowed, based on a sheet (structure composed of a grid of metallic triangles), precise and digital control in the construction of texts or volumes. This project allowed us to crystallize a dynamic way to create a three-dimensional grid, from which to understand the organic effect of the structure itself.

Fig. 4. The six-meter high "Hyposurface" module. Source: Image provided by the author.

Third, the "Clouds" product, launched in 2006 by Vitra (see Fig. 5), designed by designers Ronan & Erwan Bouroullec. An easy to install solution, developed from an assembly system with elastic bands, designed for this purpose. In this way, assembly is simple, allowing greater freedom of volumetric configurations. It is an organic language structure, composed of a combination of irregular modules, designed from the combination of triangles, which multiply at random, building chaotic environments, since the only control is the creation of colour patches and its aesthetic quality. This product allowed us to study possibilities of connecting triangles, in the construction of modules, irregular two-dimensional figures, which, when combined in a certain way, resulting in three-dimensional volumes.

Fig. 5. Possibility of "Clouds" configuration. Source: Image provided by the authors.

Starting from an A4 sheet of paper's base format, we started the artefact's entire design project (see Fig. 6). The first step started by weakening the A4 sheet through an isometric grid, formed by three sets of lines at angles of 60° to each other. Using a weakening tool on the isometric grid, we managed to create a regular weakening of the plane. With this simple exercise, it was possible to understand the potential of transforming a two-dimensional plane into a geometric, three-dimensional volume. With the survey of three-dimensional shapes through folds and cuts, we realized that we could have two types of approach. An approach where the result created was manifested disorganized. This visual disorientation acquires a dynamic interaction with the interlocutor, in a higher possibility and freedom of configurations. Another approach, where the result created was of a regular character, minimalist and monotonous morphology.

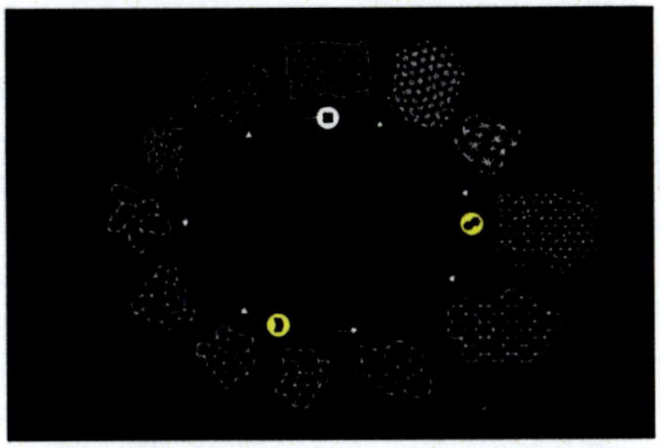

Fig. 6. Diagram with the drawn artefacts.

The development of the three-dimensional artefact for the scenographic context is related to the research itself; when analyzing the "Clouds" project, it was possible to devise a new solution with 11 distinct typologies of polygonal artefacts (see Fig. 7) that combined among themselves generate a controllable irregular three-dimensional grid, which in turn provides a volumetric-free construction. From these 11 typologies, the most versatile artefact for the workshop was selected.

Fig. 7. Diagram of 11 typologies, at the center the selected artefact.

3 Design Teaching Project

The vision of design by future designers is often seen as a solitary act of creation that depends on the vision of the author of the project [4]. Therefore, at the level of higher education, the study of product design can be much more than each student's independent creation. An effective way to help students prepare to become designers is to encourage participation in collaborative workshops, as proposed here, which encourages students to interact and work together. Making work less in-dividual, more shared among students, and exchanging different knowledge and points of view. The process of materializing the scenographic space as a team helps students to deepen their experiences.

3.1 Scenographic Context

In the challenge made by the Porto Casa da Música – Educational Services (2017/2018), the possibility arose to build a scenographic space with the students of the design degree for a children's show, "Piano Caos". The following Table 1 informs about the technical data of the show. The repertoire is composed of a set of pieces, sometimes devoid of the score where the piano in an interaction between the most diverse technologies, video and a ballerina, is the guiding thread of a show, which in the vicinity of anything goes, launches (dis)order. This minimalist or brutal character in images that shatter or ping-pong balls on a grand piano bouncing through the revival of hammers on the strings also stimulated the creation of a (dis)structured where the rhythm caused by the difference in planes allows broken projections that align with the development of the pieces.

Table 1. Technical data for the "Piano Caos" show.

Technical support	Participant
Artistic direction	Duarte Cardoso and José Alberto Gomes
Choreographic creation and interpretation	Daniela Leite Castro
Acoustic instrumentation	Duarte Cardoso
Programming, processing and synthesizers	José Alberto Gomes
Coordination of props and scenography	Bernardo Providência and Lígia Lopes
Props and set design	Students of the Product Design Course at the School of Architecture - University of Minho

3.2 Workshop Preparation

The "Piano Caos" workshop was held in the show's context with the same name to design a scenographic space for a children's show. As mentioned, and given a short time, the workshop was preceded by an investigation and mapping of design projects that were based on three significant vectors: (1) Model building solutions that can be replicated; (2) Creation of three-dimensional volumes from two-dimensional shapes; (3) Construction of dynamic forms that deconstruct regular forms in order to align the scenic space with the show's "chaos" repertoire.

This mapping served as a basis to accelerate the workshop in the first part so that students could understand the construction processes and the possible languages of similar projects. From the examples presented, the product "Clouds" by the designers Bouroullec allowed the creation of a grammar tuned by the selected artefact, a set of four triangles (or two lozenges) with flaps that, depending on the shape of the assembly, allowed the construction of different shapes, more or less accentuated, in terms of volume and orientation of the resulting part. Listed and elevated plans were distributed after the first presentation and discussion of the solutions present in architectural and design projects. Then, the students were invited to leave the workroom located on the lower floor and go up to the space where they would have to design the scenographic objects. Here the stage director presented the show's proposal, referring to the needs and limitations to be considered during the planning, as well as the characteristics correlated to the child audience.

After this brief introduction, the students, based on the elements presented, re-turned to the workroom and were provided with a set of small cardboard models, proportional to the room plan's scale, to carry out their first studies (see Fig. 8).

Fig. 8. Modeling of ideas on a small scale.

The first phase started to discuss forms and how to manipulate artefacts to control scenographic space. Assembly exercises were carried out to control plans or simple shapes, and the necessary elements for space were identified. From that moment onwards, construction of full-scale objects on corrugated cardboard was made, with dimensions that could reach seven meters in height and which would be suspended only by the cables of the web of the concert hall (see Fig. 9).

Given the pieces' size and self-supported weight, it was necessary to find resistant assembly solutions and a strategy for transporting the assembled elements to the second largest concert hall. Simultaneously, some objects allowed the construction of a visual grammar that was assumed to be the only communication language between the show, the authors and the child audience. This grammar comes from the cut and crease produced on this type of materials. It has its roots in the exercises developed by Josef Albers in the 1920s, where according to Rocha "the experimental character of free play encourages

Fig. 9. Materialization of the scenographic project.

the exchange of experiences and the simultaneous understanding of the elementary laws of form and interpretation" [5].

3.3 Musical Show "Piano Chaos"

The result developed by the students (see Fig. 10) articulates an experience on stage. That dignifies the show and aligns with the agenda's rhythm—both in its plastic expression and the interaction of the video projection on the same supports. The dynamics went through working first through a cartesian coherence supported by a mesh of regular polygons to free itself in a cosmogenic view gradually, where the dialogue with the inflexion [6] allows creating dynamic forms.

Fig. 10. Projection of video on the structures designed for the show.

4 Discussion

The initial research was preponderant so that in the first place, a range of references (i.e., visual artists, designers and architects) could be created that could be shared during the workshop. Second, justifying the artefact design applied in the workshop and the folding and assembly exercises performed during the workshop - beneficial exercises for future product designers. Here the students were able to observe the presented artefact, perceive the three-dimensional shapes created, test the idealized volumes as a team, discover possible flaws and have the opportunity to apply it to real space. In this way, it was possible to stimulate the requirements necessary for the professional practice of design, such as cooperation and coordination. The students were able to create new forms through the transformation of matter (i.e., the artefact).

5 Conclusion

In a short period, the exercise aimed to work in a real context, to value the acquisition by students of skills in the design and construction of large-scale pieces with inferior materials and given the modular typology to work on the unpredictable "chaos" where trajectories are sometimes corrected during the assembly of the pieces. According to Jackson, all designers fold (bend, shape, twist) two-dimensional materials to conceptualize three-dimensional objects [7]. He considers that most objects created (imagined, designed and produced) transform from flat materials. The folding is part of the method of transforming the material into form. Even when technical aspects are "on the table", intuition is a feature of the creative process. In this model, students sought to reach a wide variety of ideas among the differences, resulting in convergence or centralization of group ideas. As a team, the students could share and structure three-dimensional ideas from the predefined artefact, creating the scenario for the musical show "Piano Caos". This activity development shows that this workshop creation is a successful educational solution; the process or part can be replicated.

References

1. Watkins, S.M.: Using the design process to teach functional apparel design. Cloth. Text. Res. J. **7**(1), 10–14 (1988)
2. Kroes, P.: Design methodology and the nature of technical artefacts. Des. Stud. **23**(3), 287–302 (2002)
3. Liem, A., Øritsland, T., Nørstebø, C.: Introducing form and user sensitivity to mechanical engineering students through industrial design projects. In: Rodgers Hepburn, B. (eds.) Crossing Design Boundaries, pp. 348–355. Tailor & Francis Group, London (2005)
4. Montagna, G., Carvalho, C., Carvalho, H., Catarino, A.: O designer de produto como elemento de ligação nas equipas multidisciplinares. Revista Lusófona de Educação **20**, 99–108 (2012)
5. Rocha, C.S.: Plasticidade do papel e Design, 1st edn. Plátano Editora, Lisboa (2000)
6. Klee, P.: Théorie de l'art moderne, 1st edn. Gonthier Publisher, Paris (1964)
7. Jackson, P.: Techniques de pliage pour les designers, 1st edn. Dunod Editeur, Paris (2011)

Possibility of Reaction Mixture Variable Composition Identification in Semi-batch Reactor

Lubomír Macků(✉)

Department of Process Control, Faculty of Applied Informatics, Tomas Bata University in Zlin, nám. T. G. Masaryka 5555, 760 01 Zlín, Czech Republic
macku@utb.cz

Abstract. The paper shows a new procedure for the identification of the variable composition of the reaction mixture, namely determination of reaction kinetics of the chemical reaction in batch or semi-batch reactor. This procedure is based on the measurement of temperature profiles directly in the chemical reactor and makes it possible to respond to possible changes in the reactant's composition without the need for the reaction kinetics laboratory determination. This can be very advantageous, for example, in the case of recycling processes, where the quality of raw materials can vary considerably and thus there may be problems with the unsatisfactory setting of the controller for such process. Chemical batch or semi-batch reactors behave as a non-linear system in terms of control and therefore require advanced control techniques mostly. This method can also help to detect poor quality raw materials during common processes. The derivation of this method is shown and a model example of a chemical semi-batch reactor is mentioned. The issue of reaction kinetics and its influence on chemical reactors control is also discussed.

Keywords: Semi-batch reactor control · Exothermic process · Simulation · Identification · Reaction kinetics

1 Introduction

A common way of chemical reactor control preparing is by chemical reaction laboratory verification, assuming a known composition of the feed components and also analyzing the output of the reaction products. This practice is sufficient in a number of chemical processes. The reactors are mostly used for single-purpose processing of the same raw materials and the same output products are expected.

But there are situations where this procedure may be not satisfactory. A typical example could be e.g. waste materials treatment. The waste materials usually vary significantly according to the individual components proportion and also to the composition of individual components.

J. Machado et al. (Eds.): icieng 2021, LNME, pp. 246–255, 2022.
https://doi.org/10.1007/978-3-030-79165-0_24

The different composition of these secondary recycled raw materials results in different chemical reactions or a ratio of their representation in the reaction spectrum. However, this different composition can be very dangerous for the control process. Chemical batch or semi-batch reactors behave as a non-linear system in terms of control and therefore require advanced control techniques. In addition, batch or semi-batch reactors behave usually as an astatic system, i.e. they do not have a steady state.

Therefore, an individual control technique is used for a specific process and is suited to an exact process. The required quantities are measured under laboratory conditions and then results are transferred to a real system. As concerned with the exothermic reactors, the reaction kinetics and reaction heat are mainly observed. Reaction kinetics is defined by a rate constant, the so-called Arrhenius relation, which is non-linear (exhibits non-linear behavior) and is a function of temperature. The rate of reaction then depends both on the specific temperature and the concentration of the reactants. For batch processes, the entire batch (initial batch of chemicals) is placed in the reactor and the reaction is allowed to proceed through all content. Such an approach is not possible in the case of a strongly exothermic reaction, because raising the temperature above the critical limit would destroy the reactor. In such a case, the semi-batch process is suitable. The initial batch of the reaction mixture is placed into the reactor and the suitable reactant is added (dosed) gradually so that the maximum allowed temperature is not exceeded. The controlled variable here is then the temperature in the reactor and the actuating variable the added reactant. To some extent, the temperature in the reactor can be also influenced by the temperature of the cooling/heating medium in the reactor jacket. In large-volume reactors, however, the effect of the coolant temperature is considerably lower than in small-volume or laboratory reactors. In addition, if very low or very high medium temperatures in the jacket are required, the energy demand for the process increases significantly. The cooling or heating of the reactor can be to some extent influenced also by the varying flow of cooling/heating medium. However, this has also limitations and is applicable to laboratory or small-size reactors only. For economic reasons, water from commonly available sources, e.g. tanks or boreholes with a relatively stable temperature, is often used as a cooling medium. In that case, we require only the reactor walls cooling and thus the internal reaction mixture cooling as well. The in-reactor temperature control is then only possible by changing the feed rate of the selected reactant causing the chemical reaction to run. However, the temperature increase in this case is not directly proportional to the amount of reactant added, but is strongly influenced by the reaction kinetics. As was mentioned above, the rate of reaction generally depends on both the concentration of the reactants and the temperature of the reaction mixture. Moreover, the dependence of the reaction rate on temperature is exponential, i.e. the rate of reaction with temperature increases extremely.

In the next chapter, the issue of reaction kinetics will be described for a semi-batch exothermic chemical reactor. Other possibilities of reaction kinetics identification can be found in literature, for example in [1] authors try to improve a differential method, or in [2] authors deal with reaction kinetics of contaminants in water. In paper [3] authors study stirred tank chemical reactors and also mention the reaction kinetics. All these methods are complicated and/or not suitable for the system described in this paper.

2 Reaction Kinetics in a Semi Batch Exothermic Reactor

In the case of a semi-batch reactor, the initial batch of reactants is first placed into the reactor and thereafter the appropriately selected reactant is dosed in a controlled manner. The chemical mixture of the initial batch is then in excess of the added reactant. Because of the excess of the initial reactant mixture to the dosed reactant, the ongoing reaction satisfies the conditions for the first-order reaction, i.e. the rate of reaction depends only on the concentration of the reactant added. For the first-order reaction the following relation (1) can be used:

$$k = Ae^{-\frac{E}{RT[t]}} \tag{1}$$

In this relation k represents the rate constant, A is the pre-exponential factor, E is the activation energy and R is the gas constant. For the chemical reactor control, the constants A and E are usually determined experimentally.

As mentioned earlier, it is usually not possible to use classical control algorithms, e.g. PID controllers, to control the reactor. This is due to the fact that reaction kinetics is causing a delayed response to the reactant fed into the reactor.

When dosing the reactant (action variable) at the low temperature (e.g. 20 °C), the reaction rate will be significantly lower than at the high temperature (e.g. 100 °C). In addition, at the beginning of the reactant dosing, there is also a small concentration of reactant in the reactor, and the reaction rate is therefore very slow. Thus, the temperature (controlled variable) in the reactor almost does not increase and the controller tends to increase the reactant dosage. Thus, the concentration of unreacted reactant in the reactor increases without the corresponding increase in temperature. A conventional controller (without prediction) only responds when the temperature approaches the setpoint and then stops dosing. At that point, however, the amount of unreacted reactant in the reactor is already too large and the reaction runs very fast due to the high concentration and high temperature. The process cannot be stopped because of the accumulated reactant and the result may be product degradation or, ultimately, destruction of the reactor. An example of such a process with overshooting the setpoint is shown in Fig. 1. To highlight the problem, only a two-position controller is used here, the initial overshoot is more clearly visible.

As can be seen from Fig. 2, the unreacted reactant accumulated in the reactor until its concentration reached critical value which started reaction at high reaction rate what caused the temperature to exceed the setpoint. In order to limit overshoot, dosing would have to be stopped before a certain temperature was reached, i.e. the reactant concentration in the reactor had to be measured simultaneously. However, this is quite complicated and in case of the concentration sensor failure, there is again the risk of reactor destruction.

Suitable for such a system is a control based on the knowledge of the system behavior, i.e. on the model of a particular reactor. If enhanced control techniques using system behavior prediction are applied, the future conditions can be predicted and the controller stops the input reactant dosing in advance to avoid the critical reactor temperature. An example of such a procedure is given in Figs. 3 and 4. Here, the model predictive control was used for simulation. It can be seen from Figs. 4 and 5 that, in the case of predictive control, the controller terminated the reactant dosing in time based on calculations performed on the system model and thus the critical concentration in the reactor was not exceeded (Fig. 4). Thereafter, the regulator maintained approximately the same concentration value by uniform dosage until the reactor was full, which in the graphs corresponds to the time value of about 1.5×10^4 s.

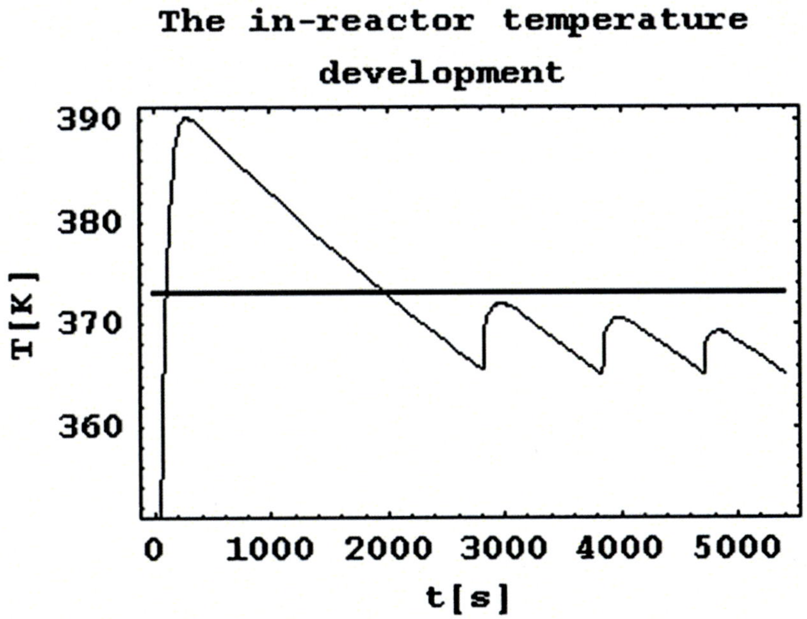

Fig. 1. Temperature overshooting caused by the accumulated input reactant.

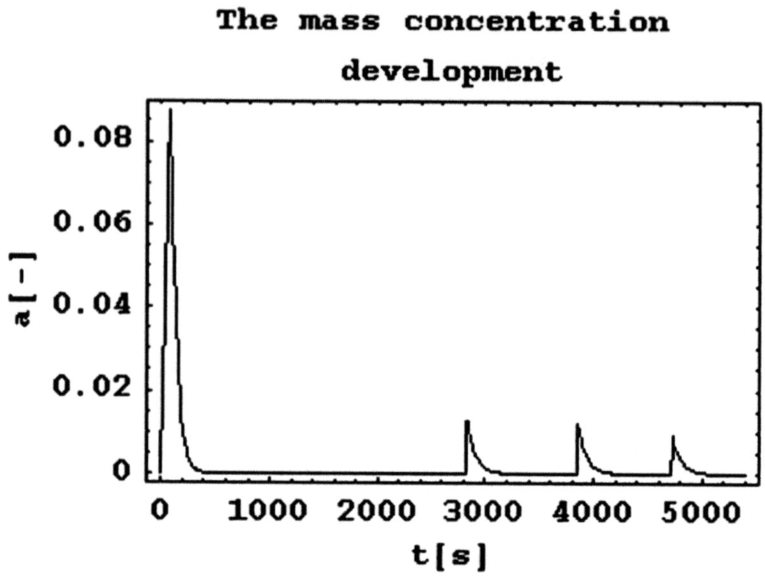

Fig. 2. In-reactor input reactant concentration causing the overshoot.

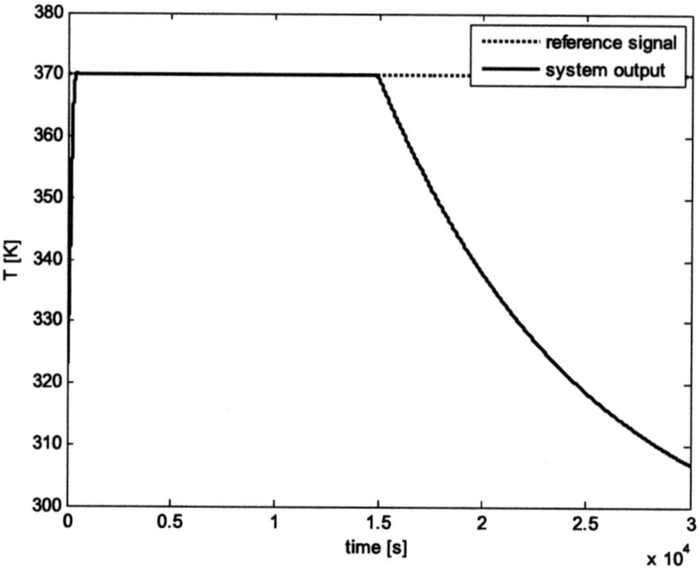

Fig. 3. In-reactor temperature development obtained from model predictive control.

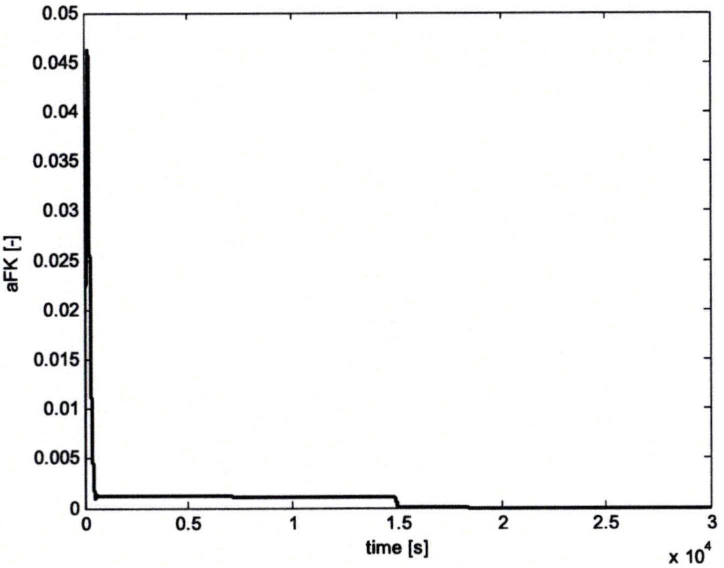

Fig. 4. In-reactor input reactant concentration obtained from model predictive control.

3 Semi-batch Exothermic Reactor Model

The chemical batch reactors are quite commonly used in the long term. An interesting view on the problem is mentioned in paper [4], which describes the old techniques used for their control.

To be able to apply prediction control, it is necessary to have a sufficiently accurate system model. The reaction kinetics and therefore the influence of parameters A and E contained in the Arrhenius equation are also included in the model equations. An example of such a model is given below, see Eqs. (2) to (5). It is a model on which the above-mentioned simulations were performed and can be found for example in [5]. The corresponding schema of the semi-batch reactor can be seen in Fig. 5.

Equations (2) and (3) were obtained from the chemical compounds mass balance, Eqs. (4) and (5) from the heat balance. The Eq. (4) involves chemical compounds heat balance while the Eq. (5) coolant medium heat balance. Individual symbols are shown in Table 1.

$$\frac{dm(t)}{dt} = F_I \tag{2}$$

$$\frac{da(t)}{dt} = \frac{F_I[1 - a(t)]}{m(t)} - A \cdot e^{-\frac{E}{R \cdot T(t)}} \cdot a(t) \tag{3}$$

$$\frac{dT(t)}{dt} = \frac{F_I \cdot c_I \cdot T_I}{m(t) \cdot c} + \frac{A \cdot e^{-\frac{E}{R \cdot T(t)}} \cdot \Delta H_r \cdot a(t)}{c} - \frac{K \cdot S \cdot T(t)}{m(t) \cdot c} + \frac{K \cdot S \cdot T_C(t)}{m(t) \cdot c} - \frac{T(t) F_I}{m(t)} \tag{4}$$

$$\frac{dT_C(t)}{dt} = \frac{F_C \cdot T_{C_I}}{m_C} + \frac{K \cdot S \cdot T(t)}{m_C \cdot c_C} - \frac{K \cdot S \cdot T_C(t)}{m_C \cdot c_C} - \frac{F_C \cdot T_C(t)}{m_C} \tag{5}$$

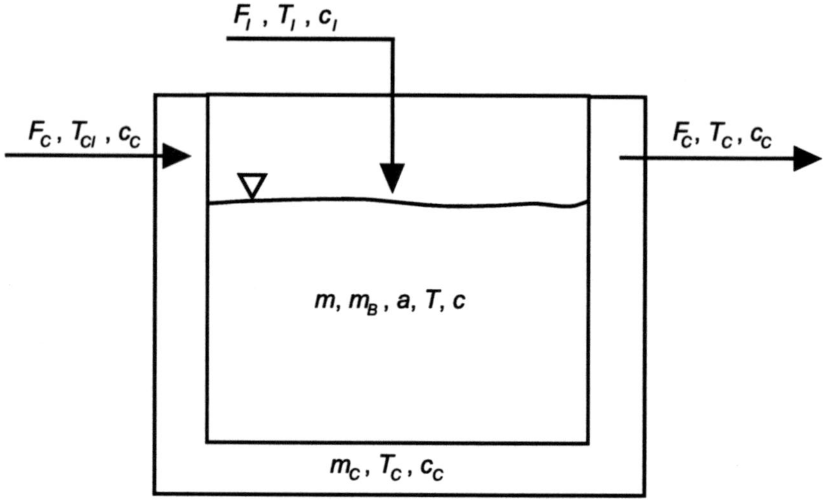

Fig. 5. Exothermic semi batch reactor scheme

Table 1. Mathematical model symbols.

Equation number	Expression	Description
(1)	$A[s^{-1}]$	Pre-exponential factor
	$E[J.mol^{-1}]$	Activation energy
	$R[J.mol^{-1}.K^{-1}]$	Gas constant
(2)	F_I $[kg.s^{-1}]$	Mass flow of the entering reaction component
	$\frac{dm(t)}{dt}$ $[kg.s^{-1}]$	Accumulation of the in-reactor content
(3)	$a(t)$[-]	Mass concentration of the reaction component
	$m(t)$[kg]	Weight of the reaction components in the system
(4)	c_I $[J.kg^{-1}.K^{-1}]$	Entering reaction component specific heat capacity
	c $[J.kg^{-1}.K^{-1}]$	Reactor content specific heat capacity
	T_I [K]	Entering reaction component temperature
	$\Delta H_r[J.kg^{-1}]$	Reaction heat

The controller adjusts the control action according to specific parameters A and E. If these parameters are constant, which for the most part of chemical production are, everything works properly. However, if these parameters change, the system model will differ from the real situation and the controller will not respond adequately.

This situation can easily occur if the reactant's composition changes, for example, if the raw material to be processed has a variable composition. That is the case of recycling processes since the quality and composition of the feedstock can vary widely and the feedstock contains various pollutants as well. In this case, the parameters *A*, *E* should be determined each time this change occurs.

The possibility of determining parameters *A* and *E* directly from the reactor, based on in-reactor temperature profiles, is described in the next chapter.

4 Parameters *A* and *E* Determination

The derivation of the relationship for determining *A* and *E* is based on the following equations:

$$Q = C\Delta T_1 \tag{6}$$

In Eq. (6) the symbol Q means heat, C is the system heat capacity and ΔT_1 is the increase of the system temperature caused by the heat generated by the reaction. As was already mentioned, the semi-batch reactor reaction kinetics can be treated as a first-order kinetic. For such case, the reaction velocity v is proportional to the dosing reactant only, as can be seen in formula (7):

$$v = -dc/dt = kc \tag{7}$$

The symbol c means a concentration of the dosing reactant, t is time and the k the velocity constant. Generated heat (Eq. 6) corresponds to the number of reacted moles and to the reaction component concentration as well. As implies from Eq. 6, increase of the temperature T is adequate to the heat:

$$dT = \frac{dQ}{C} \tag{8}$$

The heat dQ released by the chemical reaction can be written as a product of reacted moles dn and the heat of reaction ΔH_r. The number of moles also corresponds to product of concentration and volume:

$$dQ = \Delta H_r dn = \Delta H_r dcV \tag{9}$$

After replacing dQ in the Eq. (8) by Eq. (9) and dividing the result by time we obtain the relationship (10):

$$\frac{dT}{dt} = \frac{V\Delta H_r}{C}\frac{dc}{dt} \tag{10}$$

The fraction *dc/dt* here equals to reaction velocity $-v$ as implies from Eq. (7). By combining the Eqs. (7) and (10) we get the resulting relation:

$$v = -\frac{dT}{dt}\frac{C}{V\Delta H_r} \tag{11}$$

From relations (11) and (7) the equation for rate constant can be derived:

$$k = \frac{dT}{dt}\frac{C}{V\Delta H_r c} = \frac{dT}{dt}\frac{C}{\Delta H_r n} \tag{12}$$

The derivation *dT/dt* means a tangent direction which can be taken from the time-temperature development (Fig. 6), C [J.K^{-1}] is reactor thermal capacity, V [m^3] volume, c [mol.m^{-3}] molar concentration and n [mol] number of moles. To determine the rate constant, the chemical volume V and the beginning concentration of the reactant dosed to the reactor c and the reaction heat ΔH_r must be known. The tangent direction *dT/dt* can be obtained from temperature time dependence (Fig. 6) as was already mentioned. The graph in Fig. 6 can be obtained as a temperature development during a chemical reaction. The rate constant can be calculated from the expression (12) using experimentally observed data.

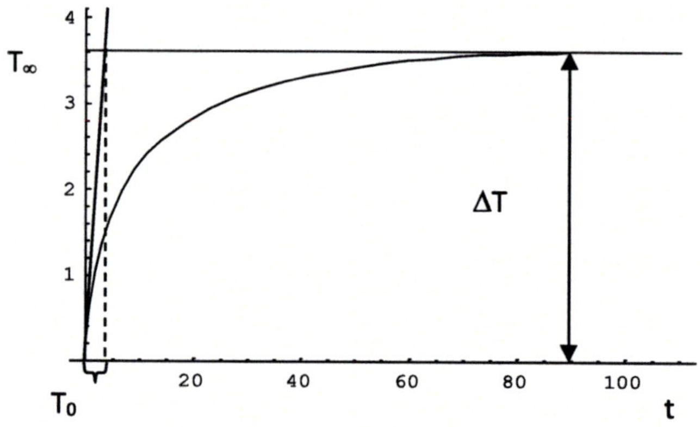

Fig. 6. An example of in-reactor temperature development.

It is suitable to identify the directive *dT/dt* at the beginning of the chemical reaction, when the initial concentration of substances is known. The initial concentration c_0 can be then used in the formula (12). The reaction mixture volume V is given by the initial chemicals volume together with the volume of reactant dosed to the reactor. The reaction heat ΔH_r can be found from the temperature differences (Eq. 6) in case that we know the reactor heat capacity C. This capacity can be measured once or can be counted mathematically.

For the preexponential factor A and the activation energy E determination, the Arrhenius relationship (Eq. 1) is transferred to a logarithmic form (13):

$$\ln k = \ln A - \frac{E}{RT} \tag{13}$$

The $\ln k$ can be assigned to y axe and the *1/T* to x axe and we obtain dependence in the form:

$$y = a + bx \tag{14}$$

The preexponential factor A and activation energy E can be than found as:

$$A = e^a \tag{15}$$

$$E = -bR \tag{16}$$

The constants A and E can be then placed to the chemical reactor mathematical model (Eqs. 1–4) for control purposes.

5 Conclusion

As was shown in this paper it is possible to determine the chemical kinetics from the in-reactor temperature profiles. This involves the Arrhenius equation parameters as are the preexponential constant A and the activation energy E. These parameters are very important for appropriate chemical reactor model description and can change with different reactant composition. This is also very important in terms of chemical reactor control because the A and E incorrect values could cause the controller failure.

To identify the Arrhenius equation parameters, we need to know the system thermal capacity, the starting reactant concentration, and the system volume. The volume and the beginning concentration are usually known. The thermal capacity can be found experimentally or counted theoretically, but only once. The procedure is based on experimentally obtained data in laboratory conditions when studying chromium sludge processing, as can be found in publication [6].

This paper also mentions the mathematical model of an exothermic semi-batch reactor, which was used for the real process simulation. Some control methods applied on this model can be found in articles [7–9]. Especially very good results were obtained with model predictive control, as can be seen in the paper [7]. Here the Self Organizing Migrating Algorithm was used to improve the chemical reactor control performance.

References

1. Bardow, A., Marquardt, W.: Incremental and simultaneous identification of reaction kinetics: methods and comparison. Chem. Eng. Sci. **59**(13), 2673–2684 (2004)
2. Alvarez-Corena, J.R., Bergendahl, J.A., Hart, F.L.: Advanced oxidation of five contaminants in water by UV/TiO2: reaction kinetics and byproducts identification. J. Environ. Manage. **181**, 544–551 (2016)
3. Benítez, M., Bermúdez, A., Rodríguez-Calo, J.F.: Adjoint method for parameter identification problems in models of stirred tank chemical reactors. Chem. Eng. Res. Des. **123**, 214–229 (2017)
4. Luyben, W.L.: Temperature setpoint-ramp control structure for batch reactors. Chem. Eng. Sci. **208**, 115–124 (2019)
5. Macků, L., Sámek, D.: Semi-batch reactor predictive control using MATLAB fmincon function compared to SOMA algorithm. WSEAS Trans. Syst. Control **13**, 466–472 (2018)
6. Macků, L.: Design of regenerate preparation control for tanning (In Czech). Zlín, Dissertation. Tomas Bata University in Zlín (2003)
7. Macků, L., Sámek, D.: Self-organizing migrating algorithm used for model predictive control of semi-batch chemical reactor. In: Silhavy, R., Senkerik, R., Oplatkova, Z.K., Silhavy, P., Prokopova, Z. (eds.) Automation Control Theory Perspectives in Intelligent Systems. AISC, vol. 466, pp. 255–265. Springer, Cham (2016). https://doi.org/10.1007/978-3-319-33389-2_25
8. Gazdoš, F., Macků, L.: Analysis of a semi-batch reactor for control purposes. In: Proceedings of 22nd European Conference on Modelling and Simulation, ECMS 2008, pp. 512–518 (2008)
9. Novosad, D., Macku, L.: Ziegler-Nichols controller with online identification versus PID controller comparison. In: Annals of DAAAM and Proceedings of the International DAAAM Symposium, pp. 1017–1018 (2010)

An Innovative Textile Product Proposal Based on Sustainability: Recycling Wastes from the Wool Industry

Regis Puppim[1,2](✉) and Ana Cristina Broega[2]

[1] Instituto Federal de Goiás (IFG), Aparecida de Goiânia Campus, Aparecida de Goiânia, Brazil
[2] Centre for Textile Science and Technology (2C2T), University of Minho, Azurém Campus, Guimarães, Portugal

Abstract. In a world where consumers are increasingly more aware of the issues on Sustainability, this scientific article presents a partial stage of the PhD research that exposes an investigation that led to the development of a new textile product for recycling clean solid waste from the wool industry. Even at an early stage, the prototypes that were achieved have a high potential, as an innovative and sustainable propose of product. The research development includes scientific methods such as Case Study, Field Research and laboratory prototyping (experimental research issue). For the study, a local and important Wool Textile Industry was chosen, for analysis, diagnostic and confirmation of the Literature Review, focusing on the industrial textile clean wastes that most express the production bottleneck. At the end, the article foresees the next steps of the research, in order to improve and characterize the product, enhancing the possibilities of sustainable aspects on it.

Keywords: Textile innovation · Textile recycling and reuse · Sustainability

1 An Introduction

Contemporarily, it is common to face different kind of products with the 'sustainable' adjective attached to, adding intangible value feelings onto the consumer, especially for those who are environmental warned. It is not different for the Textile and Fashion Industries. Nevertheless, many of those companies and brands (and even consumers) are not concerned on the sustainability core bases. Therefore, sometimes, it is possible to lead the buying act to a Greenwashing statement, that means a false manner of promoting a product and/or a service as 'sustainable', with no real meaning, by its proposal [1].

According to Veiga [2], sustainability is a term only accepted and used by the media in the 1990's, although scientists and researchers apply the term since the 1970's. Some important events helped to spread sustainability term, such as Rio 92 Conference, promoting important International Treaties, such as Agenda 21 and Kyoto's protocol, being guidelines for the environmental behaviour of the involved countries. Such movements, in order to make us re-think on environment, are still up for consideration, remembering that the 2020 Davos Economic Forum theme, that engaged sustainability as a main discussion [3].

J. Machado et al. (Eds.): icieng 2021, LNME, pp. 256–262, 2022.
https://doi.org/10.1007/978-3-030-79165-0_25

Concerning Textile and Fashion Industries, differently from many other fashion trends, that had their beginning, expanding, plenitude and decreasing moments through a short period of time [4], Sustainability is not a "common trend", it is a new way of production, use (and reuse) and consumption, re-thinking and re-defining the whole Clothing and adjacent Industries [5].

If by one side, there is a growing of a new perspective on production and consumption, within the sustainable values, by the other side, there is a strong, solid and eco unfriendly force named (Traditional) Fashion, that, constantly, order us to buy more and more, otherwise we would be outdated (out of Fashion). Thus, it is possible to remit us to Lipovetsky [6], explaining the panorama where 'Fashion' was born, as a terminology, into a modern, occidental and consumerist society. In other words, by its origins, Fashion, essentially, means to look forward, constantly, to be 'on fashion'/'on vogue', with the most possible newest trend, especially by the clothing.

In these circumstances, it seems hard to believe that Fashion and Sustainability terms can merge into one way. Therefore, for example, most of the researchers and authors do not use the term 'sustainable fashion', because it means something paradoxical, with no proper functioning as a term [7]. However, it is already possible to find interesting products and projects moving forward in this perspective, that Puppim et al. [8] already presented and find an approach to Fashion and Sustainability possibilities, divided by five subgroups that are: Materials, Processes, Consumption, End of Life Cycle, and Suitability.

In 'Materials' category the authors reveal the new technological materials, mostly resulting from recycling, and organic materials, using the less possible amount of the natural resources, such as energy, water and the raw materials itself. In Processes, it is exposed to the development of techniques and methods throughout the Fashion and Textile chains aiming the minimum (or zero/null) waste and the generation of discards. In Consumption, it is studied the practices of conscious consumption, being the Ethic a keyword. In the "End of the Life Cycle", it is shown proposals of reuse, recycles and redesign products, trying to extend the life of the product or the material. In the Suitability subgroup, it is expressed the seals and certification, that could be applied to fashion products, demonstrating the integrity and transparency of the materials, processes, and products in favour of sustainability.

It is highly demanded that materials (and by consequence, End of Life Cycle) are one of the most important topics on sustainability and fashion, which can potentiate the real 'sustainable' core meaning in a fashion and textile product, by being on the pre-product phase, marked out, also, by Thompson and Thompson [9], expressing strong references for the development of sustainable materials.

In this sense, throughout a PhD research, involving literature review, market and product research, and, particularly and most important, the Wool Textile Industry case study, focusing on discards from the manufacturing processes, we had begun the development of a new textile product, having as a investigation core the sustainable values – reusing and recycling.

2 Materials and Methods

Considered one of the most important natural fibers, specially by its market pricing [9], the Wool fiber is used for textile and clothing products since antique times, by its thermal conservation quality [10]. Thus, and adding the importance of developing textile products from natural origins [7], we took in consideration for the development of a new product. In order to achieve a better result, we studied the Wool Textile Industry manufacturing processes and its discards on each stage and chose the most appropriated textile technology.

2.1 Wool Textile Industry Process and Its Wastes

First of all, it is important to remember that the basic manufacturing process of a textile is composed by: acquisition of the raw material (depending on the type of fiber); spinning process; weaving or knitting process; and, finally, processing and textile finishing [10]. It is important to note that all stages of this production process produce wastes and discards [9]. Here, it is focused the Clean Textile Residues, contemplated by the European Waste Catalogue (EWC) codes 040221, which stands for unprocessed textile fibers waste, and 040222, for processed textile fibers waste.

The wool textile has, in particular, two major processes that are responsible for the generation of clean waste (excluding liquid effluents), where, of its total amount (100%), it counts 50% in the spinning operation (counting carding, combing and spinning, properly meaning) and 37% in warping and weaving operations (pre-phase and main phase of the weaving stage), remaining only 13% for other operations [11]. According to data from the Report of the National Institute of Engineering and Industrial Technology (INETI), it was estimated that in 1998 in Portugal, processed and unprocessed textile fiber waste was more than 80 thousand tons (Idem). Even tough wool fiber production accounts for 5%, it was explicated that the yield from the use of the raw material (which can also be perceived as part of its efficiency in the production process) has the worst result among all fibers, namely, only 67% for the raw (preparation) stage and 95% for the spinning and pre-weaving/knitting stage [11], in addition to solid waste, liquid effluents and sludge from production.

In this sense, the spinning process (with all its sub-production stages) processed over 20,000 tons and generated almost 1,100 tons of solid waste, while in the weaving and knitting process the production was over 17,000 tons and the solid waste produced was 112 tons [11].

In other words, it is possible to note, by this presented report, that wool textile manufacturing process produces a considerable quantity of waste and discard, specially the spinning and weaving/knitting stages, considering 5,50% (1,100 tons for waste on 20,000 tons for total production) and 0,65% (112 tons for waste on 17,000 tons for total production), respectively.

Although the report seems to present old numbers (more than 20 years), it is important to notice, historically, the most important residues production and its responsible processes stages.

Observing these report numbers, we chose a Portuguese Wool Textile Industry that is very important to national and European production and followed its processes stages. In this field research, we have observed, monitored and inquired many questions as possible, in order to comprehend the most relevant production bottleneck waste, considering that this industry looks forward on having recycling and reusing of its own wastes.

This investigation leads us to the combing process, that discards the small wool fibers, in order to develop a better-quality textile yarn (less roughness and pilling into the fabric). So, we have collected this material (discarded small wool fibers) with the view to study and to perform experiments. We conducted this material to a veil, which resulted fine, in order to facilitate transportation and initials experiments.

Another waste material, pointed out by the industry's engineers was the fabric selvedge, which is used to reinforce the fabric edges during the weaving process, commonly from a more strong material and different color from the fabric, that has no function after this process on the material.

2.2 Nonwoven Textile Technology

As known that small fibers, normally, does no result into good-quality textile yarns for weaving and/or knitting [10], we decided to search for nonwoven textile processes alternatives regarding the combing process wastes. Among the various types of nonwoven, this study seems to present Needle Punching process as a better way to develop our intended product, from the sectioned and collected wastes.

Needle Punching is a process for nonwoven textile that consists in a mechanical interlacing of fibers, through a repeatedly penetration by needles with protrusions into a fibrous veil [12].

This experiments on nonwoven of needle punching technology resulted according to expectation, on the veil consolidation into a 'felt kind' base material, but also on incorporating the selvedge material into this base material.

The developed process of the product was started by the consolidation of the mantle, through one passage on the Needle Punching Machine, with configurations for prototyping materials in speed of inlet and outlet conveyors at 150 cm/min and 150 mints-needling/min. The machinery used was *Automatex's Impianti Non-Woven.*

Then, within the base material consolidated, we developed an initial design for the final material (choosing colouring contrasts and design created forms) and positioned the selvedge material on it and repeated two times the needle punching pass process. This procedure was also proposed by the Rewald [12] technique.

Two different forms prototypes were developed on this stage, not resulting for one as better-quality, but for different outcomes. The first one was passing the selvedge on the back position (downside) both times, resulting in a less incorporated material but more color contrast exposed. The second, on the back and on the front (upside), each time, resulting in a more incorporated material but less color contrast exposed, because of the base material fibers passing on and through the selvedge material.

With the propose to enhance the incorporation on both prototypes, we proceeded on a felting process, which is a specially used for wool fiber textile materials on nonwoven, because of its proprieties [10]. The procedure was part handmade and part mechanical, starting with kneading the prototypes with warm water and neutral soap, by hand, for 10 min. After, we used a home washing machine on the short wash and drying process, within temperature of 60°C, that took around 1-h period.

Then, the still wet materials were positioned horizontally upon a table with plastic cover, in a room with no direct sun incidence and no wind for 12 h, for drying.

3 Results and Discussion

The achieved result for the textile product showed innovation and opened a new perspective on textile reuse and recycling, since it is used by different kinds of textile technology processes, but also, encompasses two presented, by the industry, relevant bottleneck production discards.

The felting process enhanced, even more, the consolidation of the material. Afterwards, we tested and measured the prototypes, that, then, presented a really better incorporation result, full felted, on both cases, also presenting a slight shrinkage on its dimensions, horizontally and vertically, not leaving anymore, as before, small fibers "dust" with by hand friction, and less roughness on touch. An artistical image of the results is presented (see Fig. 1).

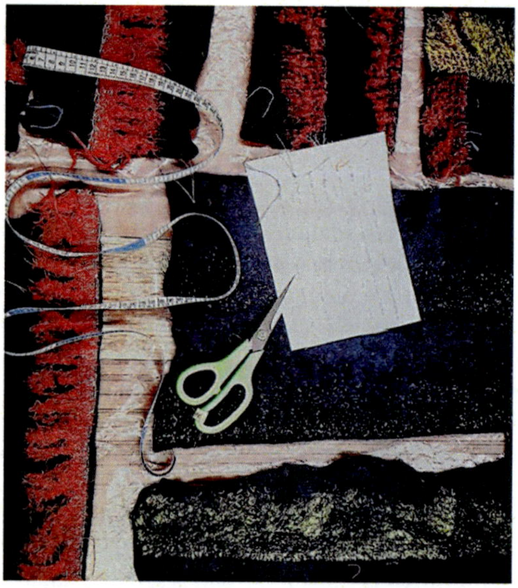

Fig. 1. Artistical presentation of the prototypes material.

This result, even on a prototyping stage, can express a good product for some home textiles, such as decorative panels or boards.

In addition, the Textile Industry used as part of the research argued that this initial result proposes a very interesting and innovative product, that represents well the research on textile waste and sustainability.

4 Conclusion and Further Studies

The next steps on the investigation and development of the textile product are on two stages, namely: 1) Design; and 2) Quality and Classification Testing.

On the Design aspect, it is important to tabulate all different possible designs for the products, concerning the disponible coloring and selvedge format disposal (See Fig. 2).

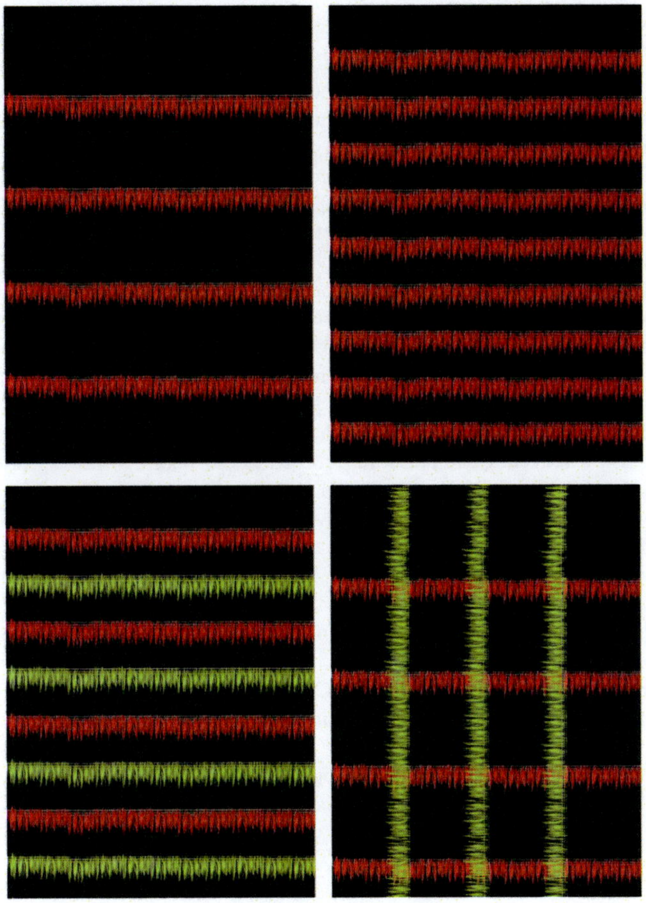

Fig. 2. Initial design studies.

On the Quality and Classification Testing perspective, it is important to elect methods on characterization for the material, intended, initially, to result for clothing destination. Some of the most important tests for achieving this result are expressed as "comfort quality", such as physical proprieties [13].

This physical method tests may suggest some adaptation and re-development for the product, reconsidering the material, the nonwoven and felting processes, and/or also finalization and textile processing, in order to achieve the expected result.

By now it is possible to express that this initial part of the research has good results, by having a innovative recycling product proposal, by the recognition of the local industry on the importance and good result for the prototype and by receiving an Honorable Mention at the International Design Awards – IDA – 2019, on the Fashion – Textile and Materials topic, in a special project developed within the core bases of this product conjoined with flexible 3D printing [14].

References

1. Manzini, E., Vezzoli, C.: O desenvolvimento de Produtos Sustentáveis: Os requisitos ambientais dos produtos industriais, 2nd edn. EDUSP, São Paulo (2011)
2. Veiga, J.E.: Sustentabilidade: A legitimação de um novo valor. SENAC-SP, São Paulo (2010)
3. Forbes Homepage. https://www.forbes.com/sites/lbsbusinessstrategyreview/2020/01/29/why-sustainability-was-the-star-at-davos-2020/#e28cdb76a387. Accessed 01 Feb 2020
4. Calderin, J.: Form, Fit and Fashion: All Details Fashion Designers Need to Know But Can Never Find. Rockpocket Publishers, Beverly (2009)
5. Brown, S.: Eco Fashion. Laurence King, London (2010)
6. Lipovetsky, G.: O Império do Efêmero: A moda e seu destino nas sociedades Modernas, 6th edn. Companhia das Letras, São Paulo (2009)
7. Berlim, L.: Moda e Sustentabilidade: uma reflexão necessária. Estação das Cores e Letras, São Paulo (2012)
8. Puppim, R., Broega, A.C., Jordão, C., Teixeira, M.A.: A case study of new products and materials from textile recycling and reuse. In: Proceedings of Textile Institute World Conference 2018, pp. 178–186. Leeds University, Leeds (2018)
9. Thompson, R., Thompson, M.: Sustainable Materials. Processes and Production. Thames and Hudson, London (2013)
10. Udale, J.: Basics Fashion Design: Textiles and Fashion. Bloomsbury Publishing, London (2014)
11. INETI: Guia Técnico - Sector Têxtil. Instituto Nacional de Engenharia, Tecnologia e Inovação (INETI), Lisboa, Portugal (2000)
12. Rewald, F.G.: Tecnologia de Nãotecidos. LCTE, São Paulo (2006)
13. Bona, M.: La Qualità Nel Tessile: Métodi fisici di controlli dei prodotti e dei processi. Paravia, Torino (1992)
14. International Design Awards – IDA Homepage. https://idesignawards.com/winners/zoom.php?eid=9-25860-19&count=0&mode=. Accessed 01 Feb 2020

Global Knowledge Generation Hotspots: An Overview from Technological Tendencies in Biotechnology

Carlos Antonio Medeiros Gambôa[1], Anapatricia Morales Vilha[1,2], Fábio Danilo Ferreira[1,2], Débora Maria Rossi de Medeiros[1,2], and Jayson Luis da Silva Ribeiro[1](✉)

[1] Federal University of ABC, Santo André, SP 09210-580, Brazil
{carlos.gamboa,anapatricia.vilha,fabio.ferreira,debora.medeiros}@ufabc.edu.br

[2] Innovation Center - Federal University of ABC, Santo André, SP 09210-580, Brazil

Abstract. This paper aims to understand the development of innovation clusters and hotspots of cutting-edge research in Biotechnology across the world, clusters understood as segments of an innovation system, gathering research institutes, universities, startups, enterprises and funding actors; and hotspots defined as clusters focused on frontier R&D, with economic high potential. The choice of Biotech as study object embeds the special characteristics of this branch of science: long term research with high uncertainty of results, huge funding and few marketable products. The analysis is based on 18 selected tendencies pointed by European Commission as Radical Innovation Breakthroughs and the frequency of these tendencies in several patent databases, focused on which countries are the main applicants since 2005. As a sub product we observed the change, in the studied range of time, of the main axis of Biotech innovation beyond USA, Europe and Japan. The analytical approach is based on literature review and quantitative screening of patent bases.

Keywords: Biotechnology innovation clusters · Innovation hotspots · Biotechnology research

1 Introduction

Biotechnology is one of the most dynamic segments of the scientific research in the world, for its huge economic potential and for its impact in the future of human health and welfare [1]. Like other knowledge dimensions, it benefits of institutional and economic ecologies with relevant aspects like geographical proximity, funding models and sources, research agents (like universities, hospital complexes and multinational enterprises), legal and institutional scenario - mainly intellectual property protection and its transfer mechanisms [2]. In the same level we need to recognize the specific characteristics of the

J. Machado et al. (Eds.): icieng 2021, LNME, pp. 263–273, 2022.
https://doi.org/10.1007/978-3-030-79165-0_26

segment: almost no products oriented to final customers, low revenue in the initial stages of projects, growth based on stock of front line research, industry secrets protection, large pay back periods [3].

Bibliographic review identifies many "loci" of concentration of institutions and companies in the Biotech segment, each of them combining, in a different way, the mentioned characteristics. This paper aims to identify, starting from a group of biotech front line research initiatives, the most important geographies of these studies, concentration and profile of these areas and its adherence to the concept of hotspots [2, 4]. The reason why of this research is the fact that the global geography of scientific and technological knowledge generation is changing, including new players of high innovative density, beyond USA, Europe and Japan.

This paper will be developed in five sections. The second section refers to Innovation Models, the third shows how innovation is produced in Biotechnology, the fourth details patent analysis, the fifth shows our aims.

2 Systemic Approach to Innovation Generation

The innovation production, from the 70s of last century, points to a concentration process in some specific areas of the world, especially in the developed countries. The innovation literature called these regions clusters, trying to identify how they started and which characteristics allowed their development.

Redman [5] defined the cluster as a geographical concentration of production chains of items similar or associated, embedded in institutional arrangements that positively act over the competitiveness of these chains. Rosenfeld [6] adds that the cluster has channels that ease communications and transactions, shared research infrastructure and access to a pool of high skilled manpower. Also important are the geographical and spatial concentration of the economic activity, vertical and horizontal relations between sectors of the industry and density of interaction between companies and individuals working in the cluster [7].

As an important share of knowledge added to innovation is tacit, demanding personal interaction, the geographical proximity is essential [8]. In addition, this proximity allows to the companies access to skilled labor coming from universities around. Metaphorically, physical proximity is improved by organizational proximity, which means organized interaction between cluster components, resulting in transaction costs reduction, improvement of the knowledge building process and access to specialized funding structures, like venture capital funds and government funding and credit lines. Finally, there is the institutional proximity, the submission of the components to the same legal, taxes and intellectual property rights regime, resulting in lower transaction costs and reduction of implementation times [2].

The globalization process of economic activities increased the range of the clusters, beyond national borders, adding density to the relationships. This increase, however, do not moved the main axis of knowledge production, only redesigns its profile, acting over the international division of work and adding new excellence centers, like China, Korea,

Israel, Taiwan and Singapore. This movement was mainly produced by multinational enterprises, which moved less technological density development activities or those that demand regional adapting, to their branches, maintaining the frontier research in their headquarters [2].

Seeming a paradox, we can see the scientific knowledge generation moving to a global reach and at the same time locally concentrated in terms of intensity [9], building Global Innovation Networks (GIN) defined as international collaboration webs between organizations (which can be universities, companies, international agencies, financial institutions, government agencies) and individuals, focused on production of knowledge and innovation [10].

Going deep, it was added to the concept of cluster the definition of hotspot, which is a group of cluster components with high growth potential e high internal competitiveness. Some factors allow the development of a hotspot: i) economies of knowledge density in the region, ii) synergy economies, due to connection with other innovative areas, and iii) linking economies, where the hotspot supply, in a specialized basis, demands evolved from the cluster where it is inserted [11]. Hotspots are mainly concentrated in the most important economic hubs of the developed countries [9]. Miguelez et al. [10] identified hotspots across the world, based on scientific papers production and patent volumes, with same conclusions of WIPO. Figures below illustrate this work (Figs. 1 and 2).

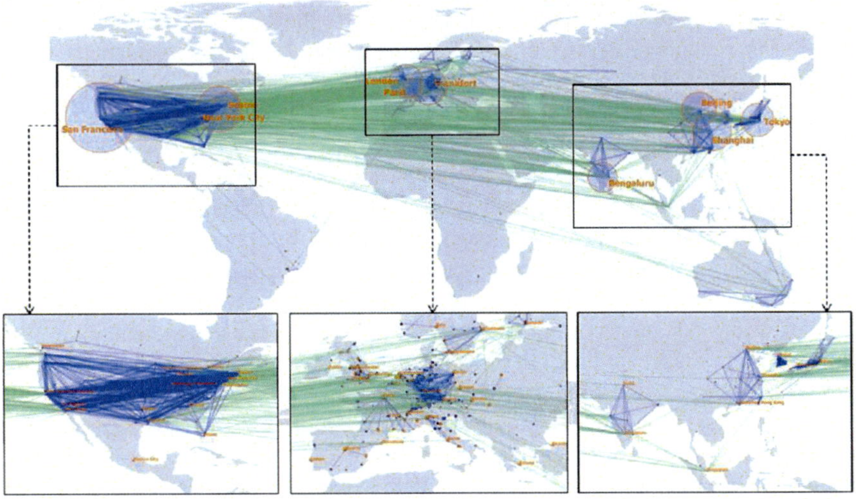

Fig. 1. Global Innovation Network, patent data 2011–2015. Source: Authors based on PATSTAT, PCT and Web of Science data (2020)

In the next section, we will expose specific characteristics of Biotechnology sector in the production of innovation and scientific knowledge.

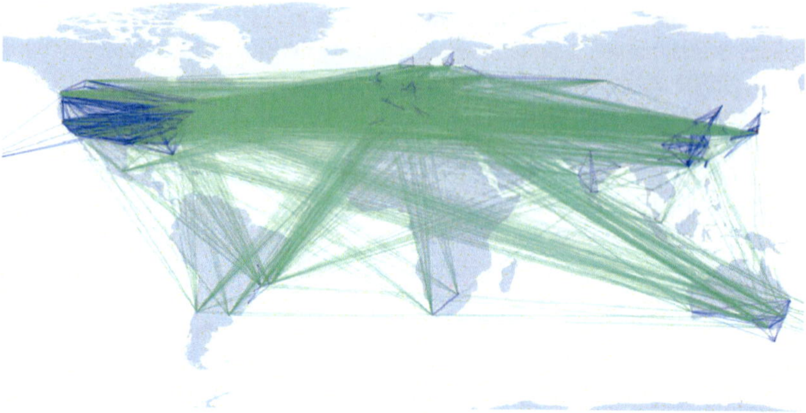

Fig. 2. Global Innovation Network, scientific publication data 2011–2015. Source: Authors based on PATSTAT, PCT and Web of Science data (2020)

3 Innovation in Biotechnology

Biotechnology, in the definition of BIO (Biotechnology Industry Organization), is a set of technologies which explores cells predicates (like their abilities to transform) and guide biological molecules (like DNA and proteins) to the benefit of mankind. In a broad concept, biotechnology is a set of technologies which use living organisms, cells and molecules to create products to benefit human health and environment and improvement of manufactured products and services, from diagnosis and therapies to agriculture and food production [12].

Biotechnology is not an industry but a set of specific activities and technologies such as biomaterials, combinatorial chemistry, DNA markers, genetic engineering, monoclonal antibodies, recombinant DNA, etc. [13]. Using these technologies, we can develop new products (i.e. artificial blood a human tissue), new processes for existing products (i.e. new methods for producing some specific protein) and new organisms for environmental cleaning or human consumption purposes. Most specialized biotechnology firms (SBFs) as well as most applications outside SBFs are now in the area of drugs for human health. The pharmaceutical industry is the main user of the biotechnology [13].

Pioneering in modern biotechnology can be seen in Boston and San Francisco areas, where we have good examples of the concepts we listed: (i) presence of universities with international first class standards, like CalTech, MIT, Harvard, Stanford; (ii) startup of companies evolved from research projects of these institutions scholars (Cetus, Genentech, Bioten, and Hybritech); (iii) funding available to start ups; (iv) suitable institutional environment, after the Supreme Court decisions that allowed to patent engineered life forms, genetic cloning and, after all, the Bayh Doyle Act, which allowed universities to license inventions funded by the government [12].

Out of United States we can see the presence of large enterprises, especially pharmaceutical and biochemical companies absorbing products and structures of SBFs and adding their integration and management capabilities to produce marketable products. These enterprises allow funding structures and connections to make viable biotech projects, which have long term paybacks, bulk costs upfront and long, complex and risky clinical tests [14]. The biggest slice of biotechnical development is made in SBFs, which are able to act in volatile and high-speed changing scenarios, and fragmented knowledge. These companies transform knowledge in go to market products and make the bridge between universities research groups and projects, and large corporations [14].

Moodysson and Jonsson [1] explain graphically the development of a Biotechnology project in the figure below (Fig. 3):

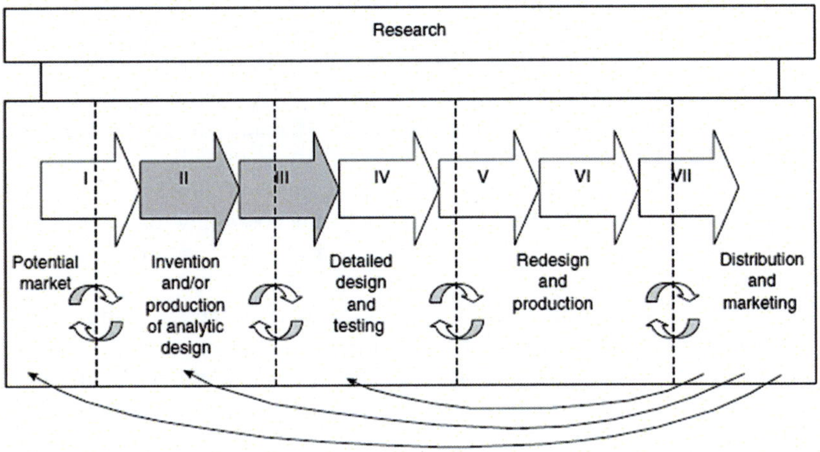

Fig. 3. Innovation Process of biotech-based-drugs. Source: Moodysson, J., Jonsson, O.: Knowledge Collaboration and Proximity - The Spatial Organization of Biotech Innovation Projects in European Urban and Regional Studies14(2): 115–131 Copyright © 2007 SAGE Publications https://doi.org/10.1177/0969776407075556

In the next session, analyzing patent basis, it will be showed the relationship between hotspots and frontier areas of Biotechnology.

4 Analytical Procedures: Patent Analysis to Detection of Conditioning Factors to Hotspots Generation in Biotechnology

In the last years patents have been used by managers working in the strategic departments of companies as well as policy makers who are interested in accessing information for their work directly from patents [15]. The information on the patent documents can reveal important indicators for the analysis of the technological behavior in specific areas.

According to Ernst [16] and Joho et al. [17], there are some main questions to be answered by patent analysis to support technological trend analysis. This includes obtaining an overview of different technology topics in a given field, identify relevant trends according to individual information needs, evaluate the importance of technological trends, observe the behavior and productivity of different players relevant to specific trends and spot new technologies related to a trend.

In the analysis of trends through patents, data can be submitted to grouping processes in order to identify characteristics of these documents that are more related to certain kinds of markets, research fields and more frequent assignees. Data mining is the process of proposing techniques for extracting useful information, patterns and trends often unknown from large amounts of data stored in databases [18].

Here we mine patents using the Google Patents database that index patents and patent applications with full text from 17 patent offices. We only include patents with a priority date from 2005 to have the landscape of the recent trends in the Biotechnology field.

The search was performed based on the most relevant words contained in the title and description of the technological areas. From the patents collected in this first search we identify the most frequently IPC code present in the results to reveal the most relevant patent class for each area. We chose to work with the maximum number of 100 patents in the result of each search to improve the results accuracy in relation to the correspondence between the collected patents and the technological area sought and eliminate potentially irrelevant results.

The data show the geographical distribution of patents and allowed the identification of the biggest players in each area, also opening up the possibility of a deeper analysis on the assignee kind (company, research institute and so forth) and the patents that were developed in partnership or individually, for example.

The researched areas were selected from a report of European Commission: 100 Radical Innovation Breakthroughs for the Future [19]. For this paper, we focused on the biotechnological innovation actions, building after them our patent database. In addition, there is a little description of each initiative and the search conditions for each one (Table 1).

Table 1. Biotechnology-radical innovation breakthroughs for the future [19].

Area	Summary
Biodegradable sensors	Electronic components with a limited lifetime subject to disappearing via hydrolysis or biochemical reactions
Lab-On-A-Chip	A lab-on-a-chip (LOC) integrates laboratory functions such as chemical analysis within a single device of small dimensions
Molecular Recognition	The study of interactions between molecules where the recognition component could be enzymes, DNA, RNA, catalytic antibodies, aptamers, and labeled biomolecules
Bioelectronics	The use of biological materials or architectures inspired by biological systems to design and build information processing machinery
Bioinformatics	Combines the methods, interests, and data of several disciplines, such as biology, mathematics, and computer science

(continued)

Table 1. (*continued*)

Area	Summary
Plant Communication	Plant communication refers to communication between plants and other organisms –and involves potential combinations with other, non-plant categories of capabilities
Gene editing	DNA is inserted, deleted, modified or replaced in an organism's genome at the targeted locations. The results are targeted mutations ('edits')
Gene Therapy	Gene-altering techniques employed for treating or preventing genetically triggered medical conditions
Antibiotic Susceptibility	These technologies enable the quick identification of therapies that are still effective for individual patients
Bioprinting (of human parts)	Particular application of 3D printing that uses polymers or genetically engineered biomaterials to produce tissues and organs, some of them implantable in the human body
Control of Gene Expression	Process by which the nucleotide sequence of a gene is used to direct protein synthesis and produce various cell structures
Drug Delivery	The administration of a healing agent or pharmaceutical complex to humans or animals in order to reach a therapeutically operative range of medication using nanomaterials
Epigenetic Change	Epigenetics refers to the heritable changes in gene function that do not entail changes in the DNA sequence
Genomic Vaccines	Inject genes that encode for the needed protein, which then cause cells to produce the protein in question
Microbiome	Microbes form microbiomes that can have both beneficial and harmful effects on human health
Regenerative Medicine	Methods to repair or replace cells, tissues, and even entire organs by using tissue engineering and cellular therapies
Reprogrammed Human Cells	Refers to either genetically reprogrammed white blood cells of the immune system or to induced pluripotent stem cells (iPSCs)
Targeting Cell Death Pathways	Targeting key regulatory molecules that trigger mechanistically distinct types of cell death might prove, a more effective, less toxic and less resistance-prone approach to cancer therapy

The Fig. 4 shows the biggest players in number of patents, considering our base of analysis, 18 categories, 100 patents each.

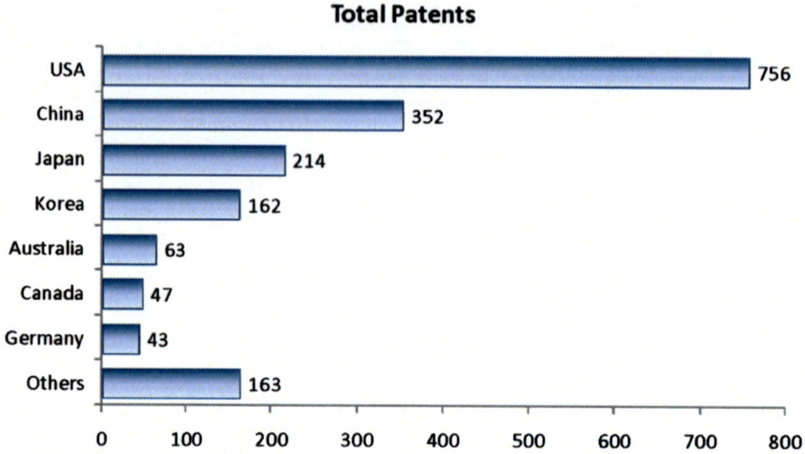

Fig. 4. Total patents by country (base 100 patents/category – 18 categories). Source: Authors analysis (2020)

The data confirm that the core of innovation is expanding, but not spraying. Following major increases in global trade and investment flows, innovation and economic development has spread to Southeast Asia, moving from Japan to the Republic of Korea and later to China [20] (Table 2).

Table 2. Biotechnology-radical innovation breakthroughs by countries. Source: Authors analysis (2020)

Area	USA	China	Japan	Korea	Australia	Canada	Others
MB	77	0	5	3	7	2	5
DD	60	5	15	8	4	3	5
BS	57	8	12	3	2	5	11
TC	53	3	16	9	4	2	13
GT	53	7	13	8	3	2	12
RH	50	13	8	9	7	2	9
AS	49	20	8	6	3	3	11
BE	45	7	22	11	0	0	11
GE	42	17	17	12	3	0	7
GV	41	19	15	7	4	3	10
CG	41	32	10	6	6	2	2
EC	40	17	16	5	7	4	11

(continued)

Table 2. (*continued*)

Area	USA	China	Japan	Korea	Australia	Canada	Others
MR	38	6	19	9	7	14	3
RM	31	13	22	6	1	2	23
LC	28	1	2	46	0	0	15
PC	23	40	8	4	3	2	7
BP	16	64	5	10	0	1	4
BI	12	80	1	0	2	0	4
Total	756	352	214	162	63	47	163

Expanding the analysis, we regrouped the 18 categories in five branches, as follows: Agriculture, Cells, Genetics, Machines and Pharmaceuticals, to detect prior interest areas of the most important players. Figures 5 present results.

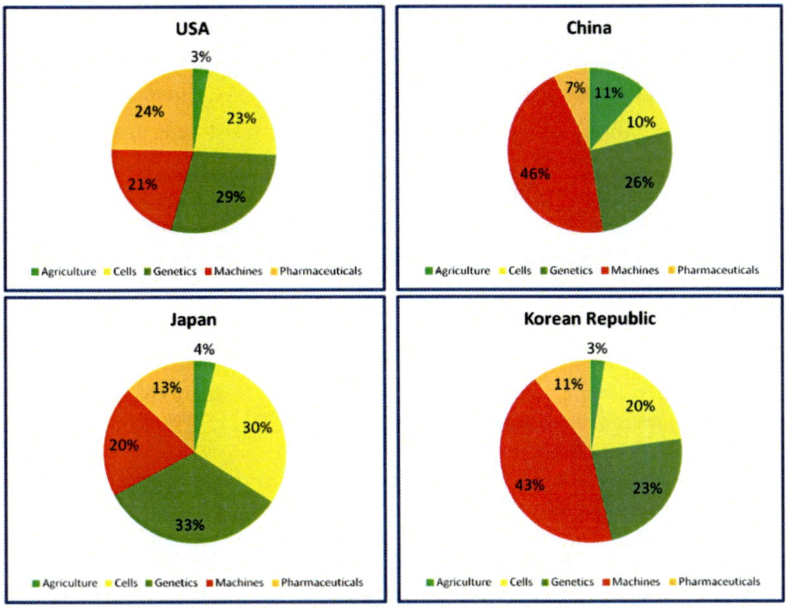

Fig. 5. Share of Biotech branches in USA, China, Japan, and Korean Republic. Source: Authors analysis (2020)

From these figures we can infer that United States have almost the same presence in all branches, except agriculture. China, Japan and Korean Republic have the lion share of investment in genetics and machines (with China adding Agriculture to the basic portfolio) using their original dynamic competencies (microelectronics, for example).

5 Final Considerations

This paper aims to understand the conditioning factors to generate global scientific, technological and innovative knowledge, especially approaching the most recent literature of Innovation Geography and Innovation Economy about Innovation Hotspots.

There are evidences that the scientific and technological development is enlarging its range beyond the historical (of the twentieth century) axis USA-Europe-Japan, including new players. This fact is especially relevant in cutting edge areas, like Biotechnology.

Based on a study of European Commission - 100 Radical Innovation Breakthroughs (RIB) for the Future [19] – we selected 18 of these RIBs with direct relation to Biotech and accessed international patent bases to explore tendencies across countries. Mining patents from 2005, based on keywords and concepts, a clear picture of the main players evolved. Examining the morphology of the patent requests in the main technological sectors which point tendencies in Biotech, we observe that even the USA still have dominance in the volume of patents, this dominance is being eroded by the new players, especially China and Korean Republic. As the American National System of Innovation is more robust and mature, the research in border areas of Biotech is diversified. In China, Japan and Korean Republic, there is focus on the most important dynamic capabilities of these countries, with expression in Genetics and electronic/mechanical devices using, for example, nanotechnology.

As we are analyzing hotspots and cutting edge technologies in Biotech, it is relevant to advance new studies to interconnect the patents overview of this paper with more variables, like technology transfer ratios, global R&D investment, map of world class universities with cutting edge research, ratio of the knowledge build up, scientific publishing related, environment (institutional, economic and scientific) and so on, to have a better understanding of the reasons that cause the birth and development of a global Biotech hotspot and its relevance in terms of border technology.

References

1. Moodysson, J., Jonsson, O.: Knowledge collaboration and proximity - the spatial organization of biotech innovation projects. Eur. Urban Reg. Stud. **14**(2), 115–131 (2007). https://doi.org/10.1177/0969776407075556
2. Crescenzi, R., et al.: The geography of innovation: local hotspots and global innovation networks. Economic Research Working Paper No. 57, November 2019
3. Coombs, J., Deeds, D.: International alliances as sources of capital: evidence from the biotechnology industry. J. High Technol. Manag. Res. **11**(2), 235–253 (2000)
4. Pouder, R., St John, C.: Hot spots and blind spots: geographical clusters of firms and innovation. Acad. Manag. Rev. **21**(4), 1192–1225 (1996)
5. Redman, J.: Understanding state economies through industry studies. Council of Governors' Policy Advisers, Washington, DC (1994)
6. Rosenfeld, S.A.: Bringing business clusters into the mainstream of economic development. Eur. Plann. Stud. **5**(1), 3–23 (1997)
7. Chiaroni, D., Chiesa, V.: Forms of creation of industrial clusters in biotechnology. Technovation **26**, 1064–1076 (2006)
8. Alcácer, J., Chung, W.: Location strategies and knowledge spillovers in management. Science **53**(5), 760–776 (2007)

9. WIPO, World Intellectual Property Organization, International Patent Classification (IPC). https://www.wipo.int/classifications/ipc/en/. Accessed 22 Jan 2020
10. Miguelez, E., et al.: Tied in the Global Network of Local Innovation. WIPO Economic Research Working Paper No. 58, November 2019
11. Kourtit, K., et al.: From islands of innovation to creative hotspots. Reg. Sci. Policy Pract. **3**(3), 145–161 (2011)
12. Koo, J., et al.: What does it take to become a biotech hot spot. Environ. Plann. Gov. Policy **27**(4), 665–683 (2009)
13. Niosi, J., Bas, T.: Biotechnology megacentres: montreal and toronto regional systems of innovation. Eur. Plann. Stud. **11**(7), 789–804 (2003)
14. Allansdotir, A., et al.: Innovation and competitiveness in European biotechnology. Enterprise Papers – No. 7 (2002)
15. Baudour, F., van de Kuilen, A.: Evolution of the patent information world–challenges of yesterday, today and tomorrow. World Pat. Inf. **40**, 4–9 (2015)
16. Ernst, H.: Patent information for strategic technology management. World Pat. Inf. **25**(3), 233–242 (2003)
17. Joho, H., et al.: A survey of patent users: an analysis of tasks, behavior, search functionality and system requirements. In: Proceedings of Symposium on Information Interaction in Context, IIiX 2010, pp. 13–24. ACM (2010)
18. Thuraisingham, B.M., Maurer, J.A.: Information survivability for evolvable and adaptable real-time command and control systems. IEEE Trans. Knowl. Data Eng. **11**(1), 228–238 (1999)
19. Warnke, P., et al.: 100 radical innovation breakthroughs for the future. In: The Radical Innovation Breakthrough Inquirer. Publications Office of the European Union, Luxembourg (2019)
20. Gurry, F., et al.: World Intellectual Property Report. The Geography of Innovation: Local Hotspots, Global Networks (2019)

Modelling of Thermal Properties and Temperature Evolution of Cork Composites During Moulding Process: Model Development

Helena Lopes[1(✉)], Susana P. Silva[2], and José Machado[1]

[1] MEtRICs Research Center, University of Minho, Campus of Azurém, 4800-058 Guimarães, Portugal
id7466@alunos.uminho.pt, jmachado@dem.uminho.pt
[2] Amorim Cork Composites, Rua Comendador Américo Ferreira Amorim, 260, 4535-186 Mozelos VFR, Portugal
susana.silva@amorim.com

Abstract. In the last years, cork-based composites proved to be able to provide new solutions for different areas like footwear or aerospace. Some of these products are the result of a combination of cork granulates with different materials - like thermosets or thermoplastics -, and its manufacture involves a thermal process. In order to simulate the manufacturing process of these types of composites, a new methodology was applied. A material composed of cork and a thermoplastic served as a case study. A model for the prediction of a cork composite mixture properties and a simulation methodology was developed to study the variation of temperature during the moulding process of cork composites. Density, thermal conductivity, and specific heat were determined based on the composite formulation and the properties of cork and the agglutinant agent, through the development of two theoretical models. The results obtained by both models were very similar. The assumptions, boundary conditions, finite volume and finite element methods formulations used for analyses are also presented.

Keywords: Temperature · Cork composites · Thermal conductivity · Specific heat · Density

1 Introduction

Cork is a natural material obtained through the harvest of *Quercus suber* L. trees. This species of oak is predominant in montado, an ecosystem that exists in different areas in the west side of Mediterranean Sea, particularly in the Alentejo region in Portugal. Classified as a cellular material, cork presents low density, almost-zero Poisson's coefficient, resilience, high energy absorption capacity, among other properties that allow its application in several fields [1]. The main productive activity of the cork industry is the production of stoppers that creates many remains that are not reintroduced in this process. The surplus of raw material can be applied mainly in the agglomerates and composites manufacturing, increasing its value [2].

J. Machado et al. (Eds.): icieng 2021, LNME, pp. 274–284, 2022.
https://doi.org/10.1007/978-3-030-79165-0_27

The production of cork composites involves diverse aspects that must be considered in order to achieve the pretended performance of the material. Nowadays, these kinds of composites are applied in many industries like automobile and aerospace, military, footwear and construction [3]. The raw material used in the composite manufacturing is cork granules, the result of trituration of cork wastes from stoppers production, virgin cork e low-quality cork. The manufacturing of the composites consists most of the times in the junction of cork granules with agglutinant products, like resins and other additives, through the combination of high temperature and pressure [3].

There are research works related to the application of new products in the development of cork composites. One of the fields with great potential is the introduction of thermoplastic as a binding agent. The use of this kind of raw materials can reduce the solvent existence and other problems involving toxicity, that are common with resins used in the agglomerates production [3, 4].

One of the manufacturing techniques consists in placing the material mixture inside a rectangular or cylindrical mould. The first step of the process is related to the mixture of the raw materials. After weighing each component according to the formulation imposed, they are introduced in a mixer to achieve a homogeneous mixture. The material is then poured into the mould and compressed until its final thickness. The mould is closed and placed inside a heating chamber to allow the agglomeration of the mixture components, due to the action of a resin thermoset or due to the melting of a thermoplastic. The processing time may variate between four and twenty-two hours with temperatures reaching 120 °C [3]. Then the mould is withdrawn from the heating chamber and placed at ambient temperature to be cooled before the material be removed from the mould. The blocks can be used as a raw material in other manufacturing processes in bulk or in laminated form [3]. Besides compression moulding, other manufacturing processes can be applied to produce cork composites such as injection moulding and extrusion [5].

The aim behind this study is to determine the temperature evolution of the material during the manufacturing process. This information could be used for optimizing process variables related to moulding process of cork composites like the temperature imposed in the heating chamber, the time spent inside the heating chamber, the amount of time necessary to cool down the mould, and the material formulation (quantities of each component in the mixture). In this article, as a first step towards the accomplishment of this goal, the mathematical formulation and numerical methods to be applied are presented as well as two different theoretical models in order to determine thermal and physical properties to be used as an input for future simulations.

2 Problem Formulation

2.1 Assumptions and Governing Equations

A schematic of the problem at study is presented in Fig. 1. For the model construction it was considered the following set of assumptions:

- Unidimensional model in the thickness direction;
- The flow of the resin/thermoplastic inside the mould is neglected;
- Only heat transfer by conduction was considered;

- Moisture content and water phase are not included in this model;
- The density of the compacted material remains the same throughout the process;
- The material was considered homogenous (the temperature in the cork particle is the same as the resin at any time);
- No deformation occurs during the compressing moulding process;
- No thermal resistance considered between the mould surfaces and the mixture.

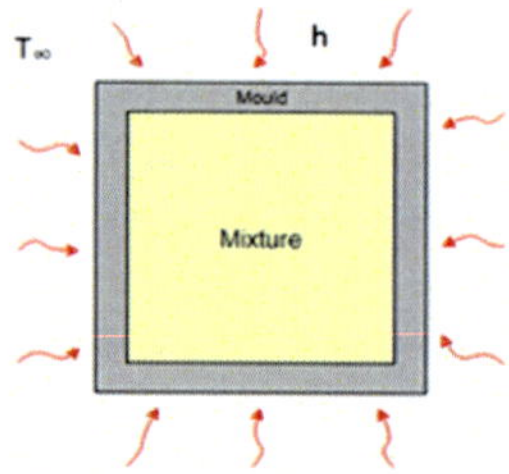

Fig. 1. Schematic of the block moulding heating process

For this problem, the governing equation in terms of cartesian coordinates (x,y,z) for transient heat transfer in a solid is given by the following equation [6]:

$$\rho c_p \frac{\partial T}{\partial t} = \left(k_z \frac{\partial T^2}{\partial^2 z} \right) \tag{1}$$

where T is temperature, t is time, ρ is density, c_p is specific heat, k is thermal conductivity and z indicates thickness direction. It was also considered the temperature influence on the thermal properties of the materials namely the thermal conductivity and specific heat.

2.2 Boundary and Initial Conditions

In this study, symmetry was considered in terms of properties and temperatures gradients at half of the thickness, to reduce computation time during problem-solving. The corresponding boundary condition is given by the following expression:

$$q|_{z=0} = 0 \tag{2}$$

where q is the heat flux [W m^{-2}].

The other boundary condition consists of convection at the mould surface, according to the following equation:

$$q|_{z=s} = h(T - T_\infty) \tag{3}$$

where s is the location of the mould surface, h is the heat transfer coefficient [W m^{-2} K^{-1}] and T_∞ is the fluid temperature.

The considered domain for the resolution of the formulated problem is depicted in Fig. 2.

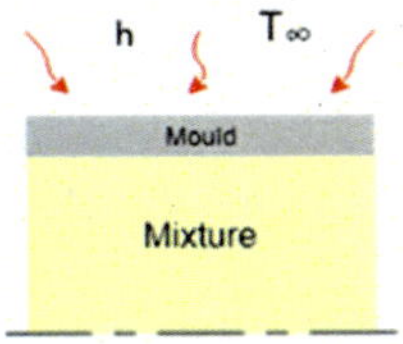

Fig. 2. Problem domain: convective heat and symmetry boundaries

3 Modelling of Thermal and Physical Properties

The input properties for the mixture were modelled according to the composition of the cork composites. It was developed models considering cork or the two phases of cork (cell walls and gaseous phase-air) to compute volume and mass fractions of each component.

In a first model (A), the material is considered as a three-component mixture before compaction (resin, cork, the air in the mixture) and a two-component mixture after compaction (resin and cork) assuming that all air between the materials is suppressed and the cork phase is compressible (Fig. 3).

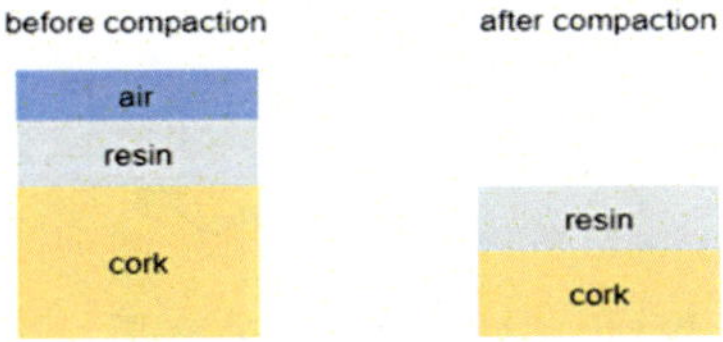

Fig. 3. Schematic for model A

The second model (B) considers cork as a combination of two different materials in a solid and a gaseous phase. The solid phase corresponds to the cell walls material and the second one is considered as air. The volume and fractions of each component of the mixture – resin, cork cell walls, and air – are calculated before and after compaction occurs, only considering the exit of the gaseous phase in the material (Fig. 4).

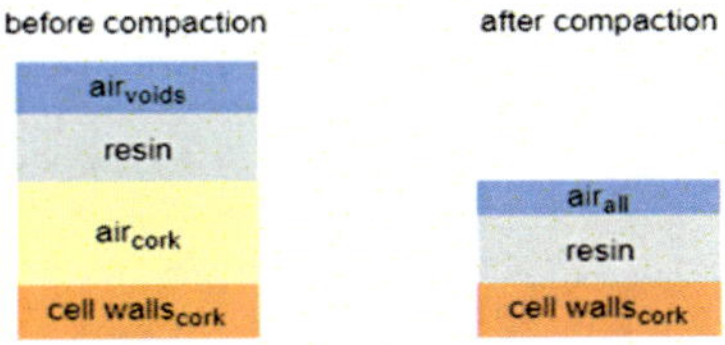

Fig. 4. Schematic for model B

3.1 Density

As no variation of volume in the mixture occurs during the heating part of the compression moulding, it was considered that the density is constant at any time of this process and is dependent only on the volume fraction of each component considered. Considering model A, the density is given by:

$$\rho = v_r\rho_r + v_c\rho_c + v_a\rho_a \tag{4}$$

Considering model B, the density is given by:

$$\rho = v_r\rho_r + v_{w_c}\rho_{w_c} + v_{a_c}\rho_{a_c} + v_a\rho_a \tag{5}$$

where v is the volume fraction after compaction, r, c and a are subscripts for resin, cork, and air from the voids, respectively, and w_c and a_c are subscripts for cork cell walls, and air inside cork, respectively.

3.2 Specific Heat

The specific heat of the mixture is calculated by the mass fraction of each component, like the following equations using model A and B respectively:

$$c_p = m_r c_{p_r} + m_c c_{p_c} + m_a c_{p_a} \tag{6}$$

$$c_p = m_r c_{p_r} + m_{w_c} c_{p_{w_c}} + m_{a_c} c_{p_{a_c}} + m_a c_{p_a} \tag{7}$$

where m is the mass fraction of a specific component after compaction given by the following expression:

$$m_1 = \frac{v_1\rho_1}{v_1\rho_1 + v_2\rho_2 + \ldots + v_n\rho_n} \tag{8}$$

where 1, 2, …, n corresponds to the component number present in the mixture.

In the present study, it was also considered the dependence of temperature on the value of specific heat. So, at each new value of temperature the specific heat of the mixture is calculated according to the following equations:

$$c_p(T) = m_r c_{p_r}(T) + m_c c_{p_c}(T) + m_a c_{p_a}(T) \tag{9}$$

$$c_p(T) = m_r c_{p_r}(T) + m_{w_c} c_{p_{w_c}}(T) + m_{a_c} c_{p_{a_c}}(T) + m_a c_{p_a}(T) \tag{10}$$

3.3 Thermal Conductivity

Model B. Regarding the thermal conductivity coefficient determination, it was considered a combination of thermal resistances pictured in Fig. 5 correspondents to the three mat components. This model is constituted by two arms, one of which are three thermal resistances in series (1) and another one where the same resistances are arranged in

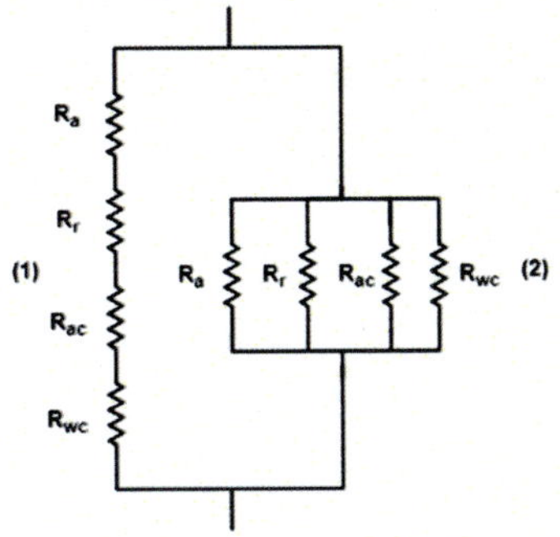

Fig. 5. Thermal resistances model used to determine the thermal conductivity coefficient

parallel (2). Once the heat flux in association in series is equal to the one across every resistance, the equivalent thermal conduction coefficient is calculated by Eq. 11. For the association in parallel, the value of the coefficient is determined by Eq. 12.

$$k_s = \left(\frac{v_a + v_{a_c}}{k_a} + \frac{v_r}{k_r} + \frac{v_{w_c}}{k_{w_c}} \right)^{-1} \tag{11}$$

$$k_p = \left(v_a + v_{a_c} \right) k_a + v_r k_r + v_{w_c} k_w \tag{12}$$

For this model, it was considered an equal part regarding the heat flux across each arm (Eq. 13).

$$k = \frac{k_s}{2} + \frac{k_p}{2} \tag{13}$$

In the present study, it was also considered the dependence of temperature on the value of the thermal conductivity coefficient. So, at each new value of temperature the thermal conductivity of the mixture is calculated according to the following equations:

$$k(T) = \frac{k_s(T)}{2} + \frac{k_p(T)}{2} \tag{14}$$

$$k(T) = \frac{1}{2}\left[\frac{v_a + v_{a_c}}{k_a(T)} + \frac{v_r}{k_r(T)} + \frac{v_{w_c}}{k_{w_c}(T)} \right]^{-1} + \frac{1}{2}\left[(v_a + v_{a_c}) k_a(T) + v_r k_r(T) + v_{w_c} k_{w_c}(T) \right] \tag{15}$$

Model A. Like the previous model, the thermal conductivity is determined considering equal parts regarding the heat flux across each arm (in series and parallel) of the three components: air from the voids, resin and cork. The following expressions indicate the equations used to determine the thermal conductivity of the coefficient considering or not a temperature dependence on this parameter.

$$k = \frac{1}{2}\left(\frac{v_a}{k_a} + \frac{v_r}{k_r} + \frac{v_c}{k_c} \right)^{-1} + \frac{1}{2}(v_a k_a + v_r k_r + v_c k_c) \tag{16}$$

$$k(T) = \frac{1}{2}\left[\frac{v_a}{k_a(T)} + \frac{v_r}{k_r(T)} + \frac{v_c}{k_c(T)} \right]^{-1} + \frac{1}{2}\left[v_a k_a(T) + v_r k_r(T) + v_c k_c(T) \right] \tag{17}$$

Temperature Dependence of Cork. Based on the Eurocode [7] it was assumed the same linear dependence of wood between 20 °C and 200 °C for cork (Fig. 6). Based on the literature value for thermal conductivity at ambient temperature (0.045 W m^{-1} K^{-1} [1]) the temperature dependence of cork was calculated by:

$$k_c = 1,67 \times 10^{-4}T + 4,17 \times 10^{-2} \tag{18}$$

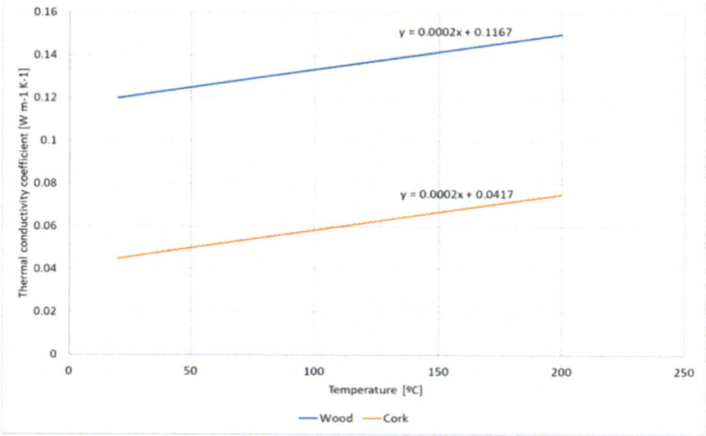

Fig. 6. Model for the temperature dependence of the thermal conductivity coefficient for cork and wood

The thermal conductivity of cork cell walls is determined based on a parallel association of a thermal resistance model of the two phases of cork:

$$k_w(T) = \frac{k_c(T) - k_a(T)(1 - x)}{x} \tag{19}$$

where x is equal to:

$$x = \frac{k_c(20) - k_a(20)}{k_w(20) - k_a(20)} \tag{20}$$

3.4 Comparative Analysis Between Models

A comparative analysis of the two volume fractions models developed was conducted. The inputs and results are presented in Table 1 and Table 2, respectively. This example considers constant values for the physical and thermal properties of the mixture.

It is possible to notice that the values obtained in the by both models are very close to each other, with highest values of thermal conductivity and specific heat for the model that considers the two phases of cork as independent components.

Table 1. Inputs for the comparative analysis between the two developed models

Input	Value	Units
Temperature	25	°C
Resin content	40%	–
Bulk density granular cork	70	$kg\ m^{-3}$
Bulk density thermoplastic powder	350	$kg\ m^{-3}$
Initial thickness	125	mm
Final thickness	35	mm
Cork density	125	$kg\ m^{-3}$
Thermoplastic density	950	$kg\ m^{-3}$
Cork cell walls density	1150	$kg\ m^{-3}$
Air density	1.2	$kg\ m^{-3}$
Cork specific heat	2000	$J\ kg^{-1}\ K^{-1}$
Thermoplastic specific heat	2300	$J\ kg^{-1}\ K^{-1}$
Air specific heat	1004.6	$J\ kg^{-1}\ K^{-1}$
Cork thermal conductivity coefficient	0.045	$W\ m^{-1}\ K^{-1}$
Thermoplastic thermal conductivity coefficient	0.35	$W\ m^{-1}\ K^{-1}$
Air thermal conductivity coefficient	0.026	$W\ m^{-1}\ K^{-1}$

Table 2. Results obtained by the two developed models

Properties	Model A	Model B
Density [$kg\ m^{-3}$]	364.59	364.59
Specific heat [$J\ kg^{-1}$ K-1]	2120.75	2123.68
Thermal conductivity coefficient [$W\ m^{-1}\ K^{-1}$]	0.0724	0.0737

4 Numerical Procedures

To solve the problem related to the temperature evolution of the block during moulding were employed two different numerical methods: finite volume and finite element. A program in MATLAB was written for solving the problem by the finite volume method, except for the case of simultaneous variables thermal conductivity and specific heat. Commercial software ANSYS 19.0 was used for finite element analyses.

4.1 Finite Volume Method

The finite volume method derives from the finite difference method formulation. However, this approach consists of the integration of the partial derivative equation in each volume control defined by central points. Assuming a linear interpolation, for the 1D problem and unsteady state conditions, the discretization equation is described by the Eq. 21 [6].

$$a_P T_P = a_W[fT_W + (1-f)T_W^0] + a_E[fT_E + (1-f)T_E^0] + [a_P^0 - (1-f)a_E - (1-f)a_W]T_P^0 \quad (21)$$

with:

$$a_E = \frac{k_e}{\delta\, x_e} \quad (22)$$

$$a_W = \frac{k_w}{\delta x_w} \quad (23)$$

$$a_P^0 = \frac{\rho c_p \Delta x}{\Delta t} \quad (24)$$

$$a_P = fa_E + fa_W + a_P^0 \quad (25)$$

in which f indicates the method to use: 0 for the explicit method and 1 for the implicit method.

The temperature at the surface of the mould exposed to heat transfer by convection by thermal flux can be calculated by the following expression:

$$a_B T_B = a_I T_I + b \quad (26)$$

where:

$$a_I = \frac{k_i}{\delta x_i} \quad (27)$$

$$a_B = a_I - S_p \Delta x + h \quad (28)$$

$$b = S_c \Delta x + hT_\infty \quad (29)$$

The other condition imposed consists in attribute the same values of temperature for each time step to equidistant points from the symmetry axis:

$$T_{x=a} = T_{x=-a} \quad (30)$$

After the definition of the discretization equation, the resulting system of equations is solved.

Constant Thermal Conductivity Coefficient. Considering a constant value for constant thermal conductivity, the value of this property on an interface is calculated using the following equation:

$$k_e = \frac{k_E k_P (\Delta x_E + \Delta x_P)}{k_E \Delta x_P + k_P \Delta x_E} \quad (31)$$

Temperature Dependence of Thermal Conductivity Coefficient. To deal with the non-linearities like a temperature dependence of thermal conductivity coefficient, it was applied the same iterative procedure for solving the discretization equation proposed by Patankar [6]:

1. Start with an estimate for the values of temperature for all grid points;
2. Calculate the values of the coefficients in the discretization equation for these estimate values of temperature;
3. Solve the set of algebraic equations to get new values for the temperature at all grid points;
4. Repeat the steps 2 and 3 until further iterations cease to produce any significant changes in the values for temperature.

4.2 Finite Element Method

The present problem was also studied recurring to the finite element method using commercial software ANSYS 19.0. After selecting transient thermal analysis, the geometry, material properties, boundary and initial conditions are introduced in the software. The problem was solved in a 2-D setup using 2-D elements.

In the finite element method for transient heat transfer problems, the temperature at nodal points is calculated by solving the following system of equations [8]:

$$[C]\{\dot{T}\} + [K]\{T\} = \{Q^a\} \tag{32}$$

where $[C]$ is the specific heat matrix, $[K]$ is the conductivity matrix, $\{\dot{T}\}$ is the nodal temperatures vector and $\{Q^a\}$ is the applied heat flow vector.

The discretization consisted is a crucial step during finite element analysis to obtain reasonable results. Besides the spatial discretization, for a transient problem is also necessary to discretize time, that is, only is calculated the temperature values for determined time steps.

The linear problem - where there is no temperature dependence of thermal properties - thermal conductivity and specific heat - can be solved by the following system of equations [8]:

$$\left(\frac{1}{\theta \Delta t}[C] + [K]\right)\{T_{n+1}\} = \{Q^a\} + [C]\left(\frac{1}{\theta \Delta t}\{T_n\} + \frac{1-\theta}{\theta}\{\dot{T}_n\}\right), \ \frac{1}{2} \leq \theta \leq 1 \tag{33}$$

where θ is the transient integration parameter and Δt is the time step.

For a nonlinear problem - accounting the temperature dependence of the thermal properties - the system of equation is solved employing the generalized trapezoidal rule and the Newton-Raphson method [8].

5 Conclusions and Future Work

This paper presents a first theoretical approach for the simulation of the temperature behaviour during moulding process of cork composites. A mathematical formulation of the problem was derived, the assumptions of the analysis and two numerical methods (finite volume and finite element) for the solution were presented. In terms of material characteristics, two analytical models were developed in order to predict the properties of a two-component mixture that includes cork granules for the process of the composite. Based on the compaction of the mixture and the data available in the literature for the constituents of the mixture, the two models presented very close values for density, thermal conductivity, and specific heat. With the properties obtained by the developed models and the numerical techniques presented, a simulated temperature profile during the moulding process of cork composites can be determined.

Future work on this topic should include the effect of the water content on the materials properties, chemical effects due to the presence of the agglutinant agent, considering the presence of a thermal contact resistance between mould surfaces and the mixture, among other options for the improvement of the model. The determination of reliable data for the properties of the components of the system should be also a matter of consideration that should be addressed.

Acknowledgments. The authors are grateful to FCT – Fundação para a Ciência e Tecnologia who financially supported this work through scholarship SFRH/BD/136700/2018 and to Amorim Cork Composites. This work has been supported by FCT – Fundação para a Ciência e Tecnologia within the R&D Units Project Scope: UIDP/04077/2020 and UIDB/04077/2020.

References

1. Silva, S.P., Sabino, M.A., Fernandes, E.M., Correlo, V.M., Boesel, L.F., Reis, R.L.: Cork: properties, capabilities and applications. Int. Mater. Rev. **50**, 345–365 (2005). https://doi.org/10.1179/174328005X41168
2. Soares, B., Reis, L., Sousa, L.: Cork composites and their role in sustainable development. Procedia Eng. **10**, 3214–3219 (2011). https://doi.org/10.1016/j.proeng.2011.04.531
3. Gil, L.: Cork composites: a review. Materials **2**, 776–789 (2009). https://doi.org/10.3390/ma2030776
4. Gil, L.: New cork-based materials and applications. Materials **8**, 625–637 (2015). https://doi.org/10.3390/ma8020625
5. Fernandes, E.M., Pires, R.A., Reis, R.L.: Cork biomass biocomposites: lightweight and sustainable materials. In: Lignocellulosic Fibre and Biomass-Based Composite Materials. pp. 365–385. Woodhead Publishing (2017). https://doi.org/10.1016/B978-0-08-100959-8.00017-2
6. Patankar, S.: Numerical Heat Transfer and Fluid Flow. CRC Press, Boca Raton (1980). https://doi.org/10.1201/9781482234213
7. EN 1995-1-2: Eurocode 5: Design of timber structures - Part 1–2: General - Structural fire design (2004)
8. Ansys 16.2.3 Documentation. https://www.sharcnet.ca/Software/Ansys/16.2.3/en-us/help/ans_thry/thy_anproc2.html. Accessed 21 Sept 2018

Pick-Up and Placement Improvement: An Industrial Case Study

Luís Silva[1](✉), José Meireles[1], Mário Pinhão[2], A. Manuela Gonçalves[3], and M. T. Malheiro[3]

[1] MEtRICs Research Center, University of Minho, 4800-058 Guimarães, Portugal
a74217@alunos.uminho.pt
[2] Bosch Car Multimedia, Rua Max Grundig, 4705-820 Braga, Portugal
[3] DMAT- Department of Mathematics, CMAT – Center of Mathematics, University of Minho, 4800-058 Guimarães, Portugal

Abstract. Surface mount technology, usually on the context of Pick-up and Placement, is used on printed circuit boards assembling. In this paper, some aspects and physical parameters related with the pick-up and placement process are analysed in detail throughout the entire work sequence, such as the different variants of these same components among the various suppliers under study. In this process, there are problems of rejection and quality. The aim of this work is to identify and analyse these types of components, as well as their differences and possible causes for their misplacement on the printed circuit boards. Measurements and analyses were performed in lab tests and the study focused more on the capacitors' assembling details. Experimental tests were carried out on the production line in order to obtain conclusive results regarding the study of nozzles and placement of components. Finally, it was concluded that nozzles 907 present a good behaviour in resistors and nozzles 925 present a good behaviour in capacitors.

Keywords: Printed circuit boards assembling · Capacitors · Components · Nozzles · Resistors · Surface mount technology

1 Introduction

Pick-up and Placement tasks require very specific equipment devoted specifically for this purpose [1]. The performance of those equipment's can be increased by the advanced development of respective control software [2], using advanced development techniques [3], as well as using accurate definition of physical solutions for handling very small size components [4–6].

Electrical applications such as mobile phones and audio devices have tended to evolve towards becoming more compact. In this regard, the reduction in the number of printed circuit boards (PCBs) in these devices, the size reduction of the components and the increase in their density in a single board are evidences. Hitherto the decrease in size of passive components such as resistors and ceramic capacitors has been huge – from the conventional 0803 (2.0 × 7.6 mm) or 0603 (1.5 × 7.6 mm) to 0402 (1 × 0.5 mm) [7], as

J. Machado et al. (Eds.): icieng 2021, LNME, pp. 285–301, 2022.
https://doi.org/10.1007/978-3-030-79165-0_28

shown in Fig. 1. However, reducing the size implies a development of its manufacturing processes and of the equipment used in its handling, to avoid compromising delivery to the customer and ensuring product quality.

Fig. 1. Size comparison of different SMT components [8].

Following the aforementioned idea, it was proposed to analyse and optimize the process of inserting 0402 components (resistors and capacitors) in PCB boards in a line of surface mount technology of an industrial company. Before optimizing the reduction of defective PCB boards, it is necessary to identify and interpret possible causes that may potentiate these consequences in the components. For this purpose, it is required to understand the structure of each of these components, their texture, roughness and dimensions. As in the insertion process, it was carried out a comprehensive analysis of the placement method, namely of the nozzles that make the closest contact in the pick-up & placement process.

This study was carried out in partnership with Bosch Car Multimedia Portugal S. A, a company located in the northern city of Braga dedicated to the development and production of communication, entertainment and instrumentation systems and safety sensors for the automotive industry [9].

1.1 SMT Process

In order to understand the context of this study, it is necessary to understand how the SMT process is carried out. Briefly, the diagram shown in Fig. 2 illustrates sequentially the process that is also applied at Bosch.

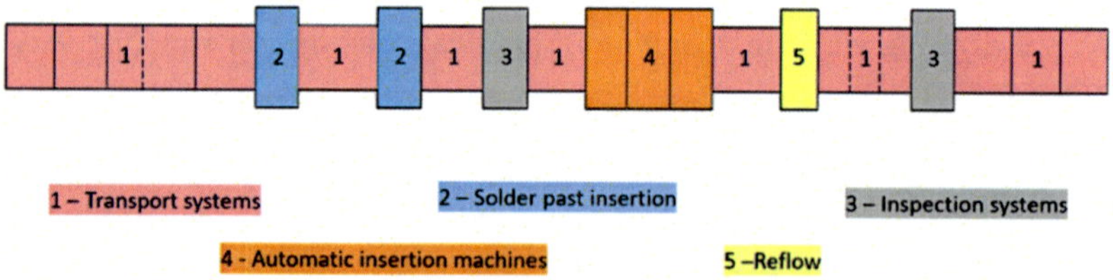

Fig. 2. Figurative example of an SMT line.

PCB boards are conveyed sequentially by the conveyor systems. The solder paste is then inserted into the PCB pads. Subsequently, this paste is inspected by an automatic vision system called SPI (Solder Paste Inspection) that checks whether the paste has been properly placed by analysing its height, width, area and volume. Then, everything being in compliance, the automatic insertion system starts to operate by inserting the components into the PCB exactly where the weld was placed. Then follows the reflow system, with the PCB going through a thermal cycle, and the paste solidifies in order to promote the adhesion of the component to the board. Finally, the AOI (Automatic Optical Inspection) vision system inspects whether all components are properly placed in their due places.

1.2 Placement Machines

Automatic insertion machines are responsible for placing the components on the PCB boards. There are different types of machines and their classification is given by different ways like design and functionality, but all of them are limited to capturing the component and placing it on the board [10]. Nowadays, all of them have a pneumatic system responsible for controlling the pick-up and placement of the components.

1.3 Nozzles

Nozzles are objects whose function is to come into direct contact with the component. They are the last object to be in contact with the component before it is inserted into the board. Its form and composition must be adequate and adjusted to the type of component in question. Generally, nozzles have cavities that allow air suction and consequently the suction of the component [11, 12].

1.4 SMD Components

There are different types of SMD components and all of them are inserted into a specification called package that safeguards their characteristics and dimensions [10]. In the case of smaller components such as the so-called flat chips, capacitors and resistors, which are rectangular components, they are easily identified by four digits (size code). The first two digits mean length (L) and the final two digits correspond to width (W). Thickness is not included in the component code and this information is usually detailed in the datasheet. It is important to note that the codes associated with the size of the component vary according to the system of units in usage, as is the case of metric and imperial systems.

For example, if the first two digits are 04, in the metric system it means 0.4 mm. However, if we proceeded working with the imperial system, it means 0.04 in., which corresponds to 1.016 mm. Thus, there is a list of measures associated with this type of components, as shown in Fig. 3.

This type of material is delivered to the client in the form of a roll of perforated tape, as shown in Fig. 4. This is the most common form nowadays, since all placement machines have mechanisms to perform the pick-up and placement of this tape: feeders, or more precisely, tape feeders.

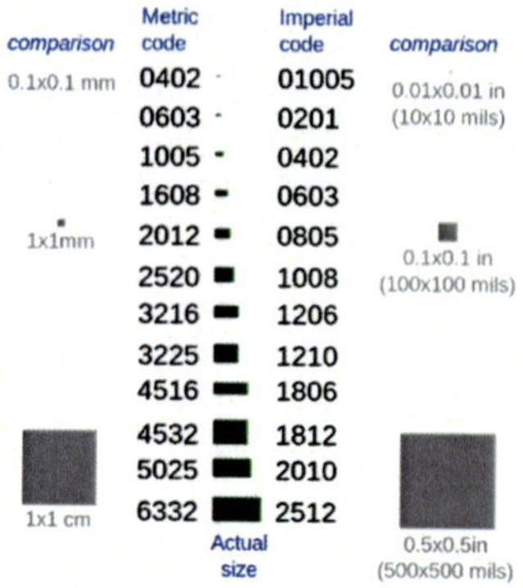

Fig. 3. Metric codes and visual representation of the respective component dimensions in both units of measurement [7].

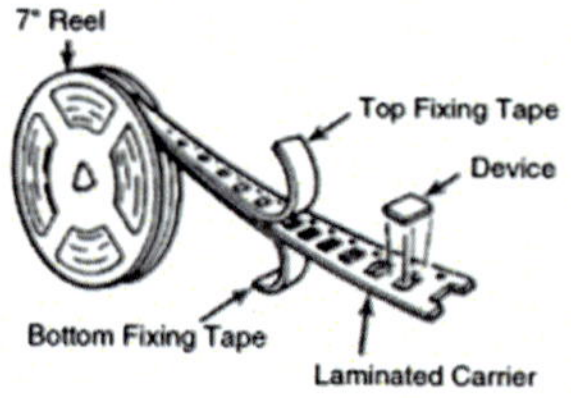

Fig. 4. Figurative example of a tape and roll set [10].

2 Physical Differences Between 0402 Capacitors and Resistors

It is known that resistors and capacitors with the same shape have similar geometric dimensions, only varying slightly in their shape. For the same part number[1] there are different suppliers, and two of them were addressed in this document, as shown in Table 1. Besides the dimensions, it is necessary to understand the influence of shape, texture and materials on the component's contact with the nozzle.

Table 1. 0402 capacitors and resistors suppliers.

Shape	0402	
Type	Capacitors (C_0402)	Resistors (R_0402)
Supplier	B	C
	A	D

Thus, in laboratory some images of the following components were captured under a microscope, firstly of the capacitors on two different sides, as shown in Figs. 5.

[1] Internal number that identifies the component.

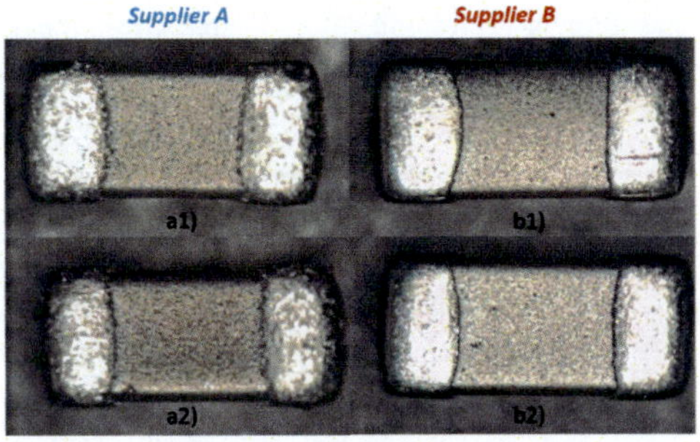

Fig. 5. 0402 Capacitor, part-number 1267370011, Supplier A and B with two different sides

Then the resistors were captured on three different sides, as shown in Figs. 6.

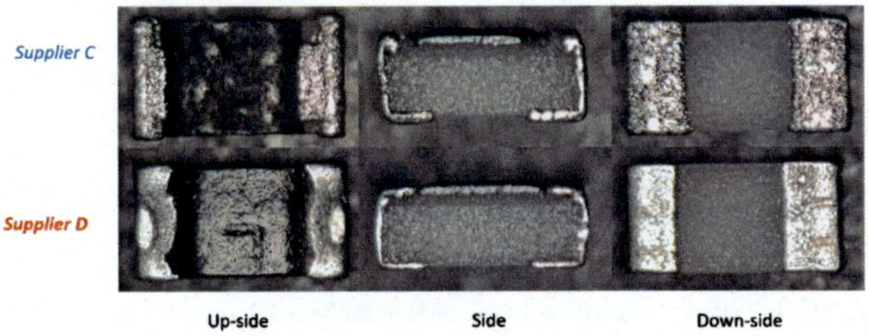

Fig. 6. 0402 Resistor, part number 1267360632, Supplier C and D with down-side, side and down-side views

We proceeded with a geometric analysis by using *ImageJ* software to measure the images captured microscopically at a 50x scale, and then the desired measurements were made by using a ratio of their pixels. Five components from different tapes and from different suppliers were analysed. We verified that these measurements agreed with their datasheet, as can be seen in an example shown in Tables 2 and 3 for components 1267370011 and 126737032, 0402 capacitor and resistor, respectively. The values for data dimensions present in the tables are obtained from the average of 5 different "mean-values" calculated for each component.

Regarding texture, it is noticeable that Supplier A's capacitors' terminals appear to be more rough, more porous and more irregular than Supplier B's, and this can cause a greater instability in the pick-up process, since the contact area between the nozzle and the component has more irregularities, and so the losses of vacuum can lead to rejection or, in certain cases, to a poor component placement on the board, thus causing a quality defect in the product, and if the product is not repairable this means that the entire board will have to be discarded.

Table 2. Capacitors' measurement values and values adapted from the datasheets of Supplier A and Supplier B, units are in mm. The specification refers to the minimum and maximum allowable values [13].

Part number 1267370011				Shape 0402		
Supplier	Supplier C			Supplier D		
Dimensions	Specification		Measurements	Specification		Measurement
L	0.95	1.05	1.03	0.95	1.05	1.00
W	0.45	0.55	0.55	0.45	0.55	0.50
T	0.45	0.55	0.55	0.45	0.55	0.50
BW/B	0.1	0.35	0.23	0.15	0.35	0.29
T1	0.39	0.48	0.44	-	-	0.39
ISO	0.50	-	0.56	0.30	-	0.35

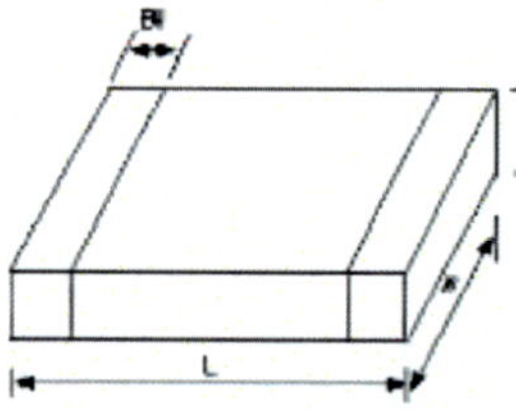

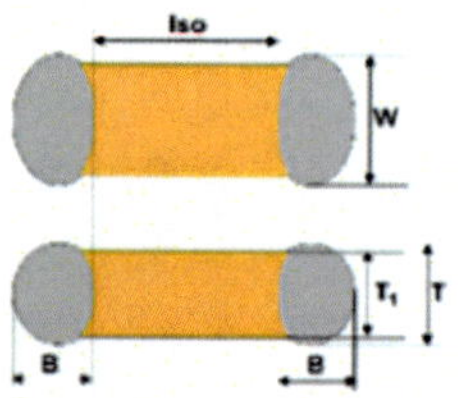

Table 3. Resistors' measurement values and values adapted from the datasheets of Supplier C and Supplier D, units are in mm. The specification refers to the minimum and maximum allowable values [14].

Part number 1267370632				Shape 0402		
Supplier	Supplier C			Supplier D		
Dimensions	Specification		Measurements	Specification		Measurement
L	0.95	1.05	1.00	0.95	1.05	1.02
W	0.45	0.55	0.50	0.45	0.55	0.55
t	0.30	0.40	0.40	0.30	0.40	0.30
a	0.10	0.30	0.12	0.15	0.30	0.15
b	0.20	0.30	0.30	0.20	0.30	0.22

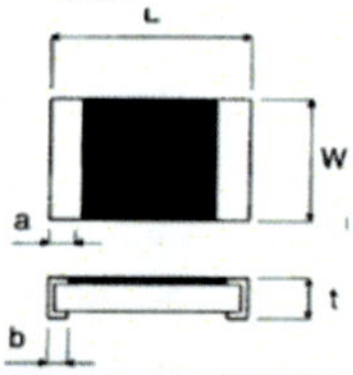

Regarding resistors, there are marked differences between Supplier C and Supplier D. At the terminal level, we confirmed that Supplier C's component base is more irregular than Supplier D's, presenting more grooves. Considering the resistor's lateral view, we can observe a more prominent protrusion in Supplier C's component. This way, it is possible to identify a declivity between the component's center and its ends. This declivity is a preponderant factor when pick-up and placement are performed, since it can cause the component's instability, thus causing eventual vacuum losses and generating a bad component's pick-up and placement. In the terminal area, Supplier C's resistor has well-defined and straight notches, whereas Supplier D's has more rounded ones.

3 Analysis and Comparison Between Different Types of Nozzles

Adapting the type of nozzle to the component in question is a very important step and must be analysed depending on material, dimensions, geometry and nozzle's suction capacity. Regarding 0402 components, we analysed Supplier E supplier's 925 and 907, whose cavities are shown in Figs. 7.

Fig. 7. Microscopic photography of Supplier E nozzle 907 and 925

Regarding dimensions, they all differ quite a lot and it should be noted that nozzle 907 is the largest of them all, with the component having a dimension approximately equal to the length and width from the first to the last cavity: 1 mm. The nozzle 907 have an air cross section bigger than 925 with 0.27 mm^2 and the nozzle 925 with 0,17 mm^2.

It is relevant to understand how the component pick-up and placement is carried out with each nozzle. Figure 8 shows a schematic representation of the possible nozzle positions when carrying out the pick-up. Pick-up is not always carried out in the component's central area and this becomes a problem if the component is not completely regular, as is the case.

Both the capacitors and the resistors have different geometries and therefore nozzle arrangement will differ from one case to another. The capacitor is higher at the terminals than at the center and in the resistors occurs the exact opposite, having more embossing in the central zone, which in a way leads to differences in the pick-ups from different types of components.

In this regard, a 2D analysis of the components and nozzles under study was carried out, as shown in Figs. 9 and 10, in order to understand how the contact was made and whether on a correct pick-up.

As previously seen in the case of capacitors, the terminals are more prominent than the central part and therefore the contact in nozzle 907 would only occur in the two lateral cavities, leaving the central part without any contact and generating losses of vacuum in that area. In the case of resistances, it would be only the central part.

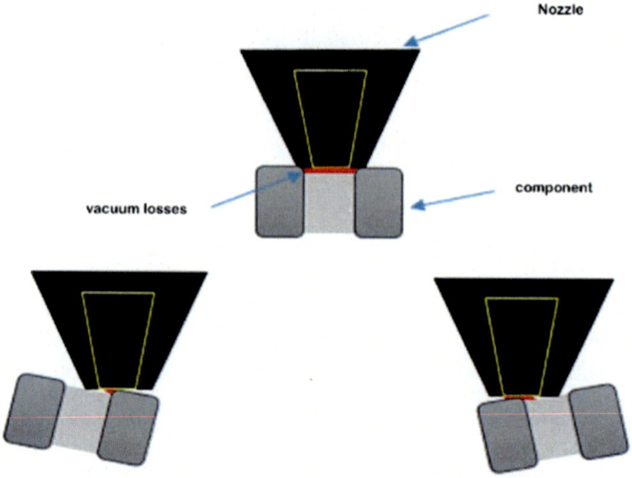

Fig. 8. Figurative example of pick-up in various positions for 0402 capacitors, in this case for a nozzle with a round cavity type 925.

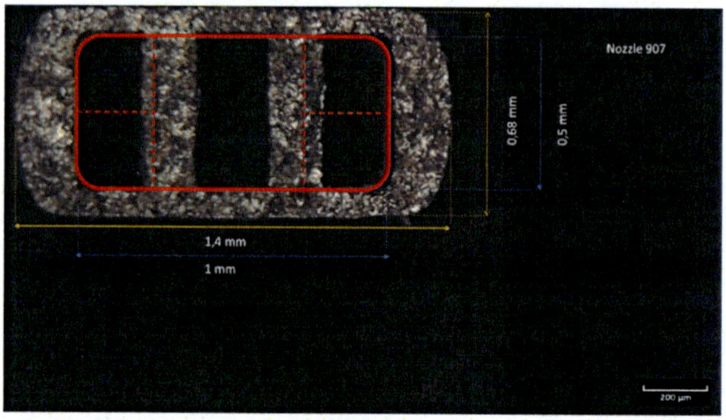

Fig. 9. Illustrative geometric comparison between nozzle 907 and component 0402.

The same would not happen with nozzle 925 because it has a smaller cavity with a circular geometry, and thus the losses would be proportionally greater since the nozzle's edge would come into contact with the component and thus cause losses in the central part. The opposite would happen in the case of resistors, and thus the pick-up would be successful.

However, when this difference is not significant – which is difficult to estimate – it can be said that suction capacity would still be able to raise the component and avoid a large part of the losses, and in this case the roles of the nozzles would be reversed. Other factors can also contribute to a correct pick-up, such as its stability: if the area is higher but irregular, the losses will be more significant, and the component will tend to

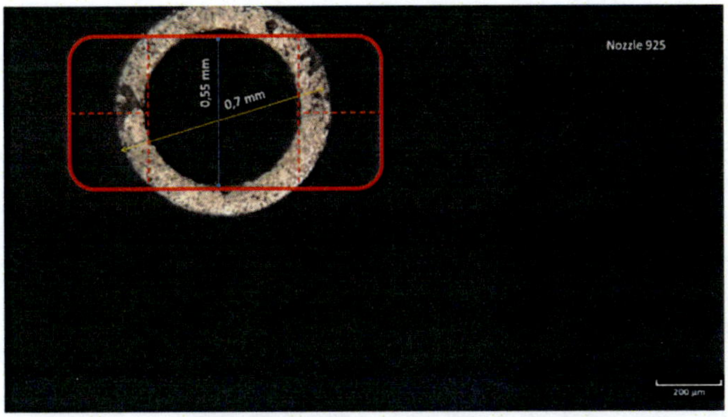

Fig. 10. Illustrative geometric comparison between nozzle 925 and component 0402.

spin or rotate and thus will be rejected. Starting from this premise, only practical tests will allow to conclude as to which is the best nozzle for each component.

4 Analysis and Comparison of 0402 Capacitors Based on Their Placement on the Tape

As previously stated, there was nothing at the dimensional level that would clash with each component's datasheet. However, an important data in the capacitors led us to scrutinize if their shape was uniform on both sides and if the measurements were different, despite meeting the required specification. This is relevant because their placement on the tape is random. So, theoretically both faces can be suitable for pick-up, since the component works in any position if the terminals are well positioned. For this reason, there should be no differences between both faces.

Therefore, at Bosch's MFT3 lab we carried out a dimensional analysis on a sample of capacitors with part number 126737011 and their respective suppliers and we have identified a tendency to arise a difference between both faces. For this reason, two sides were defined: side a, whose central part is visually more rounded, and side b, whose central part is visually more uniform, as shown in Fig. 11.

Going into more detail, some samples of these components are prepared to make some more accurate measurements and see if there are relevant findings. Thus, we tried to quantify the depression between the component's terminals and body and to prove the central body's asymmetry on both sides. This analysis was performed for both Supplier A and Supplier B.

First, the depression between the component's central body and the terminals was measured and analysed, as shown in Figs. 12 and 13.

After the analysis, we found that the depression between the component's central body and the terminals from both suppliers does not have a marked variation. In the case of Supplier, A, the depression on side a is 19.58 µm and in Supplier B is 18.67 µm.

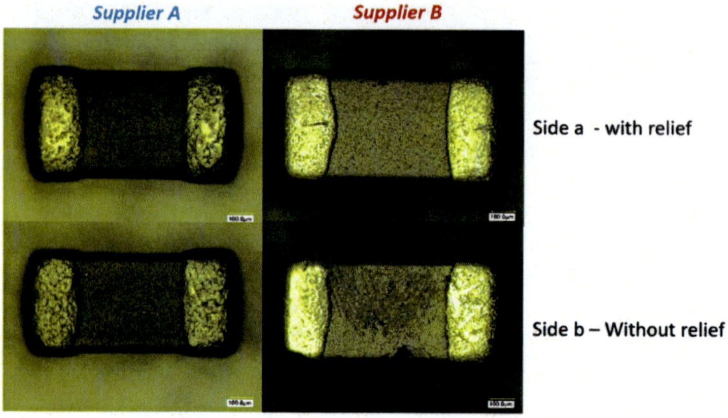

Fig. 11. Comparison of one part-number (capacitor) in two different sides with supplier A and B where was analysed the relief.

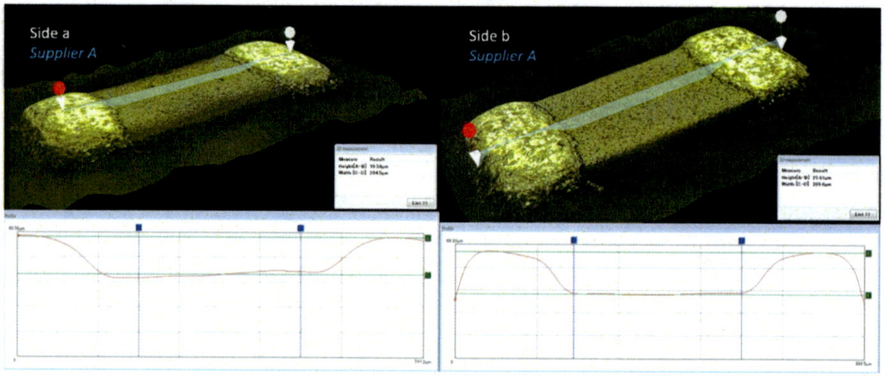

Fig. 12. Difference in different planes between the 0402 capacitor's terminals and central part from Supplier A.

On the opposite side there is a slightly different depression in both cases: Supplier A presents a depression of 25.61 μm and Supplier B a depression of 19.67 μm. Therefore, it appears that side b is the most critical one and in both cases Supplier A has the largest unevenness.

Then we carried out a last analysis on both sides a and b, to see if there would be an irregularity in the central part of the 0402 capacitors. Looking at Fig. 14, there is a noticeable difference between sides a and b is, as well as between Supplier B and Supplier A, thus proving the asymmetry.

Following the previous analysis, we concluded that side b (with a higher unevenness regarding terminals and central body) is the one with the least rounded face. It can be said that side a, being more uniform and less depressive, would be the ideal side to carry out the pick-up since the contact of the nozzle with the component would occur

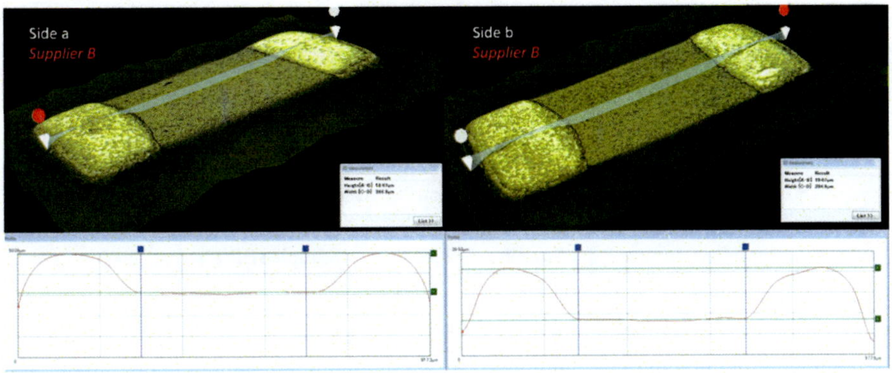

Fig. 13. Difference in different planes between the 0402 capacitor's terminals and central part from Supplier B.

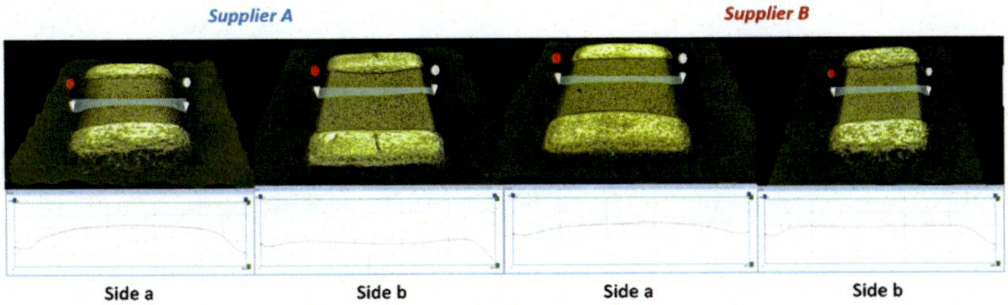

Fig. 14. Differences in different planes between the 0402 capacitors 'terminals and central part with a noticeable difference between Supplier B and Supplier A.

in a more uniform way. Thus, the component's placement on the tape may cause a poor component's placement on the board or even its very rejection.

5 Experimental Test on a Production

Following a problem extant in Supplier E assembly lines – where in several productions 0402 components were inserted with a high rejection rate and with numerous quality defects –, the need arose to analyse and understand the reasons underlying these problems. In this sense, we carried out tests to understand the impact caused by different suppliers and different nozzle configuration regarding rejection and quality of the final product on a production. A set of components was chosen in accordance with top rejection for shape 0402 and top-quality defects: 3 different capacitors and 4 different resistors were studied, among which those previously studied, together with the two Supplier E nozzles 907 and 925. It is worth noting that each of these components (except one capacitor) has two different suppliers. The test was carried out on two different assembly lines, and therefore on two different productions. These two lines were not selected randomly, but due to the high consumption of this type of material. All of these considerations can be found in Tables 4 and 5. These components are placed in the EDT which is a portion of pcb. It has several plates that are milled and divided and thus give rise to several others.

Table 4. Initial test conditions in Assembly Line 21.

Part number	Supplier	Nozzles	Sample (EDT's)
Test 1 - 08:42 - 10:30			
1267370009	A	907	450
1267370011	A		
1267360629	C		
1267360632	C		
Test 2 - 10:30 - 12:15			
1267370009	A	925	450
1267370011	A		
1267360629	C		
1267360632	C		
Test 3 - 12:15 – 13:45			
1267370009	A	925	450
1267370011	B		
1267360629	D		
1267360632	D		
Test 4 - 13:45 – 15:15			
1267370009	A	907	450
1267370011	B		
1267360629	D		
1267360632	D		

In order to ensure the test's reliability, certain initial considerations were defined:

- Ensuring that the nozzles used in the test were clean and in good condition;
- Verifying the nozzles' vacuum level before each test, ensuring that everything was within specifications.
- Quality check on solder paste printing using SPI data;
- Ensuring part-number traceability (seamless rollers were used).
- Ensuring CPk (Capability process index) and CMk (capability machine index) within the defined dates;
- Ensuring correct feeders' placement and calibration.

Table 5. Initial test conditions on Assembly Line 25.

Part number	Supplier	Nozzles	Sample (EDT's)
Test 1 – 13:55- 14:50			
1267370009	A	907	500
1267370635	B		
1267360629	C		
1267360632	C		
Test 2 – 15:19 - 16:20			
1267370009	A	925	500
1267370635	B		
1267360629	C		
1267360632	C		
Test 3 - 16:20 – 17:20			
1267370009	A	925	500
1267370635	B		
1267360629	D		
1267360632	D		
Test 4 - 17:25 – 18:05			
1267370009	A	907	500
1267370635	B		
1267360629	D		
1267360632	D		

5.1 Analysis of the Experimental Tests

Having carried out the aforementioned tests, the results obtained in both production lines were tabulated. In order to allow a more intuitive analysis, for this document we constructed Table 6, Table 7, Table 8 and Table 9 which show the evaluation performance of the components and nozzles in each test.

For this, three classifications were assigned: "Good" in case rejection is below target value 0,160% and no quality defects. "Intermediate" in case rejection is above target; and "Bad" in case quality fails regarding the components. In the case of nozzles, an average approximation of all components was made in each test. This classification prioritizes quality because the products in this production are repair-free; that is, either at client's request or at Bosch's request, no repairs are made to the product and all the EDT is discarded as waste.

Analysis of Tests on Assembly Line 21. After the test, through Table 6 and Table 7 it can be concluded that the most critical component is 1267370011, both in terms of rejection and quality.

Table 6. Classification of the results obtained after the experimental tests on Assembly Line 21 for components. The criteria are as follows: G (Good), I (Intermediate) and B (Bad).

Material	Part number	Supplier	Test 1	Test 2	Test 3	Test 4
Capacitors	1267370009	A	G	G	G	G
Capacitors	1267370011	A	B	B	–	–
Capacitors	1267370011	B	–	–	I	G
Resistors	1267360629	C	G	G	–	–
Resistors	1267360629	D	–	–	I	G
Resistors	1267360632	C	I	I	–	–
Resistors	1267360632	D	–	–	G	G

Table 7. Classification of the results obtained after the experimental tests on Assembly Line 21 for nozzles. The criteria are as follows: G (Good), I (Intermediate) and B (Bad).

Nozzle	Test1	Test2	Test3	Test4
907	I	–	–	G
925	–	I	I	–

By comparing the suppliers of this component (Supplier A and Supplier B), it is possible to verify that Supplier B presents better quality indicators, since no PCB presented defects and there was a lower rejection.

Regarding nozzles (907 and 925), 907 presents greater rejection when used with Supplier A's components and has also lower quality. Nozzle 925 did not show any significant variation between the different suppliers; however, an increase in the resistor's rejection is noted, as seen from the results from test 1 to test 2.

Analysis of Tests on Assembly Line 25. After the experimental tests were performed on Assembly Line 25, as can be seen from Table 8 and Table 9 there was no component rated "Bad", i.e., presenting quality problems.

Only when nozzle 925 was used did rejection increase, due to vacuum errors on component 1267360632 Resistor. We can therefore conclude that in this test this nozzle also displayed a bad relationship with the resistors, with Supplier D having the worst result, without dismissing Supplier C's resistor, which also performed above the target.

Test 4 using nozzle 907 performed without any errors. In terms of quality, all were perfect, with no manufacturing defects whatsoever.

Overall Analysis of the Tests Carried out on Assembly Lines 21 and 25. In general, what is evident after the 4 tests carried out on assembly lines 21 and 25 is that nozzles 925 are a great choice for capacitors since the rejection rate has remained at acceptable values and showed a better-quality performance than the 907 ones.

Table 8. Classification of the results obtained after the experimental tests on Assembly Line 25 components. The criteria are as follows: G (Good), I (Intermediate), and B (Bad).

Material	Part number	Supplier	Test 1	Test 2	Test 3	Test 4
Capacitors	1267370009	A	G	G	I	G
Capacitors	1267370635	B	G	G	G	G
Resistors	1267360629	C	G	G	–	–
Resistors	1267360629	D	–	–	G	G
Resistors	1267360632	C	G	I	–	–
Resistors	1267360632	D	–	–	I	G

Table 9. Classification of the results obtained after the experimental tests on Assembly Line 25 for nozzles. The criteria are as follows: G (Good), I (Intermediate), and B (Bad).

Nozzle	Test1	Test2	Test3	Test4
907	G	–	–	G
925	–	G	I	–

Regarding resistors, nozzles 907 presented better results in terms of rejection, showing an aggravation when we moved to the 925 ones. Regarding quality, both presented no problem.

6 Conclusions and Perspectives

In short, we noticed that at the dimensional level there is product conformity between the different suppliers. However, visually speaking, both capacitors and resistors showed differences in shape and texture, namely in their terminals.

In the study of the components with nozzles 907 and 925 we found that a bad pick-up can cause rejection, depending on how the component is caught on the tape. However, we can conclude that it is crucial that the component is caught by its terminals because it can spin and thus be rejected. Furthermore, future experimental tests should be carried out to verify which of the nozzles is suitable for each type of component. This arises from the duality in selecting the most suitable nozzle for 0402 resistors or capacitors, which can lead to vacuum losses by the correct pick-up and to losses generated by random pick-up (which is what happens in practice).

Regarding capacitors – studied here in more detail – it is evident that there is an asymmetry in several planes. It was proved that the area with the least embossment on the component's internal part and the unevenness regarding the terminals are more marked on side b and that the most critical supplier was Supplier A. Therefore, the component's side layout on the tape is probably relevant to the pick-up.

Finally, we highlighted the experimental tests which in a more conclusive way led us to the conclusion that nozzles 907 present a good behaviour in resistors and nozzles 925 present a good behaviour in capacitors.

Despite some disparities during the tests between suppliers, all of them are within specifications and therefore there is nothing to do except cancelling your purchase, which is sometimes complicated due to contractual reasons, and so the most viable option is to adjust the best nozzle to each case, resistors or capacitors, and mitigate rejection and quality problems as much as possible.

In a near future, the following remains to be ascertained:

- Behaviour in production cadence by changing nozzle 907 for resistors and nozzle 925 for capacitors.
- Quality impact regarding rejection because a single manufacturing defect causes a PCB to be discarded together with all the components inserted in it.
- The preponderance of maintenance in rejection and ascertaining the frequency of nozzle exchange.

Acknowledgements. This work is supported by: European Structural and Investment Funds in the FEDER component, through the Operational Competitiveness and Internationalization Programme (COMPETE 2020) [Project nº 39479; Funding Reference: POCI-01-0247-FEDER-39479].

References

1. Kunz, G., Perondi, E., Machado, J.: Modeling and simulating the controller behavior of an automated people mover using IEC 61850 communication requirements. In: IEEE International Conference on Industrial Informatics (INDIN), art. no. 6034947, pp. 603–608 (2011). https://doi.org/10.1109/INDIN.2011.6034947
2. Campos, J.C., Machado, J.: Pattern-based analysis of automated production systems. IFAC Proc. Volumes (IFAC-PapersOnline) **13**(PART 1), 972–977 (2009). https://doi.org/10.3182/20090603-3-RU-2001.0425
3. Kunz, G., Machado, J., Perondi, E., Vyatkin, V.: A Formal methodology for accomplishing IEC 61850 real-time communication requirements. IEEE Trans. Ind. Electron. **64**(8), art. no. 7878522, 6582–6590 (2017). https://doi.org/10.1109/TIE.2017.2682042
4. Gangala, C., Modi, M., Manupati, V.K., Varela, M.L.R., Machado, J., Trojanowska, J.: Cycle time reduction in deck roller assembly production unit with value stream mapping analysis. In: Rocha, Á., Correia, A.M., Adeli, H., Reis, L.P., Costanzo, S. (eds.) WorldCIST. AISC, vol. 571, pp. 509–518. Springer, Cham (2017). https://doi.org/10.1007/978-3-319-56541-5_52
5. Trojanowska, J., Zywicki, K., Varela, M.L.R., Machado, J.M.: Shortening changeover time - an industrial study. In: 2015 10th Iberian Conference on Information Systems and Technologies, CISTI 2015, art. no. 7170373 (2015). https://doi.org/10.1109/CISTI.2015.7170373
6. Vieira, G.G., Varela, M.L.R., Putnik, G.D., Machado, J.M., Trojanowska, J.: Integrated platform for real-time control and production and productivity monitoring and analysis. Rom. Rev. Precis. Mech. Opt. Mechatron. **2016**(50), 119–127 (2016)

7. Surface-mount technology Homepage. https://en.wikipedia.org/wiki/Surface-mount_technology#/media/File:SMT_sizes,_based_on_original_by_Zureks.svg. Accessed 18 Apr 2019
8. Kitada, T., Seki, Y.: Mounting Technique of 0402 - Sized Surface-Mount (SMD) on FPC, p. 29, Fujikura Technical Review (2011)
9. Bosch em Portugal Braga Homepage. https://www.bosch.pt/a-nossa-empresa/bosch-em-portugal/braga/. Accessed 03 June 2019
10. Yilmaz, I.: Development and Evaluation of Setup Strategies in Printed Circuit Board Assembly, 1st edn. Springer, Heidelberg (2008). https://doi.org/10.1007/978-3-8349-9872-9
11. Ayob, M., Kendall, G.: A survey of surface mount device placement machine optimisation: machine classification. Eur. J. Oper. Res. 22 (2007)
12. SMT_Nomenclature Homepage. https://www.topline.tv/SMT_Nomenclature.pdf. Accessed 15 Apr 2019
13. Samsung-Electro-Mechanics-datasheet-81461623.pdf Homepage. https://datasheet.octopart.com/CL05B104KO5NNNC-Samsung-Electro-Mechanics-datasheet-81461623.pdf. Accessed 05 Apr 2019
14. mcr-e.pdf. Homepage. https://fscdn.rohm.com/en/products/databook/datasheet/passive/resistor/chip_resistor/mcr-e.pdf. Accessed 12 Apr 2019

Acoustic Performance of Some Lined Dissipative Silencers

Marcelin Benchea, Carmen Bujoreanu(✉), and Gelu Ianus

Mechanical Engineering Faculty, Mechatronics and Robotics Department, "Gheorghe Asachi" Technical University of Iaşi, 43 Prof. Dr. Doc. D. Mangeron, 700050 Iaşi, Romania
{marcelin.benchea,cbujorea}@tuiasi.ro

Abstract. Heating ventilating and air-conditioning systems which equip our buildings/rooms represent noise sources unfavorable affecting the people decent living and working conditions. Usually, silencers are used to reduce the annoying sound from these systems. They are lined with different materials and they have various geometries in order to satisfy the consumer needs, in terms of acoustic comfort. Sound absorption and sound transmission loss characterize the materials acoustic properties, but it is not mandatory that a good absorbent material to be also an efficient one from transmission loss (attenuation) point of view. Our paper is focused on an acoustic study of different materials lining three commercial silencers of same geometry and size. We have recorded the sound data according to standards: ISO 10534-1:1998 for the sound absorption coefficient and ISO 7235:2009 for the transmission loss rating. Third octave analysis with LabView soft is used to process the sound information and then the two parameters values are calculated. The acoustic characteristics of the tested materials are discussed and features that recommend them to be used as lining materials for silencers are highlighted.

Keywords: Dissipative silencer · Sound absorption · Sound transmission loss

1 Introduction

The continuum industrial development promotes more and more powerful equipments, but their benefits are also accompanied by issues related to the human comfort. It's about the environmental pollution under all its forms: NOx, thermal, acoustic. Acoustic protection is strong linked to the sound absorption concept which quantifies the energy dissipated within the material and the transmission through it. The sound reaching the material releases an energy which can be absorbed and reflected, in relation to the material sound absorption performance. The materials acoustic behaviors are very different, depending on the configurations in which they are included and also, on the material properties [1–4].

Comfort is required in our buildings/rooms and heating ventilating and air-conditioning (HVAC) systems are important equipments for that, ensuring the proper conditions for the air quality and temperature. On the other hand, they constitute noise

J. Machado et al. (Eds.): icieng 2021, LNME, pp. 302–311, 2022.
https://doi.org/10.1007/978-3-030-79165-0_29

sources affecting not only the people comfort, but also the human health. To reduce this noise, it is usually used a silencer. Common silencers are in-duct installed on the intake and/or discharge side of the air handler. Also, they can be located on the receiver side of other noise generators (terminal boxes, valves, dampers). Literature describes the silencers types used in the sound attenuation as: reactive, active and dissipative [5]. Dissipative silencers are recommended to attenuate broadband noise [6]. Plenum boxes are dissipative ones, most widely used in ducts realizing minimal pressure drop along the silencer. They are available in various shapes and geometries according to the duct design and commonly lined with different materials in order to realize the sound attenuation. These materials are usually porous. The sound absorption performances are predicted through theoretical approaches which lead to the materials structural design from this point of view [7, 8]. The sound absorption/transmission theory is often modeled by Johnson-Champoux-Allard (JCA) approach which calculates the sound absorption coefficient taking into account of some material physical parameters such as flow resistivity, open porosity, tortuosity, and the viscous/thermal characteristic lengths [9]. Another approaches, such as Johnson-Champoux-Allard-Lafarge model and Johnson-Champoux-Allard-Pride-Lafarge modify the JCA model, introducing more parameters difficult to be analytical calculated [10]. Therefore, the theoretical models are not sufficient to predict the materials acoustic performances and the experimental investigations are needed to complete their accuracy. Our paper focuses on the acoustic performances, experimental determined, of some different lined plenum boxes in terms of sound absorption and transmission loss. These two parameters must be corroborated in order to obtain an optimized acoustic behavior, beneficial for people comfort and health.

2 Theoretical Considerations

The absorption coefficient of a material is linked to the sound frequency and also depends on the angle at which the sound wave reaches the material. Figure 1 depicts the sound wave behavior according to the energy conservation:

$$E_i = E_r + E_e \tag{1}$$

The energy E_e represents the sum of the transmitted and absorbed sound energy of the tested material.

The absorption coefficient, α, is classical defined as the ratio of all energy not reflected to incident energy [11]:

$$\alpha = 1 - \frac{E_r}{E_i} \tag{2}$$

Absorption coefficient values are settled using various standardized methods. Measurements can be made in a reverberant room, according to the ISO 354:2003 standard. Another method uses the standing wave tube technique (ISO 10534-1:1998 standard) and the impedance tube method (ISO 10534-2:1998 standard). The last method is often mentioned as the transfer-function method.

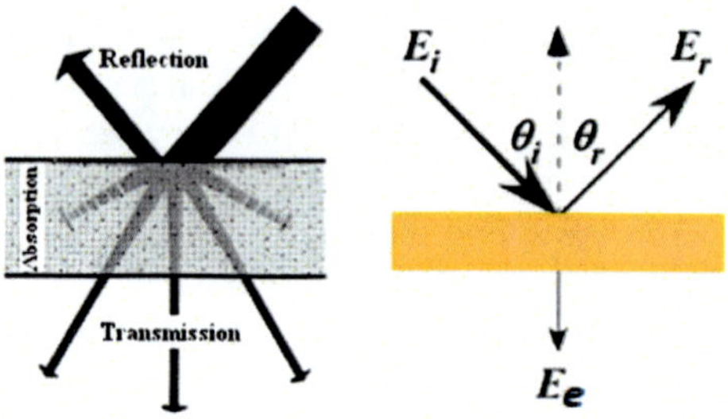

Fig. 1. Sound energy conservation [12].

ISO 354:2003 standard promotes a methodology to evaluate the reverberation time delay in diffuse sound field (reverberation room) for a tested sample placed in this room. The standard refers to the Sabine absorption coefficient.

Sabine equation is used to extract the Sabine absorption coefficient, $\alpha_{sab.}$

$$\alpha_{sab} = 0.9210\frac{V}{cS}(d_2 - d_1) + \alpha_1 \tag{3}$$

V is the room volume; c is the sound speed; S is the sample surface area; α_1 is the absorption coefficient of covered surface; d_1 and d_2 are sound decay rates (dB/s) with and without the sample. Applying this standard leads to some measurement inaccuracies, reported by literature, due to the sample room location influence and also due to its edge effects [7, 13].

ISO 10534:1998 standard presents two methods for the sound absorption coefficient determination in a normal incident sound field [14].

The standard first part describes the method for calculating the peak to minimum amplitude ratio in a standing wave tube. Thus, it is determined the magnitude and phase of the pressure reflection coefficient, and then the sound absorption coefficient α.

$$\alpha = \frac{I_1}{I_2} = \frac{|p_i|^2 - |p_r|^2}{|p_i|^2} = 1 - \left(\frac{n-1}{n+1}\right)^2 \text{ where } n = \frac{p_{max}}{p_{min}} \tag{4}$$

where I_1 and I_2 are the intensities of incident and reflected waves, respectively; p_i and p_r are the sound pressures of incident and reflected waves, respectively; n is the standing wave ratio; p_{max} and p_{min} are the maximum and minimum values of the sound pressure, respectively.

The second part of the standard uses the transfer function H concept, between two spaced microphones, and also spaced from sample. The method considers that the sound reflection coefficient, r, can be evaluated from the measured transfer function H between the two microphones. So, the pressure reflection coefficient r, then the sound absorption coefficient α, are obtained.

$$H = \frac{p_1}{p_2} \text{ and } r = |r|e^{j\varphi r} = r_r + jr_i = \frac{H - e^{-jks}}{e^{jks} + H}e^{-j2k(l+s)}, \alpha = 1 - r^2 \tag{5}$$

p_1 and p_2 are the sound pressure measured by the microphones positioned behind the sample in the tube; r_r is the real component; r_i is the imaginary component; φ_r is the phase angle of the normal incidence reflection coefficient; k is the wave number; l is the distance from the sample to the first microphone; s is the distance between the microphones.

To facilitate the comparison between materials, usually a single number is chosen to evaluate the sound absorption coefficients of each material [15]. Therefore, an indicator has been introduced, Noise Reduction Coefficient (NRC), with the value equal to the arithmetic average of the sound absorption coefficients related to frequency.

Another important parameter to characterize the silencers acoustic performance is represented by the sound transmission loss Tl. This evaluates the attenuation of the sound power level in the duct behind the silencer (as test object). It can be expressed as [16]:

$$Tl = SWL_I - SWL_{II} \text{ [dB]} \tag{6}$$

where: SWL_I - the sound power level in certain frequency band, with in-duct silencer; SWL_{II} – the sound power level in the same frequency band, without in-duct mounted silencer. The transmission loss Tl is usually measured using sound pressure levels at identical points or paths [16].

Sound absorption coefficient together with sound transmission loss values provide more complete indications of the system's performance regarding the noise sources.

3 Experimental Methodology

3.1 Method

Our method for sound absorption coefficient evaluation follows ISO 10534-1:1998 standard. We have used the acoustic interferometer technique, taking into account that in the impedance tube the sound waves are reflected from the sample, and then received by a microphone movable along the tube. So, it can be measured and recorded the incident and reflecting sound pressure. Our experimental work was conducted in an anechoic room, but it uses a modified impedance tube [2] adapted to our work space. This means that the anechoic termination of the modified impedance tube is the anechoic room itself. The whole procedure and explanations are detailed in [12, 15].

As result, the real absorption coefficient determined in the modified tube has lower values than those obtained by standard measurements (with rigid support at the end of the tube) [15].

Brief, in order to realize these measurements, our experimental setup includes a special sound-source equipment (consists of a white noise generator B&K, an amplifier and a broad band loudspeaker), impedance duct, special receiving-sound equipment including a movable microphone B&K and a soundmeter B&K connected with the microphone and NIDAQ board, data acquisition board type-NIDAQPad and laptop with LabVIEW soft.

On the other hand, the transmission loss measurements were realized according to the procedure from ISO 7235:2009, which describes, among others, the methods for determining the transmission loss, in frequency bands, of ducted silencers with and without airflow [16].

Ranging different airflows is supplied by a centrifugal fan (0 V ÷ 10 V with 1 V ratio) over which it overlaps white noise coming from the sound-source equipment. In this case the experimental setup is modified and additionally contains the above mentioned centrifugal fan, device for the flow rate (flow meter), device for the pressure drop (pressure meter), test object (the dissipative silencers); transition elements on either side of the test object. The procedure is detailed in [17].

Third-octave band analyses of recorded sound data collected, according to the above procedures (ISO 10534-1:1998 and ISO 7235:2009), were performed. The next step is to use relations (4) and (6) to calculate the sound absorption coefficient and the transmission loss (attenuation) values, respectively.

3.2 Test Samples

We have tested, from absorption point of view, three materials types lining some commercial plenum boxes.

They are denoted with “green” (glass based), “blue” (mineral based), “black” (nylon based). The lining materials thickness and porosity are the same, as specified by the manufacturer. Then we have tested, from transmission loss point of view, the three plenum boxes. They have the same width, height and length, with Φ100 mm inlet/outlet diameter for in-duct placement.

Figures 2, 3 and 4 present both the experimental stands, adapted for the anechoic room and according to standards.

Fig. 2. Sound absorption coefficient experimental setup (according to ISO 10534-1:1998).

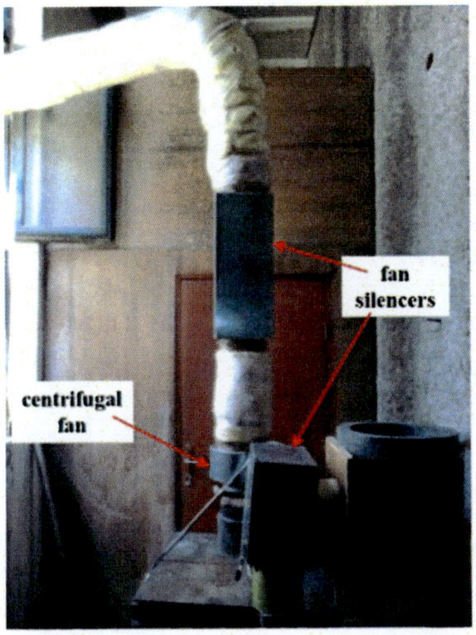

Fig. 3. Transmission loss experimental setup (outside the anechoic room).

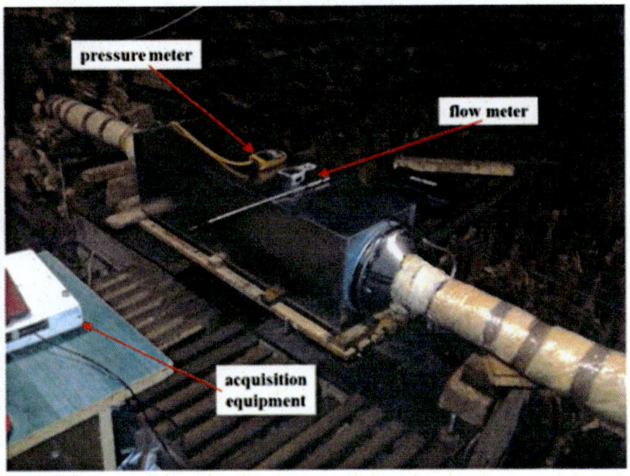

Fig. 4. Transmission loss experimental setup (inside the anechoic room).

4 Results and Discussion

As we have above explained, first, we present the sound absorption coefficients values for the three materials lining three commercial plenum boxes (Fig. 5).

An obvious increasing tendency of the absorption coefficient can be observed at frequencies higher than 1450 Hz (Fig. 5), values over 0.6 and greater for the green lining. For frequencies lower than 1450 Hz the absorption coefficient presents values under 0.6 and higher for the blue lining. Using the average absorption coefficient (NRC) concept, as absorption performance indicator, the following values were obtained: 0.50 for the green lining, 0.56 for the blue lining and 0.44 for the black lining. Obvious, the blue lining can be considered the better sound absorbent material.

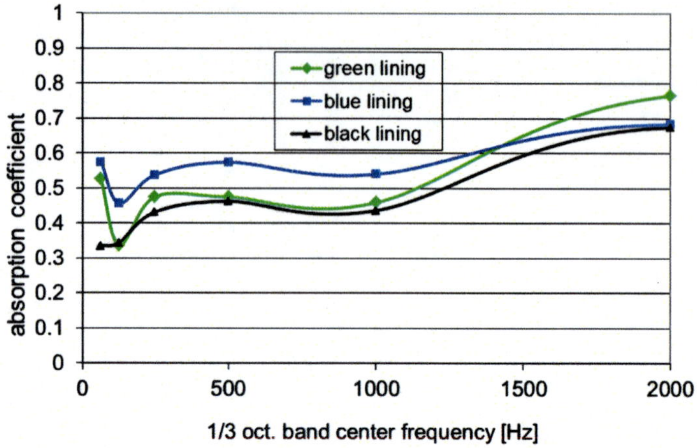

Fig. 5. Sound absorption coefficients for materials lining commercial plenum boxes.

Figures 6, 7 and 8 present the sound attenuation (transmission loss) values, recorded on the plenum boxes and then calculated, as we have above described in paragraph 2. The figures are selected for variable airflow provided by the centrifugal fan (a supply of 3 V, 7 V, 10 V). They show that the attenuation of the silencers is also influenced by the airflow in the HVAC systems, supplied by the centrifugal fan. It is observed an increasing tendency for frequencies higher than 500 Hz, with values lower than 40 dB for small airflows and maximum 15 dB for higher airflows. For small airflows, over 1450 Hz in frequencies domain of interest, the green lining has a greater attenuation compared to the others materials.

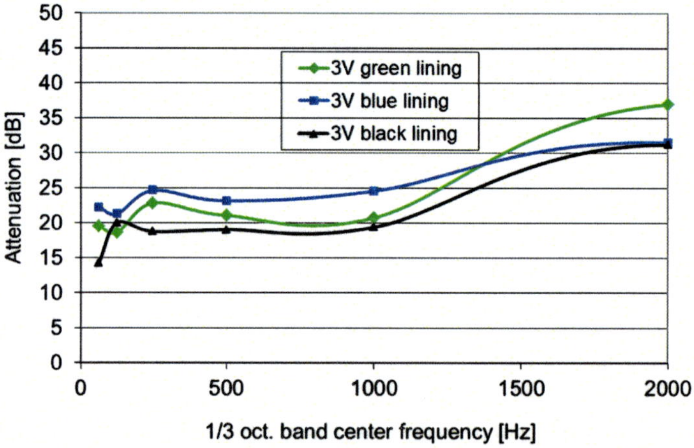

Fig. 6. Attenuation versus frequency for low airflow.

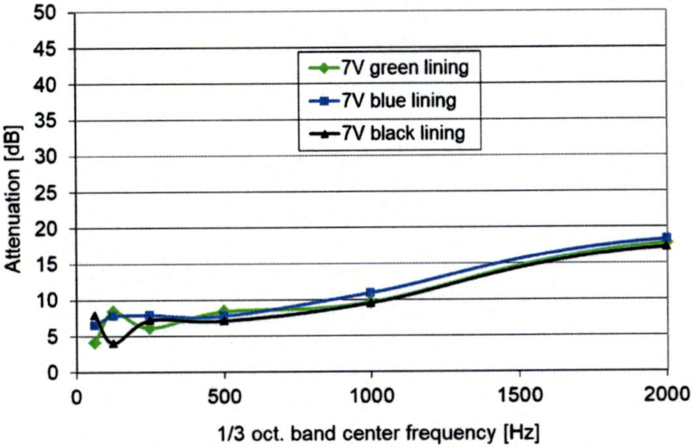

Fig. 7. Attenuation versus frequency for medium airflow.

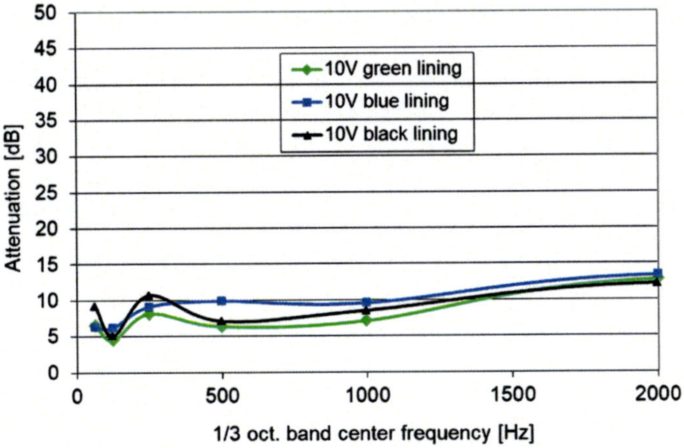

Fig. 8. Attenuation versus frequency for high airflow.

However, in the explored third octave band, at different airflows from low to high values provided by the centrifugal fan, the blue lining presents the best attenuation despite the fact that the silencer lined with this material creates a greater pressure drop in the duct's airflow path, as Fig. 9 depicts.

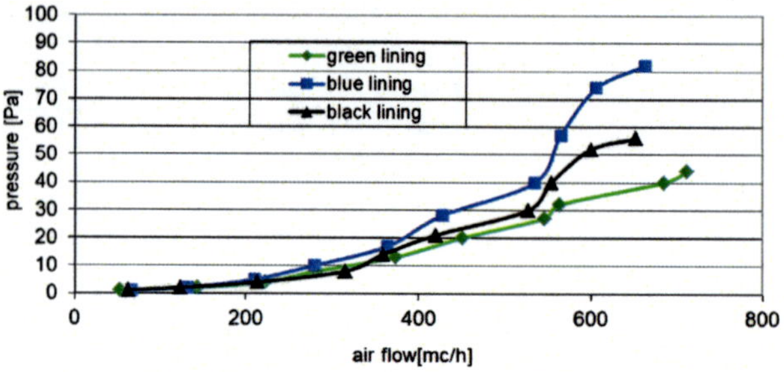

Fig. 9. Pressure versus airflow.

5 Conclusions

Materials acoustic properties are linked to their sound absorption and transmission loss characteristics. The blue lining material presents better acoustic properties under 1450 Hz frequencies according to the absorption coefficient evolution and average value, also, according to attenuation evolution. The attenuation has higher values as the air flow through the silencer is lower, because the dynamic flow of air negatively influences the noise.

The main conclusion is that the acoustical architecture takes into account of both concepts. Sound absorption, usually represented by NRC (ranging from 0 to 1), is expressed by the sound dissipation inside the material. Sound transmission loss (attenuation) is related to the sound insulation, therefore it is expressed by the reflecting sound energy. It is obvious that there is difficult to select a material and also, it's mounting system, to realize absorption and insulation at the same time.

References

1. Long, M.: Architectural Acoustics, 2nd edn. Academic Press, Cambridge (2014)
2. Leping, F.: Modified impedance tube measurements and energy dissipation inside absorptive materials. Appl. Acoust. **74**, 1480–1485 (2013)
3. Istvan, L.V., Beranek, L.L.: Noise and Vibration Control Engineering: Principles and Applications, 2nd edn. Wiley, Hoboken (2005)
4. Shen, C., Xin, F.X., Lu, T.J.: Theoretical model for sound transmission through finite sandwich structures with corrugated core. Int. J. Non-Linear Mech. **47**(10), 1066–1072 (2012)
5. Handbook Sound and Vibration Control. American Society of Heating, Refrigerating, and Air Conditioning Engineers 48, USA (2011)
6. Borelli, D., Schenone, C., Pittaluga, I.: Theoretical and numerical modelling of a parallel-baffle rectangular duct. In: Proceedings of Meetings on Acoustics, San Diego, vol. 14, p. 040004 (2012)
7. Wijnant, Y.H., Kuipers, E.R., de Boer, Ir.A.: Development and application of a new method for the in-situ measurement of sound absorption. In: Proceedings of ISMA, Leuven, Heverlee, pp. 109–122 (2010)
8. Cao, L., Fu, Q., Si, Y., Ding, B., Yu, J.: Porous materials for sound absorption. Compos. Commun. **10**, 25–35 (2018)

9. Chevillotte, F., Perrot, C.: Effect of the three-dimensional microstructure on the sound absorption of foams: a parametric study. J. Acoust. Soc. Am. **142**, 11–30 (2017)
10. Yang, X.H., Ren, S.W., Wang, W.B., Liu, X., Xin, F.X., Lu, T.J.: A simplistic unit cell model for sound absorption of cellular foams with fully/semi-open cells. Compos. Sci. Technol. **118**, 276–283 (2015)
11. Beranek, L.L., Ver, I.L.: Noise and Vibration Control Engineering: Principles and Applications. Wiley, Hoboken (1992)
12. Bujoreanu, C., Nedeff, F., Benchea, M., Agop, M.: Experimental and theoretical considerations on sound absorption performance of waste materials including the effect of backing plates. Appl. Acoust. **119**, 88–93 (2017)
13. McGrory, M., Castro, C.D., Gaussen, O., Cabrera, D.: Sound absorption coefficient measurement: Re-examining the relationship between impedance tube and reverberant room methods. In: Proceedings of Acoustics, Fremantle, Australia, pp. 1–8 (2012)
14. ISO 10534:1998, Acoustics - Determination of sound absorption coefficient and impedance in impedance tubes: Part 1: Method using standing wave ratio, Part 2: Transfer-function method
15. Oancea, I., Bujoreanu, C., Budescu, M., Benchea, M., Grădinaru, C.: Considerations on sound absorption coefficient of sustainable concrete with different waste replacements. J. Clean. Prod. **203**, 301–312 (2018)
16. ISO 7235: 2009, Acoustics- Measurement procedures for ducted silencers- Insertion loss, flow noise and total pressure
17. Bujoreanu, C., Benchea, M.: Experimental investigation of noise characteristics for HVAC silencers. In: MATEC Web of Conferences, Iasi, vol. 112, p. 07001 (2017)

Application of Advanced Co-Simulation Technology for the Analysis of Grasping

Daniele Catelani[1], Leonardo Di Paola[2], Mauro Linari[1], Erika Ottaviano[2](✉), and Pierluigi Rea[3]

[1] MSC Software Corporation, Via Santa Teresa 12, 10121 Turin, Italy
daniele.catelani@mscsoftware.com
[2] DICeM - Department of Civil and Mechanical Engineering, University of Cassino and Southern Lazio, via Di Biasio 43, 03043 Cassino, FR, Italy
ottaviano@unicas.it
[3] Department of Mechanical, Chemical and Materials Engineering (DIMCM), University of Cagliari, Via Marengo, 2, 09123 Cagliari, CA, Italy
pierluigi.rea@dimcm.unica.it

Abstract. The modern approach to the mechanical design involves advanced simulation tools that may be used in combination to the other approaches to get realistic results. In order to analyze this aspect, this paper aims to investigate the interaction between solid and soft objects involved in dynamic simulation. In the paper, we focus the attention on the mechanics of grasping considering a standard two-finger gripper interacting with a soft ball. The proposed research has a practical outcome related to the manipulation of delicate or soft objects, like horticulture products in winter house. More broad applications cover aspects of prosthetic hands, in which mechanical elements (poly-articular fingers) and soft elements (fingertips) have to be combined and co-simulated in grasping rigid and soft objects. In order to perform the task, a co-simulation approach is used. In this paper, we present preliminary results of the co-simulation between Adams and Marc software applied to a case of study.

Keywords: Simulation driven design · Multibody · Co-simulation · Education

1 Introduction

The mechatronic design of grippers and robotic hands is a very complex task, which involves several aspects of mechanics, actuation, and control. Factors such as geometry, environment, and grasping capability, strongly influence the adopted design [1], therefore, simulation tools can be very useful for the mechanics of grasping and manipulation applications. Nowadays simulation tools provide engineers with a unique, complete, and holistic performance insight by coupling multiple simulation disciplines. Everything from acoustics to MBD, to CFD, to structural analysis, and explicit crash dynamics, can be connected together using CAE solutions, as from MSC Software tools, codes and methodology [2]. For an efficient use of these tools, it is required an integration among

J. Machado et al. (Eds.): icieng 2021, LNME, pp. 312–324, 2022.
https://doi.org/10.1007/978-3-030-79165-0_30

disciplines in the engineering science, but also a great understanding of method analysis and competence with software. To achieve this goal, it is becoming quite common a close collaboration between Academia and Industry [3, 4], which can be seen as one of the motivations for this work. Among the wide range of possible engineering applications, we focus our attention on two-finger grippers, because they are widely used in automated systems and in industrial robots with a suitable mechatronic design, low-cost and easy-running components. Moreover, they are well suited to test a multidisciplinary approach to the simulation driven design, as in this paper.

Several researchers investigated different types of devices for achieving the grasping and handling of objects. Mechanical grippers with two fingers are widely used in industrial application and robotics [5, 6]. In addition, multi-fingered robotic devices and hands have been also widely investigated, as proposed for example in [7, 8].

More recently, an increasing interest has been focused on the design and control of underactuated mechanical systems, which can be defined as systems whose number of control inputs (i.e. active joints) is smaller than their DOFs. The underactuation property may have potential benefits on the dynamics of the system; cost reduction or practical purposes; actuator failure. However, the benefits of underactuation can be extended beyond a simple reduction of mechanical complexity, in particular for devices in which the distribution of wrenches is of fundamental importance, as for example grasping and manipulating objects, for which underactuation may allow adaptability to the object in grasp, as reported in [9, 10].

Another interesting application of gripper refers to the harvesting of horticulture products, which is still of great interest, although since the 70's and 90's addressed attention for automation in agriculture [11–14]. In particular, some works have been done on robotized solution [15] and particularly for the harvesting of citrus [16]. Nevertheless, specific problems for horticulture products are related to the unstructured environment, but mostly the success of the operation greatly depends on the efficiency of the gripper, which should be conveniently a dedicated device for a specific product. Indeed, a crucial aspect refers to the object being grasped, which greatly varies in not only geometry and dimension, but also because of its stiffness.

Very important relapse of this research refers to robotic prosthesis [17–19]. Nowadays, the challenge in design and construction of prosthesis hand aim to fill the gap between poly-articular multifunctional complex prostheses characterized by high performances, complex mechanisms and relatively high cost, and two or tri-digital prostheses, characterized by low-cost and ease in use, but also by limited flexibility and poor aesthetics. One possibility is to the design a robust low-cost mechanism having grasping ability and the appearance comparable to more sophisticated and realistic prosthesis with soft elements.

Therefore, in all the above-mentioned areas or research, it is of great interest to be able simulating systems taking into account the MBD together with soft /deformable elements. The so-called *Co-Simulation* strategies will be described in the next Sections.

2 *Co-simulation*: Raising the bar for CAE Accuracy, Precision, and Performance

Starting from the end of '90s, one of the trends of Computer-Aided Engineering (CAE) has been 'multiphysics' simulations, i.e. co-simulations among different physics simulation types. Multiphysics could allow, if used and applied with analyst perspective and scientific accuracy, to achieve a real-world engineering simulation. Multiphysics can be not only considered a unified discipline able to manage and solve all mathematics beyond physics, but it has evolved in finding the best way-to-work, the best methodology, the best solution to let different disciplines and codes to communicate and work together. For instance, Finite Volume Methods FVM (for fluids), Finite Element Methods FEM (for structures and acoustics) mean that for more efficient solver convergence in real world engineering problems, one or the other is chosen as the best methodology. In addition, also important, to deliver usable engineering simulation tools facilitating the data exchange between numerical models, and the definition of models, able to be easily modified and adapted to the specific discipline when partial verification and validation analysis have to be performed, or extended and completed when co-simulation has been requested to get more realistic results. Consequently, the final goal is to provide best-in-the-world class CAE co-simulation solutions to customers, users, engineers everywhere [2]. Everything from acoustics to multi body dynamics MBD, to CFD, to structural analysis, and explicit crash dynamics can be or should be connected together in a simulation portfolio. There are many examples of Multiphysics co-simulation applications that can be performed, both in two-product couplings, as well as in toolchains of product simulations that were dreams few years ago [2], see Table 1.

2.1 Multi Body Dynamics Led Co-simulation

Among the CAE disciplines, the MBD is the most suited for the *co-simulation* approach. The motivation is related to the need of simulating realistically the behavior of a complex mechanical system moving in a 3D space. Indeed, in order to perform a realistic MBD analysis, the simulation can include the flexibility and structural characteristic of the bodies (from FEM); aerodynamic, fluid dynamics force description (from CFD), controls algorithms and routines to emulate hydraulic, piezoelectric, pneumatic system as well as drivers and pilots (from Control Tools), evaluation of loads, stresses, fatigue, source of noise (FEM, Fatigue, Acoustic tools). Without neglecting particles model and routine (DEM) or advanced simulation scenario for ADAS and autonomous vehicle drive simulation (Virtual Test Drive).

Coupling MBD with FEM, CFD and other tools, allows to simulate in details highly deformable components like blades, wings, rubber door seals, tires, bushing, hydraulic actuators, avionics, battery, suspensions, landing gears, as part of complex systems like vehicles, aircrafts, robots, and so on; and to simulate complex phenomena like fuel tank sloshing, aeroelasticity coupling, thermomechanical stress [20–22].

Chaining MBD with Fatigue, Acoustic, and other tools allows recovering stresses and strains and to predict fatigue life of components working not in an isolated manner but loaded by realistic forces, and to provide fundamental information on noise radiation, thermal radiation and vibration to improve comfort.

Table 1. Possible areas of industrial applications of co-simulations.

Co-simulation multiphysics area	Industries	Applications
Fluid-Structure Interaction (FSI)	All	Aeroflutter, Valve Opening, MEMs, VIV, Thermo-Mechanical Stress, ...
Structural and Aeroacoustics	All	Cabin noise, Door Rattle, Noise and Vibration, ...
Multi-Body Dynamics and Fluids	All	Vehicle Side Wind, Vehicle Running over a Puddle, ...
Virtual Drive and Vehicle Dynamics	Automotive	ADAS Validation, Real Time Vehicle Driving Simulator, ...
Particulates and MBD and CFD	Auto, Aero, Chem, Proc	Car Stability on Surface, Filtration, Bulk Material Handling, ...
MBD and Nonlinear FEA	Automotive	Door Sag and Closing, Vehicle Extreme Load Cases, Running over an obstacle, Battery Pack Deformation
1d Systems and MBD and Controls	All	Robot Arms, Machinery, Landing Gear System, Vehicle ABS, ESC, Traction Control

The main purpose of CAE tools is to simulate the realistic behavior of a system, estimating performances and reducing time-to-market and costs, then it is crucial being able to use mature and robust *co-simulation* methodology to improve the design process. At the current state of the research, the *co-simulation* is still an on development and on-going methodology that has to be tested and verified. The collaboration between software vendors and academia plays a fundamental role not only for improving robustness of the code, increasing efficiency and speed of solvers, testing and validating methodology and routine, to arising the level of usability, but – principally – to explore and investigate new fields of applications, to certify the proposed solution and to help its development. Regarding the *co-simulation*, there are two main approaches for the development in MSC: using a standard interface, like FMI (Functional Mockup Interface), which allows physical quantities, settings, scripts to be passed between MSC tools and third parties. The other is developing a dedicated *cosim* tool, which allows easy-to-use, fast and fully embedded and integrated co-simulation analysis among MSC codes (Adams, Marc, Nastran, Cradle). In this paper, we test the second approach.

3 *Co-simulation* Architecture

Considering the variety and complexity of the systems to analyze, an enhanced co-simulation algorithm *cosim* has been developed for large multi-body system (MBS) interacting with large finite element models (FEM). In addition, to the exchange of forces and displacements data at each interaction point of the models, it computes the tangent stiffness matrix of the finite element model and send it to the controlling master code (named "*glue*" or "*cosim*" code) and share with the multi-body model. The tangent stiffness matrix represents the condensation of the global stiffness matrix (involving all displacements) into a stiffness matrix involving only the displacements at the interaction points. Forces and tangent stiffness matrix are used by the MBS to better predict the motion of the coupled systems.

This algorithm enforces the MBS model to step first (using force and stiffness of the FEM models) and later imposes the computed displacements onto the FEM models without using a fixed communication interval. Each code (MBS and FEM) is free to take a simulation step that fits well within the status of the numerical solution; it means that the communication with the *glue* code is continuous. Allowing the co-simulating models to take considerably larger simulation steps has the additional benefit of providing stable solutions and better accuracy, both for static and dynamic analyses.

3.1 Coupled Solutions and MBS-FEM *Co-simulation*

To describe coupled systems, generically multiple domains are considered, where independent and dependent variables describe different physical properties and participate in different governing equations. A MB-FEM *co-simulation* can be classified as interface variables coupling, where the coupling occurs through the governing differential equations describing different physical phenomenon, particularly dynamical and structural ones. Three processes drive the solution: 1 MB process, multiple FEM processes and 1 *cosim* (*glue*) process. In our activity, we use MSC Adams for the MB process and MSC Marc for the FEM processes because for a general-purpose application, commercially available code has the big advantage of strong and robust functionalities and validation [2].

Figure 1 illustrates the *co-simulation* process, divided into three software. The *glue* code acts as an executive routine that directs the flow of data between the MB and FEM codes. The *co-simulation* is started launching the *glue* code, which prompts the user to start the MB and FEM applications with the *glue* code.

Both the MB and FEM codes have been customized to allow the right communication each other through the *glue* code. As already said, there is no scheduled communication among the codes but they communicate with the *glue* code at the end of each integration step taken and considering the order: the MB code runs first with current values of forces and approximate compliance of the FEM models. At the end of each integration step, the MB code passes kinematic data to the *glue* code, which passes the data to the FEM code.

The implemented algorithm and the developed architecture allow every code to use the best time step needed to solve accurately their respective equations.

The MB code, which usually requests larger time step is blocked during the FEM code evaluation, waiting for the next interpolated data about loads provided by FEM code, and in turn for interpolated data of motion delivered to FEM model. In this manner, the *glue* code allows the two applications to exchange data and maintain synchronization. This co-simulation approach is used for other coupling problems, such as MB-FEM coupling,

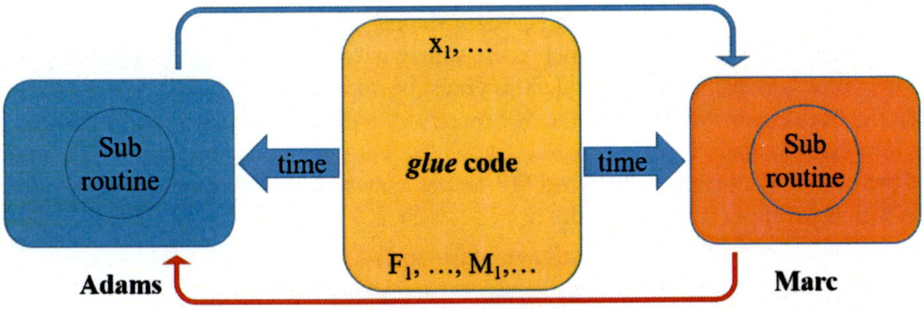

Fig. 1. A scheme to illustrate the *co-simulation* process.

MB-CFD coupling, FEM-CFD coupling and so on [23]. It could happen some instability when time step size is too big, making too small the matrix operator, again, mathematics have developed accordingly to address this problem. MB-FEM co-simulation could be time-consuming solution due to reduced number of DOFs but high number of time steps for MB model, combined with huge DOFs FEM systems. Better speed performances are achieved using techniques of analysis parallelization combined with adaptive time step algorithm, which minimize the number of stiffness matrix updates in the FEM process (on MB side computational time is smaller w.r.t. the FEM domain). It is evident, however, that it does not exist a general-purpose adaptive time stepping procedure, applicable to every MB-FEM co-simulation system.

4 Case of Study: A Two-Finger Gripper Grasping a Soft Ball

4.1 Description of the Model

The *cosim* features and performances have been tested on a two-finger gripper. The model consists of a robotic gripper that grasps a rubber ball (sphere). The claws of the gripper perform a translational movement and are driven by pneumatic cylinder (Fig. 2a). The operation of the gripper produces displacement and deformation of the rubber sphere, which is a non-linear element, modelled in FEM environment. In particular, a sphere (Fig. 2b) is used for modelling the rubber ball. Performing the co-simulation with MSC *cosim*, it is possible to obtain the grasping force variation in MB, while displacements and deformations of the sphere are obtained by in FEM environment. The MB model is composed by eight rigid bodies: actuator, base, 4 bars (up and down, right and left), two claws (right and left). They are joined with kinematic constraints. Preliminary, for the first tests, a fixed joint is used to attach the base to the ground, while a translational joint allows the movement of the actuator for claws opening and closing. Considering some compliances in the model, bushing (linear elastic connections) have been created, setting proper stiffness and damping values. The actuator is moved by a MOTION, applied to the translational joint, with a smooth STEP function. Finally, general force vectors applied to the claws are defined in the model and through their markers the communication and exchange of information between codes is defined. The location of the markers should be chosen where the contact with the object is expected (Fig. 3).

The FE model, as for the MB one, is set for the *co-simulation*. In particular, boundary conditions are set. The Marc model of the sphere is created defining the geometry and mesh properties. Modelling consideration must be applied for positioning the sphere in the right location with respect to the MB model. In addition, the selection of parameters is important to consider a realistic rubber sphere. The contact bodies are tuned in order to define 3-contact surfaces between MB model, claws and horizontal plane of the table on which the sphere is positioned.

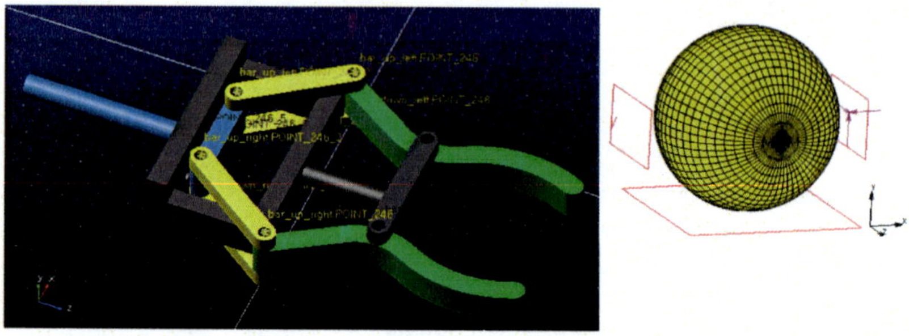

a) b)

Fig. 2. Description of the MB and FE models: a) a two-finger gripper; b) a soft ball.

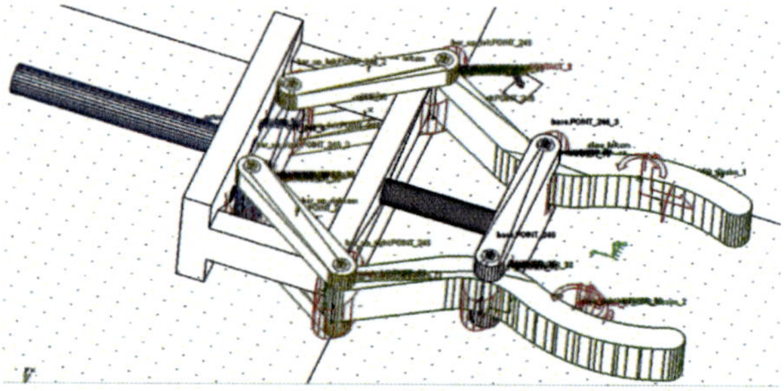

Fig. 3. Location of the markers on the model.

As already stated, the points at which a model interacts with another one (for example, points P1 and P2 in Fig. 4) are called *interactions*. At each interaction point, there must be a GFORCE in the Adams model and a NODE in the Marc model. The interaction NODEs in the Marc model must have 6 degrees of freedom and they must be position controlled by the option Co-Sim Int. Node.

In all Adams-Marc interactions, Adams passes displacements to Marc to be imposed on a NODE while Marc passes force/torque values to Adams to be used in a GFORCE at the interaction point, as shown in the scheme of Fig. 1.

Referring to the case of study, a configuration file is used by the *glue* code to:

- Assign coherently name and position of marker (MB) and nodes (FEM) where there is exchange of information (displacement from MB, forces from FE);
- Kinematics is imposed on Marc while Marc forces are applied to Adams;
- Assign name of the files of description of MB and FEM model and simulation/job;
- Definition of co-simulation parameters, like interpolation order of algorithms.

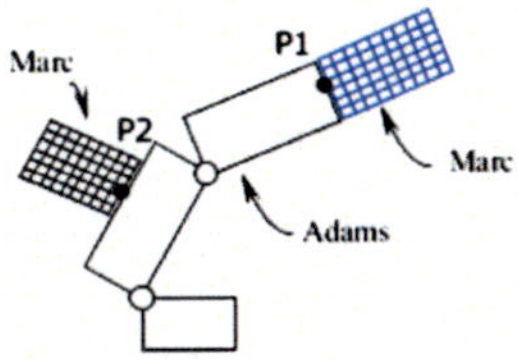

Fig. 4. Description of model *interactions*.

4.2 Simulation Results

Numerical results are shown in the following. Figure 5 and 6 show the grasping, a rotation and then a 3D movement of the gripper. In particular, Fig. 5 shows results of the grasping forces, position velocity and acceleration of the gripper CM, while Fig. 6 shows the corresponding deformation and stresses experienced during the *cosim.*

Figures 7, 8 and 9 show the results of the *cosim* for the grasping of the ball and yaw rotation of 90°. In particular, Fig. 7 shows the contact forces of the claws and the movement experienced by the system, while Fig. 8 shows the CM velocity and CM acceleration of the gripper during the simulation. Figure 9 shows the deformation and stresses experienced by the ball. The importance of these tools can be demonstrated by the inspection robot THROO, for which before prototyping a massive simulation has been carried out in several operating conditions [24].

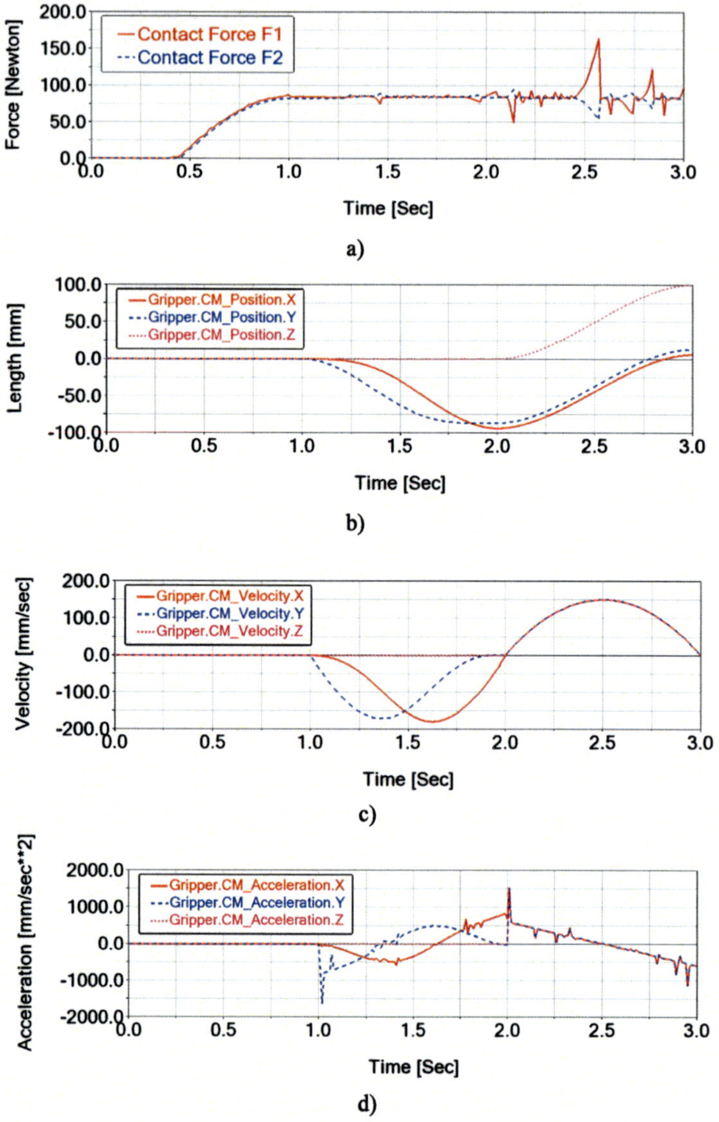

Fig. 5. Results for the simulation using *cosim* of the grasping, rotation and 3D movement of the gripper: a) contact forces of the claws; b) gripper CM position; c) CM velocity; d) CM acceleration.

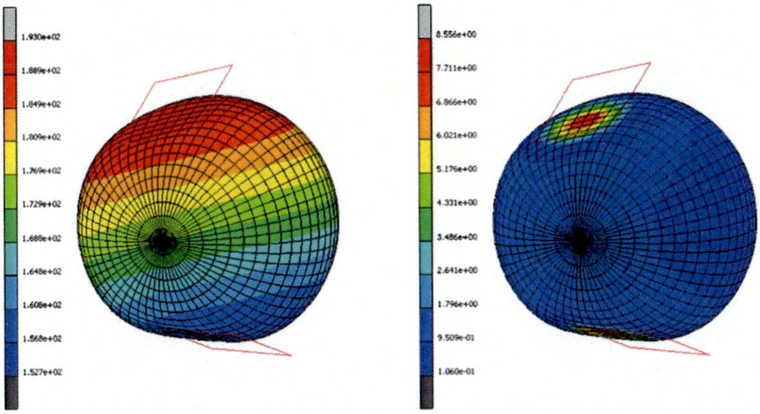

a) b)

Fig. 6. Results on the soft ball: a) deformation, b) and c) stresses.

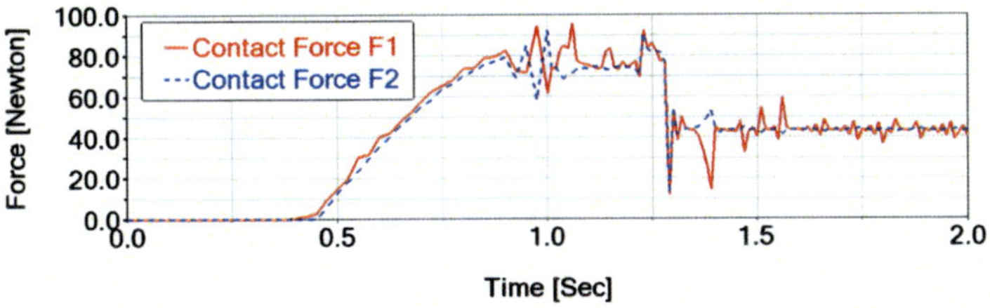

a)

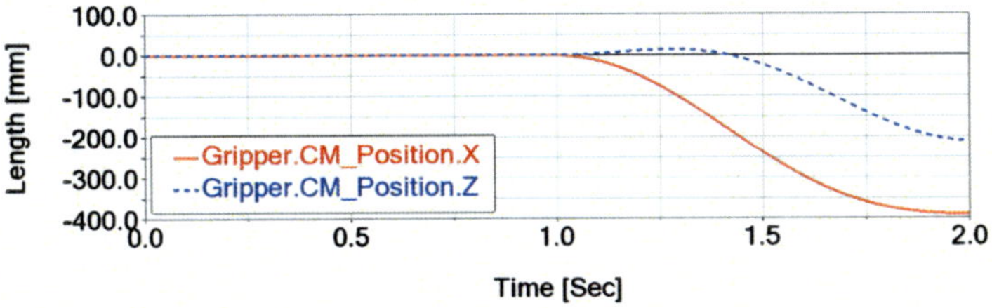

b)

Fig. 7. Results for the simulation using *cosim* of the grasping, yaw rotation 90° of the gripper: a) contact forces of the claws; b) gripper CM position.

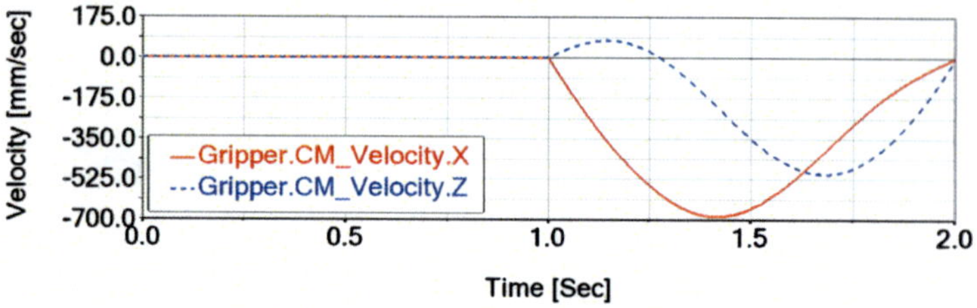

a)

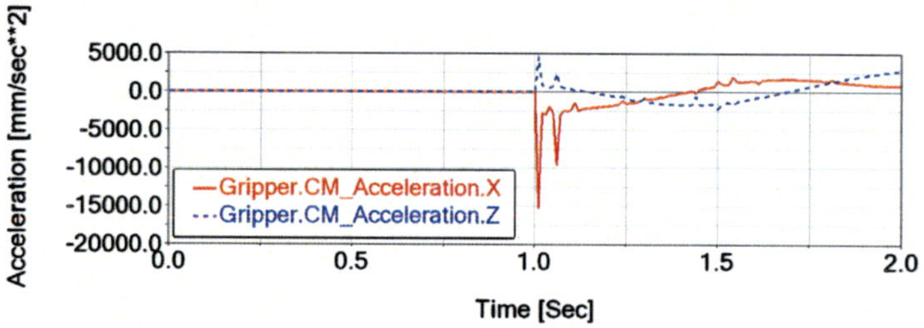

b)

Fig. 8. Results for the simulation using *cosim* of the grasping, yaw rotation 90° of the gripper: a) CM velocity; c); b) CM acceleration.

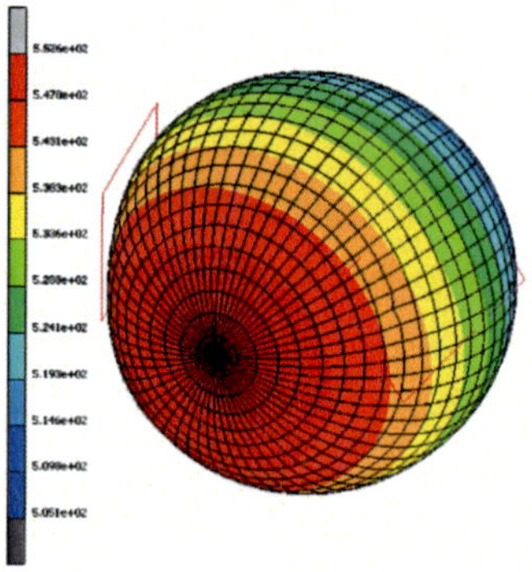

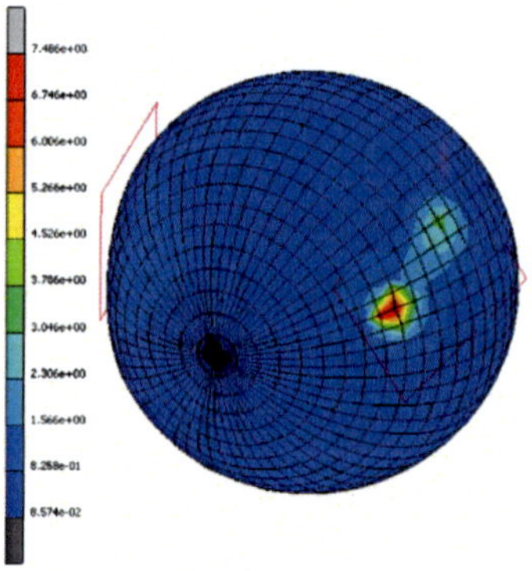

a) b)

Fig. 9. Results on the soft ball: a) deformation, b) stresses.

5 Conclusion

In this paper, we have presented some preliminary results for the *co-simulation* between Adams and Marc software. As illustrative example, a gripper composed by rigid links is simulated interacting with a soft ball simulating the grasping of a soft object. Although the simplicity of the proposed example, the co-simulation tool presented and tested here has wide range of applications in the mechanics of grasping including prosthetic robotic hands and manipulation of delicate objects as horticulture products.

References

1. Rea, P.: On the design of underactuated finger mechanisms for robotic hands. In: Martinez-Alfaro, H., (eds.) Advances in Mechatronics, pp. 131–154. InTech Europe, Rijeka (2011). https://doi.org/10.5772/24304. ISBN: 978-953-307-373-6
2. MSC software: webpage https://www.mscsoftware.com/it. Accessed 2021
3. Catelani, D.: Tools and methodology to achieve simulation driven design and the importance of leveraging numerical analysis and software knowledge from the very beginning, starting at university. In: XXV International Congress, Italian Association of Aeronautics and Astronautics, Rome (2019)
4. Petrolo, M., Governale, G., Catelani, D., Carrera, E.: Multibody models with flexible components for inflatable space structures. Adv. Aircraft Spacecraft Sci. **5**(6), 653–669 (2018). https://doi.org/10.12989/aas.2018.5.6.653
5. Raval, S., Patel, B.: A review on grasping principle and robotic grippers. Int. J. Eng. Dev. Res. **4**(1), 483–490 (2016)
6. Chen, F.Y.: Force analysis and design considerations of grippers. Ind. Robot **9**, 243–249 (1982)
7. Mason, M.T., Salisbury, J.K.: Robot Hand and the Mechanics of Manipulation. MIT Press, Cambridge (1985)
8. Hannes hand: webpage. http://rehab.iit.it/sviluppo-dispositivi-rehab-technologies-lab-iit-inail/#protesica. Accessed 2021
9. Birglen, L., Gosselin, C.M.: Kinetostatic analysis of underactuated fingers. IEEE Trans. Robot. Autom. **20**(2), 211–221 (2004)
10. Figliolini, G., Rea, P.: Overall design of Ca.U.M.Ha. robotic hand, Robotica **24**(3), 329–331 (2006)
11. Bulanon, D.M., Kataoka, T., Ota, Y., Hiroma, T.: A machine vision system for the apple-harvesting robot. Agric. Eng. Int. CIGR J. Sci. Res. Dev. **3** (2001). paper: PM 01 006
12. Ceccarelli, M., Figliolini, G., Ottaviano, E., Simon Mata, A., Jimenez, Criado, E.: Designing a robotic gripper for harvesting of horticulture products. Robotica **18**, 105–111 (2000)
13. Ottaviano, E., Toti, M., Ceccarelli, M.: Grasp force control in two-finger grippers with pneumatic actuation. In: IEEE International Conference on Robotics and Automation ICRA2000, San Francisco, Paper R0095, pp. 1976–1971 (2000)
14. Figliolini, G., Rea, P.: Ca.U.M.Ha.: Robotic hand for harvesting horticulture products. In: XXX CIOSTA-CIGR V Conference on Management and Technology Applications to Empower Agriculture and Agro-Food Systems, Turin, Italy, pp. 288–295 (2003)
15. Buemi, M., Massa, G. Sandini, F.: AGROBOT: a robotic system for greenhouse operations. In: 4th Workshop on Robotics in Agricultural the Food-Industrial, Toulouse, pp.172–184 (1995)
16. Harrell, R.C., et al.: Robotic picking of citrus. Robotica **6**, 269–278 (1990)
17. Hussain, I., Iqbal, Z., Malvezzi, M., Seneviratne1, L., Gan, D., Prattichizzo, D.: Modeling and prototyping of a soft prosthetic hand exploiting joint compliance and modularity. In: IEEE International Conference on Robotics and Biomimetics, Kuala Lumpur (2019). https://doi.org/10.1109/ROBIO.2018.8665231
18. Kulkarni, T., Uddanwadiker, R.: Overview: mechanism and control of a prosthetic arm. Mol Cell Biomech. **12**(3), 147–195 (2015). PMID: 27281955
19. Clement, R.G.E., Bugler, K.E., Oliver, C.W.: Bionic prosthetic hands: a review of present technology and future aspirations. Surgeon **9**, 336–340 (2011)
20. Elliott, A.: A highly efficient, general-purpose approach for co-simulation with ADAMS. In: 15th European ADAMS Users Conference, Rome (2000)

21. Elliot, A.: Status update on advanced general-purpose co-simulation with ADAMS. In: 2002 North American Users Conference, Scottsdale (2002)
22. FMI Co-simulation specification document at the official website. https://svn.modelica.org/fmi/branches/public/specifications/FMI_for_CoSimulation_v1.0.pdf. Accessed 2021
23. Fahlgren, E., Elliott, A., Carlson M.: Knee replacement simulation including large displacement and viscoelastic effects. In: Proceedings of BioMed2008, Irvine, 18–20 June (2018)
24. Ottaviano, E., Rea, P., Castelli, G.: THROO: a tracked hybrid rover to overpass obstacles. Adv. Robot. **28**(10) (2014). https://doi.org/10.1080/01691864.2014.891949

Chalala: Conscious Fashion Towards the Re-innovation of Santander's Weaving Tradition

Eugenia Chiara[1], Eddy Alexandra Arguello Bastos[2](✉), and Arturo Dell'Acqua Bellavitis[1]

[1] Politecnico Di Milano, Via Durando 10, 20158 Milan, Italy
{eugenia.chiara,Arturo.dellacqua}@polimi.it
[2] Viale Zara, 21, 20159 Milan, Italy

Abstract. The process of giving proper value to heritage and traditions can be carried out through fashion. Following this line of thought, this project integrates Colombian and Italian know-hows related to handmade weaving in order to re-innovate and add value to the work of the cotton weavers of Corpolienzo in Charalá (Santander, Colombia), a group of women who have dedicated themselves to rescue the textile tradition of their region. The research followed a methodology of five steps: (1) Discover; (2) Analysis and research; (3) Learn and explore; (4) Share and create; (5) Develop. The result of the project was Chalala, a womenswear collection inspired by the appreciation of craftsmanship by exploring the combination of traditional knowledge with contemporary design and the promotion of slow and premeditated processes to obtain high-quality pieces with a sophisticated aesthetic and a refined elegance.

Keywords: Womenswear · Re-innovation · Slow Fashion · Craftsmanship · Handmade weaving · Co-design

1 Introduction

In a world in which everything moves and changes at an almost imperceptible speed, the heritage and traditions are sometimes undervalued and by consequence, they are slowly disappearing. As a tangible manifestation of traditions, craft objects are the carriers of a series of symbols and meanings that are unique to the culture that produce them and able to enrich humanity's past, present, and future [1].

This project was inspired by Colombian's cultural richness, specifically, the organic cotton-weaving tradition of Charalá, a small town located in the region of Santander. The efforts to keep this tradition are sustained by a small social cooperative called Corpolienzo, who over the last 30 years has been committed to the recuperation of the labor that the Guane's indigenous started hundreds of years ago. These traditional weaving techniques combined with the Italian contemporary handmade textile making learned with the artisan Nicoletta di Gaetano, and a design-driven approach gives the possibility to obtain new and more valuable outcomes.

J. Machado et al. (Eds.): icieng 2021, LNME, pp. 325–334, 2022.
https://doi.org/10.1007/978-3-030-79165-0_31

The possibility of going back to the essential, bring back knowledge and values into a world that is immersed in the fast fashion culture is feasible thanks to the fact that some sectors in the fashion business and the consumers have realized the earth's resources are not unlimited. Leading to Slow fashion, defined not as the direct opposite of fast fashion, but as fresh vision towards the relationship between sustainability or ecological practices with transparency along the production chain, taking emphasis on quality, craftsmanship and experienced labor [2].

With the development of this project, it would be possible to continue contributing to the reaffirmation of a narrative that educates and raises consciousness in the fashion world and in the consumers about the importance of artisans and their role in the preservation of the material and immaterial identity. In the same way, craftsmen can teach societies about the preservation of the planet, using raw materials that are locally produced and employing a slow and premeditated production which is emotionally charged with their stories.

2 Territorial Identity, Multi-local Society and Interculturalism

The concept of territorial identity refers to the interactions that take place in the area inhabited by the people who give content to them, since the social being is deeply linked to the territorial segment in which it lives. Furthermore, Benedetto [3] refers to this concept as the Collective recognition, which is at the same time implicit and explicit, of a network of meanings of social fabrication. Considering ways of thinking, valuing, organizing and appropriating the environment, forming a cultural organization of the territory, whose plots are geographically limited.

Moreover, in times of globalization, societies cannot remain closed, it means they must interact with others in other to enrich knowledge and capabilities to compete. According to the aforementioned, the idea is to encourage communities to work with others under the premise of "think globally, act locally", referencing the use of local practices which are strong in their own identity, with the intention of been open and connected to other communities around the world. A Multi-local society is defined as a society based on a new relationship between local and global. According to Vezzoli and Manzini [4], the multi-local society combines a local society based on the specific features of places and their communities, with the new phenomena generated and supported world-wide by globalization and by cultural, socio-economic interconnection. Thus, local cultural practices provide more transparent production systems, often using hand skills, which can address a contemporary search for authenticity; meanwhile, the openness to the world can take them to innovation.

Following the definition of a Multi-local society, it is pertinent to describe the concept of interculturalism. This one refers to the presence and equitable interaction between culturally differentiated subjects or social groups and the possibility of generating shared cultural expressions and links, acquired through dialogue and co-work, having an attitude of respect and tolerance [5]. This concept emphasizes the possibility of producing mutual learning among the participants, based on cooperation and exchange in order to enrich both parties.

3 Design and the Actions to Innovate Within Tradition

When talking about the introduction of design into crafts, Vacca [6] states that this specific discipline acts as a connection between tradition and modernity looking forward to giving to the craftsmanship a modern perspective; while artisanship contributes with immaterial values and the know-how to produce the object. Thanks to design, it is possible to develop products that stimulate artisans to experiment with practices that can bring them closer to modernity and at the same time enrich their objects with new meanings.

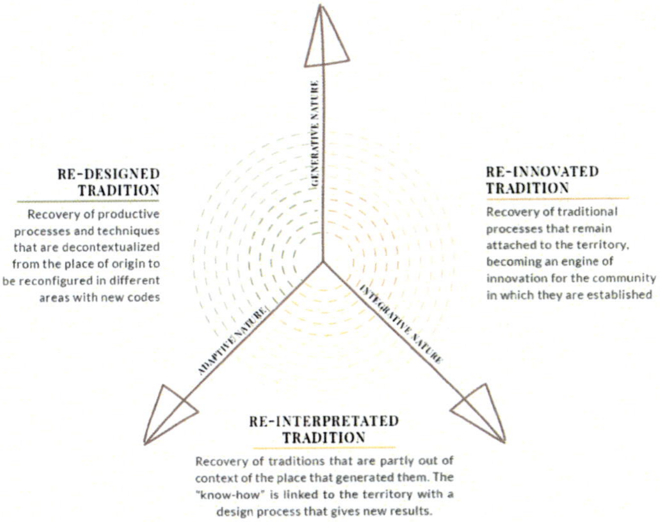

Fig. 1. The combination of innovative natures allows tradition to be Re-Designed, Re-Interpreted or Re-innovated. Image adapted by the authors from Design Sul filo della tradizione [6].

Under this perspective the author of *Design sul filo de la tradizione* [6], refers to the combination of three different whose sole purpose is to innovate within tradition:

- Adaptive nature: adjustment of external knowledge into the configuration of existing processes in different applications.
- Generative nature: development of new knowledge and skills.
- Integrative nature: integration of already existing knowledge but related to different productive sectors.

Thanks to the combination of above-mentioned concepts, three different ways in which the craft tradition can be intervened and combined with the design are generated, as shown on Fig. 1.

4 Valorization of the Identity via Fashion Design

When referring to the relationship between identity and fashion it is possible to talk about two different perspectives: One related to the identity of each individual and the other one to the identity of a local community or territory. According to Bertola [7], nowadays the fashion products are the perfect elements to define identity because they incorporate meanings and their proximity with the body make them alive, therefore they become identity vehicles.

On the other hand, the second one refers to the collaborative work between fashion and a community, its material values, people and the environment. When co-working with a certain community, the anonymity of largescale production is rejected and the human touch is immediately added, bringing with it the skills, stories, attitudes, and traditions of the ones involved; attributing an emotional charge that is impossible to deny and therefore empowering the wearer. In order to revalorize a specific territorial identity, it must be emphasized that the work should emerge from joint efforts with the local community, rather than being imposed from the outside [8]. The designer, as Parente [9] remarks, should understand the values that represent the idiosyncrasy and the diversity of a region compared to others and by doing so, set up a project which ennobles its traditions, culture, and Know-how. For instance, a way to connect fashion with a defined community is through the use of materials and processes of production, since in this way it is created a tangible link with the region. About this perspective, Clark [10] states that the use of localized physical and social resources can provide an alternative to standardization and centralization. Moreover, Aakko [11] mentions that factors such as skill, quality, materiality, aesthetics, small batch production, time and provenance must be taken into account when designing by the hand of craftsmanship.

5 Methodology

For the project development, it was employed a methodology of 5 steps: discover, analysis and research, learn and explore, share and create, and develop.

5.1 Discover

The phase comprehended the exploration of the initial context: Visit the workshop of Corpolienzo's (Charalá) artisans to learn about their history, identity, materials, techniques, and products.

5.2 Analysis and Research

The following step was the development of theoretical research about three main topics: the relationship between design and territorial identity, the development of craftsmanship related to fashion in Colombia and Italy, and sustainability and ethical practices inside the fashion industry. Here, there were also analyzed several case studies, specifically fashion brands related to each of the subjects.

5.3 Learn and Explore

The third step of the project involved a practical study of weaving techniques that took place in Italy, using raw materials made by the Colombian artisans. This step allowed a redesign of the tradition since the weaving process is done in a foreign context and the figure of the artisan and the designer were combined into one.

5.4 Share and Create

The fourth phase of the methodology aimed to make a re-interpretation and a re-innovation of Charalá's weaving tradition through the development of workshops with artisans and its subsequent results. In this case, the know-how remains linked to the territory and is supported by a design process to obtain new results that can become a source of innovation for Corpolienzo.

5.5 Develop

The last step of the project was dedicated to the development of Chalala, a womenswear collection for the season Spring/Summer 2020. The collection gave protagonist to the handmade fabrics using minimal silhouettes able to honor the feminine figure, offering garments that can be wear season after season by women with a sensibility for handmade products and that want to have intimate contact with their roots.

6 Textile Making

One of the most relevant processes of the project was the phases of the experimentation of textile making Italy and the posterior fabric co-design with the artisans of Corpolienzo using a design-driven approach which gave the possibility to obtain new and more valuable outcomes.

6.1 Experimentation as a Way of Learning

Craft objects are the result of implicit knowledge, which means that is not easy to communicate how they are developed with formal language, therefore it must be direct contact between the master and the student, and a process of experimentation in order to be transmitted: it is internalized within oneself and refined only through practice (Vacca 2013). Due to this affirmation, the process of learning how to weave with the Italian artisan Nicoletta di Gaetano was divided into theory and practice, allowing to have a better understanding of how to use theoretical concepts directly in the loom. It was possible to see that for a beginner in the art of weaving there must be constant feedback between the technical design made on paper and the following experimentation on the loom to make subsequent modifications. This step was also crucial since the knowledge acquired enabled to communicate in a technical way with the artisans of Corpolienzo and to explain the variations for the weaves and how to make technical textile design during the following workshops.

6.2 Co-designing with the Artisans Through Workshops

Throughout the project, the integration of knowledge and experience coming from craftsmanship and design was key for its development and result. In order to obtain different results, was necessary to establish a process of continuous collaboration and sharing of know-how in the ideation and development of the collection. Here, the designer's role was to act as an interpreter of artisanal traditions and ways to re-work them by the hybridization of different cultures, techniques, and technologies; while the artisanal community were the carriers of the tacit knowledge and part of ideation processes with the designer, since they interpreted the project and in their own way and assigned meaning to it.

Particularly, for the development of the project, there were executed two workshops where the two analyzed cultural scenarios, the Colombian and the Italian, were integrated through fashion design. The aims of the workshops were: help to preserve the identity as a positive heritage of artisanal culture and search for new ways of interpreting traditional elaborations with a vision that adapts to the requirements of current fashion.

Workshop A – Exploring the Identity

Exploring the identity, intended to explore the identity and tradition of the Lienzo de la Tierra and Corpolienzo to reaffirm the bases of the project and anchor them to the collection's identity.

Workshop B – Spiral of Knowledge

Workshop B, Spiral of knowledge, was divided into four phases (socialize, exteriorize, combine and interiorize) based on a model proposed by Nonaka and Takeuchi [12]. During the workshop, there were explained the fashion trends for Spring/Summer 2020, basic design concepts, and the weaves learned with Nicoletta di Gaetano. After it, there was a discussion about the possibilities of integrating the concepts that were explained further above, and finally their application in the design fabrics for garments of the collection taking as inspiration landscapes of the Santander department. In Fig. 2, one of the artisans are drawing one of the pieces for the collection based on the concepts previously explained.

Fig. 2. One of the artisans of Corpolienzo drawing her interpretation of one of the landscapes of Santander considering design principles and types of weaving.

7 Results

7.1 Results of the Workshops

Notwithstanding the principal outcome of the project was a womenswear collection, the result of the workshops was key for the final development of the research. Here, the artisans developed two different alternatives of fabrics for a kimono, as shown in Fig. 3, that later one of them was adjusted by the author of the project and included it as a piece of the collection.

Fig. 3. Final drawings developed by the artisans of Corpolienzo. In following steps, the drawing of the left was adjusted to be used as part of the final collection.

Thanks to the execution of the workshops, it was possible to confirm that there was a re-interpretation and re-innovation of Copolienzo's traditions because:

- The designer supported the artisans towards an evolution of the fabrics and together re-read their elements and the handmade techniques to construct them. Specifically, this occurred when the weaving techniques were explained during the workshop and with the continuous communication between the artisan ad the designer during the process of textile making in order to achieve the expected result.
- There was cultural mutation able to produce innovation since the artisans were able to combine the plain weaving variations, learned in first place with Nicoletta di Gaetano in Italy, with the design principles and the trends explained during the workshop for the creation of different textiles that can answer the requirements of the market.

- Taking as a starting point the techniques and the local materials and by combining them with different techniques and concepts it is possible to preserve emotions that can evoke the place in which they were created and the person in charge of its production. The final product has its own *raison d'etre* expressing the local history and traditions.

7.2 Chalala

The result of the project was a womenswear collection for the season Spring / Summer 2020 that takes its name after the Guane Cacique Chalala, who ruled in the area in which nowadays the town of Charalá is located. The collection was composed of 12 different outfits, as shown in Fig. 4, in which each one is inspired by a landscape or place in Santander (Colombia), common in the memory of the designer and of the artisans. The great majority of the fabrics selected for the garments were made using raw materials produced by Corpolienzo, designed by the author of the project and co-designed with the artisans from Charalá. They were evaluated and selected according to their weaves, functionality, finishing, texture, aesthetics, and wearability.

The collection had a clear vision towards the integration sustainability, ethical practices inside the fashion industry and restatement of territorial identity which is represented through craftsmanship combined with a design-driven approach.

Fig. 4. Final outfits of Chalala. The collection was composed of 12 looks inspired by the Santander's landscapes, common in the memory of the artisans and the designer.

8 Conclusions

During the line of the development of this project, the three actions to innovate within tradition, re-design, re-interpretation, and re-innovation, were all encountered.

The workshops with the Italian artisan, Nicoletta di Gaetano, lead to in sizes for Re-designing the tradition, giving a new perspective to the use of the raw materials coming from Colombia. In this case, the cotton threads from Charalá were partially decontextualized from their place of origin and were used by the designer/artisan for the creation of new textiles. This followed by the development of a workshop with the Colombian artisans where they got engaged with basic design principles and plain weaving variations previously acquired by the author of the project. This communication between the designer and the artisans allowed for re-interpreting traditions, resulting in the development of textiles that remained attached to the territory without losing their identity properties but with a new interpretation thanks to a contemporary criterion. The practical application of the learned concepts led to an improvement of the manual techniques of the artisan who was in charge of weaving, which shows the presence of the concept of re-innovated tradition.

The final result was the opportunity to gather know-how of different countries and disciplines to contribute to the preservation of the weaving traditions in Charalá, by innovating and enriching the creation of fabrics during the phase of weaving. These are the roots of Chalala, the resultant collection of the research.

9 Limitations and Further Work

Due to the timing constraints of the project, it was not possible to co-create all the fabrics and garments that were proposed as part of the collection. In the future, the work between designer and artisans will be extended for continuing the re-innovation of the weaving tradition in Charalá by being present in Corpolienzo and working closely with the artisans during the experimentation and design of textiles and pieces. Additionally, including the artisans into other processes of manufacturing as pattern making and sewing, such that the product is all locally produced, which should allow for an improvement of the overall process.

Acknowledgements. We thank the artisan Nicoletta di Gaetano for her guidance in the third phase of the methodology and her support for the correct development of the workshops. We would like to express our complete gratefulness to the artisans of Corpolienzo, especially to Graciela Sanabria and Liliana Ardila. The aims of the research and the result of the project were achieved thanks to their collaboration and interest. We are looking forward to working once again with them.

References

1. UNESCO: Patrimonio. In: Indicadores UNESCO de cultura para ell desarrollo, pp 109–118. Organización de las Naciones Unidas para la Educación, Paris (2014)
2. Fletcher, K.: Sustainable Fashion and Textiles: Design Journeys, second edi. Routledge, New York (2008)
3. Benedetto, A.: Nuevas alternativas para pensar el desarrollo de los territorios rurales. Cuadernos de Desarrollo Rural **57**, 101–131 (2006)
4. Vezzoli, C., Manzini, E.: Design for sustainable consumption. In: Perspectives on Radical Changes to Sustainable Consumption and Production (SCP). SCORE!, Copenhagen, pp 167–199 (2006)
5. Giménez, R.C.: Pluralismo, Multiculturalismo e Interculturalidad. Propuesta de clarificación y apuntes educativos. Educación y futuro: revista de investigación aplicada y experiencias educativas **8**, 11–20 (2003)
6. Vacca, F.: Design sul filo della tradizione. Pitagora, Bologna (2013)
7. Bertola, P.: La moda progettata. Le (sette meno una) vie del design. Pitagora, Bologna (2009)
8. Fletcher, K., Grose, L.: Fashion and sustainability: design for change. Laurence King Publishing, London (2012)

9. Parente, M.: Oltre il tangibile: il design per l'identità e la crescita dei territori. In: D4T–Design Per I Territori. Approcci, metodi, esperienze. LISt Lab, Milan (2018)
10. Clark, H.: SLOW + FASHION—an oxymoron—or a promise for the future …? Fash. Theory **12**(4), 427–446 (2008)
11. Aakko, M.: Unfolding artisanal fashion. Fashion Theory J. Dress Body Cult. **23**(4–5), 531–552 (2018)
12. Nonaka, I., Hirotaka, T.: The Knowledge-Creating Company. Oxford University Press, Oxford (1995)

Development of a Computerized Maintenance Management Model of a Laboratory Testing Service Enterprise

Teresa Morgado[1,2,3](✉), André Pinto[4], Helena Navas[2,4], and Suzana Lampreia[3,5]

[1] IPT - Polytechnic Institute of Tomar, 2300 Tomar, Portugal
t.morgado@fct.unl.pt
[2] UNIDEMI/NOVA – Research & Development Unit for Mechanical and Industrial Engineering, Universidade NOVA de Lisboa, 1900 Lisboa, Portugal
[3] CINAV - Naval Research Centre, 2810 Almada, Portugal
[4] DEMI/NOVA – Department of Mechanical and Industrial Engineering, Universidade NOVA de Lisboa, 1900 Lisboa, Portugal
[5] ESCOLA NAVAL/CINAV – Science and Technology Department of Portuguese, Naval Academy/Naval Research Centre, 2810 Almada, Portugal

Abstract. Maintenance management and the existence of maintenance management systems are fundamental challenges for companies subject to the increased competitiveness imposed by modern societies. This type of management is complex, requiring systematic approaches where it is essential to coordinate the different interdisciplinary aspects of maintenance in public and/ or private organisations. It is crucial to seek continuous improvements and balance their benefit and cost to maximise the positive contribution of maintenance to overall profitability and sustainable expansion.

This article presents a study on improving management systems and computerisation of a private organisation's maintenance activities. This company is within the market segment of service provision in the area of laboratory testing. In this area, there is a lack of studies regarding maintenance management. The Maintenance Management proposal developed in this work integrated the company's Management Systems focused on the Quality Management vision. LEAN philosophy was used for the diagnosis and identification of existing problems. And a management model was developed to allow the company to improve maintenance activities, maintenance management and control it, thus increasing its efficiency and effectiveness. The Maintenance Management proposal is based on the Computerized Maintenance Management System. With this study, it was concluded that no sudden changes should be introduced in order to obtain results as soon as possible. The changes should be gradual, planned, programmed, and controlled to test, evaluate and validate the proposed solutions to be implemented.

Keywords: Computerized maintenance management system · LEAN · Ipinza method · ISO/IEC 17025:2017

J. Machado et al. (Eds.): icieng 2021, LNME, pp. 335–346, 2022.
https://doi.org/10.1007/978-3-030-79165-0_32

1 Introduction

The 13306 European Standard [1] defines maintenance management as a set of activities to establish the maintenance objectives, strategies, and responsibilities and implement them by maintenance planning, maintenance control and supervision, and improvement of the organisation's methods economic aspects. Stamboliska et al. [2] refer to maintenance as a complex management process that associates several processes, like production, quality, environment, risk analysis and safety. Muchiri et al. [3] concluded that reliability and maintenance have a strong impact on timely services and quality. Murthy [4] refers that maintenance management requires a multidisciplinary approach with a business perspective.

One of the most critical resources determinant to maintenance management is the information [5] with the economic and industrial evolution. Information systems to support maintenance management are Computerised Maintenance Management Systems (CMMS) [6]. According to Wienker et al. [7], the organisation must understand the CMMS roles to define which information is relevant to record and ensure compliance with the maintenance management strategy. Usually, CMMS includes Assets Management, Work Orders Management, Preventive Maintenance Management and Report Management [6, 8, 9]. These functions allow better efficiency and effectiveness of maintenance management, taking advantage of Information and Communication Technologies [9]. There are many CMMS available in the market, but some authors [10, 11] identified limitations in condition monitoring analysis, equipment failure diagnostic, limit support to resource allocation and decision analysis support.

Currently, 558 companies that use computerised maintenance management systems exhibited an average of 28.3% increase in the productivity of maintenance; 20.1% reduction in equipment downtime; 19.4% savings in the cost of materials; 17.8% decrease in inventory maintenance and repair and, 14.5 months of payback time [12].

This paper presents a study to organise the maintenance information and apply a Computerized Maintenance Management System in a Laboratory of Non-Destructive Testing of a Service Enterprise. This private company [13] has its laboratory accredited through EN ISO/IEC 17025 [14]. This research aims to obtain a maintenance management model based on a CMMS that facilitates interventions in the maintenance area, takes advantage of the company's existing know-how, and helps the company expand sustainably.

The structure of this paper is organised in five sections. Starting with state of the art and followed by the research method's presentation in the Sect. 2. The Sect. 3 is developed the model maintenance management system based on CMMS, with the survey of an accredited laboratory's organisational management through EN ISO/IEC 17025. The results and the respective discussion are presented in Sect. 4, followed by the final section's conclusion.

2 Research Method

The research method followed in this work is presented in Fig. 1. First, to understand the company business's processes and procedures was conducted brainstorming meetings with all the company collaborators, including top management, characterisation of provided services, analysis of existing procedures, study of relevant documentation and universe of existing equipment, and direct observation combined with literature survey. This step's principal objective was to obtain accurate insight into the business strategy and identify tools to develop a Maintenance Management System based on CMMS. Second, was made a diagnostic of all scenarios and decided intervention criteria. Parameters regarding the operational process and scenarios to be applied to the new model were defined. Third, the conceptual framework has been enhanced by adding external factors to predict a maintenance event used as an input to the CMMS. This resulted in a conceptual decision-making framework for a contract and improvement proposal that could constitute a transition phase between the current and future sustainable implementation of a more complex CMMS.

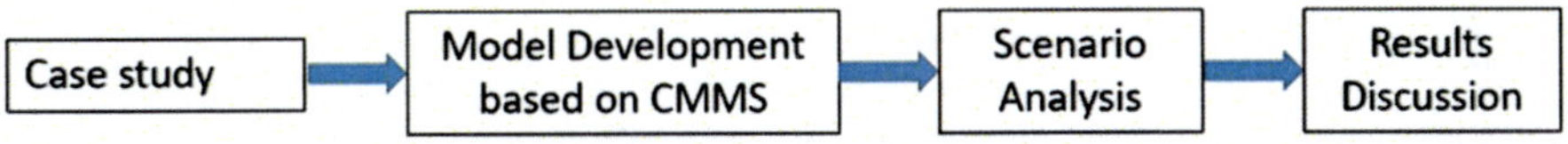

Fig. 1. Research method

3 Model Development

In this research, the case study is a laboratory accredited through EN ISO/IEC 17025 [14] of a private company.

In maintenance management, a balance between benefit and cost is sought to maximise the positive contribution of maintenance to a company's overall profitability. For the development of maintenance practices and the achievement of these goals, it is necessary to relate and coordinate different interdisciplinary aspects of a company. This type of management is complex and therefore requires a systematic approach. There should be a maintenance department/unit/module with dedicated and adequate human resources and materials for easier creation, management, and control of a maintenance management system. The inclusion of such a department will contribute to the sustainable development and expansion of the organisation.

Maintenance Planning, Programming and Control are management tools that will increase the company's maintenance's efficiency and effectiveness. Even if this type of investment in maintenance is complex, it is always worthwhile to research, analyse and organise the existing information because the most pertinent one already exists. For different reasons, its use for the benefit of maintenance is not profitable. Therefore, this work's first step was to relate the existent processes with the equipment management and its characterisation (see Fig. 2).

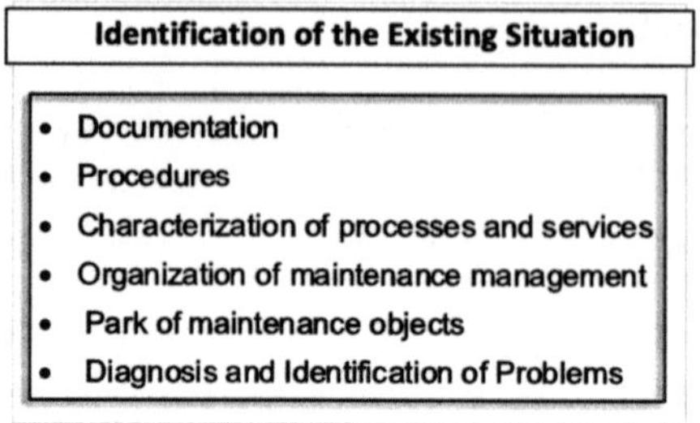

Fig. 2. First Step: Identification of the existing situation

From the survey of the existing situation, from the dialogue established essentially with the company's engineers responsible and technicians, several critical points were identified, making it possible to draw up a diagnosis of the situation. The most relevant issues used in the development of the new CMMS were:

- There is no maintenance department/unit/ module.
- There is no maintenance planning and management team.
- The Maintenance Object Park was developed and organised according to Quality Management and not according to Maintenance Management.
- Some critical documents of object park are too general and, in their current form, are not suitable to handle the Maintenance Management System.
- There is no documentation exclusively directed to maintenance planning and programming. The one that exists mixes the different phases of these processes.
- There is no library of standard preparations. The information that should be centralised in this library is scattered in several physical and computerised documents, making it difficult to consult, articulate, and possibly modify them.
- There are no work orders, so there is no coordination between them and the work reports.
- There is no documentation exclusively directed to maintenance planning and programming. The one that exists mixes the different phases of these processes.
- A document called "Note of Non-Conformity and Corrective / Preventive Action" corresponds to a work request or requisition that also tries to integrate the work order component and, in a minimal way, a work report. However, the way it is developed and the procedures that manage it do not allow effective use.
- It is not clear from the existing documentation on the work reports where the work reports originate or what kind of maintenance it is undergoing.
- There are no appropriate procedures or records that allow the calculation of performance indicators in maintenance effectively and efficiently.
- There are no qualified personnel to meet the more complex IT support needs of the company.

3.1 Scenarios Analysis

The second step consisted of structuring, planning and programming the first phase of computerisation, automatisation. Considering the CMMS areas of incidence and studying the market solutions, a Pugh decision Matrix [15, 16] was constructed based on this laboratory management's reality (see Table 1). Only three market solutions, designed by A, B and C in Pugh Matrix, had the general specifications considered in this case study. The parameters of this decision Matrix (Table 1) complied with the following assumptions:

- The user is not a computer specialist.
- The user is not a maintenance specialist.
- To test the software's suitability, it has to provide a free version.
- The free version must have clear and effective technical support for the non-specialist user.
- Ease of associating external objects and documents (e.g. photos, manuals, etc.).
- Possibility to issue and customise reports.

Table 1. Pugh decision matrix.

Criterion		Weighting factor	A		B		C	
			Simple score	Weighted score	Simple score	Weighted score	Simple score	Weighted score
1	Do not present limitations in the free version	0.15	7	1.05	7	1.05	7	1.05
2	Clarity of technical support	0.09	7	0.63	10	0.90	7	0.63
3	Ease of installation	0.14	6	0.84	10	1.40	6	0.84
4	Simplicity in adding objects	0.07	7	0.49	10	0.70	0	0.00
5	Ease of implementation	0.08	8	0.64	10	0.80	8	0.64
6	Simplicity in associating documents (Photos, Pdf)	0.07	0	0.00	10	0.70	7	0.49
7	Reporting	0.05	6	0.30	10	0.50	6	0.30
8	Ease of use	0.14	6	0.84	10	1.40	0	0.00
9	Presentation in Portuguese	0.15	0	0.00	10	0.70	7	0.49
10	Reporting customization	0.05	0	0.00	6	0.30	6	0.30
11	Independence from online operation	0.04	10	0.40	10	0.40	10	0.40
12	Easy expansion	0.05	0	0.00	10	0.50	10	0.50
	Total Score	1.00	57	5.19	113	9.35	74	5.64

Note - Software where all selection criteria were evaluated
Score Range: Simple Score - [1;10] Total Simple Score - [0;120] Total Weighted Score - [0;10]

3.2 Development of a Digital Machine Book

The reorganisation of the maintenance object park selected the equipment with more importance for the laboratory chain value. A codification and function organisational of that equipment were made.

The Ipinza method [17, 18] was used to define the maintenance type. The relation between each equipment and the score (Table 2) obtained by this method is presented in Fig. 3.

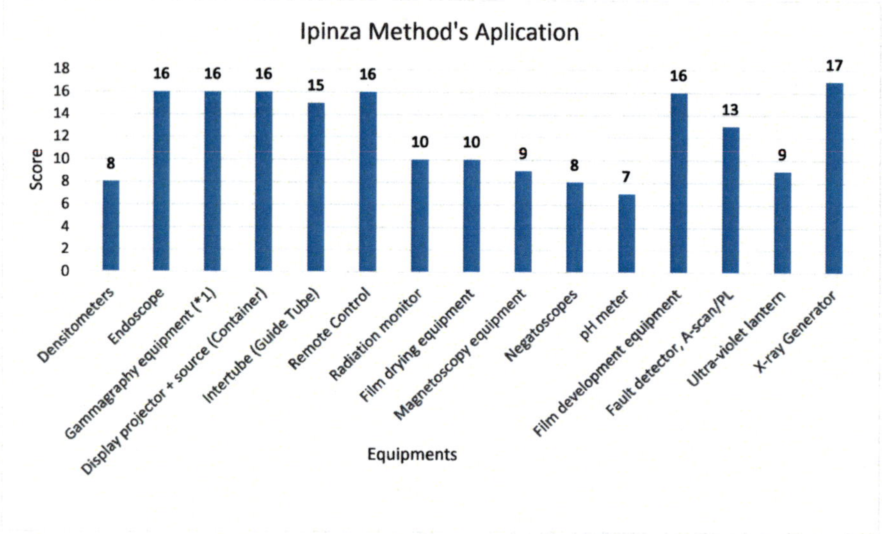

Fig. 3. Results of Ipinza method's application.

Table 2 presented the Ipinza score used in the decision matrix of Table 3. This matrix was developed based on the organisation's principal core business, in the equipment's laboratory reality and the relevant equipment of the testing value chain. After analysing the results of this study, it was possible to make decisions for the maintenance planning process and create documentation for implementing a Computerized Maintenance Management System, namely the digital machine book.

Table 2. Ipinza score.

Total score	The application of preventive maintenance is:	Application suggestion
19–22	Critical	Preventive maintenance
13–19	Important	Preventive maintenance
6–13	Convenient	Corrective maintenance
0–6	Optional	Corrective maintenance

Table 3. Decision matrix based on Ipinza method.

ASPECTS TO CONSIDER	SITUATION	SCORE
1) Effect on Production/Service	Stop	4
	Reduces	2
	Doesn't Stop	0
2) Technical Value - Equipment Economic Value	High	4
	Average	2
	Low	1
3) Damage Consequences of the breakdown:		
a) The machine itself	Yes	2
	No	0
b) The process	Yes	3
	No	0
c) Staff	Risky	1
	No risk	0
4) Logistics Dependency	Foreign	2
	Local	0
5) Labour Dependency	Third parties	2
	Own	0
6) Probability of Breakdown (Reliability)	High	1
	Low	0
7) Degree of Repair Difficulty	High	1
	Low	0
8) Flexibility and Redundancy	Simple	2
	ByPass	1
	Double	0
Total score	**16**	

3.3 Digital Machine Book

The last step consisted of organising all the Maintenance Management information, creating the digital machine book. With this aim, 33 procedures have been created: six for managing the Equipment Park; 3 for organising the Maintenance Work; 4 to control the

Technical Intervention Area; 9 to register the Objects and elaborate Preventive Maintenance Plans, 10 to Start-up the Management System and 4 to Materials Organization. Figure 4, 5, 6, 7 and 8 are presented the strategic sequence developed in the creation of the digital machine book.

The first step to organise the Equipment Park (Fig. 4) must start observing the facilities and creating the layouts, followed by the establishment of functional areas. Then these areas must be divided into systems. The cost centres were made for each system. Each object's codification and technical data were created for each system, and the objects' attributes were developed.

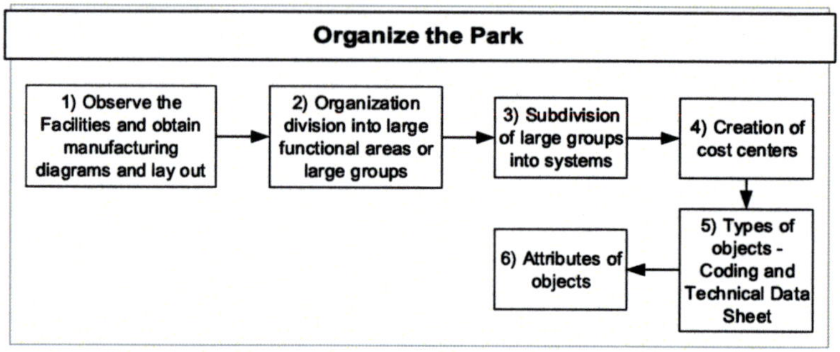

Fig. 4. The six procedures to organise equipment park.

Then, the Maintenance Work must be organised, as shown in Fig. 5. This work started with a survey and definition of maintenance type (preventive or corrective) and strategies. A standard work library was created, and finally, the fundamental description in the maintenance process was done.

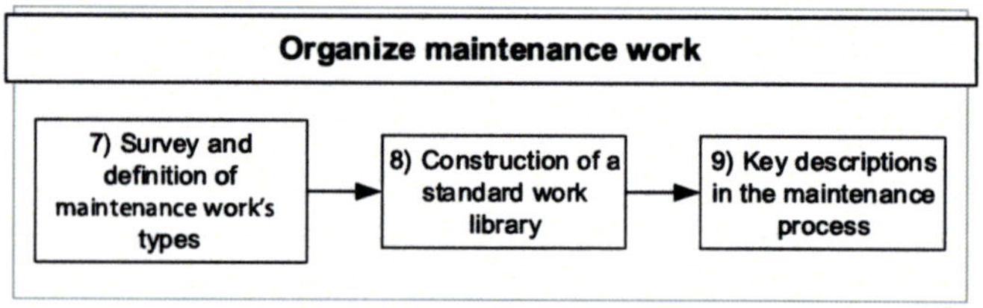

Fig. 5. Three procedures to organise the maintenance work.

After the maintenance work has been organised, the Technical Intervention Area must be developed (see Fig. 6). This step starts with defining and codifying the different levels of the technical area, following the departments. The staff and suppliers must be registered.

The Maintenance Plans must be elaborate, and the objects must be registered. This phase must follow the sequence of the procedure shown in Fig. 7. This work is very laborious and important since it started in the election of maintenance objects; registered them; established and planned the preventive maintenance plan; carried out the

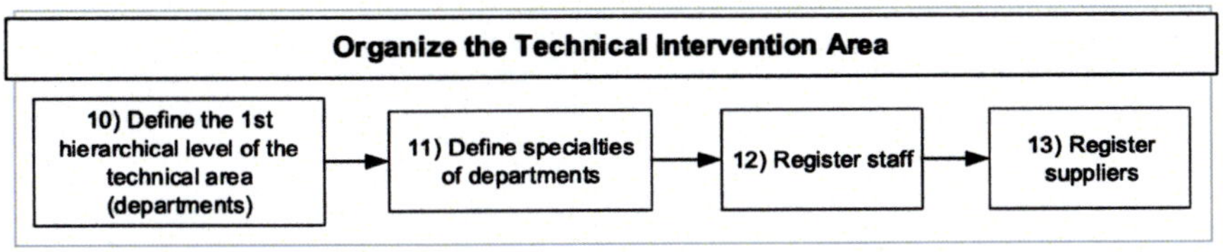

Fig. 6. Four procedures to develop the technical intervention area.

preparation work; election the components that must be included in standard work; created the material file, prepared the preventive maintenance forms; and created the maintenance inspection routines. And finish in develop maintenance plans for similar objects. These nine procedures will be repeated for the different equipment/systems existing in the enterprise. It has been decided to carry out preventive maintenance (see Table 2 and Table 3).

Nevertheless, all the enterprise equipment/systems must be included in the digital book and be part of the maintenance management system. So, at the management system's start-up (Fig. 8), all the equipment/systems must be included sequentially in the procedures manual, request forms, work reports standard, implementation of operating records, labour imputation, material and services imputations. For the equipment subjects to preventive maintenance, more procedures are needed, as shown in Fig. 8, as preparation of preventive actions, analysis of the work program, and issue the Work Orders (WOs).

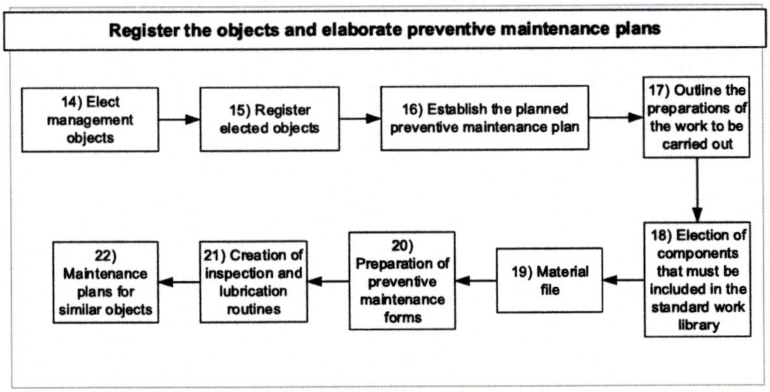

Fig. 7. Nine procedures to develop the maintenance plans.

All the maintenance material must be organised (Fig. 9) thought classes, families and subfamilies, codified and registered. These procedures are time-consuming and must be continuous work for the success of this maintenance management model.

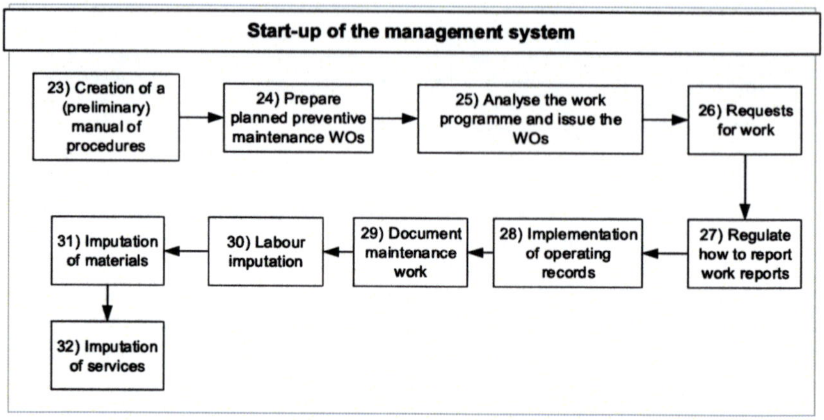

Fig. 8. Ten procedures to start-up of the management system.

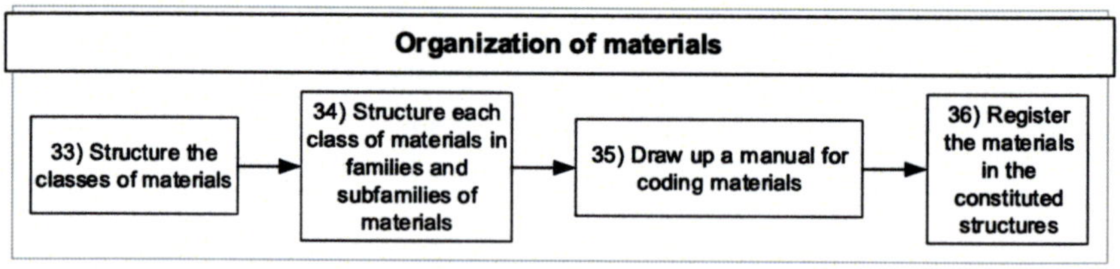

Fig. 9. Four procedures to materials organization

4 Results and Discussion

The elaboration of the flowcharts of the existing processes facilitated visualisation and highlighted the company's current procedures. It had the benefit to help structure the study of all parts of the proposed maintenance management system and leverage Continuous Improvement over time. In this work, a new Maintenance Management Model based in CMMS has an idealised intermediate level, constituting its structure's essential elements. These elements should be complemented with other systems to be developed, such as the stock management, management and prioritisation of work orders, registration systems and indicator control.

In Fig. 10, the CMMS implementation schedule is presented. The first phase of computerisation and automation are medium-term objectives, and the long-term goals include Planning and Programming Phase, Implementation Phase, Test Phase, and Entry Service. The digital-book creation belongs to the intermediate/ transitory maintenance management phase (see Fig. 10). And the Maintenance Management System supported by CMMS software can be implemented after evaluating the implemented solution and only entry to service at the end of 3 years, as shown in Fig. 10.

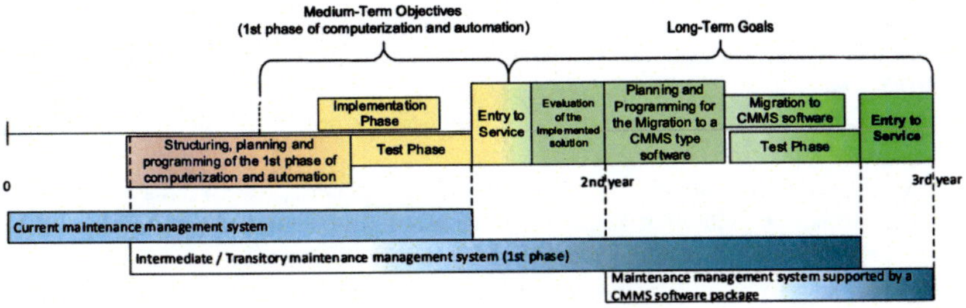

Fig. 10. CMMS implementation schedule

5 Conclusions

With this work, it is concluded that the investment in Planning, Programming and Control shows significant results in the medium and long term, causing the introduction of a mentality of continuous improvement at various company levels. However, it should be noted that this type of investment is often overlooked for not contributing directly to immediate profit and taking time to show results. It is not easy to implement without dedicated human and technical resources at all levels, from top management to specialised technicians.

It is also concluded that not should one fall into the error of introducing sudden changes with results as soon as possible. Such changes may lead to severe and unpredictable future consequences. Gradual changes should be planned, programmed, and controlled so that there is time to test, evaluate and validate the proposed solution to be implemented.

References

1. EN 13306: 2017. European Standard: Maintenance. Maintenance Terminology, December (2007).
2. Stamboliska, Z., Rusiński, E., Moczko, P.: Proactive Condition Monitoring of Low-Speed Machines, pp. 9–34. Springer, Cham (2015). https://doi.org/10.1007/978-3-319-10494-2
3. Muchiri, P., Pintelon, L., Gelders, L., Martin, H.: Development of maintenance function performance measurement framework and indicators. Int. J. Prod. Econ. **131**, 295–302 (2011)
4. Murthy, D., Atrens, A., Eccleston, J.: Strategic maintenance management. J. Qual. Maint. Eng. **8**(4), 287–305 (2002)
5. Laudon, K., Laudon, J.: Management Information Systems, vol. 6. Prentice Hall, Upper Saddle River (2000)
6. Cato, W., Mobley, R.: Computer-Managed Maintenance Systems: A Step-By-Step Guide to Effective Management Of Maintenance, Labor, and Inventory, pp. 13–55, 2nd edn. Butterworth-Heinemann (2002)
7. Wienker, M., Henderson, K., Volkerts, J.: The computerized maintenance management system - an essential tool for world class maintenance. Procedia Eng. **138**, 413–420 (2016)
8. Peters, R.: Reliable Maintenance Planning, Estimating and Scheduling. Gulf Professional Publishing. Elsiever (2015)
9. Carnero, M., Novés, J.: Selection of computerised maintenance management system by means of multicriteria methods. Prod. Plann. Control Taylor Francis **17**(4), 1335–1354 (2007)

10. Rastegari, A., Mobin, M.: Maintenance decision making, supported by computerised maintenance management system. In: Annual Reliability Maintainable Symposium, pp. 1–8. IEEE (2016)
11. Labib, A.: A decision analysis model for maintenance policy selection using a CMMS. J. Qual. Maint. Eng. **10**, 191–202 (2004)
12. IBM: Watson IoT. Understanding the impact and value of enterprise asset management - Make smarter decisions about your assets using the Internet of Things and artificial intelligence. IBM. https://www.ibm.com/downloads/cas/BX0ERPWB. Accessed 04 Feb 2020
13. Pinto, A.: Improvement Study of Management Systems Computerization of Maintenance Activities in a Service Company. Master thesis (in Portuguese), Faculdade de Ciências e Tecnologia da universidade Nova de Lisboa, Lisbon (2019)
14. ISO/IEC 17025:2017. General Requirements for the competence of testing and calibration laboratories, November 2017
15. Pugh, S., Clausing, D.: Creating innovative products using total design: the living legacy of Stuart Pugh, MA, USA, pp. 167–76. Addisen-Wesley Longman Publishing (1996)
16. Navas, H.: Evaluation of complexity for tolerancing solutions of mechanical systems using Pugh's criteria. In: AIP Conference Proceedings, vol. 1315, pp. 1511–1514 (2010)
17. Costa, J., Morgado, T., Navas, H.: Lean philosophy and management assets applied to industrial valve production company. In: Proceedings of the 1st Conference on Quality Innovation and Sustainability – ICQIS2019-Valença, Portugal, pp. 135–144, June 2019
18. Morgado, T., Santos G.: Methodological approach of maintenance management applied to a research laboratory. In: Proceedings of M2D2019, 8th International Conference on Mechanics and Materials in Design, Bologna/Italy, vol.142, pp. 4–6, September 2019

Textile Yarn Winding and Unwinding System

Filipe Pereira[1](✉), Eduardo Leite Oliveira[2], Gustavo Guedes Ferreira[2], Filipe Sousa[2], and Pedro Caldas[2]

[1] MEtRICs and Algoritmi R&D Centres, University of Minho, 4800-058 Guimarães, Portugal
[2] Mechanical Engineering Department, University of Minho, 4800-058 Guimarães, Portugal
{pg38917,pg37170}@uminho.pt, a80947@alunos.uminho.pt

Abstract. This paper presents the concepts of a Textile Yarn Unwinding and Winding System for application in the textile industry. Considering that textile industry market is constantly growing, this work aims to develop a system of unwinding and winding of textile yarn that will later be applied to a system of verification of the quality of the textile yarn through image processing. The article first presents the methodology used for prototype development. It then presents the relevant concepts and mathematical model for the development of a system that allows unwinding and winding the textile yarn as well as adequate yarn tension control. It is also shown and explained a 3D model for the developed prototype. Furthermore, a proposal for the validation criteria of the prototype system is included. Finally, the relevant variables for the prototype's main functions are identified, the system's feasibility is discussed, and the added value of this prototype is highlighted.

Keywords: Textile yarn · Unwinding · Winding · 3D model · Validation

1 Introduction

Currently, in the textile industry, there are several machines with different technologies responsible for performing the unwinding and winding process of textile yarns. These machines are mainly used for reuse of leftover lines, for combining different lines on the same coil or for splitting large coils into smaller coils.

In view of the fact that yarn quality is an increasingly important aspect for textile production, due to an increasing requirement of quality as a requirement for construction of clothing, in the market of textile industry there are several pieces of equipment able to measure characteristics of textile yarn. However, due to their high cost, many companies do not adopt such measures to verify the quality of textile yarn in their production.

Taking this into account, this article seeks to create a textile unwinding and winding system which will subsequently be applied to a low-cost system capable of verifying the quality of the textile yarn through the use of available open-source resources in terms of image processing.

The outline of this paper is as follows. In Sect. 2, Methodology Used for the Prototype Development is presented. In Sect. 3, it is described a Description of Concepts in used techniques. Section 4 presents the Developed Prototype and in Sect. 5 there are presented

J. Machado et al. (Eds.): icieng 2021, LNME, pp. 347–358, 2022.
https://doi.org/10.1007/978-3-030-79165-0_33

some topics about the validation of Developed Prototype. Finally, in Sect. 6, final remarks are enunciated.

2 Methodology Used for Prototype Development

A project consists of several stages and objectives. In order to properly identify the objectives and subobjectives that make up the winding and unwinding system of textile yarn, a methodology, classified as "rational", that guarantees a systematic and consistent approach to the solution of the problem was used.

Through this methodology, it is possible to observe (see Fig. 1) the establishment of the main and secondary objectives that are part of the system of verification of the quality of the textile yarn.

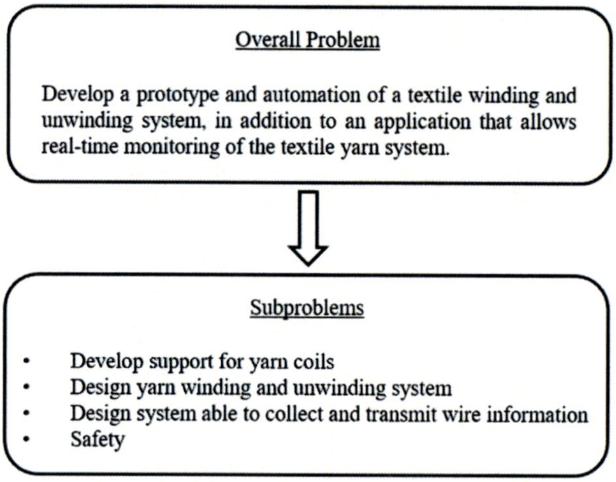

Fig. 1. Definition of project problems.

As this article is related to the yarn unwinding and winding system, and this will in turn be part of the textile yarn verification system, only the first two subproblems, "develop support for yarn coils" and "design yarn winding and unwinding system", identified above will be addressed.

2.1 Clarification and Establishment of Project Objectives

To clearly establish the objectives that are part of the project, the objective tree method was used, thus allowing a clear and useful format so that they can be addressed in the project (see Fig. 2 and Fig. 3).

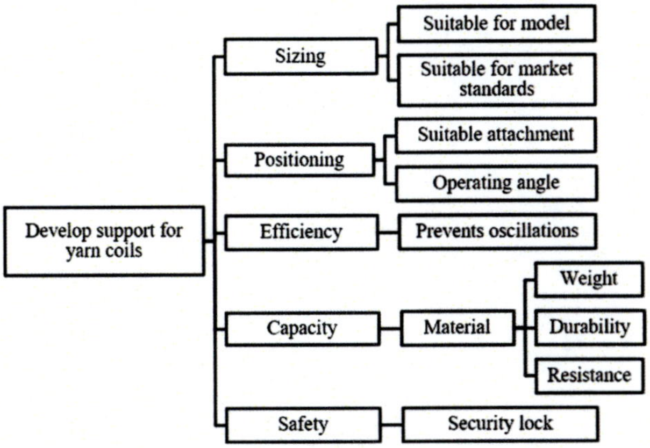

Fig. 2. Tree method for subobjectives.

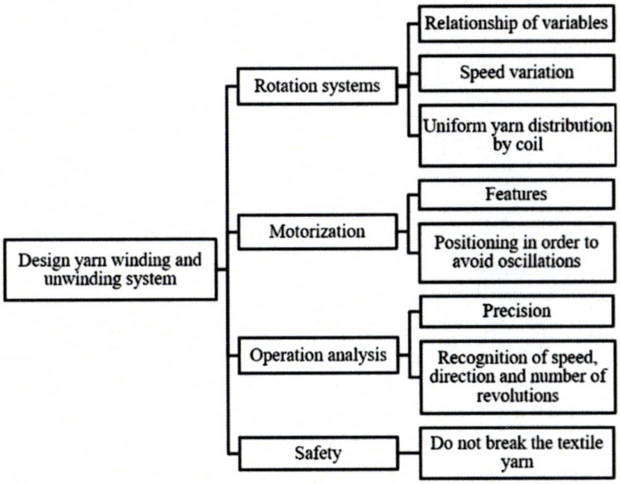

Fig. 3. Tree method for subobjectives.

3 Description of Relevant Concepts

3.1 Uniform Yarn Distribution on Coil

The distribution of the textile yarn on the coil can be made following three patterns of filament winding: helical, polar and circular according to the angle that the path makes with the axis of rotation [1]. The polar and helical windings are identical in that they cross the mandrel in length at a given angle, resulting in a coating made by alternating paths in positive and negative orientations (see Fig. 4 and Fig. 5). In the circular winding the mandrel coating is made by the deposition of the fibers perpendicularly to its axis (see Fig. 4).

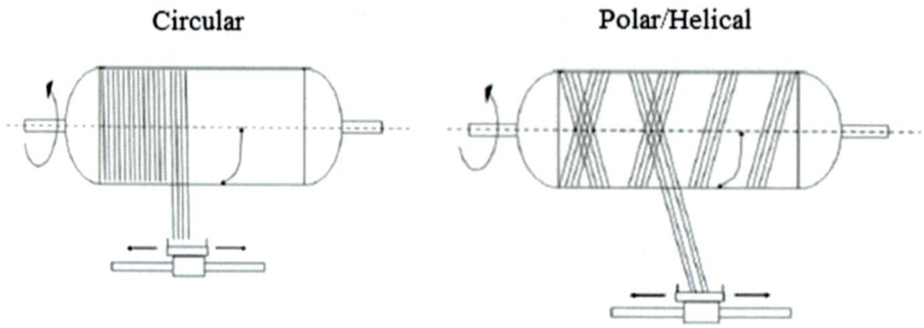

Fig. 4. Circular and polar/helical windings [1].

Fig. 5. Geodetic winding (Helical) and flat winding (Polar) [1].

3.2 Winding Techniques

There are several ways of achieving the yarn winding according to the patterns described above. These ways are classified by two principles of winding: drum-driven or random winders and spindle-driven or precision winders [2].

In drum-driven winder, the package is driven by a cylinder by surface or frictional contact. Traverse of yarn is given either by the grooves cut on the drum or by a reciprocating guide. In case of grooved drum, the drum performs the dual functions of rotating the package by surface contact and performing the traverse. However, when plain drums are used, it just rotates the package and traverse is performed by reciprocating guide (see Fig. 6). A characteristic of random winding is that since the package surface speed remains constant as its radius changes, and the traverse speed also stays constant, the coil angle[1] will, therefore, also remain constant throughout the entire package. However, traverse ratio[2] will continuously decrease with increasing package radius [3].

In spindle-driven winder, the package is mounted on a spindle which is driven positively by a gear system. If the r.p.m. of the spindle is constant, then the surface speed of the package will increase with the increase in package diameter. Therefore, principle

[1] Coil angle – the angle between a plane perpendicular to package axis and the yarn direction at the winding point [3].

[2] Traverse ratio – number of coils laid on package per each completed traverse (from one side to the other and back) [3].

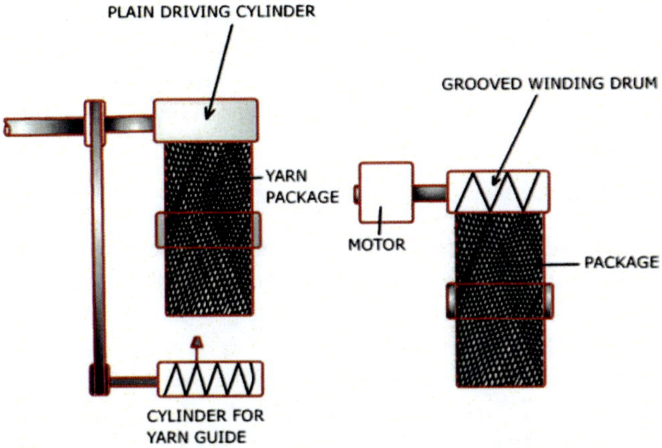

Fig. 6. Grooved and plain drum-driven winders [2].

wise there could be two types of spindle-driven winders: constant r.p.m. spindle winders and variable r.p.m. spindle winders (see Fig. 7).

In case of the latter, the spindle r.p.m. is reduced with the increase in package diameter in such a manner that the winding speed remains constant.

Spindle-driven winders are also known as precision winders as a precise ratio is maintained between the r.p.m. of spindle and r.p.m. of traversing mechanism. Precision winders ensure a constant value of traverse ratio during package building, as package radius increases coil angle will decrease [3].

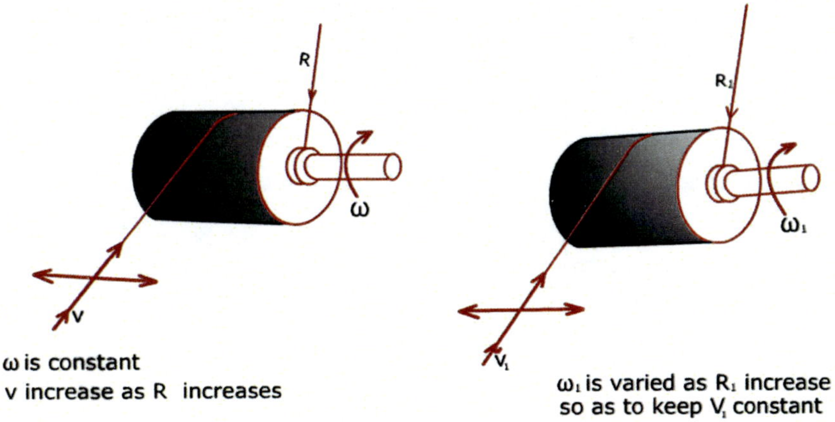

Fig. 7. Constant and variable spindle-driven winders [2].

A third winding technique, often known as step-precision winding or hybrid winding, will combine the previous winding methods, overcoming drawbacks from each one. It has different stages during winding (steps), where it will keep the same traverse ratio during each stage until the coil angle varies a determined amount. At this point a new step begins, with a new traverse ratio bringing the coil angle to the value it began with. This method avoids the patterning effect cause by certain traverse ratios in random winding, while preventing coil angle from variating as much as in precise winding [3].

3.3 Unwinding Techniques

There are two types of yarn withdrawal, side and over end withdrawal. In the side withdrawal the yarn is unwound from the side with package rotational movement, without changing yarn twist. The over end withdrawal changes the yarn twist and allows yarn unwinding without package movement [3].

In this scope we will consider the over end withdrawal since it allows higher yarn feeding velocities.

This kind of unwinding system is normally associated with a yarn guide positioned above the package with the same central axis. Here the unwinding point shifts around the package generating a centrifugal force in the yarn. With high shifting unwinding point velocity caused by high unwind velocities or reduced package diameter, the formation of a characteristic yarn balloon occurs [3].

The unwinding system needs to minimize yarn breakages. This phenomenon occurs due to two main reasons, the tension peaks and slough off [3].

The tension peaks are caused by yarn lack of freedom during unwinding, commonly associated with reduced cone taper. Since there is an increase of contact between unwound yarn and the package causing an increase of friction between the two, it raises the probability of yarn sticking to the package. This phenomenon can cause a yarn break when associated with high yarn twist and the inexistence of a significant yarn balloon [3].

The slough off consists of simultaneous unwinding of several package coils, and it's normally related with high cone taper and high unwinding velocities [3].

Thus, it's clear the importance of cone taper, unwinding velocity, yarn balloon and twist control for the yarn unwinding success.

3.4 Mathematical Model for Yarn Unwinding and Winding System

This is a Mathematical model for the unwinding of textile yarn. Below is the representation of a coil of constant internal radius R_1 and variable outer radius R_2 as the unwinding of the textile yarn occurs (see Fig. 8) [4].

The angular velocity is proportional to the rotation according to the following equation.

$$\omega = \frac{2\pi n}{60} \tag{1}$$

The linear velocity is given by Eq. 2.

$$v = \omega R \tag{2}$$

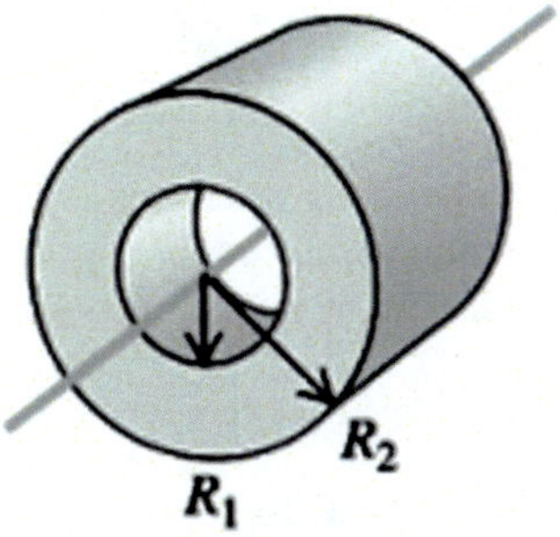

Fig. 8. Representation of the external and internal radii of a coil [4].

Considering the external radius, the substitution of Eq. 1 into Eq. 2 leads to Eq. 3 below.

$$v = \frac{2\pi n R_2}{60} \tag{3}$$

Equation 4 determines the rotation as a function of the cycle traveled. With each cycle N traversed, the outer radius of the coil R_2 is decreased by the value of the diameter of the yarn.

$$n = \frac{60v}{2\pi(R_2 - dN)} \tag{4}$$

The angular acceleration is given by Eq. 5.

$$\alpha = \frac{dw}{dt} \tag{5}$$

Knowing that n = 60f, the substitution of Eq. 1 into Eq. 5 leads to Eq. 6.

$$\alpha = \frac{2\pi df}{dt} \tag{6}$$

The frequency is the ratio of the number of cycles N by the time interval t.

$$\alpha = \frac{2\pi d\left(\frac{N}{t}\right)}{dt} \tag{7}$$

Rearranging the terms results in Eq. 8.

$$\alpha = \frac{2\pi N d\left(\frac{1}{t}\right)}{dt}$$

$$\alpha = -\frac{2\pi N}{t^2}$$

$$|\,\alpha\,| = \frac{2\pi f^2}{N}$$

$$| \alpha | = \frac{\pi n^2}{1800 N} \tag{8}$$

It will have negative signal (deceleration) when it is a winding machine, as the motor speed decreases with increasing radius, and positive signal when unwinding. The torque is the product of the moment of inertia by angular acceleration, Eq. 9 [4].

$$T = I\alpha \tag{9}$$

The moment of inertia is given by Eq. 10.

$$I = \frac{M\left(R_1^2 + R_2^2\right)}{2} \tag{10}$$

As R_2 is variable as the coil is being unwound, moment of inertia is then given by Eq. 11.

$$I = \frac{M\left[R_1^2 + (R_2 - dN)^2\right]}{2} \tag{11}$$

Since mass is the product of density by yarn volume and volume is the product of the base area by length, then the variable mass can be defined as a function of the number of cycle N.

$$M = \rho\pi L[(R_2 - dN)^2 - R_1^2] \tag{12}$$

The moment of inertia is then found as a function of the number of cycles by substituting Eq. 12 in Eq. 11.

$$I = \frac{\rho\pi L[(R_2 - dN)^2 - R_1^2][R_1^2 + (R_2 - dN)^2]}{2} \tag{13}$$

Substituting Eqs. 8 and 13 in Eq. 9, leads to the equation that represents the torque as a function of the number of cycles N.

$$T = \frac{\pi^2 n^2 \rho L[(R_2 - dN)^2 - R_1^2][R_1^2 + (R_2 - dN)^2]}{3600 N} \tag{14}$$

Below is presented the mathematical model for the winding of textile yarn. As for the winding process, with each cycle N traversed, the internal radius of the coil R_1 is increased by the value of the yarn diameter. Thus Eq. 15 determines the rotation as a function of the cycle traveled [4].

$$n = \frac{60v}{2\pi(R_1 + dN)} \tag{15}$$

At each complete cycle the external radius R_2 increases the value of a diameter d, so the moment of inertia will be represented accordingly to Eq. 16.

$$I = \frac{M\left[R_1^2 + (R_1 + dN)^2\right]}{2} \tag{16}$$

The mass of the coil also increases as the textile yarn is wound. It is defined as a function of the number of cycles N, according to Eq. 17.

$$M = \rho\pi L[(R^1 + dN)^2 - R_1^2] \tag{17}$$

Substituting Eq. 17 into Eq. 16 leads to the Eq. 18, which represents the moment of inertia of the coil as a function of the number of cycles N.

$$I = \frac{\rho\pi L[(R_1 + dN)^2 - R_1^2][R_1^2 + (R_1 + dN)^2]}{2} \tag{18}$$

Equation 19 represents the angular acceleration of the coil as a function of the number of cycles N performed.

$$|\alpha| = \frac{\pi n^2}{1800\left(\frac{R_2 - R_1}{d} - N\right)} \tag{19}$$

Substituting Eqs. 18 and 19 in Eq. 9 leads to Eq. 20. This equation represents the torque as a function of the number of cycles N.

$$T = \frac{\pi^2 n^2 \rho L[(R_1 - dN)^2 - R_1^2][R_1^2 + (R_1 + dN)^2]}{3600\left(\frac{R_2 - R_1}{d} - N\right)} \tag{20}$$

3.5 Yarn Tensioning System

During winding, it's necessary to impart proper tension to the running yarn, so that it's stable during image capture and a quality package can be achieved. This tension shouldn't surpass 1/8 of yarn breaking strength so that its properties remain unchanged.

The most common method of imparting tension to the yarn relies on applying friction to a running yarn. In this domain there are four variants of yarn tensioning available: Capstan method, Additive method, Combined method, and Automatic method.

Additionally, tension could also be applied through surface speed control of two positive feed systems. By wrapping yarn around a motorized feed roll, any yarn tension fluctuation from the unwinding process is absorbed by the feed roll. If the yarn is then passed through a second motorized feed roll, the difference in surface speed of the two feed systems will allow for effective tension regulation [3].

4 Developed Prototype

For the practical part of the project a model of the unwinding and winding system was developed. The model was developed in a CAD software, so that in the future it could be applied to the textile yarn quality verification system (see in Fig. 9 and Fig. 10).

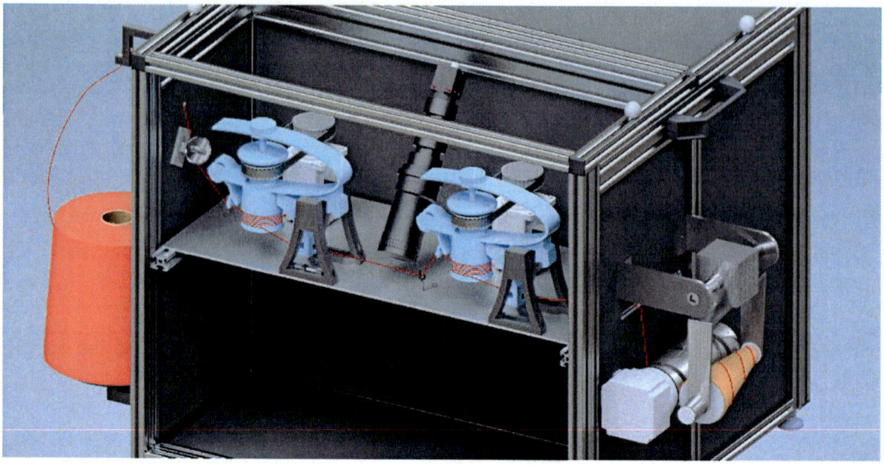

Fig. 9. Textile yarn unwinding and winding system's prototype (open view).

This prototype is formed by three subsystems, the yarn unwinding, guidance control, and winding systems. The unwinding system is based on the over end withdrawal technique described in Sect. 3.3. The guidance control system is the region where the image will be captured. This section consists of two positive yarn feeders that assure an effective yarn velocity, position, and tension control, essential for good quality image capture as described in Sect. 3.5. The winding system is designed for random winding, described in Sect. 3.2.

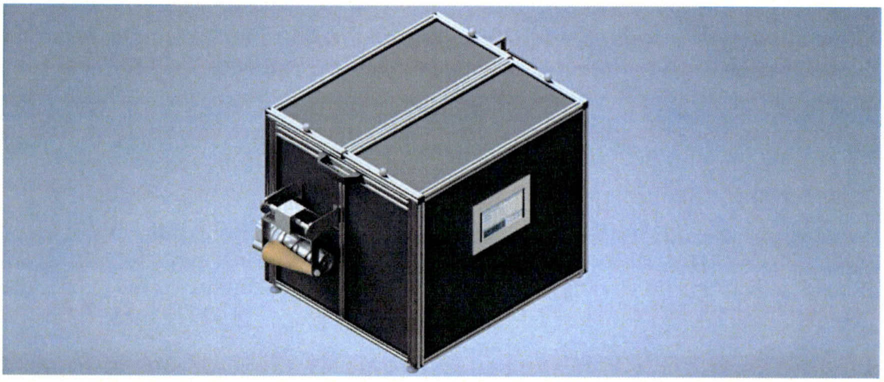

Fig. 10. Textile yarn unwinding and winding system's prototype (closed view).

5 Prototype System Validation

To properly evaluate the correct function of the system some parameters must be defined according to the objectives first identified in Sect. 2.1. For the objective of minimizing speed variation, the tolerance was set at 5% of the defined velocity. And to avoid oscillations yarn positioning must not deviate more than 1 mm from its defined path. These two criteria shall be evaluated using high frame rate image capture with dimensional references.

To meet the safety objective, and therefore not break the yarn or change its properties the maximum yarn tension should not surpass 1/8 of the yarn breaking strength, as it's explained in the Sect. 3.5. This shall be measured using a yarn tension meter.

Additionally, to confer uniform yarn distribution by coil objective some package properties such as density, evenness of yarn distribution and stability must match current industry standards.

The final objectives for package support sizing should ensure a correct fit for the selected cone sizing range.

This way, companies will be able for improving their performance and capacity for answering to issues related with suppliers, as well as issues related with clients [21].

6 Final Remarks

In this scope was possible to understand the effect of various parameters on yarn unwinding, positioning, and winding, that are essential for the system design success. The unwinding process was found to be dependent on package cone taper, unwinding velocity, yarn balloon and twist control. The image capture and wound package quality are also dependent on tension control.

Seeing as this work's three subsystems all exist and are present, separate or otherwise, in other machines, it stands to reason that the combination of all three has good probability of success, as they have all been previously proven.

Finally, the prototype's simplicity, portability and inexpensiveness when compared to similar systems in the market like USTER TESTER 6, which are heavy, complex, and expensive machines shows a promising opportunity for this system to generalize access to yarn quality control in textile industry.

Acknowledgements. This work has been supported by FCT – Fundação para a Ciência e Tecnologia within the R&D Units Project Scope: UIDP/04077/2020 and UIDB/04077/2020.

References

1. Barbosa, A.S.: Robotized Fiber Application with Quick Manipulator Programming and Quality Control. Integrated Masters Dissertation, University of Minho, Portugal (2016)
2. Cooray, T., Fernando, E.: Mathematical modeling of over-end yarn withdrawl, and the development/design of a device for uniform unwinding tension. In: 85th Textile Institute World Conference, Colombo, Srilanka (2007)
3. Koranne, M.: Fundamentals of Yarn Winding, 1st edn. Woodhead Publishing India in Textiles (WPI), New Delhi (2013)
4. Pušnik, N., Praček, S.: The effect of winding angle on unwinding yarn. Trans. FAMENA **40**, 29–42 (2016)

5. Bhattacharya, S.: Value Added Textile Yarns-Manufacturing Techniques and Its Uses, 1st edn. OM SAITECH Books, India (2020)
6. Banerjee, P.: Principles of Fabric Formation, 1st edn. CRC Press, New York (2015)
7. Praček, S., Možina, K., Sluga, F.: Yarn motion during unwinding from packages. In: Troch, I. (ed.) Mathematical and Computer Modelling of Dynamical Systems, vol. 18, pp. 553–569. Taylor & Francis (2012)
8. Kyosev, Y.: Braiding Technology for Textiles. 1st edn. Woodhead Publishing (2015)
9. Praček, S., Pušnik, N.: Simulations of Yarn Unwinding from Packages. In: Uddin, F. (ed.) Textile Manufacturing Processes, IntechOpen (2019)
10. Nuriyev, M.N., Veliyev, F.A., Hamidov, H.I., Sailov, Mahamad-Seydaliyev, I., Jabbarova, G.Z.: Development of a device for continuous monitoring parameters of the winding structure of textile bobbins. In: Ingeniería Solidaria, vol. 14, no. 24, p. 10 (2018)
11. Patent No. cn 201410607590. "Control System for Textile Winding Machine Convenient for Loading of Bobbin." ipc: B65H54/547, B65H63/00, b65H67/04. Li L. – No. CN104386539 A (2015)
12. Nuriyev, M., Seydaliyev, I., Recebov, I., Dadashova, K., Musayeva, T.: Determining the dependences for calculating a conversion scale of profile height of the controlled packing surface. Eastern-Eur. J. Enterp. Technol. **2, 1**(86), 58–62 (2017)
13. Maag, F.: Spinnspulen mit der Stufenpräzissionswicklung. Chemiefasern/Textilindustrie **6**, 416–420 (1985)
14. Nuriyev, M.N., Musayeva, T.T.: Development of algorithms surface recognition forging cross winding. Bull. NTU "KhPI" Ser. Mech. Technol. Syst. Complexes **49**, 52–55 (2016)
15. Nuriyev, M.N., Imanova, G.M.: Mechanical and Optical methods and means of controlling geometric sizes and shapes of textile packages. Bull. Sci. Pract. **5**, 65–74 (2016)
16. Pinto, R., Pereira, F., Carvalho, V., Soares, F., Vasconcelos, R.: Yarn linear mass determination using image processing: first insights. In: IECON 2019 - 45th Annual Conference of the IEEE Industrial Electronics Society, Portugal, pp. 198–203. IEEE (2019)
17. Carvalho, V., Cardoso, P., Belsley, M., Vasconcelos, R., Soares, F.: Yarn diameter measurements using coherent optical signal processing. IEEE Sensors J. **8**, 1785–1793 (2008)
18. Pereira, F., Carvalho, V., Soares, F. O., Vasconcelos, R. M., Machado, J.: Computer vision techniques for detecting yarn defects. In: Wong, C. (ed.) Appl. Comput. Vis. Fashion Textiles, Amsterdam, vol. 22, pp. 123–145. Elsevier (2018)
19. Mendes, A.F.: Design and Development of a Monitoring and Control System for a Multiaxial Loom. Master Thesis (2012–2013)
20. NPTEL. https://nptel.ac.in. Accessed 22 Jun 2019
21. Gangala, C., Modi, M., Manupati, V.K., Varela, M.L.R., Machado, J., Trojanowska, J.: Cycle time reduction in deck roller assembly production unit with value stream mapping analysis. In: Rocha, Á., Correia, A.M., Adeli, H., Reis, L.P., Costanzo, S. (eds.) WorldCIST 2017. AISC, vol. 571, pp. 509–518. Springer, Cham (2017). https://doi.org/10.1007/978-3-319-56541-5_52

TAB-Med: Automated Pill Dispenser in Residential Environments

Nuno Fernandes[1](✉), Ana Rita Amorim[1], Bárbara Silva[1], Joana Freitas[1], and João Pedro Mendonça[2]

[1] University of Minho, 4710-057 Braga, Portugal
a74165@alunos.uminho.pt
[2] MEtRICs Research Center, University of Minho, 4800-058 Guimarães, Portugal

Abstract. This paper presents an intelligent system capable of dispensing pills after the programming of the daily doses by the user. This system has been studied and designed for elderly people and their caregiver. In order to add value to the product, the equipment has an improved recharge system and user interface. Thus, it uses trays, and each one of them takes a different type of medication, taking advantage of an easy-to-use touch screen for the programming of the doses, a buzzer for sound signaling and employing a weight sensor to detect the presence of the collection cup. Alongside that, the equipment has a communication system to warn the caregiver if the patient hasn't taken their pills, if it is necessary to reload the trays or even if the patient forgot to put back the cup.

Keywords: Pills dispenser · Medication management · Intelligent system · Healthcare

1 Introduction

1.1 Motivation

Polypharmacy – the act of consuming various types of medication – carries extremely serious risks and consequences for the patient, being responsible for many internments. Due to the development of the medical industry, populational ageing is becoming a concerning problem in the XXI century. In Portugal, the number of elderly people is increasing, while the young population is decreasing. As such it is expected that by 2080 there will be 317 elders per 100 teenagers [1–3].

The medical community associates this phenomenon with memory loss, body adaption difficulty, psychological deterioration, cognitive struggle and a greater susceptibility to diseases. Older adults with intellectual disability consume 10 or more medicines. To fight these tendencies, there is a greater consumption of medication by this populational group [2–4].

Considering the lack of independency of the elder populace, they commonly mix their medication, take it with the wrong dosage, disrespect the time schedule or take it in excess, reproducing unwanted effects. This non-compliance of the patient not only

J. Machado et al. (Eds.): icieng 2021, LNME, pp. 359–370, 2022.
https://doi.org/10.1007/978-3-030-79165-0_34

becomes a problem for them, but also for the doctor, which might cause him to wrongfully question the treatment applied to the patient. As such, devices that manage the dosage are becoming increasingly more popular, since they aim to organize and facilitate medication ingestion, giving back some autonomy and tranquility to the elders [2, 5, 6].

1.2 State of the Art

Medication management devices come with many features and span a wide range of prices. The simplest type of these devices is a manual organizing medication box (Fig. 1–a). They differ in form, shape and compartments number. Their design provides them with ease of use however, it only addresses medication storage, not alerting the user when the time of ingestion comes, and it requires a large amount of time to be prepared. Since each compartment contains a dose (which contains various types of medication), mixing pills among them is a common issue [6].

Medelert (Fig. 1-b) is an example of a variation of the manual pill box with an alarming system. It has a cylinder shape with 28 compartments, holding up to 18 aspirin sized pills. It can be programmed for up to 4 doses a day, 7 days a week, with a visual and sound alarm that lasts up to 30 min [7].

The Philips Medication Dispenser Resource Center (Fig. 1–c) takes a new approach to medication management. The user must put each dosage in a covered plastic cup and insert it into the device. It has a capacity for 60 doses and can be programmed to deliver them six times a day. It can also exhibit an alarm for medication that is not stored on the device. Additionally, it may also be synchronized with a cell phone, alerting if one of the doses was not taken, or if the user is away to take it in the scheduled time [8].

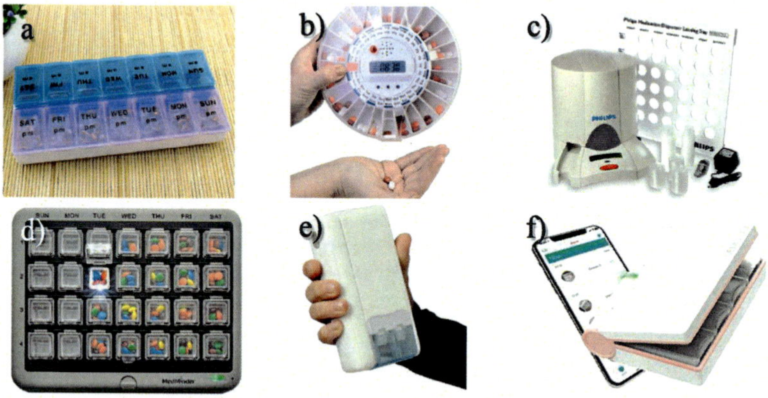

Fig. 1. Most used medication management devices: a – manual organizing medication box, b - med e-lert, c - Philips Medication Dispenser Resource Center, d – MedMinder, e – Medimi, f - Memo Box.

MedMinder™ rather than just providing a device, developed a service. The device itself (Fig. 1–d) is a box with 28 compartments that can store up to 4 doses a day, for 7 days. It has got a simple interface, without screens or digital buttons, having the capacity for sound and luminous alerts. However, the programming and communications are handled by the company, giving the user only the responsibility of filling the trays. There is also an option to ask the company to fill the trays [9].

Medimi (Fig. 1–e), was a device developed with portability in mind. It can withdraw pills up to 15 times a day but can only hold two types of pills. It has been embedded with vibration, sound and luminous alarm [10].

Lastly, another portable solution is Memo Box (Fig. 1-f). Its functionalities are managed through a smartphone app, in which the user can configure how the alarms will work and manage information about the medication. The device itself is made up of a pack of seven boxes, one for each day of the week, with 2–4 compartments each. It's also equipped with visual and sound alarms that can repeat from 10 to 10 min and can have a duration of up to 90 min [11].

2 System Architecture

After analyzing the available equipment on the market, its characteristics, settling on the target demographic, and in accordance with the goals subsequently established, an equipment was created from scratch based on some functionalities of the existing equipment, but simpler and more practical to facilitate its operation by the users [12].

Thus, to increase the capacity of stored pills and facilitate the recharging and preparation, a device with 10 circular trays was designed, in which each tray is responsible for storing one type of pill. The same tray has 15 individual compartments so that each pill is individualized and there is no interference with other pills.

The device will have an interface where the user in question will schedule when each pill is taken. When the time of each dose comes, the device will activate a beep that will notify the user. The interface will have a warning system that will tell the user to place the cup in the outlet of the equipment. When it is placed, it will release the pills of each tray programmed at that hour. If the user does not take the medication, a message will be sent to the responsible person's cell phone.

Figure 2 shows the internal block diagram of the device. At an electronic level, it is made up of the following blocks: power module, processing unit, communication module, user interface and mechanical system.

2.1 Processing Unit

The processing unit is the module responsible for managing the entire device, having the features and functionalities necessary for the device to operate under the specified conditions. The requirements for this module are to have enough memory to store all the necessary data, to have the processing capacity needed to handle the various real-time situations, and to have a real-time clock unit (RTC) [13].

The microcontroller Raspberry Pi 3 is thought to be ideal for this type of operations, since it has a 64-bit 4-core processor, operates at a frequency of 1.2 GHz, has 1 GB of RAM, a 3D VideoCore IV graphics processor, 4 USB ports, 40-pin GPIO, a full HDMI port, low power Bluetooth, slot for a micro-SD card, WiFi module, Display Interface (DSI), among other features, all packed in a space the size of a credit card [18, 19]. To work in this specific application, Raspberry Pi 3 must have additionally integrated an 8 GB SD card, a 2 A micro-USB power supply and a DSI tape cable (to connect the tablet to the microprocessor) [14].

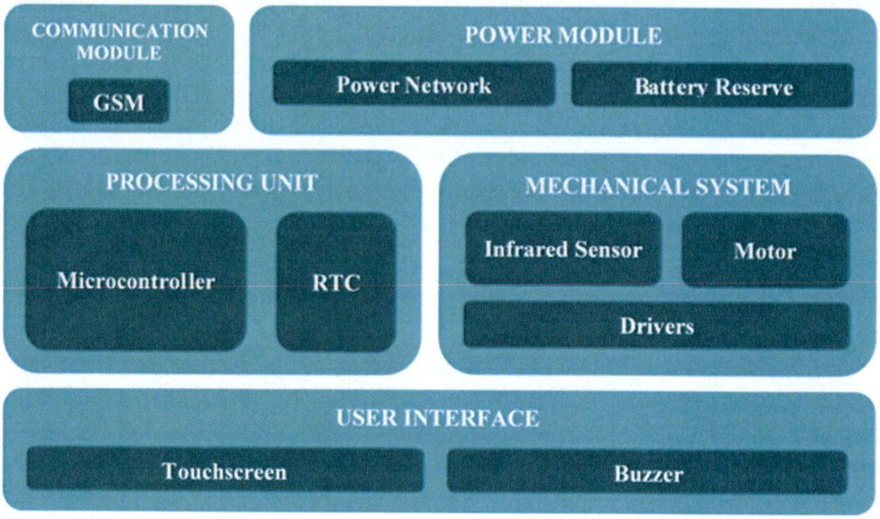

Fig. 2. Device block diagram.

For storing and providing complete data information such as day of the week, day of the month, month and year a RTC was selected. The model RTC DS3231 can accurately track hours, minutes and seconds (in formats of 12 or 24 h) and it can guarantee the release of a certain drug occurs at the correct time. The adjustment of months of less than 31 days and leap years is done automatically [15].

2.2 Communication Module

The communication module has the function of allowing wireless communication between the device and the person responsible for supervising the pills dosage.

In this project, it is meant to allow communication between the control board of the device and the mobile phone of the person in charge. For this purpose, GSM communication was chosen, which does not require user configurations, it is wireless and does not require the user to have any infrastructure beyond the power supply [16].

Thus, the SIEMENS TC35 board was chosen which is a GSM module based on standard AT commands via serial port (RS232 / TTL communication) [11].

2.3 User Interface (UI)

The user interface can be considered one of the most important components of a device since it is responsible for bridging the user and the designed system.

To ensure correct functioning of the various modules that work in conjunction with the microcontroller, it must have a correct firmware and drivers. The firmware is a low-level programming that allows the control of the hardware, being responsible for bridging the drivers and the hardware. The device driver forms a software interface that allows the user to transmit their instructions [17]. All modules have been chosen with a firmware compatible with Raspberry Pi and drivers for easy programming.

The input and display of data by the user is designed to be done through a touch screen. The screen chosen was the Raspberry Pi 7 "touchscreen produced by the Raspberry Pi Foundation.

In this screen, the user will schedule the frequency of each tablet. In Fig. 3 there is a schematic which was the most simple and intuitive interface.

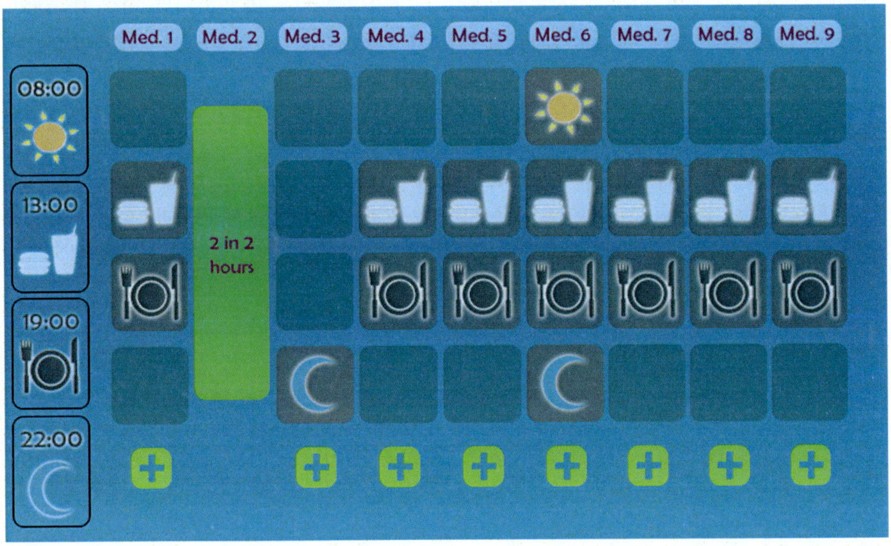

Fig. 3. User Interface Proposal: On the left, the user can choose the most common times of the day in which he must consume his pills and associate a symbol at that time; In the grid on the right, the user can choose which of these times should be scheduled for each pill. If the user must take medication at an irregular time, he can schedule the time by clicking "+ ".

When the time indicated for the pill intake arrives, a beep will sound. This sonorous signal will be performed using a buzzer. A buzzer is an electrical audio signaling device that can be mechanical, electromechanical or piezoelectric. It was decided to use a mechanical buzzer which uses a magnet to quickly move a pick, emitting a lower pitch buzz. Since many people using the equipment may be hearing impaired, a buzzer with a low sound frequency should be preferred.

After the audible signal, the user should place a cup into the pill release zone. In the area where the cup will be placed, a weight sensor will be present, which will detect the presence of the cup. It was considered that the HX711 weight sensor would be ideal, having to be calibrated for a certain weight sensitivity, considering that a calibration up to 5 kg would be enough for the intended purpose [18]. Thus, after the cup is detected, the pills are dispensed. After all pills fall, another beep sounds until the weight sensor stops feeling the weight of the cup.

2.4 Mechanical System

To guarantee that the trays achieve the desired movement, it is necessary to use motors. They are an electromechanical device that converts electrical pulses to torque, which generates discrete angular variations [19]. Since the trays need to perform a circular and precise movement, stepper motors were used. Each tray contains different types of pills, and as such, the movement each one performs must be independent and programmed for distinct times, thus it is necessary to use an individual pitch motor for each.

To choose the appropriate stepper motor, the Torque (τ, in Nm) required to rotate a compartment corresponding to an angle of 24° (= 360° (1 turn) / 15 compartments) calculation was carried out according to (1).

$$\tau = \mathrm{rFsin}\theta = \mathrm{Isin}\theta = \mathrm{mr}^2\mathrm{sin}\theta \qquad (1)$$

In (1), I represents the moment of inertia given by multiplying the mass of the tray (m), in kilograms, by its radius (r), in meters, and θ is the angle per compartment, in degrees.

Thus, knowing the radius of the tray is 0.123 m and its mass is approximately 1.012 kg (estimated by the Inventor software), a torque of 0.00627 Nm was obtained.

After a market analysis, the NEMA 23 57 mm 3 phase 1.2° hybrid stepper motor was chosen, the model being 57BYGH350A.

To activate a stepper motor, a specific hardware is required, i.e. a driver, which acts as an intermediary of the pulse signal output from the controller to the motor.

Still within the mechanical system, to detect the drop of the pills, sensors were placed at the end of the collection funnel. For the short-range sensor, the best solution is the one that makes use of infrared radiation (IR). The use of IR sensors is especially recommended when working in a noisy environment, where it is difficult to implement other systems. This type of sensor can count on inexpensive transmitters and receivers and does not require complex circuits or special adjustments, thus facilitating the elaboration of projects [20]. In this project it was decided to opt for an interrupt detection sensor (photo-interrupter) whose method of operation consists of an IR radiation emitter and a receiver installed in the same direction with opposite senses and thus the actuators only come into action when the infrared signal is interrupted by some object.

2.5 Power Module

The microcontroller uses as a source of energy a DC current with 5 V in a micro-USB input. The other modules also require power, which does not need to come fully from the microcontroller. Therefore, one more module was designed to ensure that all components receive the correct power.

For this project, the domestic power network is intended to be used as the main power source. As the voltage in the Portuguese network is alternated and has between 220 to 240 V with a frequency of 50 Hz, it is necessary to use an AC/DC transformer. This transformer not only changes the alternating current to continuous but can also limit the voltage.

A backup battery should be installed to ensure that, in the event of a power failure, this device can operate autonomously for some time. However, to perform its function, it is necessary that it be charged by the current, and it is necessary that this module also has this capacity.

The representative block diagram of the feed module is shown in Fig. 4.

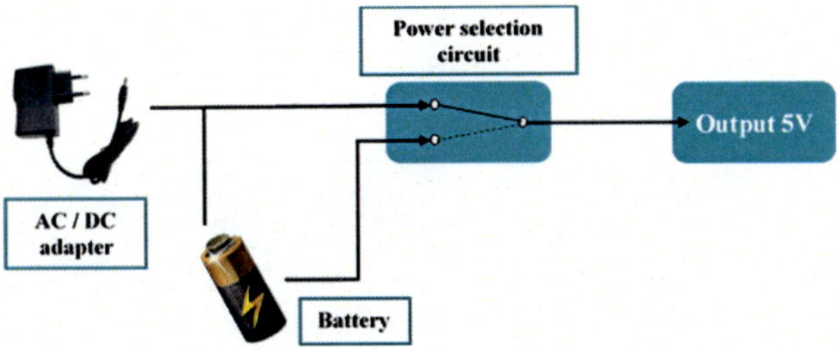

Fig. 4. Block diagram of the power module.

3 Final Product

3.1 Product Design

The design of the final product involves the conception of all components, from the design of the interior of the equipment to the design of the exterior. The equipment exterior is important since it is the perception the user will have of the system. The interior, although not visible, is what allows the equipment to behave properly.

In the exterior (Fig. 5), the device TAB-med features a hinged lid that allows the user, or the person responsible, to open the device and access the trays to load them with the proper pills. The device also features a touch screen, which allows the user to select the daily doses of medication and the respective times of the doses. For the collection of the pills, there is a platform where a cup will be placed, and the pills will be discharged.

The material selected for the device exterior construction was polypropylene (PP) since it is a low-cost material, has chemical resistance, is easy to mould, non-toxic, has low density and low moisture absorption [21].

The equipment interior is made up by the electrical devices mentioned in the previous section, by the trays, its compartments, its mechanical elements (essential for the system's mechanical function), and lastly by other mechanical components that support and unite all the equipment parts.

This system has 10 circular trays, each with a different type of pill. To avoid interactions and reactions between different pills, each one is stored individually. Each tray has 15 individual compartments that can hold up inside a 30 mm sphere. Figure 6 depicts the prototype of the trays drawn using the Autodesk Inventor software.

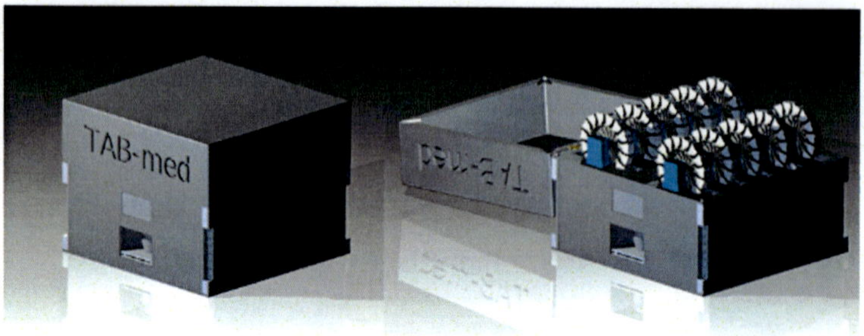

Fig. 5. TAB-med exterior.

Fig. 6. Tray prototype of the equipment.

According to Infarmed and the European Medicines Agency, the environmental conditions and the materials in contact with the pills are important aspects that must be considered. In the present project, the material selected for the compartments is in direct contact with the pills and is where they will be stored. As such, the material commonly associated with this condition is polyvinyl chloride (PVC) [23] [24]. The material selected for the remaining parts the tray was polypropylene (PP).

For the pill delivery system, an extension spring (shown in Fig. 6, on the right) was used. This spring is the driving force of the pill mechanism. The mechanism of the tray keeps the spring at a high-tension point. When the brake is released (to release the pill) the spring returns to its normal state, reducing the tension. The support of the extension spring must be made of a resistant material. Thus, the metal selected to constitute this piece was stainless steel 304. This metal has high resistance to oxidation and corrosion [22].

Regarding the opening of the compartment doors, a torsion spring was used. This type of spring was selected because it is intended for it to create tension when the door is closed. This tendency will be counteracted by a brake, which will be positioned at the opposite end of where the spring is positioned. When the brake is released, the energy stored in the spring causes the door to open.

Given that each tray has 15 compartments and there are 10 trays, 150 extension springs and 150 torsion springs are required.

The brake has the function of preventing the compartments from opening. The stainless steel 304 was also selected as the constituent material.

On the brake there is a protrusion turned to the outside of the trays. This protrusion will allow the brake to move when it touches another part during the rotational movement of the tray and thus the compartment opens.

There is a collection funnel located beneath all the trays whose function is simply to direct all the pills to the cup, as displayed in Fig. 7.

Fig. 7. Funnel responsible for the pill collection.

In addition to the elements mentioned above, there are other essential components which will allow the connection and operation of the parts, namely the screws and the hinges.

3.2 Mechanical Behaviour of the System

Apart from the electronical components, the mechanical design for this project is an essential part of its functioning, being responsible for the release of the medication. As such, the mechanics can be divided into four steps:

1. After the trays assigned for pill release at that time are identified, their stepper motors will be sequentially activated. Their motors will rotate 24° (the angular distance between each compartment).
2. Each compartment is closed by a brake that prevents its lid from opening. This brake will be released by a piece that allocates itself on the position of the pill release. Since the extension spring is in a high-tension state when the compartment is closed, once it opens, the pill inside is expelled outside.

3. The released pill is collected in a hopper that covers all the trays, located immediately below.
4. The hopper directs all the pills to a platform that contains the cup placed by the user.

3.3 Break Release System

To remove the break from the lid of the compartment a mechanical detachment platform was conceived. For this mechanism to work it is important to note that the brake has a cylinder structure that extrudes itself from its body. These structures are represented in Fig. 8 as yellow dots. Their outer trajectory is described in it as well with a blue dashed line. On the pill release zone, there is a structure (represented by a white dashed line) that disturbs this trajectory, causing the brake to move. Its movement will cause the compartment lid to open, since the torsion spring will go to its relaxed position. After the lid has opened, the compartment will then slide and release the pill inside, since the extension spring will go its relaxed form.

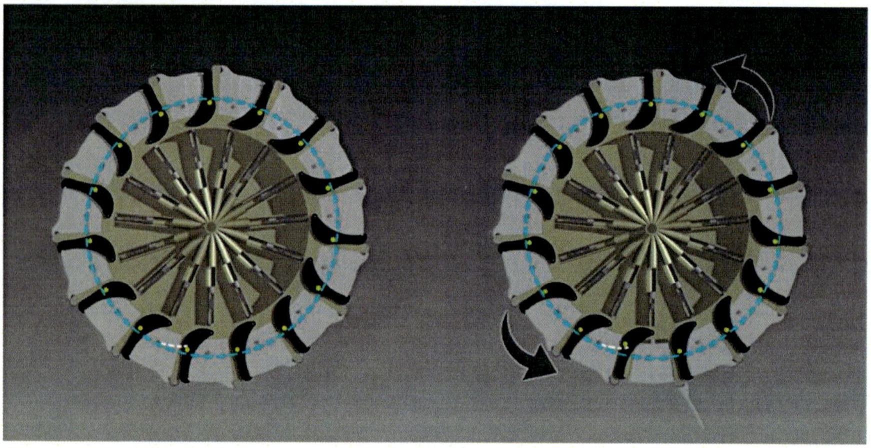

Fig. 8. Depiction of the break mechanism.

3.4 Algorithm

To execute all the tasks in the designed project, two algorithms were conceived. To guarantee their intended purpose it is advised to create an object for each type of pill. This object could be classified according to its intake schedule (if it is supposed to be one of the predefined times or a time interval). If it is supposed to be ingested in one of the predefined times, it should keep in its attributes which of them were selected. If it stands on the latter case, the first intake time and time interval should be kept in its attributes. Lastly, there should be an attribute that keeps track of how many times a type of pill has been released. The user will program and keep track of this through the U.I.

The first algorithm is conceived to oversee the medication time. Its main functionality is to trigger the pill release at the right time. Firstly, it is necessary to iteratively check the type of schedule each pill has. If it is scheduled to be in one of the predefined day times, a simple verification with the RTC module should suffice. If that is not the case, it is implied that the medication in case as an irregular time. In this case, the time elapsed between the first intake and the current time is calculated. That time is then divided by the time intervals in which the pill should be taken. If the result is an integer, it means that the time to release a pill of that medication has come. In both cases, if they return that it is time to take a pill, the pill trigger algorithm should be triggered. Since the Raspberry Pi has enough processing power, executing this algorithm once every minute should suffice.

The last algorithm is responsible for the pill release. The first step is to ring the buzzer. If the buzzer keeps ringing for an hour, a message is sent to the responsible person. The buzzer will stop ringing once the user places the cup in the designated place. After that, all the trays marked for discharge will be iterated. In each iteration, the current tray will rotate 24°. Once it is rotated, the number of times that pill has been release will be increased. If that number is greater than 10, an alert will show up in U.I. saying that that type of medication is almost over. If the IR sensor did not detect the pill release, a new rotation will occur, otherwise, the next iteration will occur. After all the iterations have occurred, the buzzer will ring until the user lifts the cup. Once the cup is lifted, the algorithm ends.

4 Conclusion

As aforementioned, this project had the main objective of developing a device that solves the polypharmacy management issue. Our prototype has the capacity to deliver the right type of medication at the right time with sound signalling. Not only that, there is also an alarm system that lets the responsible person know for the lack of medication. Even though all the components and systems necessary were thought of, some elements still need a more in-depth approach, namely the electronic systems.

Just like any other device can be improved, so can TAB-med. One of the biggest flaws it has is the space it occupies and its weight. To improve these aspects, some components should be reduced, like the number of springs or screws, using hollow walls or using other types of materials.

Acknowledgments. This work has been supported by FCT – Fundação para a Ciência e Tecnologia within the R&D Units Project Scope: UIDP/04077/2020 and UIDB/04077/2020.

References

1. Lima, F., Coimbra, C., Zilhão, M.: Mantém-se o agravamento do envelhecimento demográfico, em Portugal, que só tenderá a estabilizar daqui a cerca de 40 anos. Statistics Portugal (2017)
2. Problemas Relacionados com Medicamento (PRM) no Idoso. ACSS, Portugal (2008)
3. Masnoon, N., Shakib, S., Kalisch-Ellett, L., Caughey, G. E.: What is polypharmacy? a systematic review of definitions. BMC Geriatr. **17**(230), 1–10 (2017)

4. Aiolfi, C.R., Alvarenga, M.R.M., Moura, C.S., Renovato, R.D.: Adesão ao uso de medicamentos entre idosos hipertensos. Rev Bras Geriatr Gerontol **18**, 397–404 (2015)
5. Vinagres, N.: Desenvolvimento de um Dispensador Automático de Medicamentos. Instituto Superior de Engenharia de Coimbra (2013)
6. GMS, MED-E-LERT 28 Day Automatic Pill Dispenser. http://www.medelert.net/. Accessed 04 May 2018
7. Philips Lifeline, Automated Medication Dispensing Service. https://www.lifeline.philips.com/pill-dispenser/health-mdp.html. Accessed 05 May 2018
8. MedMinder, Maya pill dispenser. https://www.medminder.com/pill-dispensers-2/maya-pill-dispenser/. Accessed 08 May 2018
9. e-pill Medication Reminders, Portable Automatic Pill Dispenser e-pill medimi. https://www.epill.com/medimi.html. Accessed 05 May 2018
10. Tinylogics, Memo Box Smart PillBox. https://pillbox.tinylogics.com/. Accessed 07 May 2018
11. Barros, C., Leão, C.P., Soares, F., Minas, G., Machado, J.: RePhyS: a multidisciplinary experience in remote physiological systems laboratory. Int. J. Online Eng. **9** (SPL.ISSUE5), 21–24 (2013). https://doi.org/10.3991/ijoe.v9iS5.2756
12. Raspeberry Pi Learning Resources, Raspeberry Pi. https://www.raspberrypi.org/learning/hardware-guide/components/raspberry-pi/. Accessed 15 May 2018
13. Raspeberry Pi Foundation, About us. https://www.raspberrypi.org/about/. Accessed 15 May 2018
14. [Digi-Key Eletronics, Clock/Timing - Real time Clocks]. https://www.digikAvailableey.pt/products/en/integrated-circuits-ics/clock-timing-real-timeclocks/690?&utm_adgroup=Integrated+Circuits&mkwid=sitjk7rX4&pcrid=216003843098&pkw=%2Breal%20%2Btime%20%2Bclock&pmt=b&pdv=c&productid=&slid=&gclid=EAIaIQobChMIqLyrjaON2wIVTflRCh1o6gv0EAAYASAAEgJ7s_D_BwE. Accessed 15 May 2018
15. BoxEletrónica, GSM TC35 SIEMENS, https://www.boxelectronica.com/pt/gsm-gprs-gps/177-gsm-tc35-siemens.html. Accessed 20 May 2018
16. Raspeberry Pi, How to assemble the RPi touchscreen display. https://www.raspberrypi.org/forums/viewtopic.php?t=178443. Accessed 20 May 2018
17. Brites, F.G., de Almeida Santos, V.P.: Motor de Passo. Niterói. Universidade Federal Fluminense (2008)
18. RoboCore. Controlando motores: Motor de passo. https://www.robocore.net/tutorials/97.html. Accessed 19 May 2018
19. Sensor infravermelho – Funcionamento e Aplicações. Saber Elétrica. https://www.sabereletrica.com.br/sensor-infravermelho-funcionamento/. Accessed 21 May 2018
20. Linden, D.: CHAPTER 1 - Basic Concepts. McGraw-Hill, Handbook of batteries -Third Edition (2002)
21. Os Tipos de Aço Inox e Inoxidável - Conheça eles. Arinox - Distribuidora de Aço Inox. http://arinox.com.br/blog/os-tipos-de-aco-inox/. Accessed 23 May 2018
22. Embalagens de medicamentos: quais são os tipos existentes e os descartes possíveis?. http://eco4planet.com/blog/embalagens-de-medicamentos-quais-sao-os-tipos-existentes-e-os-descartes-possiveis/. Accessed 20 May 2018
23. Plastic primary packaging materials - European Medicines Agency - Orphan designation - Orphan incentives. http://www.ema.europa.eu/ema/index.jsp?curl=pages/regulation/general/general_content_000743.jsp&mid=WC0b01ac05800294ab. Accessed 21 May 2018

The Damping System in Crutches: Development of New Model

Rita Pereira[1], Amanda Carvalho[1], Vânia Costa[1], Joel Galvão[1,2], and Ana Matos[1(✉)]

[1] School of Engineering, University of Minho, 4800-058 Guimarães, Portugal
[2] METRICs Research Center, University of Minho, 4800-058 Guimarães, Portugal

Abstract. The damping systems design is a fundamental feature in a crutch because it decreases the forces in the upper members of the users and consequently reduces the chance of eventual injuries coming up. However, the damping system present in the currently marketable crutches cause discomfort to their users. Thus, the main object of this project was to develop a new damping system approach in order to minimize the problem previously stated.

Initially, a study of the existing crutches was carried out and then, a biomechanical characterization of the gait assisted by crutches was performed. Considering the results of the gait analysis and the morphology of the crutch, a hydraulic damping system was scaled and designed in the Inventor software for a specific target audience which was settled to include children between the ages of 5 and 10. This target audience was selected because usually, children don't have the required strength to effectively operate with this kind of devices and they adapt more easily to new ones.

Keywords: Assisted gait · Biomechanics · Crutch · Hydraulic damping system · 3D CAD modelling

1 Introduction

To improve the mobility of individuals with disabilities, it has been used, over the years, several devices known as mobility aids. Also, some dedicated devices are developed in the context of rehabilitation [1]. The most well-known are the crutches but it may be alternatively used other devices like walkers and walking sticks. Generally, all these devices improve the balance of their users as well as reduce the pain in their lower members since most of the load can be removed. However, they require a huge muscle strength in the upper members [2]. Therefore, it becomes important to evaluate certain parameters of each patient, like age and muscle strength, to ensure that the equipment is the most suitable.

After discussion with specialized professionals, it was concluded that the crutches are incommodious and uncomfortable since most users complain about these devices. The source of the problem is the damping system currently used on crutches.

Thus, the main goal of this work is to develop a new damping system model that ensure the comfort of the patients using crutches. The selected target audience includes

J. Machado et al. (Eds.): icieng 2021, LNME, pp. 371–387, 2022.
https://doi.org/10.1007/978-3-030-79165-0_35

children between 5 and 10 years old since they do not have the required strength to operate the device. Also, they are young which increases the adaptation capability to a new device.

2 Methodology

To carry out this project, extensive research was performed and as an outcome, it was selected the most relevant articles.

In the first stage, the literature review was assessed in order to verify the types of dampers currently used in crutches.

Afterwards, scientific articles and anthropometric studies were used to analyze gait biomechanics. In the last part of this project, Inventor, a 3D design software, was used to design a 3D model for the conceptual damping system based on previous studies and calculations.

3 Crutches

There are two main types of crutches (see Fig. 1). Axilla or underarm crutches (Fig. 1 (a)) are placed beneath the axillae and they present a typical structure including a padded axilla bar and a handpiece. They are cost-effective however they can cause pain in the axilla region which in a long term can lead to the appearance of a specific paralysis called crutch palsy. On the other hand, forearm crutches (see Fig. 1 (b)) present a forearm cuff and a padded handpiece and, therefore, they are more comfortable. Moreover, several authors considerer that this type of crutches offers more stability than the axilla crutches [3–5].

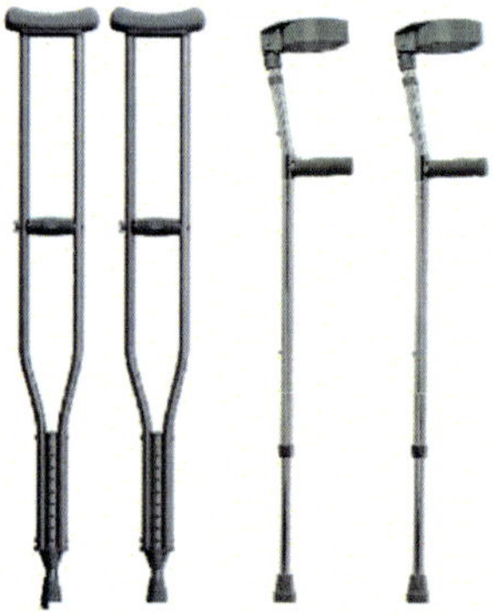

(a) (b)

Fig. 1. Main types of crutches: (a) axilla or underarm crutches and (b) forearm crutches [6].

For individuals with a height between 1.50 m and 2.00 m, standard crutches are generally made up of a rigid handpiece, forearm cuff, a leg that normally has an adjustable height depending on the height of the user and, finally, a tip (see Fig. 2).

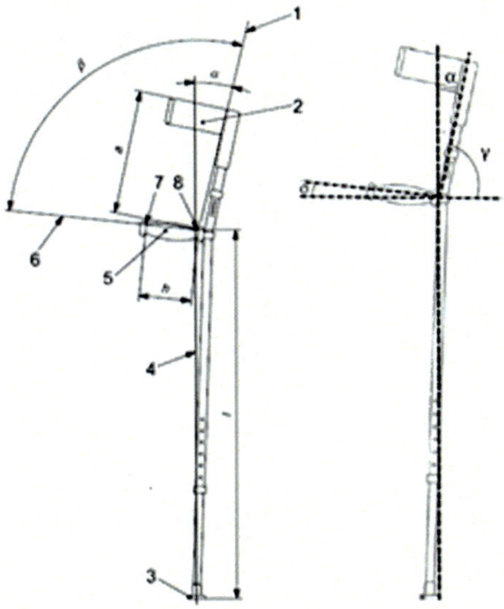

Fig. 2. Regular crutch based on the ISO 11334-1:2007 standard [7]: 1) support line of the forearm support; 2) forearm support; 3) rubber tip; 4) axis of the leg section; 5) fist; 6) wrist support line; 7) front handle reference point; 8) rear handle reference point; a) length of the arm section; h) length of the wrist; l) leg length; α) 15°; β) 97°; σ) 7° and γ) 75°.

Children crutches are usually gentle and resistant because they are mostly constituted by aluminum. Parameters like longevity, comfort and looking appearance are very important for this audience and therefore these are common features in this type of crutches [8]. Some other features like ergonomically handles may be also included providing comfort and stability which contributes to the decrease of hands blistering issue [9].

Ergoactives develop a crutch specific for children that presents a sophisticated design with shock absorbers on the handle and the tip of the crutch (see Fig. 3). These crutches are suitable for this audience since they offer an effective and comfortable mobility and at the same time, reduce the possibility of secondary injuries appearance associated with the use of these devices [10].

The Ergodyanmic model present in Fig. 4 was the commercial model acquired for this study which enabled the analysis of the complex structure of the damping system and the consideration for a new approach. This crutch has a patented structure of high-quality aluminum tubes to obtain maximum resistance, with minimum weight. The protection of these tubes is made with epoxy lacquer and they are also equipped with a flexible base, to prevent slipping. To provide maximum safety and comfort for users, all models have replaceable handles, which can be easily removed. The forearm cuff is very soft, with an ergonomic design and air ventilation for additional comfort. The greatest interest of this model and what sets it apart from the common models is on the integrated shock absorption mechanism, which helps in avoiding pain in the hands, wrists, elbows and shoulders. This mechanism is composed of a helical conical spring (Fig. 4). The peculiar

Fig. 3. Ergobaum® Junior Adults & kids forearm scock absorver crutches [10].

shape of this type of spring, one side wider than the other, allows it to accumulate forces while undergoing compression and exerting backpressure [11].

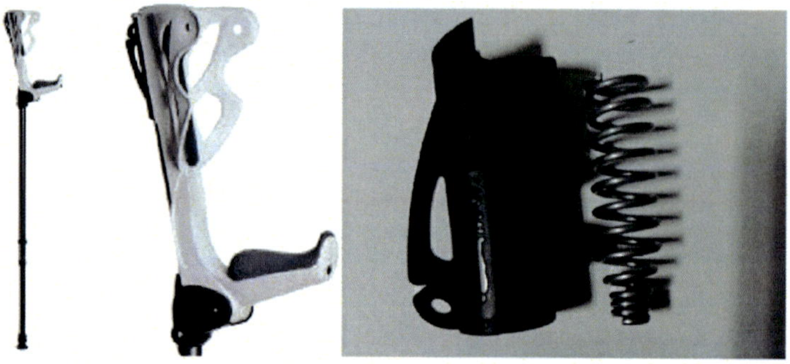

Fig. 4. Ergodyanmic® crutch and the respective damping system.

4 Conceptual Project

Before starting the detailed project, the most significant objectives for the evaluation of the medical device were described. Subsequently, the list of objectives was ordered in order of importance:

1. Security
2. Utility
3. Reliability
4. Comfort
5. Cost of material
6. Easy maintenance
7. Service life

After that, to define the solutions to be incorporated in the device, a more detailed study of each one was carried out. All solutions adopted agreed with the defined objectives.

The essential product functions were initially listed to incorporate the essential components into the final product. These components are expressed in an abstract form in terms of product requirements or functions.

The chosen components analyzed by this method were the forearm support, handle, damping system, crutch tube and tip.

For the forearm support, three different options were considered: a normal plastic holder, a velcro holder and an adjustable tape. Due to the focus of the project, it is important to maintain the crutch attached to the children arm. For this reason, the velcro tape was chosen, since it provides greater safety during the use of the crutch. Besides, it is more comfortable and easier to adjust the velcro to the forearm compared to the adjustable tape. Although the velcro suffers greater wear with its use (it loses the force of adhesion), this one can easily be acquired in the market, for a low price, and replaced in the device.

For the handle of the crutch, three hypotheses were considered such as an anti-slip and anti-breathable, handle padded with washable velcro and hand-adaptable tape. The selected solution was the handle padded with washable velcro since it provides greater comfort and greater hygiene as it is removable for washing. Although the hand-adaptable tape provides greater security, it can injure the child's fingers.

Regarding the damping system, it was decided that it would be a hydraulic damping system.

Another important component of the crutch is the tube. Once the children can get hurt or hurt someone with the crutch, it was decided to apply a coating on the tube. Between the sponge shell and the cushioned fabric, the latter was chosen for aesthetic reasons.

Finally, for the crutch tip, the weighted alternatives were anti-slip, cylindrical and conical tip. The anti-slip was chosen because it allows absorbing the asperities of the surface, which does not happen with the other two solutions.

5 Detailed Project

5.1 Adapted Child Model

As previously mentioned, this project aims to develop a crutch with a damping system for children between 5 and 10 years old. Therefore, the device was dimensioned considering anthropometric data for children in this age group. The data presented in Fig. 5 was based on the Brazilian Association of Technical Standards (ABNT) standard whose objective is to establish a system that presents the corporal measurements of children and young people [12]. Although this standard is intended for support the production of clothing, it was considered that this could be applied in the present project, since the system is based on the measures of the body and not on the measures of the clothing. Moreover, the European ergonomic scale only presents the height values and, for sizing the crutch, the values of the elbow height for the children were necessary. Due to that, the ABNT standard was used.

Thus, based on data from the ABNT standard, it is concluded that for children of 5 years and 10 years old, the model should present 0.79 m and 1 m height, respectively. Besides, the incorporation of an adjustable system, with a minimum adjustment of 2.5 cm, has been maintained, as described in the tradable crutch standard [7].

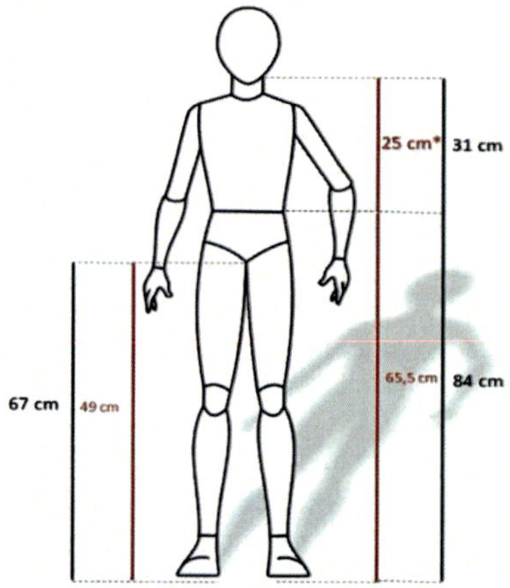

Fig. 5. Anthropometric data for 5 years (red) and 10 years (black). *Obtained by calculating the arithmetic mean of the data for 4 years and 6 years.

5.2 Biomechanical Analysis of Gait

The correct parametrization of the gait was performed using an analytical study of the simple swing gait during the support phase of the crutch. This parameterization was based on gait velocity, step length, axilla height, crutch height and the individual mass (see Table 1). These data were used to quantify the load force and the impulsive force in the support phase.

Table 1. Anthropometric data used in gait parametrization

Age group	5 years	10 years
Gait velocity	1.55 m/s	1.55 m/s
Step length	0.77 m	1.06 m
Crutch height	0.79 m	1 m
Individual mass	17.7 kg	32.5 kg

The walking speed of 1.55 m/s was selected based on the study of Rogers et al. [13].

The step length was calculated for a child of 5 and 10 years, obtaining the value of 77.42 cm and 10.86 cm, respectively. The mass values of a 5-year-old and 10-year-old child were taken from the weight and height tables for the 50th percentile. Thus, the mass of a 5-year-old and 10-year-old individual is respectively 17.7 kg and 32.5 kg [14].

5.2.1 Mass Center Movement

Tracking a person's movement on an electric walking path it is observed that the center of mass of the individual describes an approximately circular motion. When the individual walks on the horizontal floor, the circular movement of the center of mass is summed with the translation movement, and a movement of the cycloid type is obtained [15].

The parametric equations, horizontal (*t*) and vertical (*t*), for this motion, are given by Eq. 1,

$$\mathrm{x(t)} = \mathrm{a}(\omega \mathrm{t} - \mathrm{sen}(\omega \mathrm{t})) + \mathrm{v_d t y(t)} = \mathrm{a}(1 - \cos\cos(\omega \mathrm{t})) + \mathrm{h_{CM}} \tag{1}$$

where ω is the torsional velocity of the circle, a is the radius of the circle, v_d is a component of the horizontal displacement velocity, h_{CM} is the height of the center of mass and t is the time.

5.2.2 Loading Force in the Support Phase

The force applied to the shoulder joint of the crutch user is shown in Fig. 6.

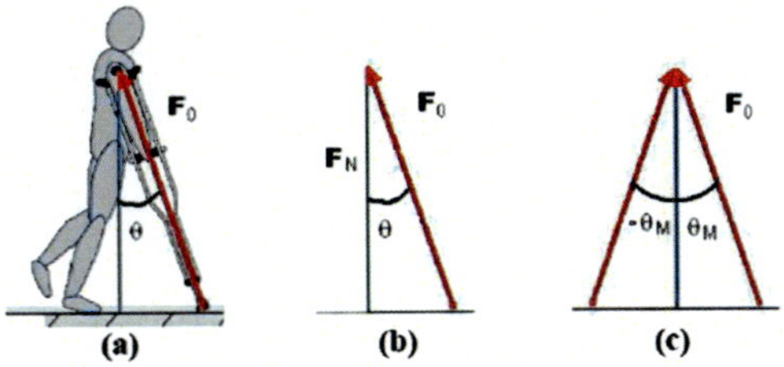

Fig. 6. Vertical force. (a) Force applied to the shoulder joint; (b) Force diagram; (c) Variation of the vertical force from the minimum angle (left) to the maximum angle (right) [15].

To simplify the calculations, the normal force F_N was considered. As shown above, the normal force value varies according to the angle θ_M. Through the trigonometric relations, the angle value is obtained for 5 years and for 10 years is 0.47 rad and 0.49 rad, respectively.

The translation speed v_T, which is equal to 1.55 m/s, and the distance travelled in a complete cycle, which is the length of the stride, was used to calculate the time that a 5-year-old children takes to complete a cycle of march, which is 0.49 s. In turn, the time that a 10-year-old child takes to run a walking cycle is 0.68 s.

The total weight of the child is distributed by the two crutches, being 50% of the weight supported by each one. Thus, the force supported by each of the crutches is given by Eq. 2.

$$F_0 = \frac{pg}{2} \tag{2}$$

Thus, for 5 and 10-years-old children, the value of F_0 is, respectively, 87 N and 160 N.

The equation of the normal force on the shoulder joint along the gait cycle is given by Eq. 3.

$$F_N = F_0 cos\left[\theta_M\left(1 - sen\left(\frac{\omega t}{2}\right)\right)\right] \tag{3}$$

The result of this equation for children of both ages is expressed in Fig. 7.

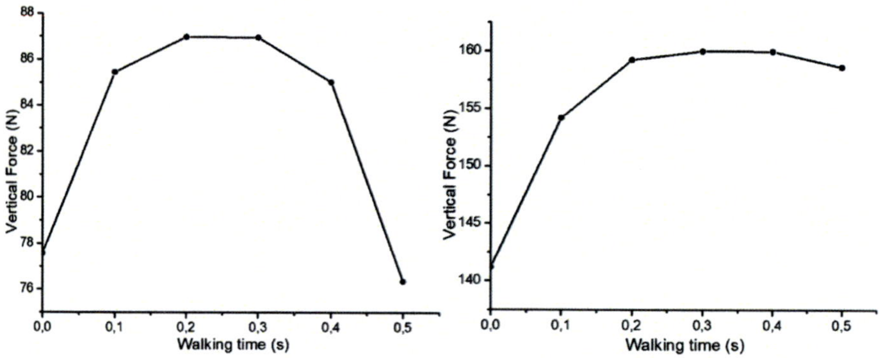

Fig. 7. Graph of the loading force in the support phase for children with 5 years (left) and 10 years (right).

5.2.3 Impulsive Force

The impulsive force is the collision force, F_c, of the crutch with the ground and it is explained in Fig. 8.

The duration of impact is $\delta t \approx 0.2$ s [15].

At the moment of impact, an individual projects about 67% of his body mass on the crutches.

The moment value P_f transferred from the crutch to the ground for a child of 5 years and 10 years is respectively, 40.46 kg.m/s and 71.91 kg.m/s, and is given by Eq. 4. The impact force of the crutch on the ground is obtained by Eq. 5. For a child with 5 years of age, the value of f_m is 202 N. For a 10-year-old child, the value of f_m is 360 N.

$$P_f = \frac{mv}{sen\theta} \tag{4}$$

$$f_m = \frac{\delta_p}{\delta_t} \tag{5}$$

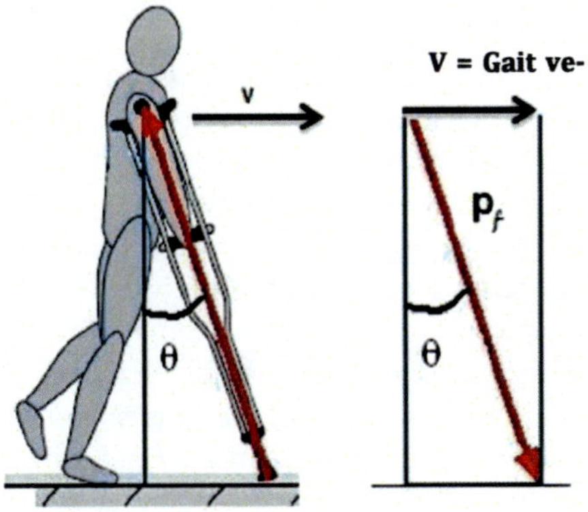

Fig. 8. Collision force of the crutch with the ground [2].

The equation that allows the modulation of the impulsive force on the impact of the crutch with the ground is given by the mathematical expression of Eq. 6.

$$F_c(t) = f_m exp\left[-100t^2\right] \tag{6}$$

The result of this equation for children of both ages is expressed in Fig. 9.

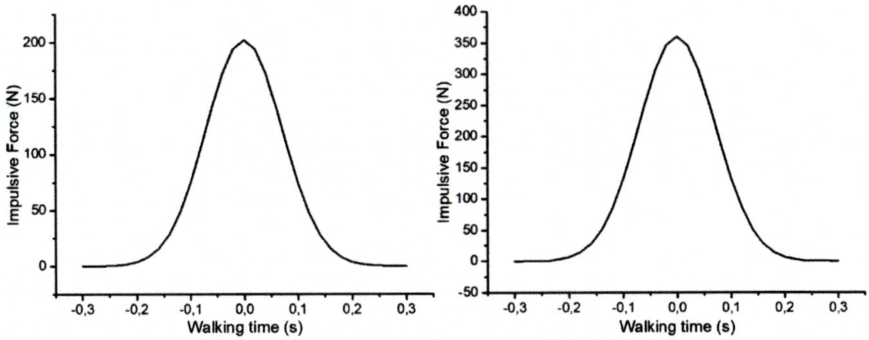

Fig. 9. Graph of the impulsive force for children with 5 years (left) and 10 years (right).

5.3 Dimensioning of the Damping System

Damping is a physical phenomenon observed in some mechanical systems in which mechanical energy dissipation occurs in the form of heat, noise, viscous or dry friction. This phenomenon occurs due to the presence of a non-conservative force that performs work, with loss of mechanical energy and, therefore, damping occurs. A damping system is composed of three essential components: piston, spring and compressible fluid. Besides, this system is composed of another component that contains the fluid, denominated body.

5.3.1 Dynamics of the System on the Crutch

It is well established that using a hydraulic system on a crutch, the shock absorber performs two movements during the assisted walking cycle: the compression movement when the child forces on the handle and the shock reduces its length with the displacement of the rod into the body; and the extension movement, when the child raises the crutch of the ground, occurring the reverse as described above [16]. In this type of system, for the damper to work correctly, the piston must offer a minimum resistance to the compression movement and great resistance to the extension movement. In the execution of these movements, the damper chamber will always be divided in two parts by the piston.

In the dimensioning of the damper, for subsequent application to the crutch, it becomes necessary to evaluate the behaviour of the fluid (oil) inside the system. Thus, calculations based on fluids mechanics are exposed in the following sections and these will allow to select correctly the work of the damper and to size the orifice of the piston that will be responsible for the damping.

5.3.2 Fluid Selection

The selected fluid was AMSOIL Synthetic Hydraulic Oil, which is an oil composed of synthetic esters and has a superior performance compared to others. The selection criteria were mainly focused on its biodegradability and, therefore, in case of spillage, no environmental damages are expected [17]. The respective oil has a specific mass (ρ) of 860 kg/m^3 and a kinematic viscosity at 40 °C of 45 $\frac{m^2}{s}$. For this damper, the specific mass of said oil is in an ideal range, since, it does not generate very high or very low damping forces. The kinematic viscosity value has been considered because it is a fundamental property in the damper performance due to its influence on the Reynolds number, which will subsequently be used to size the piston orifice. On the other hand, its high viscosity index allows excellent performance in all environments, reducing the need for seasonal changes [17].

5.3.3 Dimensioning the Orifice

The orifice present in the damper piston is primarily responsible for the damping coefficient since it generates a loss of charge in the fluid flowing from the compression chamber into the decompression chamber. This loss of load results largely from the inlet and outlet leakage losses, which vary depending on the geometry and shape of the inlet and outlet edges, the diameter and length of the orifice and on the geometry of the tube. In this way, the Mass Conservation Law and the Bernoulli Equation were used to size the piston orifice.

The Law of Conservation of Mass is required to determine the velocity within the orifice channel and hence the Reynolds number of the flow in that region. Mathematically, the Conservation Law can be represented by Eq. 7.

$$\sum_i \; (V_i A_i)_{entrada} = \sum_i \; (V_i A_i)_{saída} \tag{7}$$

The control volume was delimited between the lower and upper faces of the piston and the orifice channel. When the damping system is actuated under the influence of a force applied on the handle, it undergoes a displacement of 3.2 cm and the system is re-established in 0.2 s. The fluid inlet speed is 0.16 m/s. Considering the law of conservation of mass, the fluid inlet speed is equal to the fluid outlet speed. The Bernoulli equation (Eq. 8) expresses the conservation of energy in the flow of an incompressible fluid along a current line.

$$P_1 + \frac{1}{2}\rho V_1^2 + \rho g h_1 = P_2 + \frac{1}{2}\rho V_2^2 + \rho g h_2 + h_l + h_{lm} \tag{8}$$

This expression demonstrates that there is a loss of mechanical energy during the passage of the fluid through the orifice, represented by h_l (distributed losses) and h_{lm} (localized losses).

After the calculation of the Reynolds number (Re), it is possible to confirm that the fluid in this damping system present a laminar flow since Re in the extension is 7,82 × 10^{-5} and in the compression is 3,91 × 10^{-5}.

The friction factor depends on the Reynolds number and Eq. 9 was used to calculate this factor.

$$f = \frac{64}{Re} \tag{9}$$

Results showed that during extension and compression there is a friction factor of 3,91 × 10^5 and of 1,64 × 10^6, respectively.

The inlet losses are caused by the separation of the flow as it passes through the edges of the orifice, forming small vortices in the flow of the fluid. The appearance of these vortices results in a reduction of the effective diameter of the flow in the orifice, forcing the fluid to accelerate passing through that reduced area, known as the 'contraction vein'. The load coefficient K depends on the geometry of the orifice inlet [16].

To scale the orifice, an input load coefficient (K_{input}) of 0.04 and output load coefficient (K_{output}) of 1 was considered during compression. During the extension, an input load coefficient (K_{input}) of 0.5 and output load coefficient (K_{output}) of 1 was considered. The output load losses are due to the uncontrolled expansion of the fluid as it exits a small diameter orifice into a large reservoir, causing almost complete dissipation of the kinetic energy of the fluid. For the case under study, the loss coefficient for the output will assume the unit value. Thus, the Bernoulli equation can be rewritten in Eq. 10, where n is the number of orifices present in the piston.

Table 2 shows the results of the calculations performed for the orifice dimensioning.

$$\Delta P = P_1 - P_2 = \rho\left[\left(K_{output}\frac{(V_{orifice})^2}{2} + K_{output}\frac{(V_{orifice})^2}{2} + h_l\right)n + \frac{V_{output}^2}{2} - \frac{V_{input}^2}{2}\right] \tag{10}$$

As previously mentioned, for the correct functioning of the damper, the piston must offer little resistance to the compression movement (downward movement of the piston) and higher resistance to the extension movement (upward movement of the piston). For that, it was thought to add 4 orifices to the piston to decrease the force during compression.

Table 2. Results performed for the orifice dimensioning.

	During compression	During extension
Orifice area	$1{,}9 \times 10^{-4}$	$9{,}50 \times 10^{-5}$
Fluid velocity in the orifices	0,32 m/s	0,64 m/s
Variation of losses on entry and exit	$8{,}53 \times 10^{5}$ Pa	$1{,}71 \times 10^{6}$ Pa
Behaviour	1012 N.s/m	3037 N.s/m

Regarding the orifice geometry, the entrances incorporated present a round shape, with a radius of 1 mm, in the face of compression and a living corner in the face of extension [16].

The diameter of the orifice was chosen considering a reference piston speed of 1 m/s. In order to observe the influence of the diameter on the extension force and the compression force, the diameter of the orifice was varied between 4 and 8 mm.

Analyzing the data obtained it was found that for an orifice diameter of 5.5 mm, the compressive force (162 N) is about 67% less than the extension force (486 N). Thus, considering the issues and objectives discussed previously, it was concluded that this would be the ideal orifice diameter for the damping system.

5.3.4 Dimensioning the Body

The damper body is designed as a closed pressure cylinder and, therefore, must withstand the pressure differences generated during piston displacement. Thus, calculations were made in order to evaluate the state of stresses in cylinders subject to a certain pressure. For calculations, the maximum pressure inside the pipe (P_{int}), the atmospheric pressure (P_{atm}), the wall thickness of the pipe (t) of 5 mm and the inner diameter of the tube (D_{int}) of 17 mm were considered.

The radial stress in the tube, the tangential stress and the longitudinal tension can be calculated by Eqs. 11, 12 and 13, respectively.

$$\sigma_r = \left\{ \frac{\left[P_{int}\pi\left(\frac{D_{int}}{2}\right)^2\right] - \left[P_{atm}\pi\left(\frac{D_{int}}{2}+t\right)^2\right]}{\left(\frac{D_{int}}{2}+t\right)^2 - \left(\frac{D_{int}}{2}\right)^2} \right\} - \frac{(P_{int} - P_{atm})\left\{ \frac{\left(\frac{D_{int}}{2}+t\right)^2\left(\frac{D_{int}}{2}\right)^2}{\left[\left(\frac{D_{int}}{2}+t\right)^2-\left(\frac{D_{int}}{2}\right)^2\right]} \right\}}{\left(\frac{D_{int}}{2}\right)^2} \tag{11}$$

$$\sigma_\theta = \left\{ \frac{\left[P_{int}\pi\left(\frac{D_{int}}{2}\right)^2\right] - \left[P_{atm}\pi\left(\frac{D_{int}}{2}+t\right)^2\right]}{\left(\frac{D_{int}}{2}+t\right)^2 - \left(\frac{D_{int}}{2}\right)^2} \right\} + \frac{(P_{int} - P_{atm})\left\{ \frac{\left(\frac{D_{int}}{2}+t\right)^2\left(\frac{D_{int}}{2}\right)^2}{\left[\left(\frac{D_{int}}{2}+t\right)^2-\left(\frac{D_{int}}{2}\right)^2\right]} \right\}}{\left(\frac{D_{int}}{2}\right)^2} \tag{12}$$

$$\sigma_z = \frac{\left[P_{int}\pi\left(\frac{D_{int}}{2}\right)^2\right]}{\left[\pi\left(\frac{D_{int}}{2}+t\right)^2\right] - \left[\pi\left(\frac{D_{int}}{2}\right)^2\right]} \tag{13}$$

To determine the maximum internal pressure of the body, it was considered that during the cycle of assisted walking, more precisely, in the balance phase of the crutch, it will be suspended at a maximum height of 0.5 m. This generates a piston velocity of 3.13 m/s, obtained by Eq. 14. In order to determine the pressure felt inside the tube, Eq. 15 was used. The tensile strength of the material is represented by *Rc*, the outside diameter of the pipe in mm is represented by D, the inner diameter of the tube is represented by d and the safety coefficient is represented by λ.

$$V = \sqrt{2gh} \tag{14}$$

$$P_{int} = \frac{R_c\left(D^2 - d^2\right)}{\lambda\left(D^2 + d^2\right)} \tag{15}$$

Considering the use of a 6061-T6 aluminum tube with an external diameter of 22 mm and an internal diameter of 17 mm, an internal pressure of 33.71 MPa was obtained. The results for the stresses felt in the tube were as follows: $\sigma_r = 35{,}94\,\text{MPa}$, $\sigma_\theta = 102{,}44\,\text{MPa}$ and $\sigma_z = 22{,}14\,\text{MPa}$

5.3.5 Stem Dimensioning

The damper rod is designed considering that at one side it allows the piston to be assembled and at the other side it is the whole of the crutch. The rod undergoes tensile traction and flexion at different movements of the damper during compression. Equation 16 allowed the calculation of the tensile tension of the rod being defined by the ratio between the force during the extension (F) and the minimum cross-sectional area of the rod. Equation 17 allowed the calculation of the critical flexural stress (σcr), considering the modulus of elasticity of the material (E), the length of the rod (L) and the radius of the rod (r).

$$\sigma_T = \frac{F}{A} \tag{16}$$

$$\sigma_{cr} = \frac{\pi^2 E}{\left(\frac{L}{r}\right)^2} \tag{17}$$

For this damper element it was thought to use SAE 1045 Steel (σ_y= 530 MPa / E = 207 GPa) and therefore, the safety coefficient obtained was 11 for the traction. There was obtained a tensile stress of $2{,}1 \times 10^7$ MPa and a critical flexural stress of 190×10^3 MPa

5.3.6 Sizing of the Expansion Chamber/Floating Piston

The correct function of the damper depends on the expansion chamber that must have sufficient volume to accommodate the volume of the rod as it moves into the body. Thus, to size the camera was used Eq. 18.

$$P_i V_i = P_f V_f \tag{18}$$

The rod shifts 32 mm and the final volume of the camera has been sized to reduce 25% of its original volume. The pressure inside the chamber increased to about 101 MPa. In turn, the floating piston is designed to separate the damping fluid from the air inside the expansion chamber. Attached to the floating piston, the system presents a disc which operates a seal to provide two points of support during the displacement of the piston, preventing the misaligned work of the system.

5.3.7 Piston Design

In order to verify the resistance of the piston to the forces that will actuate on it, a computational simulation was performed in the Inventor software (Fig. 10).

After this analysis, it was found that the minimum safety factor is 15, having the same value as the maximum safety factor. As this value is high it has been concluded that the piston will withstand the forces to which it will be subjected.

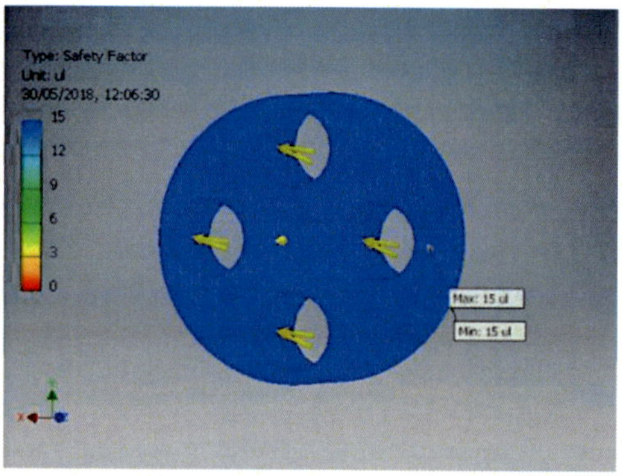

Fig. 10. Distribution of the safety factor on the piston after computational simulation.

5.3.8 Sizing of the Spring

To scale the spring, it was decided to consider the elastic constant (k), which was determined using the loading force of a 10-year-old child (159 N).

Thus, the total force that the spring should restore was calculated. In this case, this force is the sum of the loading force (F_{load}) that the individual exerts on the crutch with the compressive force F_{load}) that the oil exerts on the spring (Eq. 19).

$$F_{total} = F_{load} + F_{comp} \tag{19}$$

The total value of the force exerted on the spring is 321 N.

Through the value of the total force and the dimensions referred to in the previous section, the compression spring sizing software The Spring Store was used. The parameters of the springs selected for each of the ages are shown in Table 3.

Table 3. Spring parameters selected by The Spring Store software.

D_e [mm]	t [mm]	l_0 [mm]	F [N]	k [N/mm]
20	2.85	15	327	124

6 Final Model of the Damping System

As was already reported, the damping system currently used in crutches is uncomfortable for their users since it requires high muscle forces to make the system functional. Thus, the purpose of this work was to develop a new damping system to implement in crutches for children between 5 and 10 years old. This system needs to be comfortable and functional for all users within these range of ages.

The designed damping system features a relatively simple assembly that leads to a rather complex manufacturing process. The Inventor software was used to construct the 3D model represented in Fig. 11.

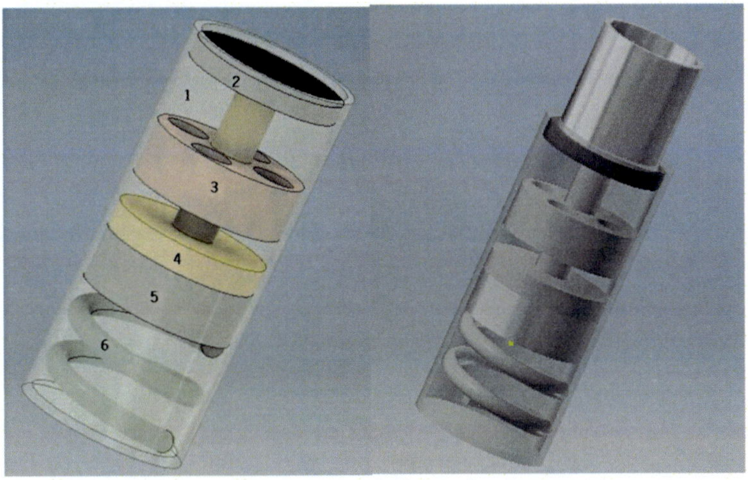

Fig. 11. Final model of the damping system with different components (on the left) and inserted in the crutch (on the right): 1) body; 2) sealant; 3) piston; 4) rod connected to a cylinder; 5) floating piston and 6) helical spring.

The system developed throughout this project is more comfortable and functional for all patients, as verified during the dimensioning of the damper. Moreover, in order to improve the stability of the system during the use of crutches, the damper was implemented in the crutch tube. Thereby, it is also expected a decrease of injuries in patients using crutches.

7 Conclusion

The main goal of this project was to remodel a new damping system to the crutches currently available in the market since these are uncomfortable and cause more tiredness and effort on the part of the upper limbs. However, since the habituation of children to new devices is easier, 5 and 10-year-old children were defined as the target audience.

Through the study of the anthropometric measures of the children, it was concluded that the crutch should have a minimum height of 0.79 m and a maximum height of 1m. After this analysis, the biomechanical characterization of the assisted gait was carried out in order to calculate the value of the forces exerted on the crutch during the support phase. Thus, it was concluded that the force exerted on a crutch by a 5-year-old child is 87 N and by a 10-year-old child is 159 N.

The damping system was developed according to the parameters and requirements of the proposed application. It is therefore considered that the system designed to be integrated into a crutch will be effective in damping the impact with the ground and the forces exerted during the assisted walking cycle of children between 5 and 10 years. Furthermore, this damping system provides more comfortable use without requiring so much effort from the upper limbs.

Acknowledgements. This work has been supported by FCT – Fundação para a Ciência e Tecnologia within the R&D Units Project Scope: UIDP/04077/2020 and UIDB/04077/2020.

References

1. Barros, C., Leão, C.P., Soares, F., Minas, G., Machado, J.: RePhyS: a multidisciplinary experience in remote physiological systems laboratory. Int. J. Online Eng. **9**, 21–24 (2013). https://doi.org/10.3991/ijoe.v9iS5.2756
2. Westerhoff, P., et al.: In vivo measurement of shoulder joint loads during walking with crutches. Clin. Biomech. **27**, 711–718 (2012). https://doi.org/10.1016/j.clinbiomech.2012.03.004
3. Goehring, M., Kenyon, L.: Axillary or forearm crutches: comparing postural sway in single leg stance and perceptions of safety. J. Acute Care Phys. Ther. **2**, 72 (2011)
4. Maguire, C., Sieben, J.M., Scheidhauer, H., Romkes, J., Suica, Z., de Bie, R.A.: The effect of crutches, an orthosis TheraTogs, and no walking aids on the recovery of gait in a patient with delayed healing post hip fracture: a case report. Physiother. Theory Pract. **32**, 69–81 (2016)
5. Cassel, C.K., Walsh, J.R.: Geriatric Medicine: Volume II: Fundamentals of Geriatric Care (2012)
6. Ordem dos Enfermeiros: Cuidados à Pessoas com Alteração da Mobilidade - Posisionamento, Trasferências e Treino de Deambulação (2013)
7. ISO 11334–1:2007 - Assistive products for walking manipulated by one. https://www.iso.org/standard/36054.html. Accessed 14 Apr 2018
8. Crutches for Children Aged 6–10 Years. https://www.completecareshop.co.uk/mobility-aids/crutches-children/crutches-for-children-aged-6-10-years. Accessed 21 May 2018
9. Childrens Crutches. https://www.completecareshop.co.uk/mobility-aids/crutches-children/childrens-crutches. Accessed 21 May 2018
10. Ergoactives: Crutches for kids and junior adults - Ergobaum Junior (pair). https://www.ergoactives.com/products/ergobaum-junior-crutches. Accessed 25 May 2018

11. Adriano, R.: Molas I. https://pt.slideshare.net/RenaldoSilva/aula-21-molas-i. Accessed 04 May 2018
12. ABNT/CB-17: Vestuário – Referenciais de medidas do corpo humano – Vestibilidade para Bebê e Infanto – Juvenil. https://epoca.globo.com/edic/662/662_vidautil_norma.pdf. Accessed 04 May 2018
13. Rogers, E.: Analysis of force distribution on upper body limbs during ambulation with crutches (2014)
14. SBP: Tabela de percentil 50. http://www.sbp.com.br/fileadmin/user_upload/img/documentos/valores_referencia.pdf. Accessed 18 May 2018
15. Oliveira, J.H.: Proposta de amortecimento adaptável para muleta canadense (2017)
16. Oliveira, F.S.D.: Projeto de um amortecedor para protótipo de veículo off-road (2014)
17. AMSOIL: Biodegradable Synthetic Hydraulic Oil. 1–2 (2017)

Use of Virtual Mirror in Design of Individualized Orthopedic Supplies

Filip Gorski(✉), Pawel Bun, and Kaja Stefanska

Poznań University of Technology, Piotrowo 3 STR, 61-138 Poznan, Poland
{filip.gorski,pawel.bun}@put.poznan.pl,
kaja.stefanska@student.put.poznan.pl

Abstract. The paper presents studies on a prototype of a virtual mirror application, intended to aid the design process of individualized orthopedic supplies, such as limb orthoses or prostheses. The application is a part of an automated design and manufacturing system, relying on 3D scanning and 3D printing for manufacturing of supplies adjusted to specific patients. The application itself is divided into two parts: the configurator and the mirror. The paper presents briefly a concept, a prototype application based on Kinect sensor and studies on a small group of patients, as well as results of a survey study. Certain inaccuracies of Kinect sensor were detected in the tests, preventing smooth user experience with the mirror. However, the presented approach was viewed in a favorable way by the test group – the concept of a virtual mirror is a promising solution that should be utilized in the design process of individualized orthopedic supplies. On the basis of observations and surveys, recommendations for future development were formulated.

Keywords: Augmented reality · Virtual mirror · Orthopaedic supplies

1 Introduction

Orthopedic supplies are used for healing and rehabilitation of persons with diseases and malfunctions of human locomotory organs. A traditional process of manufacturing of individualized supplies (e.g. prostheses or orthoses of limbs) is time-consuming and arduous, as well as costly [1]. Modern computer techniques enable shortening of time and reduction of cost of production of such products. Patient's comfort is increased both before use – during manufacturing – as well as during product operation. This is possible thanks to use of new technologies, such as 3D scanners, 3D printers and CAD systems, as mentioned in [1] and [2]. Augmented and virtual reality (AR and VR), on the other hand, can aid the design process and make it more attractive [3].

A concept of virtual mirror, i.e. AR mirror, consists in displaying or representing a human silhouette and its movement, enhanced and/or augmented of additional objects. Although the technique itself is quite popular, there are not many literature examples of its professional use. The existing ones concern application in commerce, e.g. for shoes

J. Machado et al. (Eds.): icieng 2021, LNME, pp. 388–394, 2022.
https://doi.org/10.1007/978-3-030-79165-0_36

[4] as well ass regular clothes [5] and (rarely) in medicine, mostly in rehabilitation [6] or pain therapy [7].

The first medical example of virtual mirror is a system presented in [6]. In the system, a realistic avatar is displayed, following user's movements in all planes and in real time. The system contains real-time modulation of visual feedback on human body movement, used to gather data in the rehabilitation process. The second example is a complex system, involving electromyography for movement sensing, as well as machine learning for better reaction for user movement. The result is a realistic experience, significantly helping to reduce phantom pains [7].

Use of virtual mirror concept as a process helping to customize a prosthesis or orthosis is a concept which is practically non-existent in literature, as well as in commercial applications. There are, however, applications in prosthetics – in paper [8], the authors proposed a dedicated augmented reality mirror for teaching patients how to operate a myoelectric prosthesis, with promising results.

A prominent medical use of augmented reality mirrors is the Magic Mirror – a system for anatomy learning, developed at Technical University of Munich [9]. This system utilizes Microsoft Kinect to show human internal tissues and organs on their own mirror reflection, for educational purposes. It was developed and successfully implemented, and the authors claim that (as of 2019) it is the only AR system to date that to be successfully integrated into such a large scale, educational setting for anatomy learning.

In opinion of authors of this paper, after analyzing available literature, the AR technology can be characterized as having high potential for use in the design process of customized prosthetic or orthopedic equipment. Both medical and commercial examples of such systems can be found. However, implementation of virtual mirror for prosthesis or orthosis visualization was previously utilized only for training or healing purposes – not in the design process.

The authors proposed and built their own solution – results of the studies on a prototype with a small group of patients are presented in this paper.

2 Materials and Methods

2.1 System for Automated Design and 3D Printing of Orthopedic Supplies

The solution presented in this paper is a part of a project focused on rapid manufacturing of orthoses and prostheses, named "Automation of design and rapid manufacturing of individualized orthopedic and prosthetic products on the basis of data from anthropometric measurement". A concept of the AutoMedPrint system, which is being developed in the project, is shown in Fig. 1.

Fig. 1. Concept of the AutoMedPrint system

The user of the system first gets its limb 3D scanned by a contactless optical 3D scanner. Then, the automated algorithm processes the scan and an intelligent CAD model is used to generate an invidualized variant of an orthosis or a prosthesis. When it is generated, it is shown to the user via the virtual mirror, while the user can alter some minor features (material, color, etc.). Then the product is manufactured using 3D printing technology. The whole process is short (patient engagement is less than 45 min in total) and allows generating cheaper products in shorter time, as shown in earlier works [10]. Detailed description of system operation, as well as technical details of subsequent stages are explored in other papers, such as [10] and [11].

2.2 Concept of the Virtual Mirror

The concept of the virtual mirror and configurator application assumes that the orthosis/prosthesis model generated automatically for a given patient (after 3D scanning of their limb) will be displayed to the user, which will allow to verify its correctness and change configuration features. The Unity 3D environment was selected to create the application functionality, and the Kinect device was selected as the motion sensor. Unity is an universal 3D application development platform – a popular game engine. It was selected because of its versatility and hardware compatibility. The version used was 2018.2.20 – it was selected due to backward compatibility issues. A 50" TV was selected to display the application, and a gyro mouse was used for user interaction with the application. The solution assumes two separate user views that can be freely switched between. One of them is used for configuration and the other for "trying on" orthopedic devices. The input data were previously automatically generated models of orthoses and prostheses.

The motion tracking system area was set so that the sensor detects a user immediately and to ensure the camera's field of view covers most or all of the user silhouette. The Kinect sensor was set at a height of about 0.8 m. The initial distance from which the user takes the "T" position was set at 2 m. After detecting the silhouette, the person using the application can walk approx. 1 m forward to better see the screen. Slight sideways movements are also possible, but to a greater extent they cause a noticeable shift of the generated frame relative to the user's image from the RGB camera. Figure 2a shows the application screen used to configure the selected device - WHO (wrist hand orthosis), and 2b - a user using the application.

2.3 Methodology of Experimental Studies

The research was carried out on a sample of 9 users aged 20–49 years, for whom individual orthoses had been previously 3D printed. Before using the application, users were informed about its functionality and the purpose of the study. The use time of the system was on average from 1 to 3 min (it was not strictly limited). The sample size was deliberately small – it consisted only of people (patients) who were previously 3D scanned, and for whom the real, fitting orthoses were created using 3D printing. The purpose of this particular study was to make some decisions on development of the prototype solution, hence the sample size was not increased (the authors plan testing much more users using the second prototype).

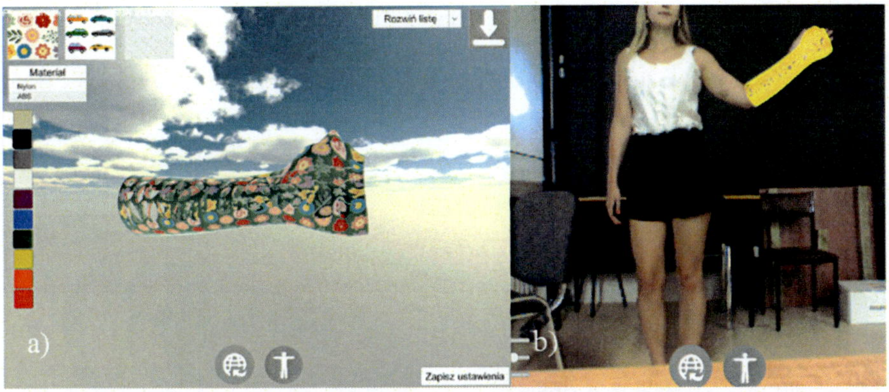

Fig. 2. Virtual mirror – a) wrist hand orthosis 3D configurator, b) user image with superimposed orthosis model

Immediately after the test, each user completed a specially created survey. The survey consisted of 13 questions. The first eleven questions consisted in assessing the individual features of the application on a five-point Likert scale, while the last two - selecting two out of five answers. Questions have been grouped into questions about the configurator, virtual mirror and the entire application.

3 Results

3.1 Survey Results

Results of the survey studies are presented in Table 1.

Question 12 and 13 concerned the best and the worst thing about the application, respectively. The best thing, as the users answered, was easiness of use, with the bad fitting of virtual model to camera image being the worst.

In general, the configurator was evaluated very positively, gaining high scores except graphics quality and loading times – these two areas were highlighted as needing improvements. The virtual mirror was evaluated poorly, mostly due to fitting problems, directly translating into losing the feeling of "realness" of the experience.

3.2 Observations

The following critical observations were made during the tests, regarding the:

- The skeleton is well fit to the user image only when the user is located frontally against the sensor,
- Rapid movement or being too close to the sensor makes the system not recognize the skeleton properly,
- Depending on the limb position, the product model fits it differently, what may influence a general "realness feeling" score, that is also related to the sensor not recognizing hand rotation,

Table 1. Results of the survey study on Kinect virtual mirror prototype

Category	Question	Average score (1–5)
Configurator	1. 3D model loading time	4.2
	2. 3D graphics quality	3.8
	3a User interface – clarity	4.6
	3b User interface – quality	4.1
	3c User interface – arrangement	4.6
	4. Intuitiveness of use	4.6
	5. Fluency of operation	4.7
Virtual mirror	6. Fitting of virtual model to real image	2.2
	7. Fluency of movement response	2.6
	8. Intuitiveness of use	4.3
	9. "Realness" of feeling	2.2
Entire application	10. Usefulness in making design decisions	3.8
	11. Increasing engagement and interest in the product	4.1

- Dynamic loading of 3D models is too long and graphics quality should be improved.

No user spent more than 2 min using the virtual mirror. However, they decided to spent considerable time with the configurator instead (3 min and higher), playing with possible designs. While using the mirror, the users were observed to pay more attention to their own movements and their reflection, rather than to the product and its features – their attention should be drawn to it, which would be achieved if the fluency and fitting of the image were ensured.

4 Conclusions

The novelty of the presented approach lies in the innovative use of previously known virtual mirror approaches, used in both commerce and medicine, and mixing them to obtain a system of helping the patients design their own medical device. To authors' knowledge, no such system is currently being implemented.

The prototype of the virtual mirror was tested and was generally assessed positively by the users. However, certain features are unacceptable, mostly due to technical issues with the sensor, influencing poor user experience and evaluation. Therefore, the concept was changed in the second prototype. The Kinect sensor was ditched and a number of alternative solutions were being tested (Intel RealSense SDK, Vive Tracker sensors and others).

Based on the observations and evaluation by the users, very important decisions were made in the development process of the whole AutoMedPrint system. The part of the design process requiring patient's engagement was divided into two separate stages: the configuration and the mirror. This helped with reducing loading times, as well as obtaining a general better impression by a next test batch of users. The graphics quality, as well as user interfaces were improved, basing on detailed guidelines provided by several test users. The improved product configurator in usage is presented in Fig. 3. For the virtual mirror, the RealSense sensor was decided to be used – the application is still in development.

Fig. 3. The improved prototype of orthosis configurator

It is noteworthy, that the virtual mirror is a solution that has much potential and is generally interesting for the potential patients, so it is worth developing – the sensor shortcomings could be overcome in subsequent iterations. However, it is not a crucial stage in the whole process of design of individualized orthopedic equipment, so it was given a less priority and the users can choose to skip it altogether. Future studies will involve a considerable (50) number of patients and the readiness and usefulness of the solution will be thoroughly tested.

Acknowledgments. The studies were realized with a support from Polish National Center for Research and Development, in the scope of the "LIDER" program (grant agreement no. LIDER/14/0078/L-8/16/NCBR/2017).

References

1. Cha, H.Y., et al.: Ankle-Foot Orthosis Made by 3D Printing Technique and Automated Design Software. Republic of Korea, Hindawi (2017)
2. Buonamici, F., et al.: A CAD-based procedure for designing 3D printable arm-wrist-hand cast. Comput.-Aided Des. Appl. **16**(1), 25–34 (2019)

3. Jimeno, A., Puerta, A.: State of the art of the virtual reality applied to design and manufacturing processes. Int. J. Adv. Manuf. Technol. **33**(9–10), 866–874 (2007)
4. Redaelli, C., Pellegrini, R., Mottura, S., Sacco, M.: Shoe customers' behaviour with new technologies: the Magic Mirror case. In: 2009 IEEE International Technology Management Conference (ICE), Leiden, Netherlands, pp. 1–10 (2009)
5. Erra, U., Scaniello, G., Colonnese, V.: Exploring the effectiveness of an augmented reality dressing room. Multimedia Tools Appl. **77**, 25077 (2018)
6. Roosink, M., et al.: Real-time modulation of visual feedback on human full-body movements in a virtual mirror: development and proof-of-concept. J. NeuroEngineering Rehabil. **12**, 2 (2015)
7. Perry, B.N., Armiger, R.S., Wolde, M., McFarland, K.A.: Clinical trial of the virtual integration environment to treat phantom limb pain with upper extremity amputation. Front. Neurol. **9**, 770 (2018)
8. Anderson, F., Bischof, W.: Augmented reality improves myoelectric prosthesis training. Int. J. Disabil. Hum. Devel. **13**(3), 349–354 (2014)
9. Bork, F., et al.: The benefits of an augmented reality magic mirror system for integrated radiology teaching in gross anatomy. Anat. Sci. Educ. **12**(6), 585–598 (2019)
10. Górski, F., et al.: Automated design of customized 3D-printed wrist orthoses on the basis of 3D scanning. In: Okada, H., Atluri, S.N. (eds.) ICCES 2019. MMS, vol. 75, pp. 1133–1143. Springer, Cham (2020). https://doi.org/10.1007/978-3-030-27053-7_97
11. Górski, F., Wichniarek, R., Kuczko, W., Żukowska, M., Lulkiewicz, M.: Experimental studies on 3D printing of automatically designed customized wrist-hand orthoses. Materials **13**(18), 4091 (2020)

Stapler Anvil Groove Profile Optimization

João Veiga[1(✉)], Carlos Ventura[2], and João Pedro Mendonça[1]

[1] MEtRICs Research Center, University of Minho, Campus of Azurém, 4800-058 Guimarães, Portugal
joaoveiga@dem.uminho.pt

[2] ACCO Brands Portuguesa, Zona Industrial de Paçô, 4970-249 Arcos de Valdevez, Portugal

Abstract. Staplers are tools that commonly use staples to perforate and join sheets of paper or fabric. The staple, first, punches the paper or fabric and then folds in contact with a solid surface of an anvil. This work aims to develop an optimized rigid anvil to reduce the maximum force required during staple folding for stapling between 2 and 25 sheets of paper. A commercial stapler is used as a case study. Force curves for stapling 0 to 25 sheets of paper in intervals of 5 sheets were measured for this stapler and the maximum forces are always found to occur during the folding of the staple legs. Finite Element Analysis (FEA) models were developed and fitted to real results to be used in an optimization job of the anvil groove profile. An asymmetric anvil is proposed capable of stapling between 2 and 25 sheets of paper in good conditions and indicates that an asymmetric anvil has potential to reduce maximum forces during folding. Results show a reduction of 32% on maximum force for stapling 25 sheets when compared with the commercial stapler anvil and maximum forces are only 8% higher than the maximum forces for paper punching. For fewer sheets the reduction in maximum force lessens and stapling 2 sheets of paper with the proposed anvil results in an increase of 23% in maximum force. However, the maximum forces are still less the less sheets are stapler, therefore, the proposed anvil reduces the overall maximum forces required for the entire system.

Keywords: Staplers · Anvil · Staple · Asymmetric anvil

1 Introduction

Staplers are hand-held or electric-powered devices used to join sheets of paper or fabric typically through the deformation of a staple. Stapleless staplers exist which join paper together without the use of a stapler as in [1], usually offering less sheet capacity, however the first type is far more common in either paper joining applications and medical applications.

A common staple to join sheets of paper or fabric is a two-legged fastener with a U shape. During stapling the two legs are driven through sheets of paper or fabric and then folded to enclosure them. On most staplers, the staple's legs are folded by being pushed against a metal component with a grove at the base of the stapler called anvil. The simplest staplers use rigid anvils, more complex staplers can use anvils with movable

J. Machado et al. (Eds.): icieng 2021, LNME, pp. 395–409, 2022.
https://doi.org/10.1007/978-3-030-79165-0_37

clinchers [2–4]. There is evidence that, when using rigid anvils, the forces to fold the staple legs are usually higher than the forces required to punch the paper or fabric, [5, 6] show evidence of this for surgical staplers, and Sect. 3.1 proves this behavior in this case study. However, a more global study of stapler applications is necessary to take reliable and confident conclusions.

There is a wide range of staplers with variable maximum capacities and geometry. This work focuses on hand-held staplers with a common design as case study. These staplers are composed by a base and an upper part that rotates relative to the base through a revolute joint without transformation of the force performed by the user. The user inputs force at the stapler cap that directly actuates the driver blade – component responsible to push the staple against the paper and anvil – the staple punches the paper and is deformed in a rigid anvil.

The objective of this work is to reduce the maximum necessary force to fold the staple legs for stapling between 2 and 25 sheets of paper by optimizing the anvil's groove geometry. Since forces are higher during the folding of the staple, reducing this force is the same as reducing the overall maximum force and has a secondary advantage by smoothing the "force - driver blade displacement" curve, improves the usability of hand-held staples and also reduces the required power of the actuators in mechatronic devices without increasing the part-count.

This paper is divided in 3 sections. Section 2 introduces the staplers and problem under study, as well as a small literature review of solutions for reducing the stapling required force and the apparatus used for the experimental tests. Section 3 shows the work developed from experimental tests to finite element models, an analysis of the behavior of the system and methodology to find a solution for the specific problem. Finally, Sect. 4 presents the obtained solution and a comparison of its performance against the results for the commercial stapler.

2 Problem Specification

The staple used for this work is the common commercial staple recommended for the stapler under study and the paper is bond paper of 80 g/m^2.

Figure 1 shows an exploded view of the CAD parts composing the staplers under study and presents the nomenclature used for the components.

To execute the stapling process correctly (in good conditions) the final geometry of the system must respect some specifications: At the end of the stapling process the staple tip should be at a maximum of 0.4 mm away from the bottom surface of the paper and should be facing the paper, i.e. the end direction in which the tip points should have rotated more than 90° from its original position, which is downwards, and be pointing towards the paper.

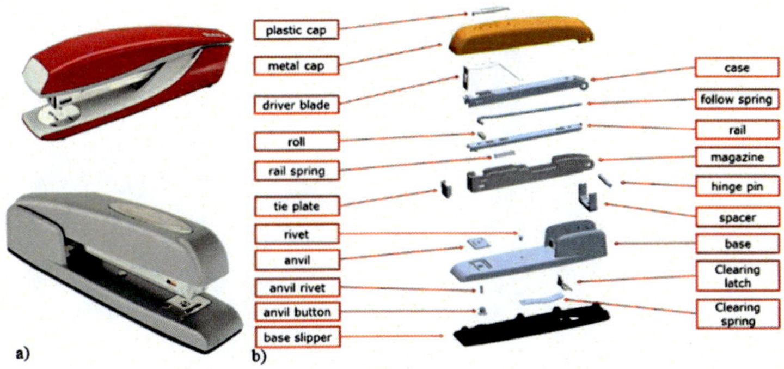

Fig. 1. Staplers under study. a) Staplers, c) Exploded view and nomenclature.

2.1 Literature Solutions for Force Reduction

Reducing the necessary force for the stapling process has been tackled in many ways. Most of these ways consist of a creation of a lever effect. A way to create a lever effect is simply to extend the stapler cap, or respective handle component where the user inputs force, beyond the position of the anvil opposed to the side of the pivot axis of the handle relative to the component that actuates the driver blade as in [7]. A four-bar mechanism can also be used inside the staple body as shown in [8].

Other solutions use a spring to store the energy inputted by the user, then release this energy in a stapling blow. A lever effect can also be used where the user does a motion larger than the motion performed by the spring and driver blade components. Moreover, the force-displacement curves provided by the user are related to spring deformation and can be smoother than the force-displacement curves for a more conventional stapler system, avoiding the peak forces of the stapling process and further reducing the necessary maximum force [9].

These solutions reduce the necessary force to fold the staple as well as the necessary force to punch the paper. However, increases the stapler's part-count and complexity.

The work present on literature, relating to the improvement of anvil geometry, is scarce.

2.2 Experimental Apparatus

Experimental tests were performed at Portuguese ACCO Brands at in.Cubo facilities using a Mecmesin MultiTest 2.5-dv force tester and a support apparatus, Fig. 2. The driver blade component is cut and mounted on a metal structure which has a vertical translational degree of freedom with respect to the base. The force gauge acts on this metal structure. At the base is fixed a magazine – component that stores the staples – and the anvil aligned with each other. A stack of paper is placed between the magazine and anvil, and the force gauge can be set to move at a constant speed pushing the staple down and measuring the force during the stapling process.

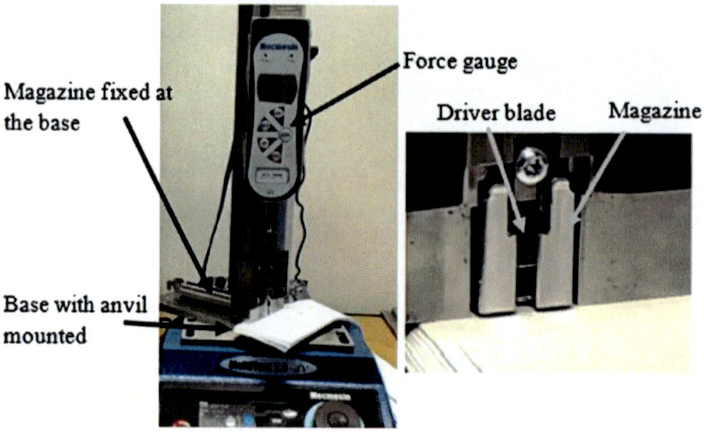

Fig. 2. Equipment used for experimental tests. Mecmesin MultiTest 2.5-dv and support for anvil, magazine and driver blade.

3 Developed Work

3.1 Experimental Results

The results presented in this section are relative to experimental tests of 80 g/m^2 bond paper sheets stapling with the anvil of the commercial stapler under study at 300 mm/min (vertical downwards velocity of the force gauge/driver blade).

Figure 3 shows five samples of force-displacement curve for stapling 25 sheets of paper which represent the highest maximum forces for the cases under study. The vertical axis shows the necessary force, in percentage, and the horizontal axis shows the downwards displacement of the driver blade (equal to the displacement of the force gauge).

The curves can be divided into three regions. As an example, sample 55 was divided in such regions: the detachment of the staple from the staple set represented by the blue region; then the punching of the paper during the green region; and finally the staple folding represented by the red region. The three relative maximums of this curve are highlighted by the vertical dashed lines.

These three regions are common to all curves for stapling 2 to 25 sheets of paper and the maximum force always occurs during the folding region. For 25 sheets the maximum force for folding the staple is 59% higher relative to the force value to punch the paper. The maximum force during detachment is much smaller than the other maximum forces.

For 25 sheets of paper, at the end of the test, the measured force drastically increases. Such happens because either the test apparatus reached its displacement limit, or the staple upper part already reached the paper causing the apparatus to be compressing the paper and staple. For fewer sheets of paper some increase in force can be measured at the end due to paper re-penetration as well (after folding, the staple legs punch the paper from the bottom surface).

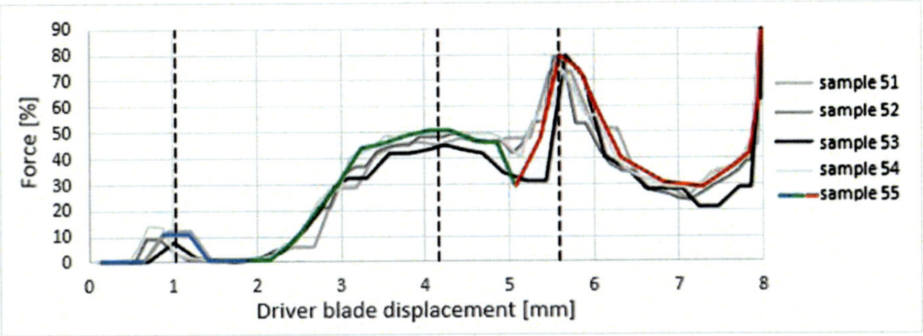

Fig. 3. Force-displacement curves of experimental tests for 25 sheets of paper on an anvil of the commercial stapler.

Figure 4 shows 5 sample of force-displacement curve for stapling 15 and 20 sheets of paper.

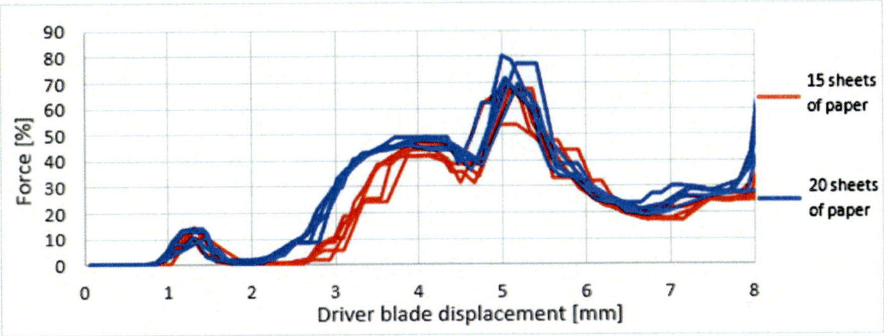

Fig. 4. Force-displacement curves of experimental tests for stapling 15 and 20 sheets of paper on an anvil of the commercial stapler.

And Fig. 5 shows 5 sample of force-displacement curve for stapling 5 and 10 sheets of paper. The paper punching region is less and less noticeable as the number of sheets is reduced due to the thin thickness of the pad of paper but still exists.

Force-displacement curves were also obtained for stapling without paper, Fig. 6. The main objective of this tests was to get real data without the influence of paper to adjust the theoretical curves from FEA to the real curves by estimating the coefficient of friction between the staple and the anvil and to evaluate the staple model.

Reducing the number of sheets of paper reduces the relative maximum forces required not only to punch through the paper but also to fold the staple. The absolute maximum force always occurs during the staple folding region. The maximum force for stapling 5 sheets of paper is on average 44% of the maximum force required to staple 25 sheets of paper.

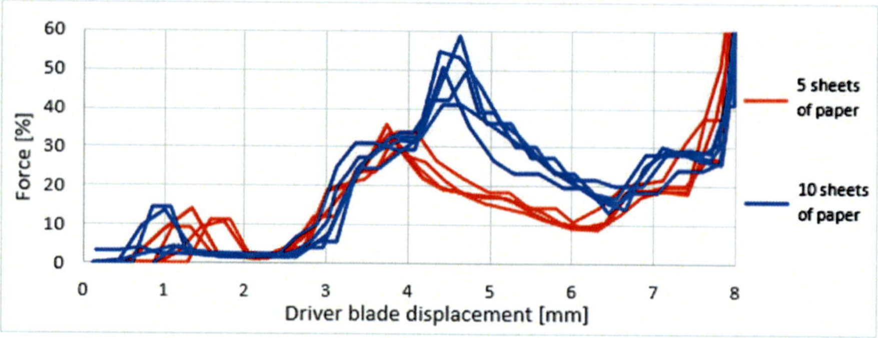

Fig. 5. Force-displacement curves of experimental tests for stapling 5 and 10 sheets of paper on an anvil of the commercial stapler.

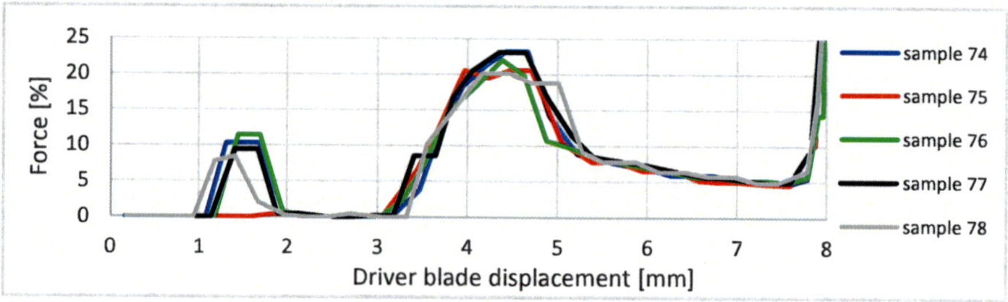

Fig. 6. Force-displacement curves for stapling without paper.

To enable a better comparison between the curves for the different numbers of sheets of paper, Fig. 7 shows the last sample of each set of tests for each number of sheets of paper. The curves are representative of the behavior and force required but some variability exist between each sample for the same number of sheets of paper as can be seen by Figs. 3, 4, 5 and 6.

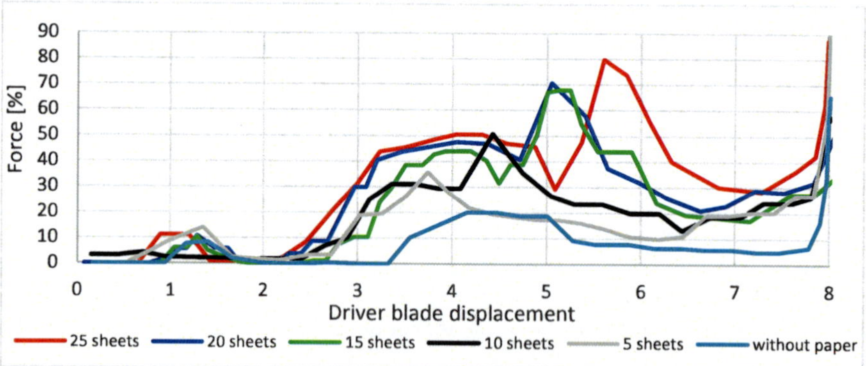

Fig. 7. Force-displacement curves of the last sample of each set of tests for each number of sheets of paper.

3.2 Stapling Process Model and Assumptions

The stapling process model consist of a FEA model of the staple, paper, anvil and relevant parts of the driver blade and magazine. The paper punching is not modelled, and the paper already has its holes at the beginning of the simulation.

The staple material model is in accordance with data from tensile tests provided by ACCO Brands Portugal.

Paper is an orthotropic fibrous material. It shows an oriented distribution of mechanical properties due to fiber orientation. Its main directions are usually defined as machine direction (MD) - direction in which fibers are usually aligned - cross direction (CD) and out of plane direction (ZD). Young Modulus and stiffness are usually higher in the MD direction than in the CD direction [10, 11]. Its properties vary greatly depending in the type of paper, manufacturing method, basis weight and other factors. In the literature several material models have been proposed to describe the elastic-plastic behavior of paper. However, in this work a simplified paper model is used since we are only interested in modelling paper as bodies that produce resistance to the staple motion causing it to deform. The model used corresponds to data for the MD direction of 70 g/m^2 Kraft paper form [12] which gave sufficiently approximated results in terms of force-displacement curves and staple deformation for this particular case. The paper was assumed isotropic with linear elasticity and multilinear isotropic hardening and is able to deform past its fracture point since it is known that the paper tears in some regions near the holes during stapling, especially when stapling few sheets. The sheet thickness is approximately 0.1 mm.

In an effort to balance model cost with accuracy, during preliminary work the set of paper sheets were assumed as a block drastically reducing the number of contacts involved and the simulation time. However, for final results, as those presented in chapter 4, each individual sheet of paper and their contacts are modelled.

The remaining bodies are assumed rigid.

When used the commercial theoretical anvil – based on the technical drawings – only a quarter model of the system is modelled, since the anvil is theoretically symmetric. When simulating a real profile or any asymmetric profile a half-model is used. The quarter model profile imposes that the staple is always centered with the anvil due to its boundary conditions.

Figure 8 a) shows a quarter model with symmetry along the xy and zy-planes and with the paper modelled as one body. Figure 8 b) shows a half model with symmetry along the xy-plane with each sheet of paper modelled. The system start position is as shown in Fig. 8, during the simulations the driver blade moves downwards at constant speed until the staple upper part touches the paper upper surface.

The magazine surfaces and the anvil are fixed during the downward motion of the driver blade and there is no restriction on the movement of the paper and staple except that imposed by the conditions of symmetry and contacts with the driver blade, magazine and anvil.

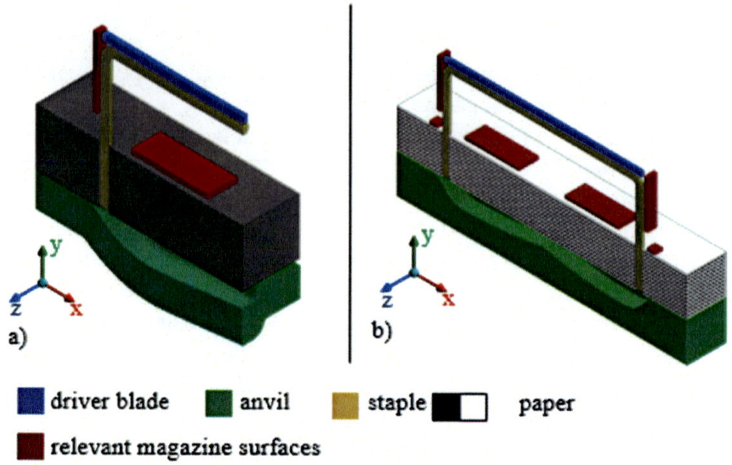

Fig. 8. Models of the stapling process. a) quarter model and paper modelled as one solid. b) half-model and all sheets of paper modelled independently.

After the driver blade downward motion, the driver blade becomes fixed, the anvil moves downwards allowing the staple to elastically recover. For stability, during the motion of the anvil, the nodes belonging to the faces parallel to the zy plane of the paper (excluding the face with symmetry condition if existent) are also considered fixed maintaining their deformation from the last point before the anvil starts to move. Such prevents the paper and staple to move in the x direction while the staple is recovering.

3.3 Fitting the Models to Test Results

The coefficients of friction between anvil and staple and between staple and paper are important parameters to better approximate the FEA models to the real behavior. These parameters were estimated by fitting the force-displacement curves as result of FEA simulation to the real force-displacement curves from experimental testing.

Figure 9 shows the FEA model results together with the real test data, shown in Fig. 6, for stapling without paper (no paper influence, only coefficient of friction between staple and anvil is evaluated). The coefficient of friction estimated between the anvil and staple is approximately 0.2. The FEA models compared use a theoretical anvil geometry and a real anvil geometry. The theoretical anvil geometry corresponds to the nominal dimensions of the commercial anvil as in its technical drawing. A real anvil of a commercial stapler was cut along its groove in order to obtain images of their profiles through a scanning electron microscope (SEM). From the SEM images, various points of the groove profile were measure which, together with splines that connect such points, form an approximation of what corresponds to a real profile.

Figure 10 shows the FEA model results together with the real test data, shown in Fig. 3, for stapling 25 sheets of paper. Together with the estimation of the previous coefficient of friction, a coefficient of friction of 0.15 between the staple and paper was found to give a good approximation to the real results.

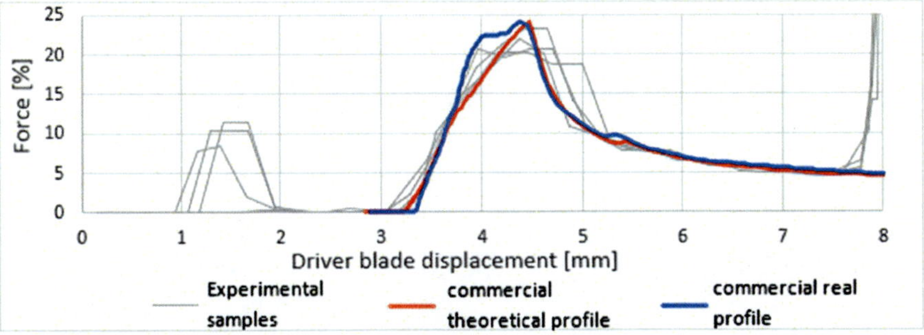

Fig. 9. Comparison force-displacement curves for stapling without paper between real test data and FEA simulation model considering a coefficient of friction of 0.2 between the staple and anvil for the theoretical anvil and a profile measured from a real anvil of the commercial stapler.

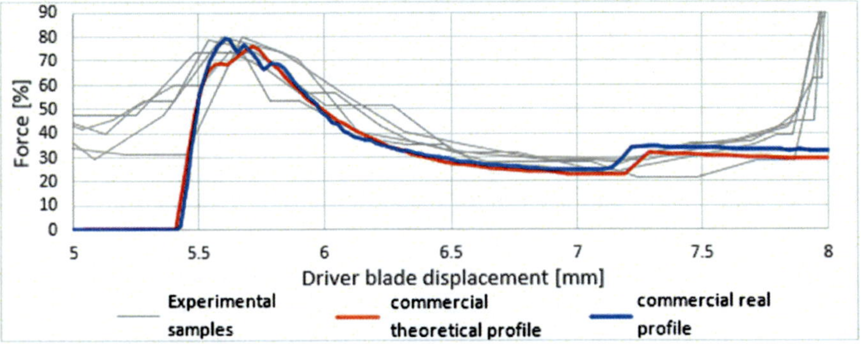

Fig. 10. Comparison force-displacement curves for stapling 25 sheets of paper between real test data and FEA simulation model considering a coefficient of friction of 0.2 between the staple and anvil and a coefficient of friction of 0.15 between the paper and staple for the theoretical anvil geometry and a profile from a real anvil of the commercial stapler.

3.4 General Behavior of Stapling Process

As seen the maximum force occurs during staple folding. Simulation results suggest that such maximum force occurs during the initial deformation of the staple legs. Most of the staple leg deformation occurs on the outside regions of the groove with the more inwards regions being only used when stapling very few sheets of paper or without paper. Figure 11 shows 3 moments of the stapling process when stapling 2 and 25 sheets of paper. At the maximum force, the contact point of the staple tip with the anvil has only a displacement in the x-direction of 0.48 mm when stapling 2 sheets and 0.30 mm when stapling 25 sheets relative to the initial x position of the same point of the staple tip. In reality and without considering yz-symmetry (real profile and asymmetric anvils), the staple can move (without deforming/rigid body motion) slightly to the left and right (x-direction) but is limited by the magazine, in this case, for nominal dimensions, it is only possible a motion of 0.12 mm for either side.

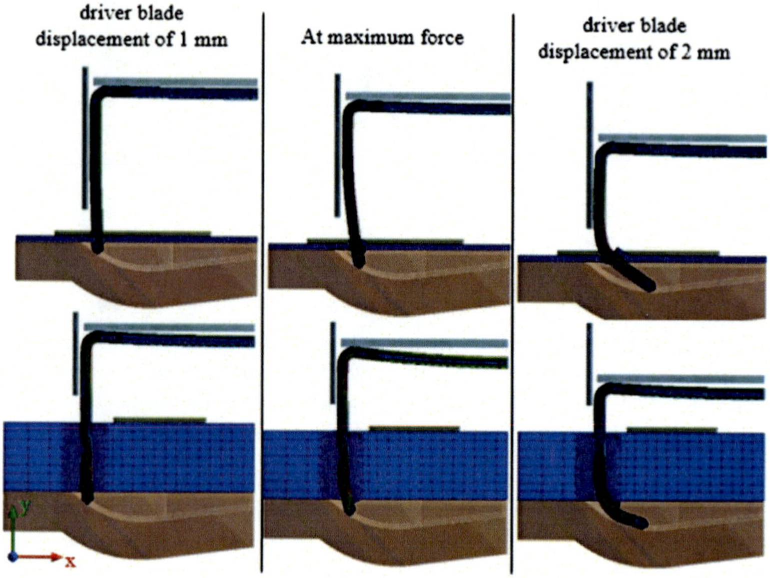

Fig. 11. Staple deformation sequence for 2 sheets of paper (upper row) and 25 sheets of paper (bottom row) for driver blade displacements of 1 mm, at maximum force of respective force-displacement curves, and for driver blade displacements of 2 mm.

In Fig. 11, red regions are yielding regions of the staple. When stapling fewer sheets, the staple yields in an upper region of its legs and the contact point with the anvil moves slightly more inwards on the anvil groove that when stapling more sheets. The contact point also changes to the side of the staple leg sooner when stapling more sheets.

3.5 Proposed Solution

The anvil groove can be characterized by its depth – vertical distance of the profile to the upper surface of the anvil – and slope – angle of the profile relative to the horizontal – at any given point of its profile. According with a preliminary study, a greater depth of the profile with the same slope at the region where maximum force occurs generally decreases the required force as the distance from the force acting point to the yielding region of the staple increases. However, if greater depth is maintained, such results in a less folded staple which is problematic for stapling many sheets. Changing the slope of the groove so that the groove profile starts rising sooner – closer to the staple first contact point with the anvil – usually makes the staple fold more solving the problem mentioned previously for stapling many sheets of paper. However this usually not only results in an increase in necessary force for stapling few sheets of paper as it can cause too much staple re-penetration from the paper backside when stapling fewer sheets. Moreover, staple jamming can occur since for fewer sheets the staple-anvil contact point moves more inwards in the anvil groove and if the profile rises to fast the staple can either stick to the anvil due to friction and/or become blocked. Increasing the slope near the first contact point maintaining the depth usually also reduces the maximum force, but such means that the depth of the profile increases rapidly causing a similar behavior with the mentioned above. The optimization of this system is a balance between the reduction of maximum force and deformation of the staple, where improvements in performance for stapling many sheets can worsen performance for stapling fewer sheets.

An optimization work using the Nonlinear programming by quadratic Lagrangian (NLPQL) method was carried out with the objective to optimize an asymmetric anvil geometry to reduce the maximum required force for stapling 25 sheets of paper. An asymmetric anvil is assumed with the idea of offsetting the required maximum forces to fold each individual leg along the driver blade displacement. The goal is to initially deform more one leg than the other reaching the respective peak forces at different times/driver blade's displacements.

Figure 12 shows the parameterization of one side of the anvil and some constant dimensions. A minimum number of parameters were chosen that give good control of the groove profile at the outside region. The point p1 is the point of first contact between the tip of the staple and the anvil and p2 is the point of highest depth of the profile. The parameters are: *py1* – the depth of p1; *Ap* – the slope at p1; *py2* – maximum depth of the groove; *px* – distance in the horizontal direction between p1 and p2. The other side of the groove has the same parameterization, thus making 8 parameters in total. The constant dimensions are based on the anvil of the commercial stapler. To join the end of the groove with p1, and p1 with p2, splines are used. And to join p2 with the middle of the groove a circumference arc is used.

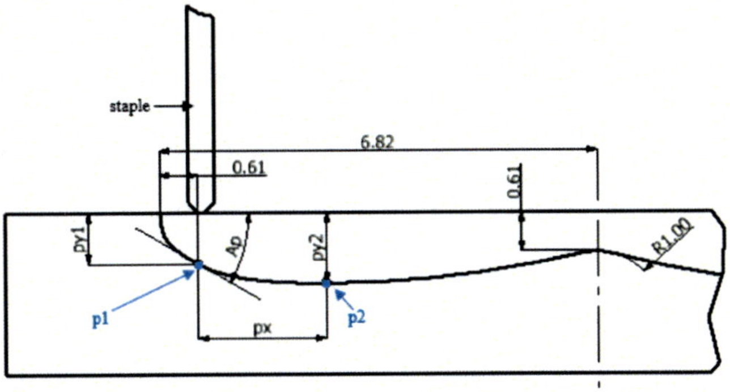

Fig. 12. Anvil parameterization for optimization.

For each side, *py1* can vary between 0.5 mm and 1.6 mm and *px* can vary between 1.2 mm and 3 mm. The maximum and minimum values for *Ap* are function of p1 and the maximum and minimum values of *py2* are function of *py1* and *Ap*. The minimum value of *py1* ensures that the space between the paper and the first contact point with the anvil is slightly larger than the thickness of the staple and the maximum value is an exaggerated value for the depth of the first contact point. p2 is always lower than p1 by a minimum difference function of *px* and *Ap* and together with the minimum value of *px* ensures that the staple does not get stuck when stapling 2 sheets or without sheets of paper, according with preliminary work.

4 Results

From the optimization a candidate profile was chosen. The final parameters for the left side, in mm, are: py1 = 0.5, py2 = 1.12, px = 1.2 and Ap = 44°; and for the right side are: py1 = 0.8, py2 = 1.17, px = 1.2 and Ap = 47°. The splines were changed to arcs of circumference and straight lines that accurately match the splines giving the final groove geometry characterized by Fig. 13.

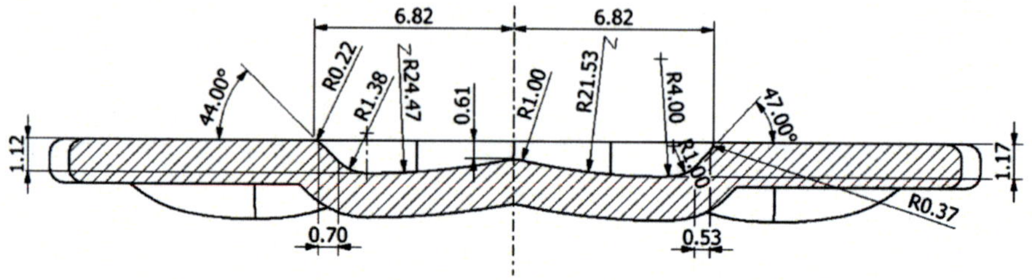

Fig. 13. Groove profile of the proposed new anvil.

The maximum force for the proposed new anvil groove profile is 32% less than the required force for the anvil theoretical groove profile of the commercial stapler. The necessary force curve is more spread out along the driver blade motion, Fig. 14. Comparing with the mean maximum value to punch the paper, the maximum force for the proposed new anvil is 8% greater which is far less than the 59% difference for the anvil of the commercial stapler.

By discretizing the sheets of paper, the force-displacement curve for stapling 25 sheets of paper with the theoretical anvil of the commercial stapler changes slightly, but the general behavior is very similar. Some relative motion between the sheets of paper exists.

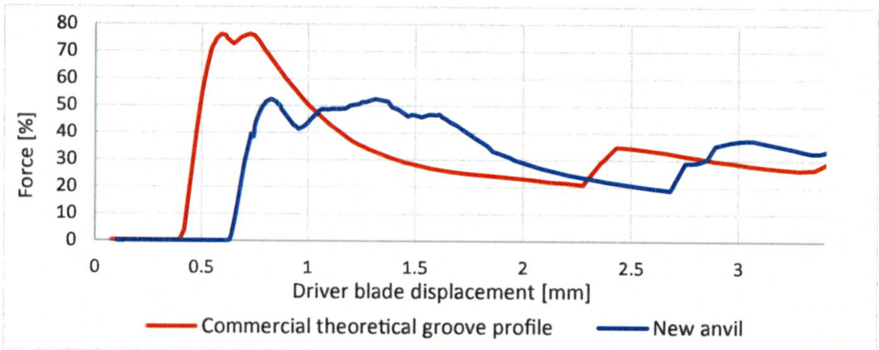

Fig. 14. Force-displacement curves comparing the anvil of the commercial stapler with the new anvil geometry for stapling 25 sheets of paper using a discretized model of the paper (all sheets modelled individually).

Regarding the final deformation of the staple, at the end of the simulation the anvil moves downwards while the driver blade remains fixed at the end position. This allows the staple legs to elastically recover and results in the final expected geometry of the system, Fig. 15. Both legs are sufficiently folded and the tips of the staple point to the paper. Regarding the requirement of distance to the paper, the right leg is expected to stay at 0.32 mm to the paper and the left leg at 0.35 mm being both less than the 0.4 mm requirement with a small gap.

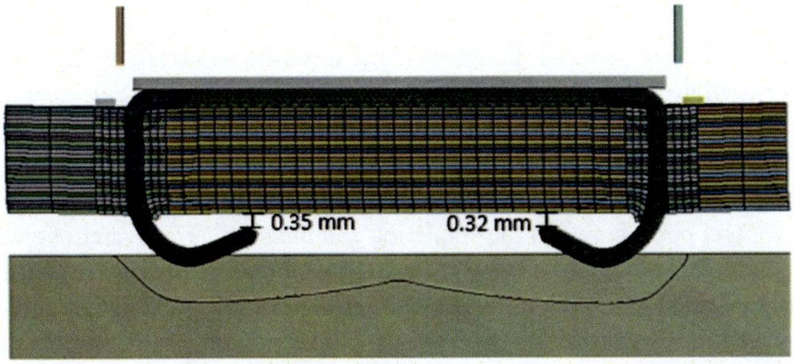

Fig. 15. Final geometry of the system and distance of the tips to the paper surface for stapling 25 sheets of paper with the new anvil geometry.

As the number of sheets of paper is reduced the reduction in maximum force diminishes. For stapling 15 sheets, middle duty, the reduction in maximum force is 29% and the maximum force is approximately the same as the maximum force for stapling 25 sheets of paper but slightly less.

For two sheets of paper the stapling processes occurs successfully but it results in an increase of 23% of maximum force when compared to the results for an anvil of the commercial stapler, Fig. 16.

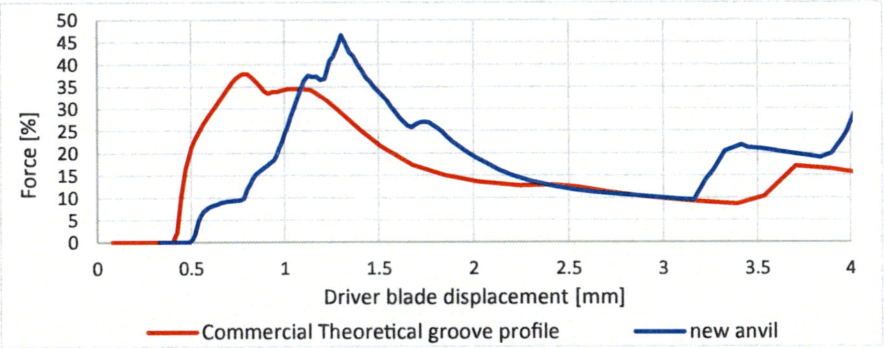

Fig. 16. Force displacement curves comparing the anvil of the commercial stapler with the new anvil geometry for stapling 25 sheets of paper using a discretized model of the paper.

Taking into account the obtained results, and respective actuating force reduction, it is important to highlight that it is possible to think in obtaining a global mechatronic stapler even for common applications. The actuating devices needed for this purpose are relatively easy to obtain and it is possible, also, to think in developing a dependable device connected with other information systems belonging to the office in which it will be used [13–16].

5 Conclusion

In the studied stapling process, the maximum force occurs during the staple folding process and is higher the more sheets are stapled. A study was carried out with the objective to reduce the maximum forces for folding the staple when stapling between 2 and 25 sheets of paper and thus reduce the overall maximum forces.

The maximum forces to fold the staple occur in a region of the groove profile near to the first contact point between the staple leg and the anvil. The system behavior when stapling many sheets of paper and few sheets of paper is slightly different with the staple yielding in upper a region of its legs when stapling fewer sheets of paper. The staple-anvil contact point also moves more inwards in the groove profile and the staple is in contact with the anvil by its tip for longer when stapling fewer sheets of paper. Therefore, optimizing the anvil's groove profile is a balance between performance for maximum number of sheets stapling and minimum number of sheets stapling.

An optimization analyses was carried out with the proposition of an asymmetric anvil. Results show that asymmetric anvils are good candidates for reducing maximum forces. For this case study, the proposed anvil results in a reduction in maximum force of 32% when compared with the anvil of the commercial stapler, and the maximum force is only 8% higher than the required maximum for paper punching smoothing the force curve. For few sheets of paper its performance is worse than for the anvil of the commercial stapler resulting in an increase of 23% in the maximum required force for stapling 2 sheets of paper, however the maximum forces are still higher the more sheets of paper stapled, meaning that the overall maximum force required to actuate the stapler is reduced.

The force curves can further be reduced by using this solution together with solution found in the literature that produce lever effects and/or use spring mechanisms.

More work is required to evaluate this solution to a broader set of stapler applications. The objective consists of being able for producing similar reductions in maximum force for different ranges of minimum and maximum number of sheets, staple sizes and other relevant system parameters.

Acknowledgements. The research leading to these results has received funding from the European Regional Development Fund under the Operational Programme "Competitiveness and Internationalization" - Research and Development (R&D) and Innovation to SMEs – R&D Individuals Projects with grant agreement n° 038397, correspondent to the project entitled STITCHED - Innovative Stitching Solutions: advanced stapling solutions and devices.

References

1. Kastner, A.: Stapleless stapler. U.S. patent 5 024 643, 18 June 1991
2. Hagemann, C.W.: Bypass clincher for stitching machine. U.S. patent 4 593 847, 10 June 1986
3. Marks, J.S.: Flat clinchstapler anvil assembly. U.S. patent 2014/0203060, 24 July 2014
4. Manabe, K.: Staple clinching mechanism in stapler. U.S. patent 6 036 074, 14 March 2000
5. Green, D.T., King, P.W., Soltanoff, L.: Surgical stapler for stitching body organs. U.S. patent 3 494 533, 10 October 1996
6. Plyley, A.K., Vidal, C.A., Redmond, R.J., Minck, J.L., Solyntjes, A.J.: Surgical stapler with mechanisms for reducing the fring force. U.S. patent 5 489 058, 6 February 1996
7. Jairam, A., Sohi, B.: Heavy duty stapler. U.S. patent 6 776 321, 17 August 2004
8. Kamdasamy, B.: Paper processing tool with force reducing drive arrangement. U.S. patent 7 942 298, 17 May 2011
9. Marks, J., Spring energized desktopstapler. U.S. 7 178 709, 20 February 2007
10. Szewczyk, W., Marynowski, K., Tarnawski, W.: An analysis of Young's modulus distribution in the paper plane. Fibres Text. East. Euro. **14**, 91–94 (2006)
11. Gary, A. B.: The Elastic Properties of Paper: A Review. IPCT Technical Paper Series, 1–21 (1984)
12. Tryding, J.: In-Plane Fracture of Paper. Phd Thesis, Division of Structural Mechanics, Lund University, Lund, Sweden (1996)
13. Leão, C.P., Soares, F.O., Machado, J.M., Seabra, E., Rodrigues, H.: Design and development of an industrial network laboratory. Int. J. Emerg. Technol. Learn. **6** (SPECIAL ISSUE.2), 21–26 (2011). https://doi.org/10.3991/ijet.v6iS1.1615
14. Kunz, G., Perondi, E., Machado, J.: Modeling and simulating the controller behavior of an Automated People Mover using IEC 61850 communication requirements. In: IEEE International Conference on Industrial Informatics (INDIN), 6034947, pp. 603–608. (2011). https://doi.org/10.1109/INDIN.2011.6034947
15. Silva, M., Pereira, F., Soares, F., Leão, C.P., Machado, J., Carvalho, V.: An overview of industrial communication networks. In: Flores, P., Viadero, F. (eds.) New Trends in Mechanism and Machine Science. MMS, vol. 24, pp. 933–940. Springer, Cham (2015). https://doi.org/10.1007/978-3-319-09411-3_97
16. Kunz, G., Machado, J., Perondi, E., Vyatkin, V.: A formal methodology for accomplishing IEC 61850 real-time communication requirements. IEEE Trans. Ind. Electron. **64**(8), 7878522, 6582–6590 (2017). https://doi.org/10.1109/TIE.2017.2682042

Environmental and Socio-economic Impact Assessment of the Switchgrass Production in Heavy Metals Contaminated Soils

Leandro Augusto Gomes[1], Jorge Costa[1,2], Fernando Santos[3], and Ana Luísa Fernando[1](✉)

[1] MEtRICs, Departamento de Ciências e Tecnologia da Biomassa, NOVA School of Science and Technology|FCT NOVA, Universidade NOVA de Lisboa, Campus de Caparica, 2829-516 Caparica, Portugal
ala@fct.unl.pt
[2] Escola de Gestão, Engenharia e Aeronáutica, Instituto Superior de Educação e Ciências Lisboa, Alameda das Linhas de Torres 179, 1750-142 Lisboa, Portugal
[3] Universidade Estadual ddo Rio Grande do Sul, Avenida Bento Gonçalves 8855, Porto Alegre 91540-000, Brazil

Abstract. The increased demand for biomass to produce bioenergy is arousing prices and land-use concurrence. These conflicts may be solved by producing dedicated crops for energy on surplus land that cannot be used for food, feed, nature conservation, or urbanization, reducing the indirect land-use change (ILUC) problems. This work aims to evaluate the environmental and socio-economic impact of switchgrass production in heavy metals contaminated soils. To determine ecological, social and economic sustainability, different categories were studied: energy balance, gases emission, land use, biological and landscape diversity, cost savings/losses, costs of CO_2 abatement, consumers/producer's acceptance and potential employment creation. Overall results suggest that switchgrass production in heavy metals contaminated soils has positive aspects and others less positive over switchgrass production in non-contaminated soils. The productivity loss in Cu and Zn contaminated soils reduces the energy, costs, and greenhouse savings but may contribute to improve the biological and landscape diversity and the soil and waters quality. In Pb and Cd contaminated soils, there was no adverse effect on the productivity, and after that, no effects on the environmental and socio-economic aspects, compared to non-contaminated soils. Yet, in Cr contaminated soils, the toxicity affected the switchgrass significantly, and no productivity was observed.

Keywords: Perennial crops · Switchgrass · Phytoremediation · Heavy metals · Soil contamination · Environmental impact assessment · Socio-economic impact

1 Introduction

Industrial crops are sustainable alternative feedstocks that can replace petroleum in energy, biofuels, and materials production [1–3]. However its cultivation must be sustainable. Indeed, the utilization of these biomasses is generating increased land prices

J. Machado et al. (Eds.): icieng 2021, LNME, pp. 410–419, 2022.
https://doi.org/10.1007/978-3-030-79165-0_38

and competition against land for food production [4]. One alternative is the utilization of lands that can not be used for food production, usually due to contaminants [5–7]. Soils contaminated with heavy metals, the land degraded for mining and intense agricultural activity are options for industrial crop cultivation [8–10]. Moreover, most of the industrial crops were already identified as tolerant to heavy metals contaminated land, and able to stabilize or accumulate those contaminants [11–16].

Switchgrass is a perennial crop with a lifespan of over 15 years [17], that can be used to produce biofuels, bioenergy, and bioproducts [18]. Despite switchgrass is not a hyperaccumulator, it can be used in the decontamination of soils - phytoremediation, due to its tolerance to soils contaminates with heavy metals. Indeed, switchgrass tolerance and productivity were also studied under several heavy metals contaminated soils, such as Cd, Pb. Cr, Cu and Zn [19–22]. Considering switchgrass versatility, bioenergy and phytoremediation potential, this work aimed to evaluate the environmental and socio-economic impact of the production of switchgrass in Zn, Pb, Cr, Cu, Ni, and Cd contaminated soils, to integrate them into a sustainable agriculture development, in Portugal and the Mediterranean.

2 Materials and Methods

In this study, it was assumed that switchgrass was used as solid fuel (biomass being harvested, crushed, dried and pelletized and then transported to be used in combined heat and power plant). To estimate the environmental and socio-economic impact on the production of switchgrass in heavy metals contaminated soils, switchgrass results obtained in the work of Gomes et al. [23, 24] were the basis for the assessment. In this work, switchgrass was cultivated in different heavy metal contaminated soils: Zn450, Pb450, Cr300, Cd4, Ni110, and Cu200 (respectively, 450 mg $Zn.kg^{-1}$, 450 mg $Pb.kg^{-1}$, 300 mg $Cr.kg^{-1}$, 4 mg $Cd.kg^{-1}$, 110 mg $Ni.kg^{-1}$, and 200 mg $Cu.kg^{-1}$). The concentrations of the contaminants were based on the limit value for each element in soil according to the Portuguese Decree-Law [25]. In order to prevent water stress, irrigation with 950 mm of tap water was applied along the growing cycle. Control soils were also tested with non-contaminated soil.

According to the methodology developed and applied by Schmidt et al. [18] and Fernando et al. [26, 27], the study focused on several categories for the environmental impact assessment. Energy savings were calculated by subtracting the energy input from the potential energy produced by the biomass's combustion being produced. Energy balance results obtained from Schmidt et al. [18] were used to estimate energy savings. The reduction of greenhouse gas emissions was calculated based on the same work [18]. Socio-Economic analysis was based in the work of Khanna et al. [28].

3 Results and Discussion

3.1 Productivity of Switchgrass

Figure 1 shows the relative yield of switchgrass cultivated in heavy metals contaminated soils compared to control soils. According to these results, it was observed that production of switchgrass in Zn and Cu contaminated soils produced less biomass than control, although this reduction was not statistically significant. Zn soils produced 82% of the control production and Cu soils produced 84% of the control production. Soils contaminated with Pb, Cd and Ni did not affect the yields compared to the control soils, but in Cr contaminated soils, there was no biomass production, indicating that this metal affected the growth of switchgrass significantly.

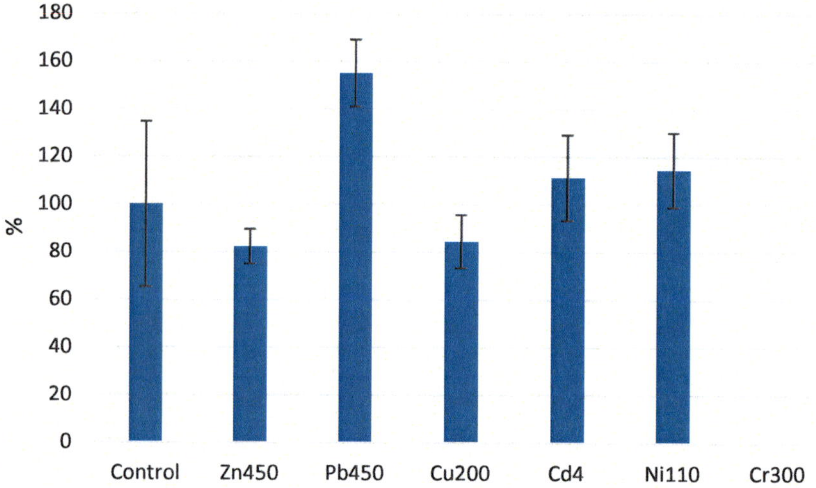

Fig. 1. Yield of switchgrass in heavy metals contaminated soils compared with the control (%).

The obtained outcomes (except for the chromium trial) show the potential for switchgrass production in contaminated soils with heavy metal contaminated land. In the work of Barbosa et al. [15], it was found that soils contaminated with Zn, affected Miscanthus but not giant reed, another two perennial crops that can be used for bioenergy, and lead did not also affect the growth of giant reed, as observed in this work with switchgrass. Chromium contamination also affected the growth of giant reed, although it was still possible to obtain some biomass production, something that was not possible in this work with switchgrass [15].

3.2 Energy Balance and Potential Reduction of Greenhouse Gases Emissions

The energy balance estimated for switchgrass production and use in heavy metal contaminated soils is presented in Fig. 2. It is possible to observe that the energy balance is statistically the same for all contaminations (in this case, the estimates in Cr soils were disregarded). This balance highlights that the energy potential of the crop was not affected by the contaminants studied (in the applied concentrations). According to the scenario presented, control show an energy balance of 197 GJ/ha, Zn and Cu showed lower energy balance than the control (154 and 159 GJ/ha, respectively), and Pb, Cd, and Ni showed higher energy balance than the control (310, 220 and 227 GJ/ha, respectively). In the contaminated soils, the energy balance follows the yields pattern, being lower, when yields are lower, and higher, when yields are higher. It is possible to see that in all the contaminated soils (except in Cr), the production and use of switchgrass for bioenergy through combustion will allow to save a significant amount of fossil energy.

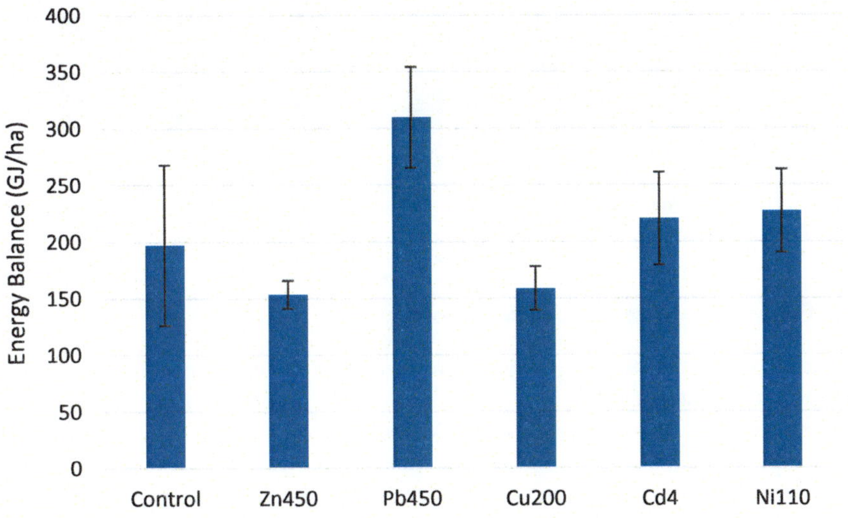

Fig. 2. Estimated energy balance (GJ/ha) for switchgrass production and use.

Figure 3 presents the potential reduction of greenhouse gases emissions (GHG) due to switchgrass production and use. According to the results estimated, the use of switchgrass as solid fuel is relevant to reduce GHG emissions, even when biomass is obtained from heavy metal contaminated soils (in this case, the estimates in Cr soils were disregarded). The experiment showed that the contamination does not significantly impact the amount of saved GHG, with results obtained in Pb450 soils, presenting even a significant increase compared with the control.

The reduction in yields affected also negatively the energy balance and the greenhouse gasses emission in the work of Schmidt et al. [18], that tested switchgrass in marginal soils of Europe.

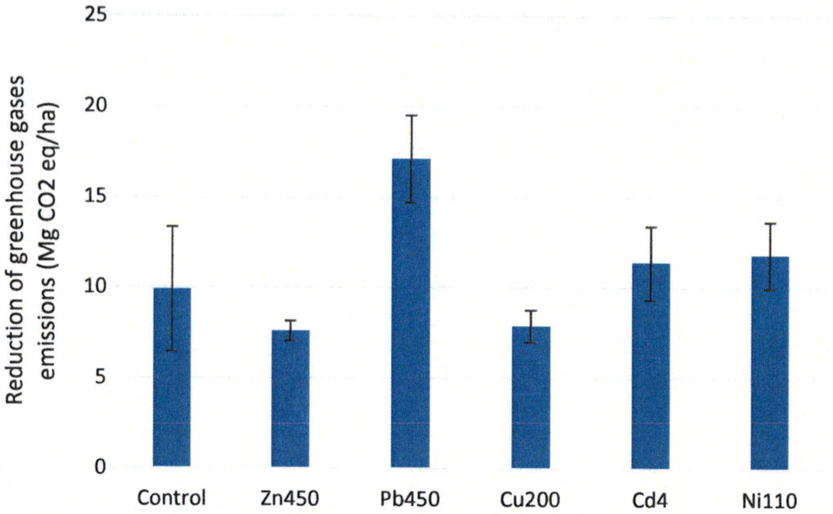

Fig. 3. Estimated reduction of greenhouse gases emissions (Mg CO_2 eq/ha) due to switchgrass production and use as solid fuel.

3.3 Other Environmental Considerations

A comparison of the land use of switchgrass in heavy metals contaminated soils, with control, is presented in Fig. 4.

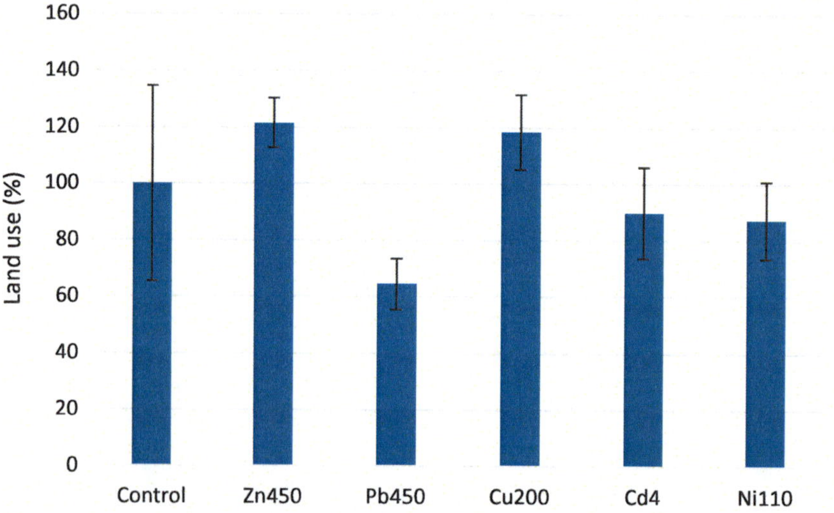

Fig. 4. Land use of switchgrass in heavy metals contaminated soils compared with a control (%).

It is possible to observe that the contamination of the soil did not affect significantly the area needed for switchgrass cultivation (in this case, the estimates in Cr soils were disregarded). Less area was needed when yields were higher than control (case of Pb soils) and higher area was needed to produce the same quantity of biomass for bioenergy when yields were lower (case of Zn and Cu), but differences to control soils were not significant. The same was also observed in the work of Boléo et al. [29]. In this work, lower yields of Miscanthus were obtained in Zn contaminated soils and therefore, higher land area was needed to obtain the same amount of material for energy.

This result indicates that switchgrass is a good option as an energy crop cultivated in soils contaminated with heavy metals. In fact, using contaminated land to produce biomass for biobased materials, biofuels, heat and power, can avoid problems derived by land use competition for food and feed. Also, considering that the land is degraded, this change in land use may be beneficial for the environment and for landowners due to the possible restoration of the functions and services of the soil ecosystem. The possibility to produce a biomass in a degraded soil will contribute to provide a cover for wildlife and will enrich landscape, namely aesthetics or structural heterogeneity, thus contributing to the landscape and biological diversity. Moreover, perennial grasses like switchgrass require reduced use of agrochemicals and soil tillage. These plants have a high aerial biomass and an extended belowground biomass, and consequently, it contributes to increase soil organic matter content and litter deposition and its presence in the soil contributes to reduce the erosion risk. These conditions favor occurrence of soil fauna and microfauna, especially decomposers, contributing to biodiversity. Moreover, a late harvest, e.g., end of January, may provide a shelter for small mammals and birds and site for invertebrates during wintering. In addition, switchgrass fields gain in heterogeneity if planted in smaller plots instead of as a wide landscape monoculture. For example, the case of postmining land, where the remediation areas are not contiguous [1, 2, 4, 13, 27].

3.4 Economic and Social Considerations

The breakeven delivered price for switchgrass production, resulting from the operating costs at farm gate plus transportation is presented in Fig. 5. Production of switchgrass in heavy metals contaminated soils increases significantly the breakeven delivered price of the biomass when yields are lower than in non-contaminated soils. In control, the breakeven price is about 20 €/Mg biomass. The breakeven delivery price for switchgrass in Zn and Cu contaminated soils showed an increase of around 20%, while for switchgrass in lead, nickel, and cadmium contaminated soils the value is similar or even lower (as in the case of Pb soils). Therefore, the increase in the cost makes switchgrass from zinc and copper contaminated soils a less viable alternative. On the other hand, lead, nickel, and cadmium already represent a viable alternative to the biomass production, making this an up-and-coming option for Pb, Ni and Cd contaminated soils landowners. The cost of switchgrass for bioenergy, even in the control, is noticeably higher than the cost of fossil-based energy (coal energy-equivalent biomass price, 15 €/Mg) [28] and is not yet economically attractive. Nevertheless, if grants and subsidies are credited to the production and conversion of biomass to energy and if a supplementary compensation for CO_2 abatement or other benefits are attributed, then this result can be minimized. On the other hand, if the cost of contaminated land remediation is included, switchgrass production in contaminated land may become economically sustainable.

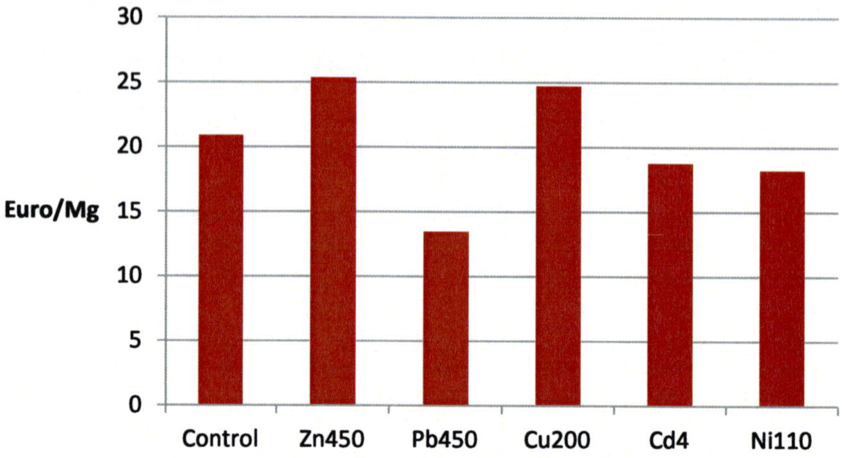

Fig. 5. Breakeven delivered price (€/Mg) of the switchgrass production and transportation.

Results of the costs of the abated emission of CO_2 when switchgrass is used as solid fuel are presented in Fig. 6. In this assessment cost of the conversion were not included, thus results presented are well below the real ones. The costs of the abated CO_2 emission show the efficiency of the use of switchgrass for energy to reduce greenhouse gas emissions. These costs enable us to evaluate the cost effectiveness of the switchgrass to energy option to reduce CO_2 emissions. According to Fig. 6, with a premium for CO_2 abatement above 9 €/Mg (CO_2 eq.), the use of switchgrass from non-contaminated soils, and from soils contaminated with Pb, Cd and Ni as solid fuel is feasible. But not from Zn and Cu contaminated soils. In this case it would be necessary a higher premium, namely 16 €/Mg (CO_2 eq.), for the soils contaminated with Zn, and 15 €/Mg (CO_2 eq.), for the soils contaminated with Cu, due to lower yields. Lead-contaminated biomass's conversion process goes further, showing a profit potential when considering the costs for the CO_2 uptake and no need for a premium for CO_2 abatement.

Production of switchgrass in contaminated soils can provide to human communities many social benefits. The use of degraded land for energy crops production still comprises much debate, and not always is socially accepted, but simultaneously this approach involves new opportunities, especially with a non-food crop that may have an economical income. Production of switchgrass on heavy metals contaminated soils, contributes to reduce the contamination of these soils, reestablishing its ecosystem function and services, and thus contributing to reduce environment and human exposure to pollutants. Also, the production of switchgrass in heavy metals contaminated soils, contributes to reduce human exposure to the effects on health of GHG emissions and the environment exposure to these emissions. Switchgrass production and use present also a positive improvement in terms of employment creation in small and medium-size enterprises and in rural areas, especially in less productive areas due to contamination. It also contributes positively to a more balanced rural development and to prevent rural exodus. Labour requirements per hectare for the production, in the farm, of switchgrass is 9 h per hectare per year [30]. Cultivation in contaminated soils doesn't represent an

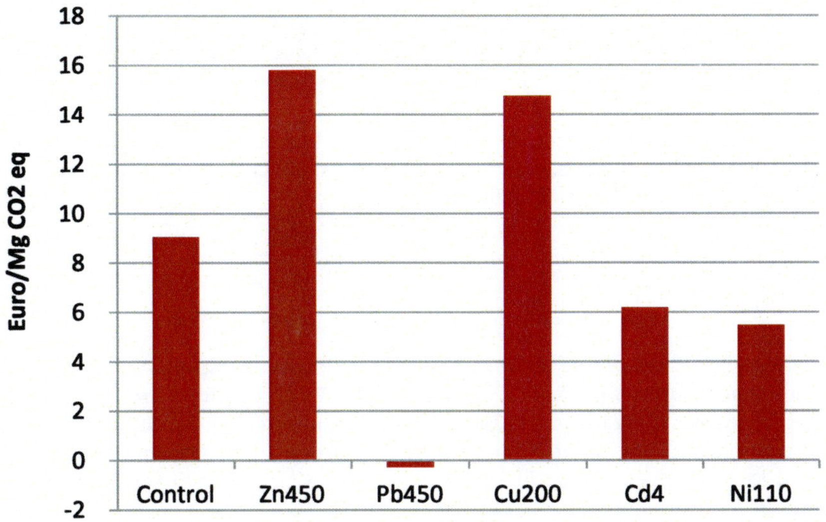

Fig. 6. Costs of the abated CO2 emission of switchgrass use for energy (€/Mg CO_2 eq).

increase in agricultural activities, by comparison with control, and so, does not lead to an increase in employment. But, if we compare with permanent degraded land (0–2 h per hectare per year) [30], than there is a considerable higher increase in employment.

4 Conclusions

One of the biggest constraints for the development of the bioenergy sector is the need to optimize and balance its economic, environmental and the social sustainability. The phytoremediation of heavy metals contaminated soils with switchgrass arises as a good option to consider compared to the abandonment of a heavy metals contaminated soil or to the traditional physical and chemical remediation techniques. Both from an environmental standpoint with the restoration of the degraded soil ecosystem and its services, or from an economic standpoint, either is from the bioenergy revenues to land owners or from the lower cost of phytoremediation compared to the traditional remediation processes. Overall results advocate that the production of switchgrass in heavy metals contaminated soils have positive characteristics and others that are less positive over switchgrass production in non-contaminated soils. The productivity loss in Zn and Cu contaminated soils reduces the greenhouse and the energy savings but it may contribute to enrich the biological and landscape diversity of those soils and the quality of waters and soils. Moreover, in Pb, Cd and Ni soils, results are promising compared with results obtained in non-contaminated soils, making switchgrass an alternative to be produced and used as solid fuel, helping to tackle the greenhouse effect and decontaminating at the same time that generates energy and profits for the owners of the heavy metals contaminated soils. Overall results suggest that for lead-contaminated soils in which the concentration is the limit allowed by the Portuguese legislation, the benefits of switchgrass production increases when compared to control, being even more attractive for investors. On the opposite, switchgrass production in Cr contaminated soils, at the level studied, revealed that this option should not be considered for phytoremediation and bioenergy production.

Acknowledgments. This research was supported by the Mechanical Engineering and Resource Sustainability Center-MEtRICs, which is financed by national funds from FCT/MCTES (UIDB/04077/2020 and UIDP/04077/2020).

References

1. Oliveira, J.S., Duarte, M.P., Christian, D.G., Eppel-Hotz, A., Fernando, A.L.: Environmental aspects of Miscanthus production. In: Jones, M.B., Walsh, M. (eds.) Miscanthus for Energy and Fibre, pp. 172–178. James & James (Science Publishers) Ltd, London (2001)
2. Fernando, A.L., Rettenmaier, N., Soldatos, P., Panoutsou, C.: Sustainability of perennial crops production for bioenergy and bioproducts. In: Alexopoulou, E. (ed.) Perennial Grasses for Bioenergy and Bioproducts, pp. 245–283. Academic Press, Elsevier Inc., UK (2018)
3. Fernando, A.L., Godovikova V., Oliveira J.F.S.: Miscanthus x giganteus: contribution to a sustainable agriculture of a future/present – oriented biomaterial. In: Materials Science Forum, Advanced Materials Forum II, vol. 455–456, pp. 437–441 (2004)
4. Dauber, J., et al.: Bioenergy from "surplus" land: environmental and socio-economic implications. BioRisk **7**, 5–50 (2012)
5. Abreu, M., et al.: Evaluation of the potential of biomass to energy in Portugal-conclusions from the CONVERTE project. Energies **13**(4), 937 (2020)
6. Von Cossel, M., et al.: Prospects of bioenergy cropping systems for a more social-ecologically sound bioeconomy. Agronomy **9**(10), 605 (2019)
7. Von Cossel, M., et al.: Marginal agricultural land low-input systems for biomass production. Energies **12**(16), 3123 (2019)
8. Barbosa, B., Fernando, A.L.: Aided phytostabilization of mine waste. In: Prasad, M.N.V., Favas, P.J.C., Maiti, S.K. (eds.) Bio-Geotechnologies for Mine Site Rehabilitation, pp. 147–158. Elsevier Inc., UK (2018)
9. Barbosa, B., Costa, J., Boléo, S., Duarte, M.P., Fernando, A.L.: Phytoremediation of inorganic compounds. In: Ribeiro, A.B., Mateus, E.P., Couto, N. (eds.) Electrokinetics Across Disciplines and Continents - New Strategies for Sustainable Development, pp. 373–400. Springer, Cham (2016). https://doi.org/10.1007/978-3-319-20179-5_19
10. Fernando, A.L., Boléo, S., Barbosa, B., Costa, J., Duarte, M.P., Monti, A.: Perennial grass production opportunities on marginal Mediterranean land. BioEnergy Res. **8**, 1523–1537 (2015)
11. Fernando, A.L., Barbosa, B., Costa, J., Papazoglou, E.G.: Giant reed (Arundo donax L.): a multipurpose crop bridging phytoremediation with sustainable bio-economy. In: Prasad, M.N.V. (ed.) Bioremediation and Bioeconomy, pp. 77–95. Elsevier Inc., UK (2016)
12. Sidella, S., Barbosa, B., Costa, J., Cosentino, S.L., Fernando, A.L.: Screening of giant reed clones for phytoremediation of lead contaminated soils. In: Barth, S., Murphy-Bokern, D., Kalinina, O., Taylor, G., Jones, M. (eds.) Perennial Biomass Crops for a Resource-Constrained World, pp. 191–197. Springer, Cham (2016). https://doi.org/10.1007/978-3-319-44530-4_16
13. Barbosa, B., Costa, J., Fernando, A.L.: Production of energy crops in heavy metals contaminated land: opportunities and risks. In: Li, R., Monti, A. (eds.) Land Allocation for Biomass Crops, pp. 83–102. Springer, Cham (2018). https://doi.org/10.1007/978-3-319-74536-7_5

14. Barbosa, B., Costa, J., Fernando, A.L., Papazoglou, E.G.: Wastewater reuse for fiber crops cultivation as a strategy to mitigate desertification. Ind. Crops Prod. **68**, 17–23 (2015)
15. Barbosa, B., et al.: Phytoremediation of heavy metal-contaminated soils using the perennial energy crops Miscanthus spp. and Arundo donax L. BioEnergy Res. **8**, 1500–1511 (2015)
16. Papazoglou, E.G., Fernando, A.L.: Preliminary studies on the growth, tolerance and phytoremediation ability of sugarbeet (Beta vulgaris L.) grown on heavy metal contaminated soil. Ind. Crops Prod. **107**, 463–471 (2017)
17. Bouton, J.: Improvement of switchgrass as a bioenergy crop. In: Vermerris, W. (ed.) Genetic Improvement of Bioenergy Crops, pp. 309–345. Springer, New York (2008). https://doi.org/10.1007/978-0-387-70805-8_11
18. Schmidt, T., Fernando, A.L., Monti, A., Rettenmaier, N.: Life cycle assessment of bioenergy and bio-based products from perennial grasses cultivated on marginal land in the Mediterranean region. BioEnergy Res. **8**, 1548–1561 (2015)
19. Arora, K., Sharma, S., Monti, A.: Bioremediation of Pb and Cd polluted soils by switchgrass: a case study in India. Int. J. Phytorem. **18**, 704–709 (2016)
20. Li, C., Wang, Q.H., Xiao, B., Li, Y.F.: Phytoremediation potential of switchgrass (Panicum virgatum L.) for Cr-polluted soil. In: Proceedings of International Symposium on Water Resource and Environmental Protection, vol. 3, pp. 1731–1734 (2011)
21. Juang, K.W., Lai, H.Y., Chen, B.C.: Coupling bioaccumulation and phytotoxicity to predict copper removal by switchgrass grown hydroponically. Ecotoxicology **20**, 827–835 (2011)
22. Novak, J.M., Ippolito, J.A., Watts, D.W., Sigua, G.C., Ducey, T.F., Johnson, M.G.: Biochar compost blends facilitate switchgrass growth in mine soils by reducing Cd and Zn bioavailability. Biochar **1**(1), 97–114 (2019). https://doi.org/10.1007/s42773-019-00004-7
23. Gomes, L., et al.: Phytoremediation potential of the perennial crops giant reed and switchgrass to soils contaminated with heavy metals. In: Proceedings of the 27th European Biomass Conference and Exhibition, EUBCE 2019, pp. 175–177. ETA-Florence Renewable Energies, Florence (2019)
24. Gomes, L., et al.: Phytoremediation potential of perennial crops in soils contaminated with heavy metals. In: Vilarinho, C., Castro, F., Gonçalves, M., Fernando A. L. (eds.) Book of Proceedings of the 5th International Conference WASTES: Solutions, Treatments and Opportunities. CVR - Centro para a Valorização de Resíduos, Guimarães (2019)
25. Decreto-Lei N. 276/2009 Regime jurídico de utilização agrícola das lamas de depuração em solos agrícolas. Diário da República 192, Série I, pp. 7154–7165 (2009)
26. Fernando, A.L., Duarte, M.P., Almeida, J., Boléo, S., Mendes, B.: Environmental impact assessment of energy crops cultivation in Europe. Biofuels Bioprod. Biorefin. **4**, 594–604 (2010)
27. Fernando, A.L., Costa, J., Barbosa, B., Monti, A., Rettenmaier, N.: Environmental impact assessment of perennial crops cultivation on marginal soils in the Mediterranean Region. Biomass Bioenergy **111**, 174–186 (2018)
28. Khanna, M., Dhungana, B., Clifton-Brown, J.: Costs of producing Miscanthus and switchgrass for bioenergy in Illinois. Biomass Bioenergy **32**, 482–493 (2008)
29. Boléo, S., Fernando, A.L., Duarte, M.P., Mendes, B.: Environmental and socio-economic impact assessment of the Miscanthus production in Zn contaminated soils. In: Castro, F., Vilarinho, C., Carvalho, J., Castro, A., Araújo, J., Pedro, A. (eds.) Book of Proceedings 2nd International Conference: Wastes: Solutions, Treatments and Opportunities, pp. 657–662. CVR, Centro de Valorização de Resíduos, Guimarães (2013)
30. Fernando, A.L., Duarte, M.P., di Virgilio, N., Mendes, B.: Is Miscanthus a sustainable land-use alternative for the energy market in Portugal? In: De Santi, G.F., Dallemand, J.F., Ossenbrink, H., Grassi, A., Helm, P. (eds.) Proceedings of the 17th European Biomass Conference and Exhibition, From Research to Industry and Markets, pp. 2270–2272. ETA-Renewable Energies and WIP-Renewable Energies (2009)

Influence of Hand Sanitisers on the Friction Properties of the Finger Skin Amid the COVID-19 Pandemic

Vlad Cârlescu(✉), Cezara Măriuca Oprişan, Bogdan Chiriac, Gelu Ianuş, and Dumitru N. Olaru

Department of Mechanical Engineering, Mechatronics and Robotics, The "Gheorghe Asachi" Technical University of Iasi, 700050 Iasi, România
vlad.carlescu@tuiasi.ro

Abstract. The recent COVID-19 pandemic involved in increased hand hygiene to prevent virus transmission. There are a lot of hand hygiene products available on the market but if are frequently used they may alter skin barrier integrity and function. This aspect can be important in daily living activities that involve gripping, feeling and manipulating objects. In this paper the authors studied the influence of the frequently hand disinfection on the friction behavior of the finger skin. Commercially available alcohol-based hand sanitisers were used to perform repeated skin disinfection and the coefficient of friction (COF) was measured on three healthy subjects by using a steel cylinder laterally sliding on the finger tip skin. The preliminary results showed that for all test subjects the coefficient of friction significantly decrease, up to 50%, when frequently hand sanitisers are used compared to daily "dry" skin condition.

Keywords: COVID-19 · Hand disinfection · Skin friction

1 Introduction

The world is facing a medical crisis amid the COVID-19 pandemic and the role of adequate hygiene is inevitable in controlling the spread of infection in public places and healthcare institutions. World Health Organization (WHO) and Centers for Disease Prevention and Control (CDC) recommends use of Personal Protection Equipment (PPE) and frequent hand sanitizing as one of the ancillary strategies to reduce transmission of the newly identified virus [1, 2]. Healthcare workers (HCWs) are the most exposed to a variety of irritants and allergens due to PPE and hand hygiene measures, as well as face mask induced pressure-related skin damage [3–5].

Frequent hand sanitizing can have a broad spectrum of potential adverse effects, ranging from cutaneous xerosis (abnormally dry skin) to severe allergic or irritant reactions [5–10]. Thus, hand hygiene products may alter skin barrier integrity and function that can be promising in daily living activities that involve rubbing against surfaces,

J. Machado et al. (Eds.): icieng 2021, LNME, pp. 420–428, 2022.
https://doi.org/10.1007/978-3-030-79165-0_39

feeling, gripping or manipulating objects. Moreover, the manipulating of objects by fingers for excessive sweating with difficulties in safety grip [11] can be influenced by the frequently hand disinfection.

The objective of this paper is to demonstrate that the frequently hand disinfection can influence the coefficient of friction (COF) of the skin fingers generally by an important decreasing of the COF, depending on the type of disinfectant.

2 Hand Disinfection

In order to prevent virus transmission hand hygiene can be realized by usually handwashing, antiseptic handwashing, and antiseptic hand sanitisation, generally alcohol-based hand sanitisers commonly named "alcohol-based hand rub (ABHR)" [12, 13]. Also, alcohol-free hand rubs were used for hand disinfection during the COVID-19 pandemic but they are not recommended by the health organizations and should therefore be avoided [14, 15].

The ABHRs mainly contain a type of alcohol like ethanol (65% to 95% v/v) as antiseptic agent and various excipients including viscosity enhancers, emollients, buffers, preservatives, colorants and fragrances, depending on the type of the formulation. Gel formulations are more portable and convenient to dispense on-the-go due to their ease of use and low risk of spillage compared to liquid-based products. Gel-based formulations also reduce the evaporation rate of alcohol and help alcohol to spread and penetrate through contaminating organisms. Disinfectant effectiveness in the ABHRs depends on type of alcohol, concentration, quantity applied on hands and time of exposure [12, 16–19]. Frequent exposure to alcohol can cause skin dryness. Emollients like glycerin (0.5% to 0.73% v/v), as well as other skin conditioners, have been shown to decrease the drying effect of alcohol on the skin [20–23].

Public health organizations recommended a time of 10 s to 60 s for hands disinfection. It has been argued that rubbing time is significantly affected by various other factors such as the dose of the sanitiser, the hand size (application surface area), and the formulation itself [24–26].

Moreover, hand sanitisation will remain at the forefront of infection prevention measures and it is reasonable to speculate that the current awareness of the general public of the importance of hand disinfection will remain assimilated and will become an integral part of people's hygiene practices, even post-COVID-19 era [27].

3 Experimental Investigations

3.1 Participants

In this work the experiments have been realized in vivo on the middle finger skin of three healthy subjects without any hand dysfunction or neurological disorders. Three persons participated at experiments: male 1 (36 years-old), male 2 (29 years-old) and a 29 years-old female.

3.2 Hand Sanitisers Formulation Tested

In the present study two ABHRs formulation were used for hand disinfection:

- liquid alcohol-based hand rub (liquid ABHR): ethanol (75%), benzalkonium chloride (0.1%), glycerin, purified water;
- gel alcohol-based hand rub (gel ABHR): ethanol (70%), water (28.5%), glycerin (0.5%), 2-propanol (0.5%), aminomethyl propanol (0.2%).

The liquid ABHR was and is still used in our faculty during the pandemic and gel product is commercially available.

3.3 Hand Disinfection Procedure

In this experiments hand disinfection was performed by using a spray container for liquid-based product and the available recipient of commercial gel formulation. The hand disinfection was realized in short-term application of 10 times repeated dose in room temperature condition. The hand sanitisers doses applied on the hand were approximatively 2 ml of liquid ABHR and 1 ml of gel product on each application step. Hands were rubbed together after each sanitiser application and wait up to 60 s until start to feel dry.

3.4 Friction Tests and Equipment

Friction tests were realized by using a Tribometer UMT-2 (Bruker) in a reciprocating-sliding fashion to evaluate the influence of hand sanitisers on the friction properties of the finger skin.

From our findings there are no studies to reveal the influence of hand sanitisers on the friction behavior of the skin. Plum et al. [28] evaluate skin barrier response of the forearms exposed to either water immersion or occlusion followed by repeated short-term applications of ABHR assessed by measurement of transepidermal water loss (TEWL), electrical conductance, pH, and erythema.

In this study the authors have used an experimental testing methodology previously reported [29, 30]. With the proposed methodology the authors obtained both the variation of tangential force and COF in the contact of a steel cylinder and middle finger skin, for various normal loads. In Fig. 1 is presented the Tribometer UMT-2 and a detailed of cylinder probe in contact with the finger tip.

Friction test were performed by using ABHRs products described in Sect. 3.1 and 10 sanitiser application step procedure was adopted. Firstly, for all test subjects the coefficient of friction was evaluated in “dry” skin condition. The hand was positioned on the linear table of the Tribometer as is presented in Fig. 1. The hand is relaxed and not moved during the tests. Before start the test, the cylinder was positioned like his middle is perpendicular to the finger tip. Through the steel cylinder a normal load of 1 N was applied and laterally moved in a reciprocating-sliding fashion on a total stoke

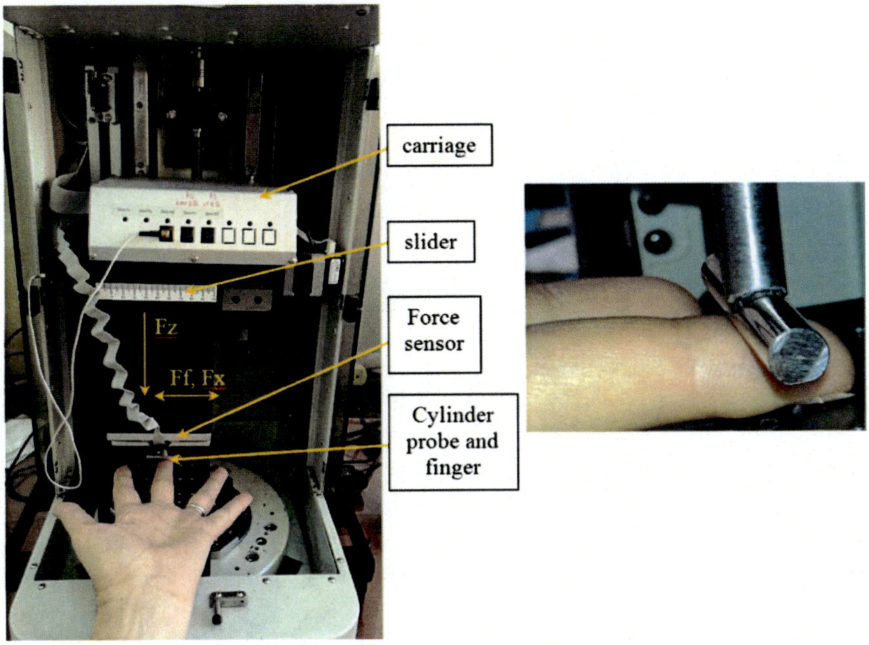

Fig. 1. Experimental setup for friction measurements

of 20 mm and sliding velocity of 1 mm/s. The friction test parameters such as normal load (F_Z), tangential force (F_X), horizontal displacement (X), time (T), velocity (V3) and COF have been monitories by the Tribometer. During the tests a data rate of 10 Hz was used.

4 Results

4.1 Skin Friction

In Fig. 2 is presented variation of the tangential force (F_X) as function of the horizontal displacement of cylinder along the stroke with 10 mm in one direction and 10 mm in the opposite direction. It can be observed successive "stick" phases with lateral deformation of the finger tissue and "slip" phases with a relative constant tangential force.

The skin friction behavior during the experimental tests can be explained as follows, beginning from the start point:

- point A is the reference point when the middle of the cylinder is perpendicular to the finger and the initial normal load is applied by the carriage;
- when the initial load is reach the set value, the slider moves the cylinder in one direction in order to positioning to the maximum point of stroke (C). In this step the elastic deformation of the skin occurs (AB) that correspond to a "stick" phase and the friction force increase until the cylinder start to slide to the skin.

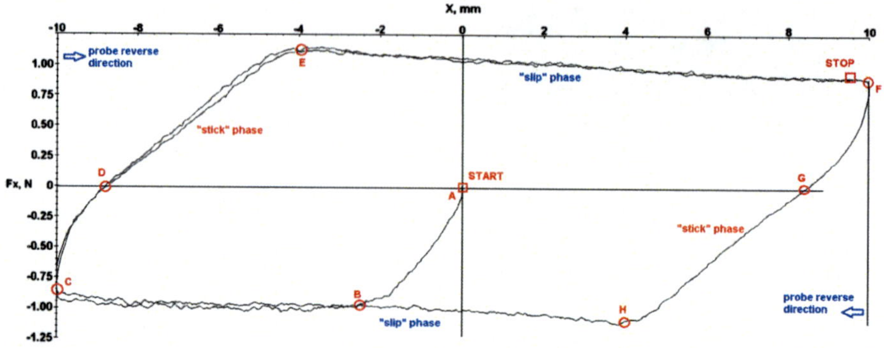

Fig. 2. Tangential force as function of probe position on middle finger in "dry" skin condition for a normal force of 1 N and sliding velocity of 1 mm/s

- the test itself begins from point C when the skin is already strained elastically in one direction (AB). At this point the probe is moved in the reverse direction and the finger skin start to relax (CD), the friction force decreases, passes through zero and increases again but with reversed sign as the elastic deformation of the skin is occurring in the opposite direction (DE).
- the "slip" phase occur and the cylinder slides on the finger skin until reach the end of the stroke (F). In this point the probe reverse direction, the skin that was strained (DE) begins to relax (FG) and strained (GH) in the opposite direction. Again, when the friction force exceeds the adhesion force the probe start to slides and the friction process continue.

The skin friction process presented above can also be observed in Fig. 3 that illustrates the typical variation of the coefficient of friction as function of the cylinder probe position along the stroke.

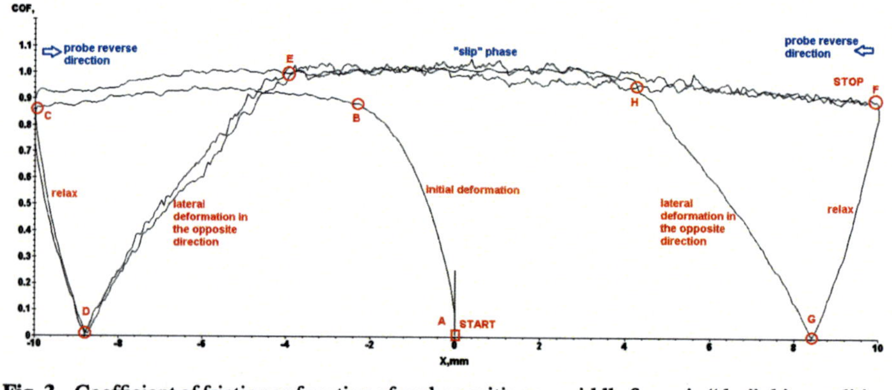

Fig. 3. Coefficient of friction as function of probe position on middle finger in "dry" skin condition for a normal force of 1 N and sliding velocity of 1 mm/s

Figure 4 illustrates the variation of the normal load (F_Z) during the friction tests. The normal force reveals some variations due to the elastic behavior of the skin and blood vessels from skin layers structure. In Fig. 4 is also plotted the sliding velocity (V3) of the cylinder probe.

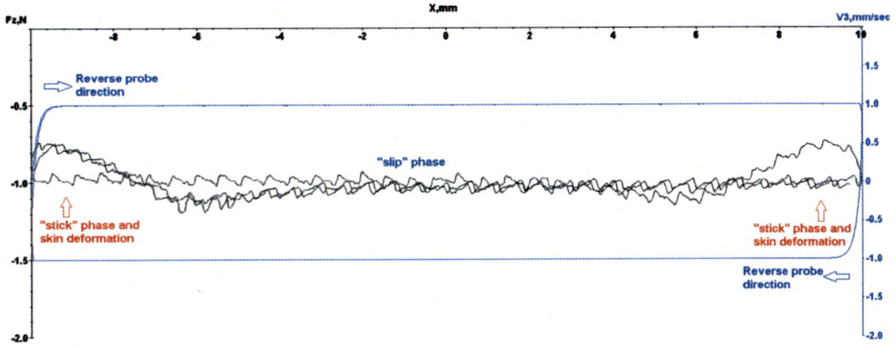

Fig. 4. Variation of normal force (F_Z) and horizontal sliding velocity (V3) in friction tests

4.2 The Coefficient of Friction on Disinfected Skin

Friction tests were performed on the middle finger of left hand adopting 10 repeated sanitisers application. In Fig. 5 are presented the variation of the coefficient of friction for test subjects in both "dry" and disinfected skin with liquid ABHR.

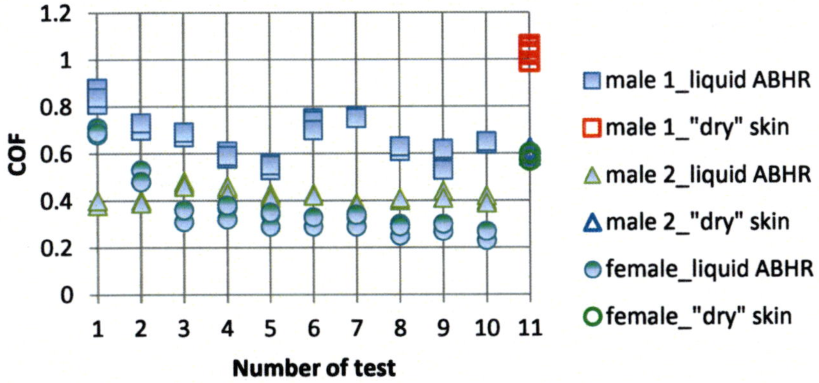

Fig. 5. COF variation during the 10 times hand disinfection with liquid ABHR

It can be observed that for all test subjects the coefficient of friction significantly decreases when liquid ABHR sanitiser is used compared with "dry" skin condition. For female subject the COF decreases more than 50%, from 0.6 in "dry" skin condition down to 0.23 when hand was disinfected with liquid ABHR.

Figure 6 illustrate the variation of the coefficient of friction for test subjects for both "dry" and disinfected skin with gel ABHR. Again, it can be observed a decrease of COF with about 50% for female subject from 0.8 to 0.4.

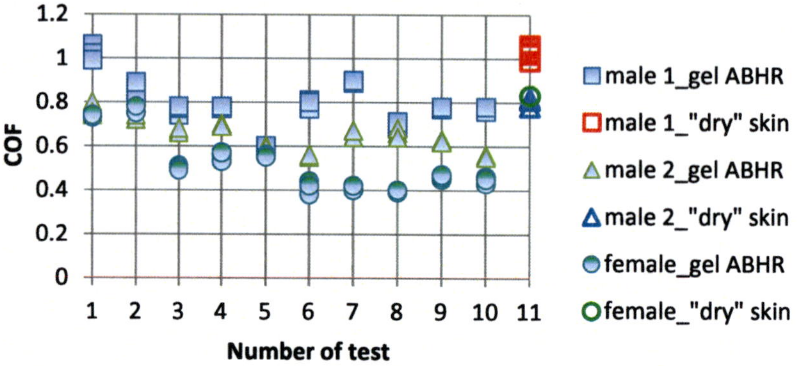

Fig. 6. COF variation during the 10 times hand disinfection with gel ABHR

From Figs. 5 and 6 it can be observed that the female subject has the lowest values for COF during the hand disinfection. It can be explained by the fact that the skin friction properties are dependent on many factors such as gender, age, moisture etc. [31].

5 Conclusions

This paper presents an experimental study regarding the influence of frequently hand disinfection in the COVID 19 pandemic. The variation in skin friction properties is important in skin damage affections like contact dermatitis and can influence the daily live activities such as sensing, rubbing, feeling or manipulating objects.

The authors have used a particular methodology to evaluate the coefficient of friction of the finger skin by reciprocating-sliding a steel cylinder perpendicular to the middle finger tip of left hand for three test subjects. The friction tests were realized by a repeated application of alcohol-based hand rub sanitisers commercially available. The preliminary results reveal that the coefficient of friction (COF) showed a significantly decreases of more than 50% for female subjects when ABHR sanitisers are used compared with "dry" skin condition. The experimental methodology and the results can be useful in studying the hand sanitisers and their influence on the skin barrier function especially for healthcare workers (HCWs) that use Personal Protection Equipment (PPE) and frequent hand sanitizing being exposed to a variety of irritants and allergens. Further work can be done by testing many other subjects with particular skin condition and various sanitisers formulation and correlating the skin friction behavior with the skin layer properties and environmental factors.

References

1. CDC: Show me the science – when & how to use hand sanitizer in community settings (2020). https://www.cdc.gov/handwashing/show-me-the-science-hand-sanitizer.html. Accessed 3 Feb 2021
2. World Health Organization. https://www.who.int/publications/i/item/10665-331495. Accessed 3 Feb 2021
3. Alpalhão, M., Filipe, P.: The show must go on. The impacts of SARS-CoV-2 pandemic on cutaneous allergology and patch testing. Eur. Ann. Allergy Clin. Immunol. **52**(6), 280–281 (2020)
4. Blicharz, L., et al.: Hand eczema—a growing dermatological concern during the COVID-19 pandemic and possible treatments. Dermatol. Ther. **33**, e13545 (2020)
5. Darlenski, R., Tsankov, N.: COVID-19 pandemic and the skin: what should dermatologists know? Clin. Dermatol. **38**, 785–787 (2020)
6. Birnbach, D.J., McKenty, N.T., Rosen, L.F., Arheart, K.L., Everett-Thomas, R., Lindsey, S.F.: Does adherence to World Health Organization hand hygiene protocols in the operating room have the potential to produce irritant contact dermatitis in anesthesia providers? Anesth. Analg. **129**(6), e182–e184 (2019)
7. Spence, N.Z., Lu, M.E., Larson, A.R., Ortega, R.: COVID-19 and occupational skin hazards for anaesthetists. Br. J. Anaesth. **125**(6), e476–e478 (2020)
8. Bhatia, R., et al.: Iatrogenic dermatitis in times of COVID-19: a pandemic within a pandemic. JEADV **34**, e563–e566 (2020)
9. Drenovska, K., Schmidt, E., Vassileva, S.: Covid-19 pandemic and the skin. Int. J. Dermatol. **59**, 1312–1319 (2020)
10. Kiely, L.F., Moloney, E., O'Sullivan, G., Eustace, J.A., Gallagher, J., Bourke, J.F.: Irritant contact dermatitis in healthcare workers as a result of the COVID-19 pandemic: a cross-sectional study. Clin. Exp. Dermatol. **46**, 142–144 (2021)
11. Zackrisson, T., Eriksson, B., Hosseini, N., Johnels, B., Krogstad, A.L.: Patients with hyperhidrosis have changed grip force, coefficient of friction and safety margin. Acta Neurol. Scand. **117**, 279–284 (2008)
12. Todd, E.C.D., Michaels, B.S., Holah, J., Smith, D., Greig, J.D., Bartleson, C.A.: Outbreaks where food workers have been implicated in the spread of foodborne disease. Part 10. Alcohol-based antiseptics for hand disinfection and a comparison of their effectiveness with soaps. J. Food Prot. **73**, 2128–2140 (2010)
13. Edmonds, S.L., et al.: Comparative efficacy of commercially available alcohol-based hand rubs and World Health Organization-recommended hand rubs: formulation matters. Am. J. Infect. Control **40**, 521–525 (2012)
14. Jing, J.L.J., Pei Yi, T., Bose, R.J.C., McCarthy, J.R., Tharmalingam, N., Madheswaran, T.: Hand sanitizers: a review on formulation aspects, adverse effects and regulations. Int. J. Environ. Res. Public Health **17**(9), 3326 (2020)
15. Dear, K., Grayson, L., Nixon, R.: Potential methanol toxicity and the importance of using a standardized alcohol-based hand rub formulation in the era of COVID-19. Antimicrob. Resist. Infect. Control **9**, 129 (2020)
16. Berardi, A., et al.: Hand sanitisers amid COVID-19: a critical review of alcohol-based products on the market and formulation approaches to respond to increasing demand. Int. J. Pharm. **584**, 119431 (2020)
17. Rundle, C.W., et al.: Hand hygiene during COVID-19: recommendations from the American Contact Dermatitis Society. J. Am. Acad. Dermatol. **83**(6), 1730–1737 (2020)

18. Boyce, J.M., Pittet, D.: Guideline for hand hygiene in health-care settings: recommendations of the Healthcare Infection Control Practices Advisory Committee and the HICPAC/SHEA/APIC/IDSA Hand Hygiene Task Force. Infect. Control Hosp. Epidemiol. **23**, S3–S40 (2002)
19. Kampf, G.: Efficacy of ethanol against viruses in hand disinfection. J. Hosp. Infect. **98**, 331–338 (2018)
20. Ahmed-Lecheheb, D., Cunat, L., Hartemann, P., Hautemaniere, A.: Prospective observational study to assess hand skin condition after application of alcohol-based hand rub solutions. Am. J. Infect. Control **40**, 160–164 (2012)
21. Harbarth, S., et al.: Interventional study to evaluate the impact of an alcohol based hand gel in improving hand hygiene compliance. Pediatr. Infect. Dis. J. **21**, 489–495 (2002)
22. Kramer, A., Bernig, T., Kampf, G.: Clinical double-blind trial on the dermal tolerance and user acceptability of six alcohol-based hand disinfectants for hygienic hand disinfection. J. Hosp. Infect. **51**, 114–120 (2002)
23. Houben, E., De. Paepe, K., Rogiers, V.: Skin condition associated with intensive use of alcoholic gels for hand disinfection: a combination of biophysical and sensorial data. Contact Dermatitis **54**, 261–267 (2006)
24. WHO: WHO guidelines on hand hygiene in health care: a summary (2009)
25. Wilkinson, M.A.C., Ormandy, K., Bradley, C.R., Fraise, A.P., Hines, J.: Dose considerations for alcohol-based hand rubs. J. Hosp. Infect. **95**, 175–182 (2017)
26. Kenters, N., Eikelenboom-Boskamp, A., Hines, J., McGeer, A., Huijskens, E.G.W., Voss, A.: Product dose considerations for real-world hand sanitiser efficacy. Am. J. Infect. Control **48**, 503–506 (2020)
27. Dindarloo, K., et al.: Pattern of disinfectants use and their adverse effects on the consumers after COVID-19 outbreak. J. Environ. Health Sci. Eng. **18**, 1301–1310 (2020)
28. Plum, F., Yüksel, Y.T., Agner, T., Nørreslet, L.B.: Skin barrier function after repeated short-term application of alcohol-based hand rub following intervention with water immersion or occlusion. Contact Dermatitis **83**, 215–219 (2020)
29. Oprişan, C., Cârlescu, V., Barnea, A., Prisacaru, G., Olaru, D.N., Plesu, G.: Experimental determination of the Young's modulus for the fingers with application in prehension systems for small cylindrical objects. IOP Conf. Ser. Mater. Sci. Eng. **147**(1), 012058 (2016)
30. Cârlescu, V., Oprişan, C.M., Ianuş, G., Olaru, D.N.: Evaluation of friction behaviour on human finger skin considering precision grip task. IOP Conf. Ser. Mater. Sci. Eng. **997**, 012007 (2020)
31. Derler, S., Gerhardt, L.-C.: Tribology of skin: review and analysis of experimental results for the friction coefficient of human skin. Tribol. Lett. **45**, 1–27 (2012)

Examination of Adhesion Strength of D-Gun Sprayed Coatings Based on Tungsten and Chromium Carbides

Yuriy Kharlamov, Volodymyr Sokolov, Oleg Krol(✉), and Oleksiy Romanchenko

Volodymyr Dahl East Ukrainian National University, 59-a Central Pr., Severodonetsk 93400, Ukraine
krolos@i.ua

Abstract. The results of experimental research of the adhesion strength of D-Gun sprayed coatings based on tungsten and chromium carbides are presented. The basic structural and technological factors that affecting the adhesion strength of coatings are systematized. It is proved that the adhesion strength of D-Gun sprayed coatings depends on combination of materials in system «coating-basic material», roughness of sprayed surface, composition, dispersion and method of powder manufacturing, etc. It is shown that D-Gun spraying due to high velocity of sprayed particles increases the role of mechanical activation of contacting materials and makes it possible to obtain coatings with high adhesion strength with reduction of requirements for preparation of spraying surface. As known ideas about mechanisms of adhesion allow to obtain only qualitative estimates of adhesion strength of D-Gun sprayed coatings, therefore, the development of models that take into account a set of structural and technological factors affecting the adhesion strength is required.

Keywords: D-Gun sprayed coatings · Adhesion strength · Spraying distance · Tungsten carbide · Chromium carbide · Hardness

1 Introduction

Modern science and technique is characterized by extensive application of various protective and functional coatings [1–5]. In the relevant field – surface engineering [6] – numerous methods of obtaining coatings and films are widely used. Among them, the technique and technology of thermal spraying [7–11] are intensively developing. If at first the methods of thermal spraying were based on mechanism of thermal activation of interacting materials, then at the present stage of development of science and technology, high – velocity methods in which mechanical mechanism of activation plays a predominant role are actively developing. Such methods include D-Gun spraying [1, 12–16]. Among the main advantages of D-Gun sprayed coatings are following: possibility of obtaining durable coatings during spraying; wide possibilities of regulation of thermal cycle of formed coating and product; high growth rate of coating thickness; reduced

J. Machado et al. (Eds.): icieng 2021, LNME, pp. 429–440, 2022.
https://doi.org/10.1007/978-3-030-79165-0_40

requirements for quality of preparation of sprayed surface; relative simplicity of designs and high reliability of technological equipment; high density of resulting coatings, etc.

2 Literature Review

The application and development of D-Gun sprayed coatings is associated with the requirements of strong and dense coatings. In most publications on this issue, much attention is paid to such an important property of coatings as adhesion strength [12, 16–18]. However, to date, the mechanisms of formation and formation of strong coatings for D-Gun spraying have not yet been sufficiently researched, which complicates the development of technological processes and designs of coated products, as well as corresponding technological equipment [19–24]. The quality and performance characteristics of thermal sprayed coatings, regardless of spraying method, strongly depend on adhesion strength of connection between coating and basic material of product, since failure of coating leads to destruction of «coating-basic material system». As a rule, this is an unacceptable failure. However, the prediction and control of adhesion strength of coating is difficult because it depends on many factors [12, 20, 25, 26]: method of thermal spraying and technological modes of coating and their subsequent processing; operating conditions of coated product and methods of adhesion strength determination; composition and type of starting material of coating, distribution of powder particles by size and morphology; basic material of product for spraying and the condition of sprayed surface, including presence and condition of oxide films (composition and thickness of oxide); roughness (the height of projections relative to average size of splats, and distance between projections of base roughness, which is characterized by root mean square roughness $R\Delta q$, that is, root mean square of ordinates of roughness profile); cleanliness (removal of surface impurities and abrasive residues); temperature and preheating method before spraying at temperature sufficient to remove adsorbents and condensates, etc.; residual stresses formed in «coating-basic material» system both during spraying process and during further processing and operation. These stresses also depend on many factors, including combination of materials in «coating-basic material» system, thermal spraying conditions and average surface temperature of substrate, and so on.; environmental conditions, such as temperature and humidity, ambient vibration transmitted to sample during test, and so on.

In addition, the structure of thermal sprayed coatings differs from the structure of compact materials, as they are usually composed from single particles (splats), the real contact surface of which is from about 15 to 60% of their surface: non – molten particles, pores of different shape, cracks and so on. The main mechanisms of adhesion between thermal spraying coating and lining are highly dependent on actual contacts between splats and between splats and substrate. They are divided into three main groups: (1) mechanical (or anchor) adhesion; (2) metal – metal connection (diffusion phenomenon); (3) chemical compound (formation of intermetallic compound with substrate).

The character of coating failure from substrate may be adhesive, cohesive or mixed. The adhesion strength can be determined by fracture mechanics [15, 20], which takes into account energy required for origin or propagation of cracks and evaluates adhesion in «coating-basic material» system in terms of fracture toughness. It is necessary to experimentally establish conditions of equilibrium under which elastic energy created by an external force (determined by geometry of sample and applied load) is balanced by propagation of a stable crack. At the critical value of rate of deformation energy release G_c, J/m^2, the crack propagates and, consequently, adhesion of «coating-basic material» system breaks down. However, this requires application of special laboratory equipment and samples of coated materials, as well as appropriate methods of analysis [20, 27–30].

Objective. The objective of this article is to research adhesion strength of D-Gun sprayed coatings based on tungsten and chromium carbides, as well as to analyze the existing ideas about formation mechanisms of strong adhesion of D-Gun sprayed coatings.

3 Research of Adhesion Strength

Adhesion strength is a complex characteristic of adhesive and cohesive strength of coatings, regardless of the method and technology of their application. Research of adhesion strength of tungsten carbide coatings with titanium alloys VT9 (85,9–91%Ti) and VT3-1 (85,9–91%Ti) were performed by method of failure of a conical pin with a diameter of 1,5 to 1,6 mm according to method described in [12, 20]. The thickness of coatings of standard powders of alloys VK8 (WC-8%Co) and VK15 (WC-15%Co) was from 0,2 to 0,3 mm. The maximum adhesion strength was achieved at a spraying distance of 150 mm (Fig. 1). The surface of samples was subjected to finishing treatment on turning machine tool before spraying of coating. The significant variation of adhesion strength values is due to presence of defects in structure and above all porosity of coatings. When using fine powders (from 2 to 5 μm) reasons of pore formation are: intensive coagulation of powder particles with formation of large conglomerates; formation of powder accumulations on walls of barrel due to subsidence and subsequent accidental breakdowns and transfer to surface of part, etc. When large pores are situated near boundary surface «coating-basic material» in area of the pin end measured local adhesion strength is much lower than the average.

From these results, it follows that values of adhesion strength and their variation are slightly lower for coatings with a higher cobalt content, which can be explained by higher proportion of metal connections in contact area of «coating-basic material» system.

The adhesion strength of D-Gun sprayed coatings depends on method of surface preparation of samples. The criterion R_z, μm, is not sufficient to predict adhesion strength, thus, it is necessary to consider other parameters that characterizing sprayed surface. When coating from VK15 (WC-15% Co) alloy is spraying on titanium alloy VT3-1 (85,9–91%Ti), maximum strength is achieved with light abrasive grit blasting of surface by corundum (Table 1).

When surface is treated by electrocorundum by method of abrasive grit blasting, surface of part is exposed to cold-hardening with formation of a developed micro relief, favorable for creation of high values of pressure at micro protrusions due to influence of radially spreading layers of particles. Increasing of microhardness of surface layer does not change the material's sensitivity to deformation velocity and to stress relaxation, as evidenced about dependence of hardness of deformation velocity that obtained by kinetic microhardness method. Higher hardness of surface layer of part is favorable for localization of plastic displacement in area of contact of particles with surface of part.

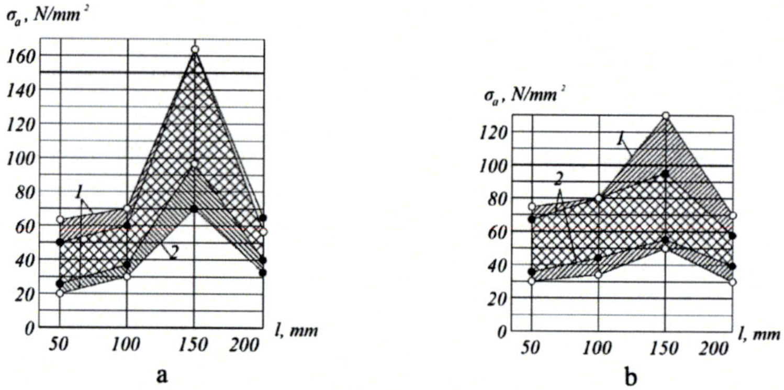

Fig. 1. Dependence of adhesion strength on failure of coatings from mechanical mixtures of powders of tungsten carbide with cobalt with samples of titanium alloys: (a) – VT9 (86,7–90,4% Ti); (b) – VT3-1 (85,9–91%Ti).Coatings with powder VK8 (WC-8%Co) – (1), VK15 (WC-15%Co) – (2)

The adhesion strength also depends on coating thickness and test temperature (Fig. 2). Temperature tests were carried out at a thickness of coatings of 0.3 to 0.5 mm. The rms deviation of tensile adhesion strength measured in MN/m^2 at temperature was: 20 °C – 13,1; 200 °C – 8,1; 300 °C – 21,4; 500 °C – 25,1. The increase of adhesion strength with increase temperature is explained by the partial removal of internal residual stresses in coatings.

The effect of spraying distance on adhesion strength of carbide coatings with titanium alloys during shear tests was researched (Fig. 3). The coatings were sprayed on cylindrical samples with a diameter of 19 mm through a mask that provides formation of an annular belt with a width of 1,5 to 3 and a thickness of 0,15 to 0,2 mm. The character of shear adhesion strength coincides with similar dependencies of tensile adhesion strength on failure, but maximum value of adhesion strength reaches at $l = 100$ mm. This is due to change of state of sprayed particles and conditions of high – velocity heterogeneous flow of samples.

The dependence of adhesion strength and other characteristics of coatings from traverse rate of sprayed surfaces were set according to methodology described in [12, 20]. As the velocity of movement increases, adhesion strength on failure and coefficient of powder application decrease (Fig. 4). This is due to increase of relative overlap area of sprayed single spots, and thereby decrease in substrate temperature in area of coating formation. The degree of influence of peripheral areas of spraying spots, contaminated with fine particles of solids, soot, etc., is also increasing. Moreover, changing of movement velocity does not affect microhardness, porosity and phase composition of coatings, but reduces surface roughness of sprayed coatings.

Table 1. The influence of preparation method of surface of the alloy VT3-1 (85,9–91%Ti) on adhesion strength with a coating from powder VK15 (WC-15%Co)

The method of surface preparation	R_Z, μm	Adhesion strength, MN/m^2
Finishing machining (turning)	10	50
	14	65
	16	120
Light sandblasting treatment by corundum with a grit of 450–980 μm at an air pressure of 0,45 MPa for 5 s	20	56
	22	101
	24	115
	27	140
Same for 120 s	36	90
	38	65
	39	35
Shot blasting treatment by steel balls of 0,5–1 mm at an air pressure of 0,45 MPa for 150 s	2°–3°	43°–72°

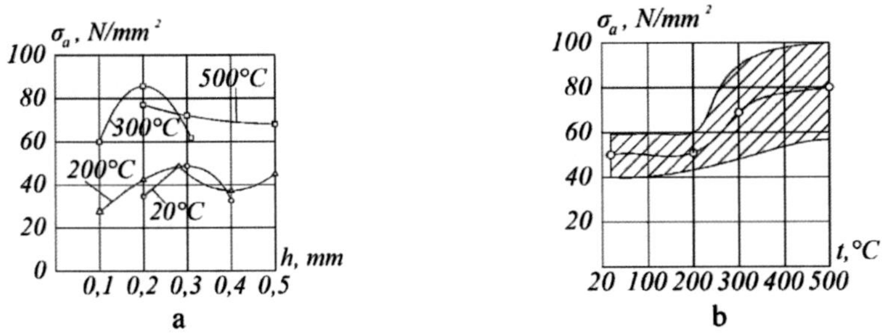

Fig. 2. Dependence of tensile adhesion strength on failure of coatings from powder VK8 of thickness (a) and test temperature (b)

Dependence of adhesion strength and other characteristics of coatings from powder VK15 (WC-15%Co) and VK18S (WC-18%Co) on movement velocity of sprayed surface: 1 – coefficient of powder application K_e; 2, 3 – adhesion strength on failure of pin powder. As hardness of steel samples increases, the adhesion strength decreases (Fig. 5). The highest adhesion strength have coatings from powder VK18S (WC-18%Co), the lowest – from powder KHN15S (3%Cr-2%C-15%Ni). Abrasive grit blasting of sprayed surface increases adhesion strength of coatings.

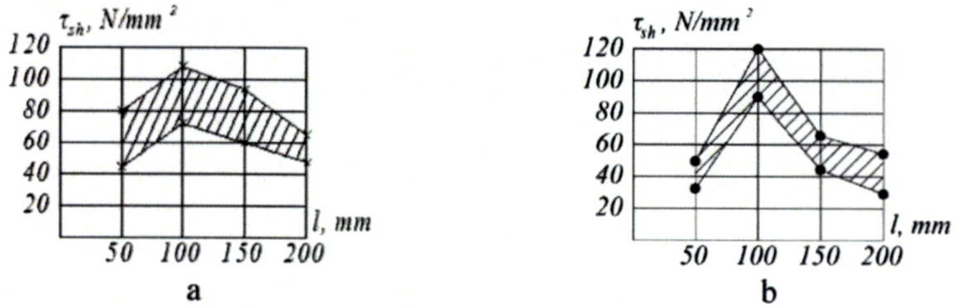

Fig. 3. Dependence of tensile adhesion strength on shear of coatings from powder VK8 (WC-8%Co) with titanium alloys VT9 (86,7–90,4% TI) (a) and VT3-1 (85,9–91%Ti) (b)

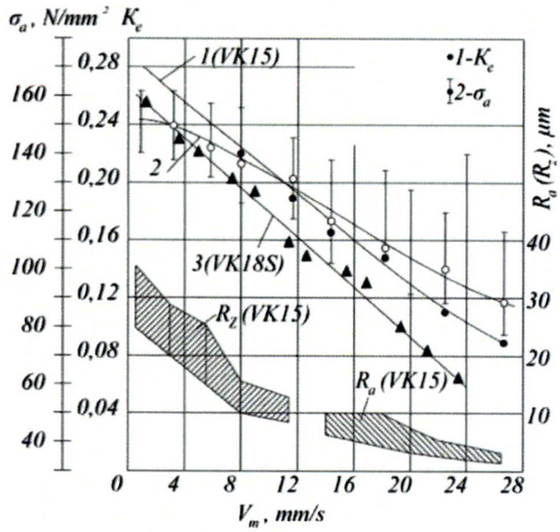

Fig. 4. Dependence of adhesion strength and other characteristics of coatings from powder VK15 (WC-15%Co) and VK18S (WC-18%Co) on movement velocity of sprayed surface: 1 – coefficient of powder application K_e; 2, 3 – adhesion strength on failure of pin powder

The only criterion (1) that determines adhesion strength is critical deformation ε_{cr}, at which integrity of contact area of substrate material and coating are broken by cracks formation [25],

$$\varepsilon_{cr} = f\left(\frac{\alpha_C}{\alpha_S}; \frac{E_C}{E_S}; \frac{G_C}{G_S}; \frac{T_{mC}}{T_{mS}}\right) \tag{1}$$

where α, E, G, T_m – are coefficient of thermal linear expansion, modulus of elasticity and shearing, melting temperature. Here the values with index C refer to coating and with index S – to main material of product. However, there is no direct dependence of adhesion strength on physico–mechanical properties of research materials of coating. Therefore, it is necessary to consider other properties of materials, such as their brittleness. Least affected to brittle fracture is tungsten carbide. This can be explained by higher values of adhesion strength of coatings based on WC. The porosity of obtained coatings (higher for coatings based on chromium carbide) and differences in conditions of thermodynamic interaction of sprayed particles with substrate also have influence.

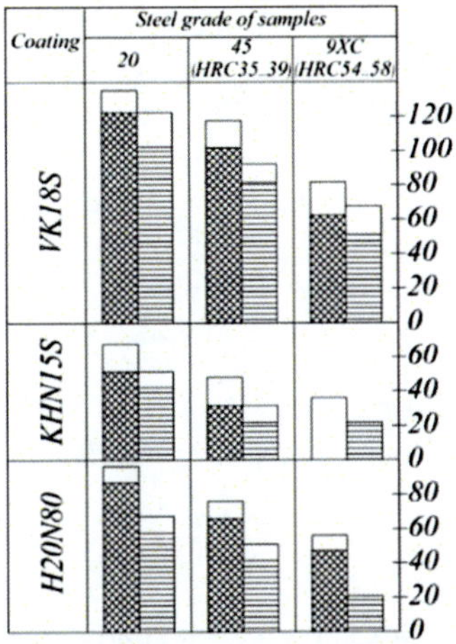

Fig. 5. Adhesion strength of D-Gun sprayed coatings. The surface of samples is prepared by: abrasive grit blasting (oblique hatching); grinding (horizontal hatching)

At the same time, the material of substrate has significant effect on adhesion strength of coatings. The character of decrease of adhesion strength of all types of coatings with increasing yield strength σ of steel substrate is approximately the same (Fig. 6). And for coatings on basis of chromium carbide there is a more gentle dependence. More clearly, the dependencies of change of adhesion strength are observed in form of dependence $\sigma_a/\sigma_{0,2}$ on boundary of yield strength of steel substrate (Fig. 7). Most significantly, the relative adhesion strength is reduced for coatings based on tungsten carbide, but at the maximum value of yield strength coatings based on tungsten carbide retains high adhesion with substrate. For Nichrome coatings (Fig. 7c) and especially on basis of chromium carbide (Fig. 7b) on samples from high carbon steel 9HS (0,85–0,95%C - 0,98–1,25% Cr - 1,2–1,6%Si) the adhesion reliability is dramatically reduced.

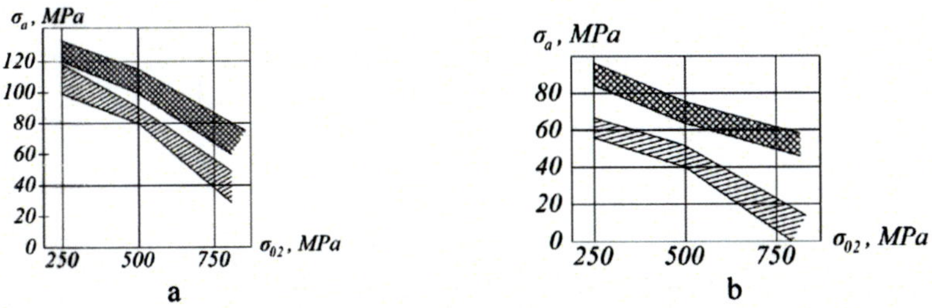

Fig. 6. Dependence of adhesion strength of D-Gun sprayed coatings from powder VK18S (WC-18%Co) – (a), KHN15S (3%Cr-2%C-15%Ni). – (b) and H20N80 (20%Cr- 80%Ni) – (c) from yield strength of steel substrate. The surface of samples is prepared by: abrasive grit blasting (oblique hatching); grinding (horizontal hatching)

For nichrome coatings application of samples after abrasive grit blasting leads to the most significant increase of adhesion strength (Figs. 6 and 7). This is also noticeable by reaction on tungsten carbide coatings.

Let's estimate the possibility of reaching in area of collision of sprayed particle with substrate of dynamic stresses that equal to yield strength of base material, that is, the condition of

$$\sigma_{dyn} \geq \sigma_p, \tag{2}$$

where $\sigma_p = \rho_p V_0^2$ – impact pressure, ρ_p – density of particle material, V_0 – velocity of impact. Then critical value of the velocity of impact is:

$$V_{0cr} = \sqrt{\sigma_{0,2} / \rho_{mp}}. \tag{3}$$

Calculations show that at a D-Gun spraying with particles velocities up to 500–800 m/s, even at maximum values of yield strength of substrate material, this condition is kept (Fig. 8). Therefore, reaching in area of collision of stress particles equal to or exceeding static yield strength of substrate material is not sufficient to form a strong adhesion with substrate.

To analyze conditions of formation of interatomic connections between materials of sprayed particle and substrate, we use experimental data about activation energy of steel in different structural state [26], where the activation energy is obtained in MJ/m^3, and influence of external force on activation energy of rupture of interatomic connection is taken into account by value $\gamma' \sigma_a^2$, here γ' in MN/m^2, and σ_a – the average tension acting on interatomic connections.

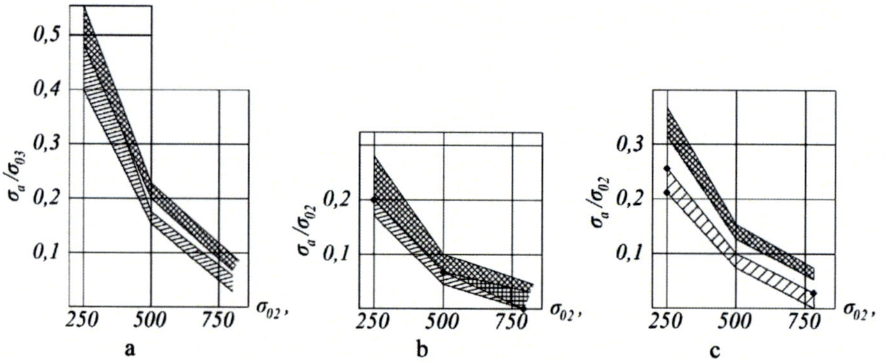

Fig. 7. Dependence of adhesion strength of coatings $\sigma_a/\sigma_{0,2}$ from yield strength of steel substrate $\sigma_{0,2}$: VK18S (WC-18%Co) – (a), KHN15S (3%Cr-2%C-15%Ni) – (b), H20N80 (20% Cr - 80% Ni) – (c) and from yield strength of steel substrate. The surface of samples is prepared by: abrasive grit blasting (oblique hatching); grinding (horizontal hatching)

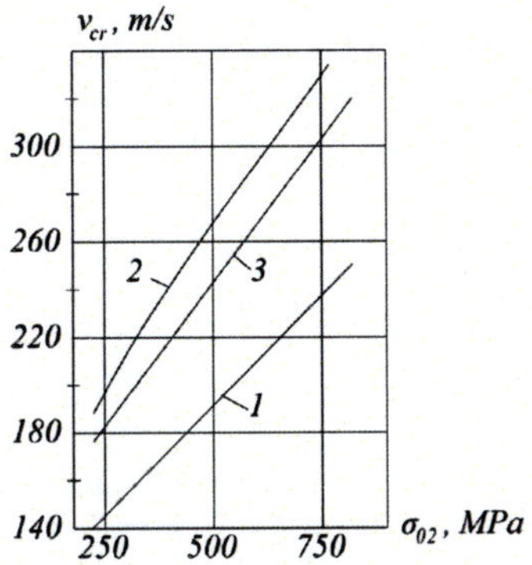

Fig. 8. Critical velocity of impact of particles of powder WC (1), Cr_3C_2 (2) and nichrome (3) to reach stresses in collision area that equal to yield strength of substrate $\sigma_{0,2}$

Transforming the known equation for relative adhesion strength of particles to the substrate [8] we will get:

$$\overline{\sigma} = 1 - \exp\left\{-\frac{\nu t_a}{\exp\left[\left(E'_a - \gamma'\sigma_a^2\right)V_a kT_c\right]}\right\}, \tag{4}$$

where V_a – is volume of atom ($V_a \approx A/N_A\rho$, where A – is atomic mass, N_A – is Avogadro number, ρ – density); ν – oscillation frequency of atoms; t_a – activation time of substrate (action of impact pressure in contact area); E'_a – energy of activation of formation process of interatomic connections (coincides with activation energy of destruction); γ' – a structurally sensitive coefficient that estimates activation volume and overtensions on interatomic connections compared to average tension in sample; p – tensions that acting in contact of particle with substrate; k – Boltzmann constant; T_c – contact temperature of «particle – substrate».

The calculations with taking into account values E'_a and γ' obtained in conditions of cyclic loading [26] show a strong dependence of relative adhesion strength on value of dynamic stresses in contact area. When spraying coatings on steel 20 (0,17–0,24% C), the increase of dynamic stresses can significantly reduce contact temperature required for formation of a strong adhesion (Fig. 9). In calculations were accepted $t_a = 1 \cdot 10^{-7}$ s, $V_a = 1,173 \cdot 10^{-29}$ m, $K = 1,3805 \cdot 10^{-29}$ MJ/deg.

When spraying coatings on steel 45 (0,42–0,5% C) the role of dynamic stresses in formation of strong adhesion is reduced (Fig. 10). To obtain equal adhesion strength of particle with steel 45 (0,42–0,5% C), when all else conditions are equal, it is necessary to obtain higher values of temperature in contact area. Moreover, with same growth compared to steel 20 (0,17–0,24% C) contact stresses, the level of required temperatures that necessary for formation of strong adhesion decreases less intensively. In both cases, an increase contact stresses reduces temperature interval between relative strength values that $\overline{\sigma}$ close to 0 and 1.

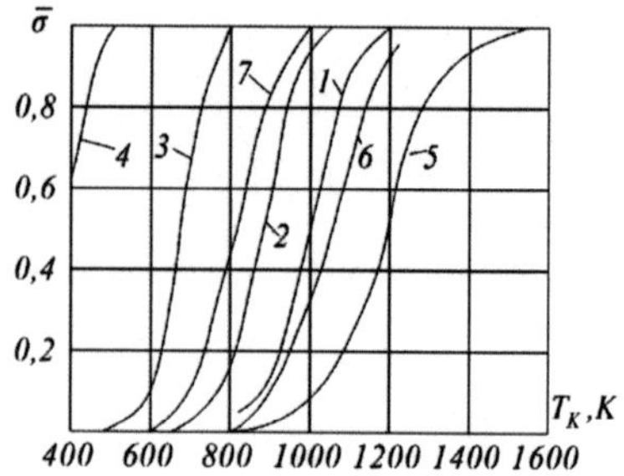

Fig. 9. Calculated dependences of relative adhesion strength of sprayed particles with steel 20 under the following conditions: $\upsilon = 10^{13}s^{-1}$ (σ_a, MPa: 100 (1), 200 (2), 300 (3), 400 (4)); $\upsilon = 10^{12}s^{-1}$ (σ_a, MPa: 100 (5), 200 (6), 300 (7))

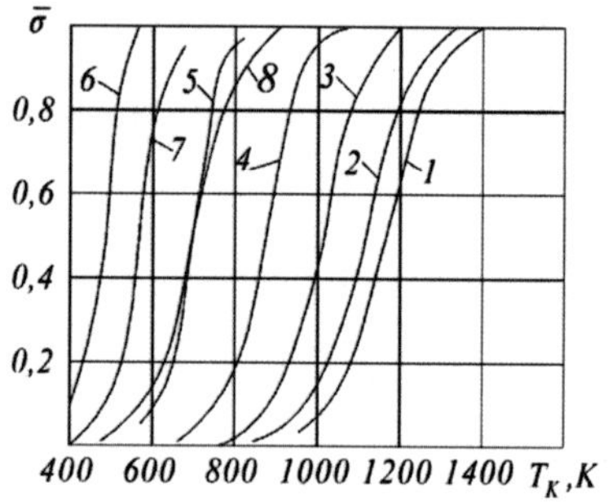

Fig. 10. Calculated dependences of relative adhesion strength of sprayed particles with improved steel 45 under the following conditions: $\upsilon = 10^{13}s^{-1}$ (σ_a, MPa: 100 (1), 200 (2), 300 (3), 400 (4), 500 (5), 600 (6)); $\upsilon = 10^{12}s^{-1}$ (σ_a, MPa: 600 (7)); $\upsilon = 10^{11}s^{-1}$ (σ_a, MPa: 600 (8))

4 Conclusions

The adhesion strength of D-Gun sprayed coatings depends on a complex of design and technological factors, including combination of materials in system «coating-basic material», roughness and other parameters of sprayed surface state, composition, dispersion and method of manufacturing of powder, etc.

D-Gun spraying due to high velocity of sprayed particles increases role of mechanical activation of contacting materials and makes it possible to obtain coatings with high adhesion strength with reduction of requirements for preparation of sprayed surface.

Known knowledges about mechanisms of adhesion allow to obtain only qualitative estimates of adhesion strength of D-Gun sprayed coatings and requires development of models that take into account a set of constructional and technological factors that affect the adhesion strength.

References

1. Kharlamov, Y.: The development of D-Gun spraying technologies. Visnik. V. Dahl EUNU 7(237), 114–132 (2017)

2. Karpus, V., Ivanov, V., Dehtiarov, I., Zajac, J., Kurochkina, V.: Technological assurance of complex parts manufacturing. In: Ivanov, V., et al. (eds.) DSMIE 2018. LNME, pp. 51–61. Springer, Cham (2019). https://doi.org/10.1007/978-3-319-93587-4_6
3. Fesenko, A., Basova, Y., Ivanov, V., Ivanova, M., Yevsiukova, F., Gasanov, M.: Increasing of equipment efficiency by intensification of technological processes. Period. Polytech. Mech. Eng. **63**(1), 67–73 (2019)
4. Krol, O., Sokolov, V.: Parametric modeling of transverse layout for machine tool gearboxes. In: Gapiński, B., Szostak, M., Ivanov, V. (eds.) MANUFACTURING 2019. LNME, pp. 122–130. Springer, Cham (2019). https://doi.org/10.1007/978-3-030-16943-5_11
5. Krol, O., Sokolov, V.: Parametric modeling of gear cutting tools. In: Gapiński, B., Szostak, M., Ivanov, V. (eds.) MANUFACTURING 2019. LNME, pp. 3–11. Springer, Cham (2019). https://doi.org/10.1007/978-3-030-16943-5_1
6. Yushchenko, K.A., Borisov, Y.S., Kuznetsov, V.D., Korzh, V.M.: Surface Engineering. Scientific Thought, Kiev (2007)
7. Ilyushchenko, A.F., Okovityj, V.A., Kundas, S.P., Formanek, B.: Formation of Thermal Sprauing Coatings: Theory and Practice. Bestprint, Minsk (2002)
8. Anciferov, V.N., Bobrov, G.V., Druzhinin, L.K.: Powder Metallurgy and Spray Coatings. In: Mitin, B.S. (ed.) Metallurgy, Moscow (1987)
9. Rogovyi, A., Khovanskyy, A.: Application of the similarity theory for vortex chamber superchargers. J. Phys. Conf. Ser. Mater. Sci. Eng. **233**, 012011 (2017)
10. Rogovyi, A.: Energy performances of the vortex chamber supercharger. Energy **163**, 52–60 (2018)
11. Pierre, L., Fauchais, J., Heberlein, V., Maher, R., Boulos, I.: Thermal Spray Fundamentals. Springer, Boston (2014). https://doi.org/10.1007/978-0-387-68991-3
12. Kharlamov, Y., Budag'janc, N.A.: D-Gun Processes in Industry. Publishing house of VUGU, Lugansk (1998)
13. Davis, J.R. (ed.): Handbook of Thermal Spray Technology. ASM International, Ohio (2004)
14. Pawlowski, L.: The Science and Engineering of Thermal Spray Coatings, 2nd edn. Wiley, Chichester (2008)
15. Heintze, G.N., McPherson, R.: Fracture toughness of plasma sprayed zirconia coatings. Surf. Coat. Technol. **134**, 15–23 (1988)
16. Movshovich, A.Y., Chernaja, Y., Ishchenko, G.I.: Investigation of influence of physico – mechanical characteristics of D-Gun sprayimg coatings on wear resistance of cutting elements of reworkable dies. Vestnik NTU «HPI» **45**, 22–28 (2011)
17. Faradzhallah, M.A., Panarin, V.E., Bys', S.S.: The problem of technological residual stresses in improving the quality of D-Gun spraying coatings. Visnik Khmelnytskyi Natl. Univ. **2**, 14–18 (2011)
18. Rjahovskij, A.V., Kosenko, V.V., Vlasenko, V.N.: Features of estimation of adhesion strength of D-Gun spraying coatings. Weapons Syst. Mil. Equip. **3**(31), 215–217 (2012)
19. Sokolov, V., Krol, O., Stepanova, O.: Automatic control system for electrohydraulic drive of production equipment. In: International Russian Automation Conference (RusAutoCon) 2018. IEEE (2018)
20. Lin, C.K., Berndt, C.C.: Measurement and analysis of adhesion strength for thermally sprayed coatings. J. Therm. Spray Technol. **3**(1), 75–104 (1994)
21. Sokolov, V., Krol, O., Stepanova, O.: Choice of correcting link for electrohydraulic servo drive of technological equipment. In: Ivanov, V., et al. (eds.) DSMIE 2019. LNME, pp. 702–710. Springer, Cham (2020). https://doi.org/10.1007/978-3-030-22365-6_70
22. Sokolov, V., Krol, O.: Determination of transfer functions for electrohydraulic servo drive of technological equipment. In: Ivanov, V., et al. (eds.) DSMIE 2018. LNME, pp. 364–373. Springer, Cham (2019). https://doi.org/10.1007/978-3-319-93587-4_38

23. Fedorovich, V., Mitsyk, A.: Mathematical simulation of kinematics of vibrating boiling granular medium at treatment in the oscillating reservoir. Key Eng. Mater. **581**, 456–461 (2014)
24. Sokolov, V., Krol, O., Stepanova, O.: Nonlinear simulation of electrohydraulic drive for technological equipment. J. Phys. Conf. Ser. **1278**, 012003 (2019)
25. Maksimovich, G.G., Shatinskij, V.F., Kopylov, V.I.: Physicochemical Processes During Plasma Spraying and Destruction of Coated Materials. Nauk. Dumka, Kiev (1983)
26. Fedorov, V.V.: Kinetics of Damage and Destruction of Solids. Fan, Tashkent (1985)
27. Pavlenko, I., Trojanowska, J., Ivanov, V., Liaposhchenko, O.: Scientific and methodological approach for the identification of mathematical models of mechanical systems by using artificial neural netpublications. In: Machado, J., Soares, F., Veiga, G. (eds.) Innovation, Engineering and Entrepreneurship. HELIX 2018. LNEE, vol. 505, pp. 299–306 (2019)
28. Pavlenko, I., Simonovskiy, V., Demianenko, M.: Dynamic analysis of centrifugal machines rotors supported on ball bearings by combined application of 3D and beam finite element models. J. Phys. Conf. Ser. Mater. Sci. Eng. **233**, 012053 (2017)
29. Pavlenko, I.: Static and dynamic analysis of the closing rotor balancing device of the multistage centrifugal pump. Appl. Mech. Mater. **630**, 248–254 (2014)
30. Kundrák, J., Mitsyk, A., Fedorovich, V., Morgan, M., Markopoulos, A.: The use of the kinetic theory of gases to simulate the physical situations on the surface of autonomously moving parts during multi – energy vibration processing. Materials **12**(19), 3054 (2019)

An Approach to Ship Equipment Maintenance Management

Suzana Lampreia[1] (✉), Teresa Morgado[1,2,3], Helena Navas[2,4], Rita Cabrita[4], and José Requeijo[2,4]

[1] CINAV-Escola Naval/Naval Research Centre-Portuguese Naval Academy, 2810 Almada, Portugal
suzana.paula.lampreia@marinha.pt, t.morgado@fct.unl.pt
[2] UNIDEMI/NOVA – Research and Development Unit for Mechanical and Industrial Engineering, Universidade NOVA de Lisboa, 1900 Lisboa, Portugal
[3] IPT - Polytechnic Institute of Tomar, 2300 Tomar, Portugal
[4] DEMI/NOVA – Department of Mechanical and Industrial Engineering, Universidade NOVA de Lisboa, 1900 Lisboa, Portugal

Abstract. Ships can navigate several months on sea. Maintenance and supply are the base for ship successful navigation and mission. Every type of equipment should have a maintenance plan. If it is a corrective maintenance plan, some risks are inherent. This study pursues a Management Maintenance System considering minimal costs, where the best equipment availability and performance is the objective. For this work, an air compressor was chosen as study case from a ship and considered three study stages. The first stage consisted of the definition of evaluation criteria and its meaningfulness. The data treatment from the first stage can provide enough information to define the second stage's maintenance methodology decision. Also, the decision-making itself based on the process is the third stage. The development of decision-making methodology for maintenance management was based on a Fuzzy method considering a Risk-Based Maintenance on ship equipment.

Keywords: Risk · Naval maintenance management · Fuzzy methodology

1 Introduction

The developments in maintenance management are in continuous changing due to complex systems and equipment [1]. Fuzzy data is being applied for modelling complex systems [2]. The decision-making processes based on risk are supported by variables that do not admit accurate values. So in this work, is developed a methodology to quantify the "quality" and try to eliminate the decision based on human sense and deductive thinking, where people "infer a conclusion from what they know" [3, 4].

Jamshidi [5], in his work, refers to "The evaluating criteria for Maintenance Strategies Selection (MSS) depend on the organisational goals and objectives and could be decided in consensus with field experts." So before starting the study, the evaluating criteria

J. Machado et al. (Eds.): icieng 2021, LNME, pp. 441–450, 2022.
https://doi.org/10.1007/978-3-030-79165-0_41

should be defined [6]. Besides, the strategies and the criteria should be adapted to the needs of the organisation and system.

On the ship's maintenance management, the decision-based in advanced industrial models can be a pillar for decision-making representatives and managers. This paper focuses on naval equipment's. Knowing ship equipment's and systems performance versus condition can be crucial for a decision-making process. This work proposes the approach of decision-making based on a risk-based maintenance (RBM) system. It can "reduce the probability of failure of equipment and the consequences of failure." [7]. For this aim knowing the equipment's lifecycle, the maintenance plan is defined accordingly and to enhance the equipment availability [8].

When a risk-based maintenance system is applied, preferentially, it should be based on a quantitative system [9]. So, this paper presents a methodology to quantify the risk of not pursuing a maintenance action even when needed. The authors believe that applying the Fuzzy models will define the criteria and decision-making process on a Risk-Based Maintenance process implemented on a system or equipment. Therefore, a Fuzzy analysis is proposed in the obtained data. Figure 1 shows the five Fuzzy used criteria for this study: the monitoring cost, efficiency, failure, safety, and if it is feasible.

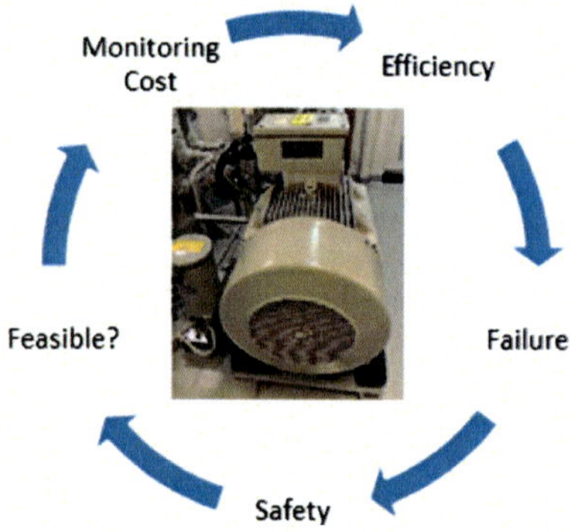

Fig. 1. Fuzzy criteria defined for the equipment

2 Condition-Based Maintenance in Ships

Ships and maritime transports have been the most critical means to take goods from their origin to destiny for more than centuries. Meanwhile, the globalisation and growing market competitiveness have changed, and still change, the way to manage firms. In this logic, the maintenance is a critical variable to reduce management costs, "whose costs account for the most cost of ships machinery maintenance" [10, 11].

Given this situation, it is crucial to establish a maintenance plan according to the data collection and analysis of equipment that should be considered too. There are two types of maintenance in theory: Corrective (CM) and Preventive Maintenance (PM). The second one divides into Systematic Maintenance and condition-based maintenance (referred to as CBM). CBM, according to Telford et al. [12], is "preventive maintenance based on performance and/or parameter monitoring and the subsequent actions." Lately has been highlighted because of its power to reduce maintenance costs effectively. This strategy recommends that maintenance actions (decisions) should be performed based on the "information collected through condition monitoring process" [13].

CBM may be used in three main methodic steps: data acquisition, data processing and maintenance decision-making [11, 14], and this data is achieved, if possible, by longs periods of monitoring systems. Once it is decided that equipment is going subjected to condition-based maintenance, the first work is to assess its decay status. If there are a set of components is vital to study their decays status, decay degrees, and the decays set direction and evaluate which one has decayed more [10]. With these data, it is profitable to schedule the next maintenance and avoid failures.

According to Hwang et al. [15], CBM has four function modules: diagnostics - that match the asses of equipment decays status and investigate the type of correspondent failures that can occur; prognostics - that follows the identification of a failure module type and then tell the Remaining Useful Life (RUL); maintenance - using diagnostics and prognostics results from it is possible to maintenance equipment and schedule it; last but not least, configuration managements function when are design standards and requirements of the equipment to prevail constant and consistent at all times.

Decays data can either be generated by a real data validated simulator, and based on that data, CBM is an important and proactive decision-making tool [16]. Thus, this strategy has the advantage of preceding warnings of stop failures and increased precision in failure prognostic. On the other hand, the main disadvantage is an investment in installing and controlling the monitoring equipment and on stipulating the decision-making systems [11, 13, 17].

CBM has been studied in the most variety of manufacturing firms, equipment or set of equipment during the past years. This paper will study if CBM is appropriate to manage and decide maintenance plans in ships, as has already been done [10, 12, 15].

3 Fuzzy Methodology in Maintenance

Ahmad and Kamaruddin [13] refer that "The main aim of diagnosis is to provide early warning signs to engineers while the monitored equipment is operating".

Maintenance plans are frequently designed by studying the maintenance tasks and then prioritisation those. There are a lot of logic methods to follow, and Fuzzy Logic is one of them. Fuzzy models have been highlighted "for measuring uncertainty in productions control fields", such as maintenance concerns [18, 19, 20].

Fuzzy modelling, as a many-valued logic, allows the utilisation of imprecise information. These models can achieve solutions with uncertain input data that standard mathematical models cannot [19]. A mathematical model is quantitative, so it may show several challenges in the implementations due to systems complexity, which is highly dimensional, nonlinearities and parametric uncertainties. In contrast, qualitative approaches receive an analysis of historical data to develop a solution [21]. Fuzzy theory's main advantage is that "deals with subjective, incomplete or unreliable knowledge bases" by having inputs linguistic variables, thus is the most appropriate to qualitative approaches [22].

Once is decided that fuzzy logic will define maintenance plans, it is crucial specify which kind of logic is appropriated: Early Warning Systems (EWS) [11], Real-time information (RTI) [18], Network Formulation (NT) [21], Multi-Criteria Decision-Making (MCDM) [23], Supporting Risk Matrix Prioritisation (SRMP) [24], and others.

This paper will use the Fuzzy Logic considering Risk-Based Maintenance (referred to as RBM) as the maintenance decision system. RBM is a crucial tool to lead failure mitigation by studying accidents probability and respective costs consequences. It also has a critical role in controlling the validity of safety investments, managing potential accidents instead of performance maintenance plans [25].

RBM aims to schedule maintenance with dynamism taking the risk of failure as a start point. First, risk should be calculated, and then maintenance is scheduled [12, 22]. Risk Assessment can be approached by Risk Index or Probability. Then, Maintenance Schedule Technique by Expert Judgement or Optimisation. Another approach is to "assess the consequences of action and prioritise maintenance tasks based on the risk of potential failures" [20].

Selecting the criteria (or multiple criteria) is possible to obtain priority weigh based on Fuzzy interval ranges. All interactions between criteria and maintenance plan alternatives will be considered uncertainty, which provides better results than conventional decision-making methods [26].

4 Methodology Development

The ship in the study is maintained through the Maintenance Management System – MMS - that the aim is to reach the balance between preventive and corrective maintenance. The goal is to monitor equipment performance. Three methodologic steps usually compose MMS (see Fig. 2): 1 - Planning; 2 - Execution; 3 - Information Treatment. Maintenance planning should consider the specifications and needs to proceed with maintenance along time. Execution is the concretisation, monitoring and due dates control of maintenance plan and risk analysis. It can be organised by sequence: first, the need for maintenance is given; second, manufacturer representatives or shipyard resources are made available; next, the required spare parts are made available; for last, the execution or not itself is coordinated. Information Treatment is started by collecting maintenance data, then analysing and treating and finally distributing this information.

Although the organisation has its own maintenance plan, it also follows the Portuguese normative [27]. The maintenance system plan is split into preventive maintenance and maintenance in the organisation, Fig. 3. In this study, the principal type of applied maintenance is corrective, but the conditioned maintenance techniques are considered to avoid total failure.

When it comes to technical means, the ship, which it was used for study, has three levels of maintenance: First Level - is when maintenance is performed by replacement of components and tools on board. In other words, when it is possible to develop the maintenance plan inside the ship with its resources (personnel and material) safely and effectively. Second Level - is when maintenance can be performed onboard and/ or on land because ship resources and capacity are not self-sufficient to execute it. Third Level - is when equipment or components should be maintained on land because their

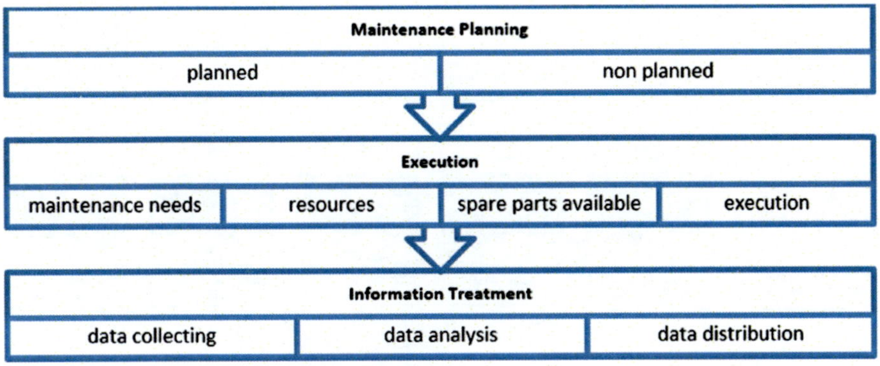

Fig. 2. Methodology scheme

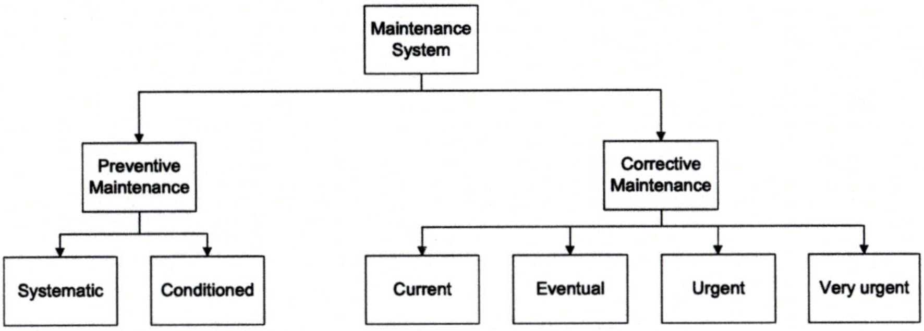

Fig. 3. Maintenance in the organisation

maintenance plan and the current situation is complex, and the space on board and its resources do not allow safe maintenance. This Third Level is executed by shipyard or by representatives of official manufacturers [28].

The Failure criteria, it was defined as the failures based on the MTBF (Mean Time to Failure). The Safety criteria were based on the personnel injuries, the equipment safety and any environmental impact. For the Efficiency criteria, it was analysed the efficiency of the personnel and material performance. For Feasible maintenance, it was considered the available employees and tools. The criteria Cost of Monitoring included personnel education, portable equipment, fixed equipment, and software costs. All the five criteria compete with a data processor, which provided the results to maintenance decision.

The Fuzzy Logic Control was applied to analyse corrective maintenance risk and the implication to do or not in the universe criteria, as showed in Fig. 4, considering other risk-based maintenance systems used to compressors [10].

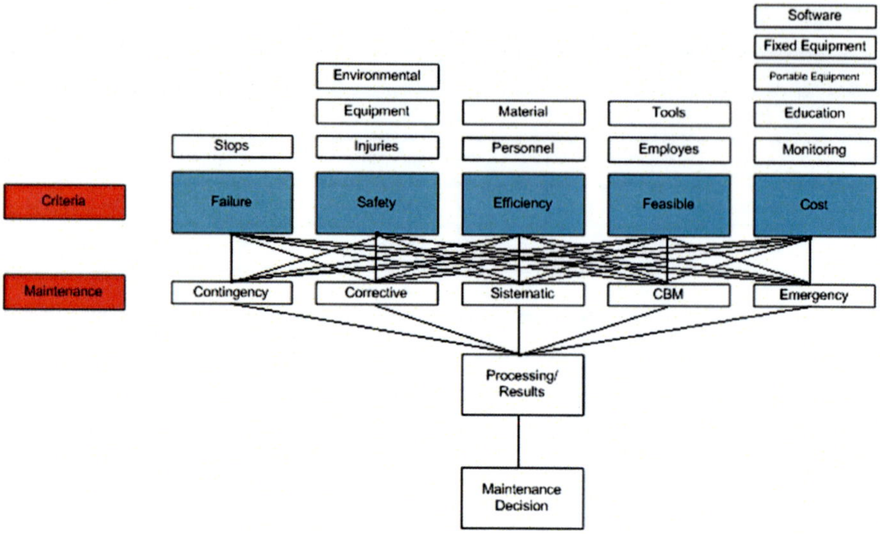

Fig. 4. Fuzzy on risk analysis

5 Case Study

An air compressor onboard it is vital for other systems functioning. Engines and power generators can depend on this for a start and even for some operations during the running. The air-compressor understudy is pursuing a bath life cycle, and it is in the middle of the lifecycle. It is believed that a risk-based maintenance system support by Fuzzy methodology can be applied [29]. The present research uses an air compressor from a ship to validate the developed system. According to the manufacturer manual, the air compressor maintenance must be based on time interval predetermined and carried out under an established time schedule or set times of use – Schedule Maintenance. An example of this schedule is Fig. 5.

An example of periodic checks on the compressor is in Fig. 6. These checks correspond to periodic maintenance defined by the manufacturer.

The idea of the developed system is, considering the air compressor maintenance defined by the manufacturer (phase 1 of the defined methodology), as shown in Fig. 7, to quantify the risk of not doing planned maintenance. It will be considered that: if the planned maintenance of 1000 h or more hours is not made, some simple maintenance will be done (for example, oil and filters change), and the risk of not doing it must be quantified. It was considered 10% of additional risk for each undoing maintenance, only the risk of the oil and filter unchanging will be exposed; this is first-level maintenance carried out by the ship. It was considered the activity *per si*, what can happen and the impact. To obtain the results and quantify, the probability, impact, and exposure were applied.

The considering maintenance cannot be feasible for the five defined criteria (see Fig. 4). The fuzzy methodology is expressed in phase 2 of the described methodology of the execution.

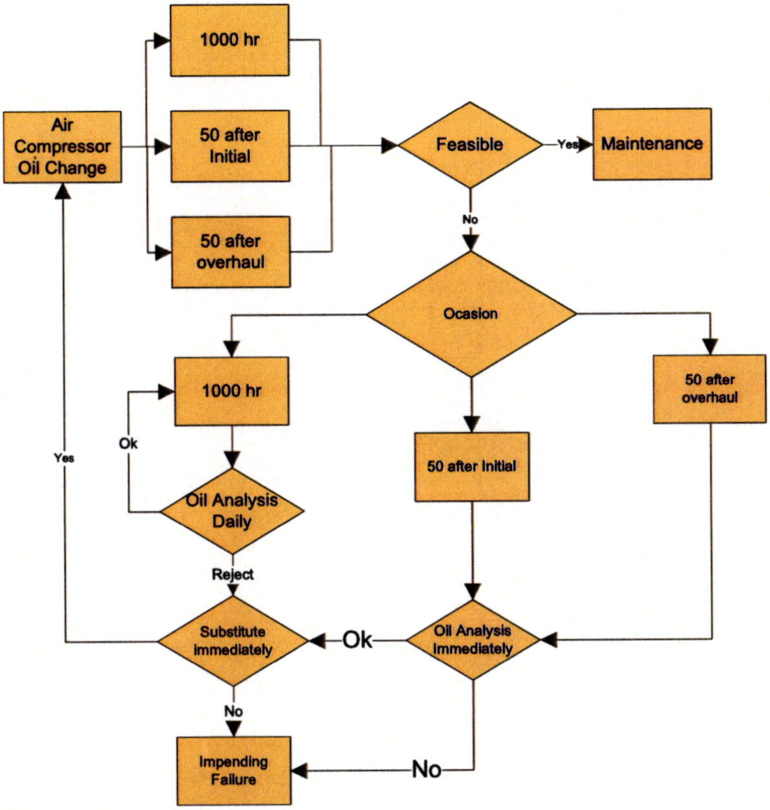

Fig. 5. Air compressor maintenance for 50 h and 1000 h [30]

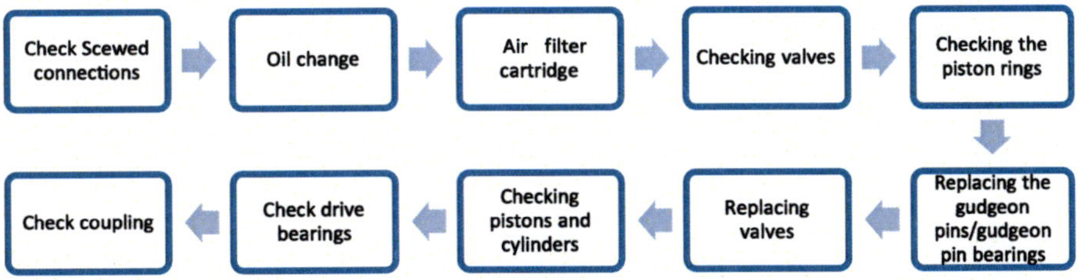

Fig. 6. Air compressor checks [30]

With undoing the maintenance, but with good oil analysis and functioning parameters, the results were positives so that the air compressor can function normally. With some loss of oil properties, it can work with some time limit. With the total loss of oil properties and the parameters out of control, the compressor should be stopped immediately, with the risk of the catastrophic anomaly. Corresponds to phase three of the defined methodology.

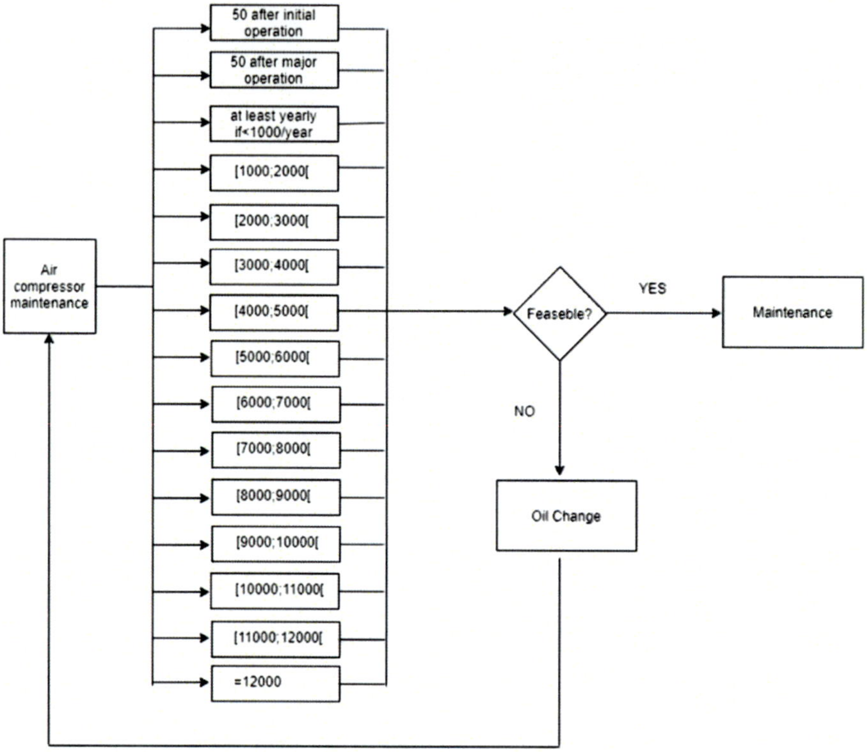

Fig. 7. Air compressor maintenance according to the manufacturer manual [30]

So, if the decision is not to proceed to maintenance, the air compressor must continue with Condition Based Maintenance (CBM), considering oil analysis, parameters check, and vibration measures. If the performance of the compressor decreases, these procedures should be made more often.

6 Conclusions

For this work, three-stages of risk-based maintenance analysis were considered. At first, the criteria were defined. Then, the way of criteria interacts and its values with software or personnel evaluation were established. In the third stage, as a result, the maintenance plan was defined.

The studied air compressor is selected equipment from a ship, and other equipment types depend on it, so the risk of undoing some maintenance can conduce to a failure. Some condition-based maintenance techniques should be used to avoid an unexpected failure, and the risk of unfulfilled maintenance is calculated.

The three-stage defined methodology can be applied in equipment monitoring.

For risk assessment, the risk should be identified, analysed, and quantified.

It is also concluded that applying the Fuzzy logic in a Maintenance System Plan is viable and can support decision making. Also, the RBM is workable for the organisation, can decrease the maintenance cost, and increase the reliability in systems.

References

1. Khan, F., Haddara, M.: Risk-based maintenance (RBM): a new approach for process plant inspection and maintenance. Process Saf. Prog. **24**(4), 252–265 (2004). Faculty of Engineering and Applied Science, Memorial University of Newfoundland, St. John's Newfoundland, A1B 3X5, Canada
2. Kaymak, U., Babuska, R., Setnes, M., Verbruggen, H., Lemke, H.: Methods for simplification of fuzzy models. In: Ruan, D. (ed.) Intelligent Hybrid Systems: Fuzzy Logic, Neural Networks, and Genetic Algorithms. Springer, Boston (1997). https://doi.org/10.1007/978-1-4615-6191-0_4
3. Rignel, D., Chenci, G., Lucas, C.: Uma Introdução à Lógica Fuzzy (An introduction to fuzzy logic). Revista Eletrônica de Sistemas de Informação e Gestão Tecnológica, Environment and Technology **1**(1), 17–28 (2015)
4. Tutorialspoint, Fuzzy Logic – Control System. https://www.tutorialspoint.com/fuzzy_logic/fuzzy_logic_control_system.htm. Accessed 16 Dec 2019
5. Jamshidi, A.: Risk-based maintenance of critical and complex systems. Ph.D. thesis, Université Laval – Québec, Canada (2015)
6. Khan, F., Haddara, M.: Risk based maintenance (RBM): a quantitative approach for maintenance/inspection scheduling and planning. Int. J. Loss Prev. Process Ind. **16**, 561–573 (2003)
7. Arunraj, N., Maiti, J.: Risk based maintenance – techniques and applications. J. Hazard. Mater. **142**, 653–661 (2006)
8. Lampreia, S., Lobo, V.: Estratégias de Manutenção num Ambiente de Escassos Recursos. 15° Congresso Nacional de Manutenção da APMI 2019, Braga, 21–22 novembro 2019
9. Ratnayake, R., Antosz, K.: Risk-based maintenance assessment in the manufacturing industry: minimisation of suboptimal priorisation. J. Manag. Prod. Eng. Rev. **8**(1), 38–45 (2017)
10. Tan, Y., Niu, C., Tian, H., Hou, L., Zhang, J.: A one-class SVM based approach for condition-based maintenance of a naval propulsion plant with limited labeled data. Ocean Eng. **193** (2019). https://doi.org/10.1016/j.oceaneng,106592
11. Vafaei, N., Ribeiro, R., Matos, L.: Fuzzy early warning systems for condition based maintenance. Comput. Ind. Eng. **128**, 736–746 (2016). https://doi.org/10.1016/j.cie.2018.12.056
12. Telford, S., Mazhar, M., Howard, I.: Condition based maintenance (CBM) in the oil and gas industry: an overview of methods and techniques. In: Proceedings of the 2011 International Conference on Industrial Engineering and Operations Management, Kuala Lumpur, Malaysia pp. 1152–1159 (2011)
13. Ahmad, R., Kamaruddin, S.: An overview of time-based and condition-based maintenance in industrial application. Comput. Ind. Eng. **63**(1), 135–149 (2012). https://doi.org/10.1016/j.cie.2012.02.002
14. Jardine, A., Lin, D., Banjevic, D.: A review on machinery diagnostics and prognostics implementing condition-based maintenance. Mech. Syst. Signal Process. **20**(7), 1483–1510 (2006). https://doi.org/10.1016/j.ymssp.2005.09.012
15. Hwang, H., Lee, J., Hwang, J., Jun, H.: A study of the development of a condition-based maintenance system for an LNG FPSO. Ocean Eng. **164**, 604–615 (2018). https://doi.org/10.1016/j.oceaneng.2018.07.004

16. Bousdekis, A., Magoutas, B., Apostolou, D., Mentzas, G.: A proactive decision making framework for condition-based maintenance. Ind. Manag. Data Syst. **115**(7), 1225–1250 (2015). https://doi.org/10.1108/IMDS-03-2015-0071
17. Kothamasu, R., Huang, S., VerDuin, W.: Handbook of maintenance management and engineering. Int. J. Adv. Manuf. Technol., 1012–1024 (2006). https://doi.org/10.1007/978-1-84882-472-0
18. Lu, K., Sy, C.: A real-time decision-making of maintenance using fuzzy agent. Expert Syst. Appl. **36**(2), Part 2 (2009). https://doi.org/10.1016/j.eswa.2008.01.087
19. Cox, E.: Fuzzy Modeling and Genetic Algorithms for Data Mining and Exploration. The Morgan Kaufmann Series in Data Management Systems (2005)
20. Ratnayake, R., Antosz, K.: Development of a risk matrix and extending the risk-based maintenance analysis with fuzzy logic. Procedia Eng. **182**, 602–610 (2017). https://doi.org/10.1016/j.proeng.2017.03.163
21. D'Angelo, M., Palhares, R., Cosme, L., Aguiar, L., Fonseca, F., Caminhas, W.: Fault detection in dynamic systems by a Fuzzy/Bayesian network formulation. Appl. Soft Comput. **21**, 647–653 (2014). https://doi.org/10.1016/j.asoc.2014.04.007
22. Gallab, M., Bouloiz, H., Alaoui, Y., Tkiouat, M.: Risk assessment of maintenance activities using fuzzy logic. Procedia Comput. Sci **148**, 226–235 (2019). https://doi.org/10.1016/j.procs.2019.01.065. The Second International Conference on Intelligent Computing in Data Sciences, ICDS2018
23. Al-Najjar, B., Alsyouf, I.: Selecting the most efficient maintenance approach using fuzzy multiple criteria decision making. Int. J. Prod. Econ. **84**(1), 85–100 (2003). https://doi.org/10.1016/S0925-5273(02)00380-8
24. Cullum, J., Binns, J., Lonsdale, M., Abbassi, R., Garaniya, V.: Risk-based maintenance scheduling with application to naval vessels and ships. Ocean Eng. **148**, 474–485 (2018). https://doi.org/10.1016/j.oceaneng.2017.11.044
25. Yazdi, M., Nedjati, A., Abbassi, R.: Fuzzy dynamic risk-based maintenance investment optimization for offshore process facilities. J. Loss Prev. Process Ind. **57**, 194–207 (2019). https://doi.org/10.1016/j.jlp.2018.11.014
26. Kumar, G., Maiti, J.: Modeling risk based maintenance using fuzzy analytic network process. Expert Syst. Appl. **39**(11), 9946–9954 (2012). https://doi.org/10.1016/j.eswa.2012.01.004
27. Norma Portuguesa: Terminologia de Manutenção. NP EN 13306:2007(2007)
28. Marinha: Manual do Sistema de Gestão da Manutenção e do Sub-sistema de Manutenção Planeada (Maintenance Management system and Planned Maintenance Subsystem Manual). ILDINAV802. Direção de Navios (1998)
29. Oliveira, N.: Risk based maintenance for compressor systems. Masther thesis, Department of Marine Technology, Norwegian University of Science and Technology (2015)
30. Marinha: Grupo Electro Compressor Ar Arranque Sauer Type WP 81 L – Manual de Operação (Electro Air Compressor Group Starter Sauer Type WP 81 L – Operational Manual). Direção de Navios

The Use of Smart Insoles for Gait Analysis: A Systematic Review

Lauriston Medeiros Paixão[1,2](✉), Misael Elias de Morais[1,2], Frederico Moreira Bublitz[1,2], Karolina Celi Tavares Bezerra[1,3], and Carlúcia Ithamar Fernandes Franco[2]

[1] Center of Strategic Technologies in Health, Campina Grande, Brazil
[2] State University of Paraiba, Campina Grande, Brazil
[3] METRICs Research Center, University of Minho, 4800-058 Guimarães, Portugal

Abstract. The intelligent insoles have pressure and inertial sensors integrated to a set of software systems, for analysis and visualization of the collected information, as well as for the application of machine learning algorithms, in order to analyze the distributions of the plantar pressure and the parameters of the march. The objective of this article is to evaluate how the intelligent insoles can be used for gait parameter analysis, which includes the following parameters: stride duration, cadence, stride length and gait speed. It is an integrative review of articles published in the last five years in the bases: SciELO, PubMed and Lilacs, with a final sample of 7 articles. The research data showed that there are several models of intelligent insole prototypes, with different sensor and software systems, and all were effective in evaluating the gait. However, the research of Zhao et al. 2020 obtained better results, due to the design of the validation study in comparison with another intelligent insole already commercialized and for presenting a software system and interface that was more effective to analyze and display diverse gait parameters.

Keywords: Smart insole · Wearable devices · Medical devices · Gait analysis

1 Introduction

The plantar region of the feet plays a fundamental role in stabilizing the moving color-po. The muscular action moves the distribution of plantar pressure, changing the rotation of the foot around the ankle and the distribution of total body weight in both feet [2]. These stability limits change constantly depending on the task, the person's biomechanics and the support surface, which can be stable or unstable, varying according to the type of footwear.

Plantar and gait pressure measurement was generally entrusted to a laboratory environment, mainly through pressure platform systems, also known as baropodometers. In recent years, intelligent in shoes or insoles systems have been designed to provide real-time monitoring of daily, work and sports activities, and can also be used for injury prevention, disease assessment and diagnosis, and rehabilitation support.

J. Machado et al. (Eds.): icieng 2021, LNME, pp. 451–458, 2022.
https://doi.org/10.1007/978-3-030-79165-0_42

The intelligent insoles have pressure and inertial sensors integrated to a set of software systems, for analysis and visualization of the collected information, as well as for the application of machine learning algorithms, in order to analyze the distributions of the plantar pressure and the gait parameters. This way, these wearable systems became much more accepted, due to their portability, low energy consumption and, mainly, by the possibility of accurately monitoring the movements of the human body [9].

Several studies with risk populations have already been carried out using intelligent insoles, in order to evaluate the plantar pressure of hemiplegics [5]; in postural control of children with Cerebral Palsy [7]; to detect when excessive pronation occurred in patients with hallux valgus [1]; in the treatment of Parkinson's disease [11]; alcohol-induced walking disorders [8]; as well as in the prevention and healing of pressure ulcers in individuals with diabetic neuropathy [10].

The objective of this article is to evaluate how intelligent insoles can be used for gait pattern analysis, which includes the following parameters: stride duration, cadence, stride length and gait speed. Walking is not only one of the most important functions and activities of daily life, but also a parameter to monitor health status.

2 Methodology

To achieve the defined objective, the integrative literature review was selected as the research method, which aims to gather and condense research results on a delimited theme or issue, in a systematic and ordered manner, contributing to the deepening of the knowledge of the investigating theme and including the analysis of relevant research that provides support for decision making and improvement of clinical practice.

The systematic review follows six stages: 1. identification of the problem and selection of the research hypothesis or question, by elaborating the guiding question and establishing the descriptors; 2. establishment of criteria for inclusion and exclusion of articles, sampling or search in the literature; 3. selection of articles; 4. definition of the information to be extracted from the reviewed works - objectives, methodology and main conclusions, as well as their analysis; 5. discussion and interpretation of results; and, finally, 6. synthesis of the knowledge evidenced in the analyzed articles/presentation of the results of the systematic review.

The guiding question of the research was based on: "What is the scientific evidence on the use of intelligent insoles for gait analysis? For the survey of articles in the literature, a search was made in the following databases: National Library of Medicine (PubMed); Scientific Electronic Library Online (SciELO); Latin American and Caribbean Literature in Health Sciences (LILACS). They were used to search for descriptive articles in Portuguese, Spanish and English combined with the boolean operator AND as follows: palmilha inteligente AND análise da marcha; plantilla inteligente AND análisis de la marcha; smart insole AND gait analysis.

The inclusion criteria were: to contain information on the use of intelligent insoles in the evaluation of walking in healthy individuals, in the period between 2010 and 2020. Those studies that did not address the subject in a clear manner and that dealt with secondary sources were excluded.

In the first database search nine results were found, all in the PubMed database. In the second phase the titles and abstracts of each journal were read and those that did not meet the inclusion criteria and were duplicated were excluded, resulting in a final quantity of nine articles. In the third phase the full nine articles were read and those that met the inclusion criteria were added to the survey. Finally, two were excluded and the study was restricted to seven journals.

After selecting the final sample of the articles, some pertinent information was extracted from each one of them for use in this review, they are: publication reference (year of publication, data base, indexed journal, authors, title of the work); research objective; methodology, focusing on the technological characteristics of the intelligent insoles; and the results obtained as to the analysis of the march by intelligent insoles.

3 Results and Discussion

The use of intelligent insoles for gait analysis in selected studies is presented in Table 1 and Table 2. Among the information extracted are: name of the authors; characteristics of the sample as quantity, age and division of groups; method or instrument of evaluation; duration and intervention and the results.

The objective of the study by Das and Kumar [4] is the development of an intelligent insole to quantify the space-time parameters and gait phases. For the prototype of the intelligent insole 7 FSR-type pressure sensors were used in selected areas to quantify the force distribution in the following regions: posterior heel (PH), anterior heel (AH), first metatarsal (M1), second metatarsal (M2), fourth metatarsal (M4), fifth metatarsal (M5) and hallux (GT) (Fig. 1). This system was able to record and identify quantitatively the phase and events of the gait, being able to detect a pre-fall condition or pathological marches.

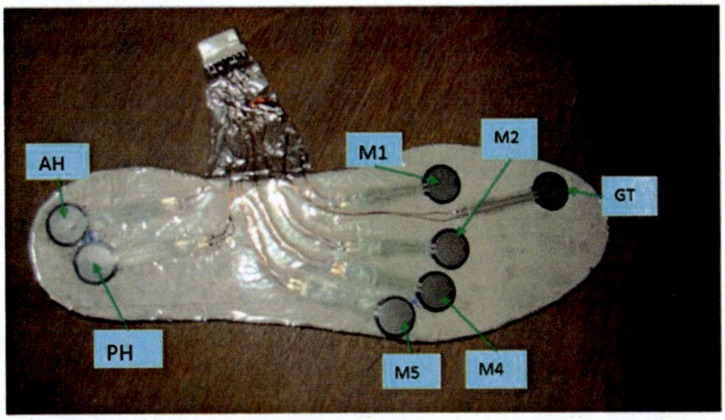

Fig. 1. Prototype of smart insole with FSR sensor. (Source: Das and Kumar [4])

In another research, it was proposed the development of an intelligent insole, called FreeWalker, capable of identifying the stages and parameters of gait in real time. The insole system was divided into three main parts: 1) insole hardware, with 8 FlexiForce type pressure sensors and 1 MPU 6050 type IMU (InvenSense) with accelerometer and gyroscope; 2) embedded system, for acquisition, SD card storage and wireless data transmission in real time; and a software and interface using LabView (National Instruments, USA), for data analysis and visualization. This new device was effective in validation tests for gait analysis and can be used for various applications such as sports, entertainment and rehabilitation [12].

The article by Wang, Kim and Min [14] aimed to develop a lightweight, durable, wireless and soft material-based smart insole (SMSI) and examine its feasibility range for real-time gait pattern analysis (Fig. 2). The insole prototype was designed with two layers of shielding and one layer composed of ten capacitive pressure sensors model W-290-PCN (A-jin Eléctron, Busan, Korea) for each foot. The data of the ten channels were sampled in 100 Hz. The data transmission was through bluetooth, and the sum of all the data collected from the ten channels was analyzed with the paired insoles at the same time. In the gait tests, channels nine and ten specifically detected the area of pressure distribution at the moment of heel touch, and channels one and two detected the moment of gait propulsion. Each heel stroke was calculated for the step count, and the time passed was defined by the time between two consecutive heel strokes on the same foot. The insole developed in this study presented equivalent performance to commercial sensors in the evaluation of the step count, stride time and for analysis of the bilateral gait coordination using the phase coordination index (PCI).

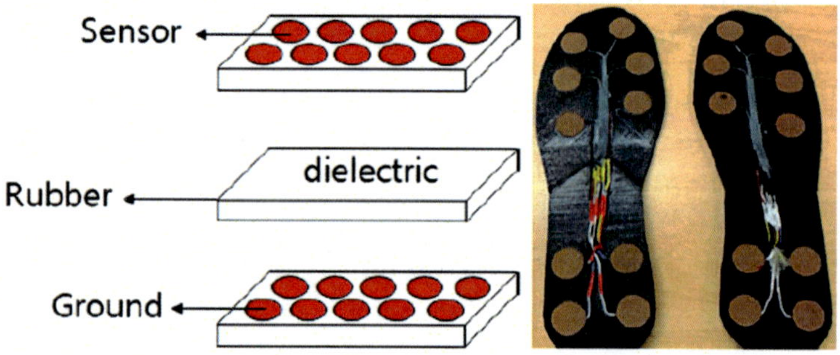

Fig. 2. The structure of the soft-material-based smart insole (SMSI) [14]

Min and contributors [6] proposed to developed a textile capacitive proximity sensor (TCPS) for a gait monitoring system. A smart insole (270 mm long) was designed consisting of three layers, including two layers of shielding and a single channel capacitive pressure sensor layer. According to the results presented in Table 2, the proposed in-shoes system can be useful in monitoring and measuring the gait as an intelligent health system. In addition, the proposed system can store long-term data to generate big data, and one can expect to monitor or detect diseases related to the musculoskeletal system and central nervous system in the initial phases.

The objective of the study by Choi and contributors [3] was to identify users using multimodal sensor data acquired through an intelligent insole, called FootLogger (Fig. 3), with eight pressure sensors and one triaxial accelerometer, and transmission via Bluetooth to smartphone. Six pressure sensors are placed on the forefoot and two on the backfoot, specifically on the heel. The sensor recovers values of 0, 1 and 2 depending on the intensity, where 0 indicates that there is no pressure, that is, the foot is out of the

Table 1. Characteristics of the selected articles.

	Article 1	Article 2	Article 3	Article 4
Author	Das, R., Kumar, N. [4]	Wang, B., et al. [12]	Wang, C.; Kim, Y.; Min, S. D. [14]	Min, S.D. et al. [6]
Sample	- Three healthy male volunteers, from 22 to 28 years old	- One healthy adult	- Fifteen healthy individuals of both sexes with a mean age of 25.1 ± 2.64	- Twelve subjects of both sexes with mean age of 24.91 ± 2.74
Methodology	- Three repetitions of running tests at normal and high speed - The ZebrisTM platform system was used to compare the data obtained in the prototype - The highest-pressure points and elapsed times were tested in the support and gait balance phases	- Test protocol for prototype viability: from steps forward, to a 'U' turn and 10 steps back to its initial position, at normal walking speed	- Walk for 3 min, on a treadmill, at fixed speed of 1.5 km/h (C1), 2.5 km/h (C2), 3.5 km/h (C3) and 4.5 km/h (C4) - The prototype was compared to a commercial F-scan device (Tekscan, USA)	- To evaluate the accuracy of the system, the step count and its error rates were simultaneously detected with the naked eye, with a ZIKTO Walk (ZIKTO Co., Korea), a pedometer HJ-203-K (Omron Co., Japan), and the time spent by the F-scan (Tekscan, USA)
Results	- The system was able to measure force within the range 0–100 N with an accuracy range of ±2 N - The PH and M4 regions generated the maximum force of 25 N - The average error compared to Zebris™ in time parameters is 0.01 s	- The FreeWalker insole proved to be effective for analysis of gait parameters such as: pressure distribution, cadence, step length, gait speed and foot orientation in space	- The step count, stride time and bilateral gait coordination analysis using the phase coordination index (PCI) between the two in-shoes systems resulted in high correlation	- The reliability of F-scan and TCPS was 99% and the correlation coefficient was 0.685 (p-value = 0.000) - TCPS showed lower error rate than ZIKTO Walk (9.1%) and pedometer (4.77%)

ground, while the values of 1 and 2 indicate increased pressure at the site of foot contact with the ground. The sampling rate was 100 Hz. The proposed method effectively extracted the individual characteristics of the measured gait data using the wearable sensors and showed excellent user identification performance. It is worth noting that walking patterns can vary even for the same user, depending on several factors, for example, walking speed usually depends on the physical and mental condition of the user.

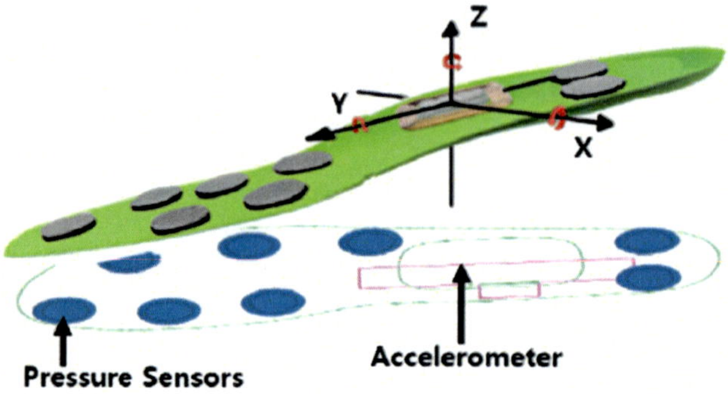

Fig. 3. Sensor structure of the smart insole, "FootLogger". (Source: Choi et al. [3])

The objective of the research of Wang and contributors [13] was to demonstrate the potential of intelligent insoles to monitor the signs of human walking. A pair of intelligent insoles integrated with piezoelectric polyvinylidene fluoride (PVDF) nanogenerators (NGs) that are manufactured to simultaneously collect energy from human motion and monitor the signs of human gait were used for testing. Multi-target magnetic sputtering technology is applied to form the layers of the aluminum electrode on the surface of the PVDF film and the self-feeding insoles are manufactured using advanced 3D seamless plane knitting technology. The output responses of the NGs are measured at different speeds of motion and a maximum value of 41 V is obtained, corresponding to an output power of 168.1 μW. The results show that it is possible to clearly distinguish claudicating, slow, normal, and rapid gait states using multiscale entropy analysis of the stride intervals.

The study by Zhao [15] designed a plantar force and pressure measurement and analysis system (WPPFMA) with a flexible film-based sensor array with sixteen piezoresistive sensors that were distributed across four columns and six rows to detect the magnitudes and distributions of force/pressure in the plantar region (Fig. 4). The system also had a wearable data acquisition device with a Bluetooth module and an interface (laptop or smartphone) and dedicated software.

In the tests of different walking tasks, the plant force/pressure remained stable in the standing and sitting positions, but varied during the dynamic activities. A greater plantar force/pressure (50%) occurred when the subject climbed stairs than when he descended. Similarly, a higher frequency of the force/pressure cycle occurred when the subject ran than when the subject walked. Higher peaks in the plantar force/pressure profile occurred in the jump position than in the standing and walking positions. The results reliably validated the applicability of the developed film-based sensor system for plantar pressure force detection under static and dynamic conditions.

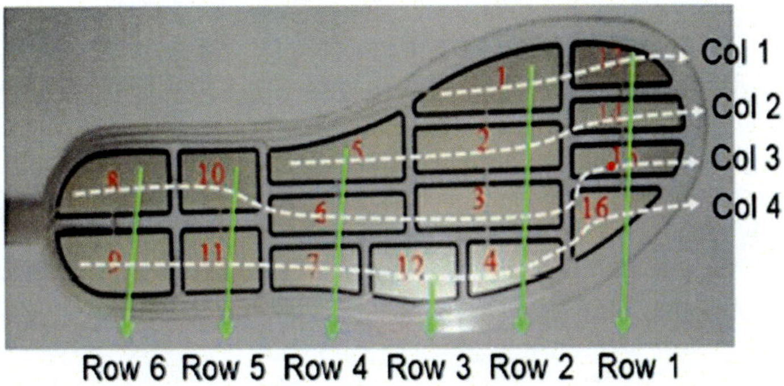

Fig. 4. Flexible sensor matrix film with 16 piezoresistive sensors. (Source: Zhao [15])

Table 2. Characteristics of the selected articles.

	Article 5	Article 6	Article 7
Author	Choi, S., et al. [3]	Wang, W., et al. [13]	Zhao, S. [15]
Sample	- Fourteen individuals of both sexes, aged between 20 and 30 years	- Two subjects	- One healthy 46 kg adult subject
Methodology	- The proposed method consists of a pre-processing stage for the extraction of discriminating characteristics and a classification stage for identifying users	- Walking on a treadmill at different speeds or limping	- Measurements of strength in the plantar region under static and dynamic conditions, including standing with one or both feet, sitting, natural walking and other daily activities - such as running, jumping and climbing stairs - Comparison with Gaitview® AFA-50 system
Results	- The proposed method effectively extracted the individual characteristics of the measured gait data using the wearable sensors and showed excellent user identification performance	- The results show that it is possible to clearly distinguish claudicating, slow, normal and rapid gait states using multiscale entropy analysis from stride intervals	- The new device had applicability to monitor plant force/pressure under static and dynamic usage conditions

4 Conclusion

The research data showed that the intelligent insoles were effective in all articles regarding the analysis of gait parameters. However, the research from Zhao [15] obtained better results, due to the design of the study with validation in comparison to another intelligent insole already commercialized, besides the characteristics of the intelligent insole, mainly of the software and interface that was able to analyze and display several parameters of the march. It is suggested more research with this type of device, with an intervention time and larger samples, in other functional or pathological conditions.

References

1. Berengueres, J., Fritschi, M., Mcclanahan, R.: A smart pressure-sensitive insole that reminds you to walk correctly: an orthotic-less treatment for over pronation. In: 36th Annual International Conference of the IEEE Engineering in Medicine and Biology Society, Chicago, pp. 2488–2491 (2014)
2. Christovão, T.C.: Effect of different insoles on postural balance: a systematic review. J. Phys. Ther. Sci. **25**(10), 1353–1356 (2013)
3. Choi, S., et al.: User identification from gait analysis using multi-modal sensors in smart insole. Sensors **19**(17), 3785–3798 (2019)
4. Das, R., Kumar, N.: Investigations on postural stability and spatiotemporal parameters of human gait using developed wearable smart insole. J. Med. Eng. Technol. **39**(1), 75–78 (2015)
5. Davies, R.J., et al.: A personalized self-management rehabilitation system for stroke survivors: a quantitative gait analysis using a smart insole. JMIR Rehabil. Assist. Technol. **3**(2), 1–11 (2016)
6. Min, S.D., et al.: Development of a textile capacitive proximity sensor and gait monitoring system for smart healthcare. J. Med. Syst. **42**(4), 76–88 (2018)
7. Neto, H.P.: Effect of posture-control insoles on function in children with cerebral palsy: randomized controlled clinical trial. BMC Musculoskelet. Disord. **13**, 193–199 (2012)
8. Park, E., et al.: Unobtrusive and continuous monitoring of alcohol-impaired gait using smart shoes. Methods Inf. Med. **56**(1), 74–82 (2017)
9. Razak, A.H.A., et al.: Foot plantar pressure measurement system: a review. Sensors (Switzerland) **12**(7), 9884–9912 (2012)
10. Telfer, S., et al.: Virtually optimized insoles for offloading the diabetic foot: a randomized crossover study. J. Biomech. **60**, 157–161 (2017)
11. Tsiouris, K.M., et al.: PD Manager: an mHealth platform for Parkinson's disease patient management. Healthc. Technol. Lett. **4**(3), 102–108 (2017)
12. Wang, B., et al.: FreeWalker: a smart insole for longitudinal gait analysis. In: Annual International Conference of the IEEE Engineering in Medicine and Biology Society, pp. 3723–3726 (2015)
13. Wang, W., et al.: Self-powered smart insole for monitoring human gait signals. Sensors **19**(24), 5336–5345 (2019)
14. Wang, C., Kim, Y., Min, S.D.: Soft-material-based smart insoles for a gait monitoring system. Materials **11**(12), 2435–2449 (2018)
15. Zhao, S.: Flexible sensor matrix film-based wearable plantar pressure force measurement and analysis system. PLoS ONE **15**(8), 1–16 (2020)

An Investigation Regarding the Impact of Running-In on Rolling Contacts Lives

Marcelin Benchea(✉) and Spiridon Crețu

Mechanical Engineering Faculty, Mechatronics and Robotics Department, "Gheorghe Asachi" Technical University of Iaşi, 43 Prof. Dr. Doc. D. Mangeron, 700050 Iaşi, Romania
marcelin.benchea@tuiasi.ro

Abstract. Microtopography of the working surfaces subjected on rolling contact can be realized by various manufacturing technologies (lapping, polishing, turning, grinding, etc.). The aim of this work is to point out the impact of the initial surface asperities on running-in phenomenon, geometry of active surface and rolling contacts durability. At first, the article exhibit the experimental tests carried out by using the two discs AMSLER machine with proper samples for pure rolling motion. The microtopography of the working surfaces have been measured before and after tests. Secondly, a numerical study is presented for the samples used on the two discs AMSLER machine, where the measured topography data and loading conditions are used as input data in the analysis model. The last part of the article exhibit a numerical investigation for the rolling bearing CARB C2318 where the modified rating life have been computed with methodology given by ISO 16281.

Keywords: Surface rugosity · Running-in process · Plastic displacement · Modified rating life

1 Introduction

When the working surfaces of two bodies are loaded for the first time the microtopography of both surfaces generally suffer. The microtopography modifies that occur from starting to stable condition are related with the running-in process [1–3] and affect the contact pressures repartitions, and stress state, especially in the shallow surface layers, and consequently the various wear types involved in the durability of contacting bodies [4–6]. Recently studies on running-in [7–10] show that changes in the surface asperities likewise alter the chemistry of surface, remanent stresses and microstructure close to the active surfaces.

The numerical models used to describe the microtopography changes [11, 12] require the input of a 3D rough surface. Because the experimental acquirement of such data does not grant enough samples of profile to be used, a main necessity for any micronumerical analysis is to produce random rough surfaces with identical or similar characteristics as the effective surfaces [13, 14].

J. Machado et al. (Eds.): icieng 2021, LNME, pp. 459–471, 2022.
https://doi.org/10.1007/978-3-030-79165-0_43

The running-in and steady-state experimental investigation were conducted on two rollers [15] and was observed the evolution of wear depth/weight and surface rugosity during the experiment. It was evidenced that the slope of wear depth and worn weight decreases as the asperities are smoothed and the number of plastically deformed asperities is reduced. In the assessment of rugosity parameters of hardened steel surfaces [16] it was pointed out that in the running-in process the most prominent asperity heights suffer plastic strain through the primary loading period.

The impact of the relative sliding movement on rolling contact fatigue life of surfaces with sinusoidal rugosity in mixed lubrication was investigated in [17] without considering the surface modification due to wear. A model for the rising of rolling contact fatigue (RCF) cracks initiated from the surface in the material of rolling bearings was presented in [18]. In this study the greatest value of the von Mises subsurface stress was considered as the critical stress for the RCF developed in the number of stress cycles which carry, finally, to critical damage.

Recently, in [19] was pointed out that in the running-in process of rolling bearings, the microgeometry of the working surfaces has two major effects: *i*) an increase in the regional clearances due to elastic deflection and, *ii*) the initiation of pressure fluctuations affect the rolling bearing life. Also, in the evaluation of the bearing life was considered the lubricate quality parameter κ as a function of the lubricate parameter λ whichever depends on the film thickness and the surface rugosity. Furthermore, in recent studies [20, 21], to compute the rolling bearing life was used Ioannides and Harris model, taking into account the surface and subsurface stress.

To model the non-linear strain rate under deformation of materials stressed in elastic-plastic field, an analysis model was build [22, 23]. The model is developed in the framework of incremental hypothesis of plasticity using the von Mises plastic flow principle and Prandtl-Reuss equations. Take into account the nonlinear kinematic and isotropic hardening laws of Lemaitre-Chaboche [24, 25] the model considers the cyclic hardening phenomenon.

The microtopography of working surfaces generate sharp peaks in pressure repartitions which disturb the von Mises equivalent stress repartition with a negative influence on fatigue life of rolling contact [26].

2 Material and Methods

2.1 Experiment

Experimental tests on the two discs machine (AMSLER), Fig. 1, have been carried out to evaluate the surface rugosity modification during the running-in process.

The two discs are rolling one versus other with constant rotating velocity rate $k = n_2/n_1 = 0.906$, where n_2 and n_1 are rotating speed of the upper and lower disc respectively.

Two discs with size dimensions of *59* mm and *53.5* mm manufactured from 42 CrMo4 steel were used. The upper disc with size dimension of *59* mm was crowned with radius of *300* mm and the lower disc with size dimension of *53.5* mm was cylindrical. The discs had *10* mm in width and were hardened and tempered resulting a hardness of *425 HB*.

Trying conditions: rotating velocity $n_1 = 181$ rpm, normal loading $F = 140$ N, running time *10* min, SAE46 (H46) lubricant, working temperature *25* °C.

Fig. 1. Two discs machine (AMSLER).

2.2 Methods

Rough Surface Simulation. All surfaces have some form of structure microtopography that change according to its textures and the way it have been made. Corresponding to [5] there are three types of parameters for the rugosity contour: magnitude parameters (the normal characteristics of the surface deflections), array parameters (the straight characteristics of the surface deflections) and composite parameters.

Rugosity Amplitude Parameters. The experimental acquirement of the rugosity contours is time overwhelming, so that, the various characters of the rough surfaces have to be established analytically or numerically, in order to study their role upon the evolution of a particular rolling contact. Majority of the statistical properties of a rough surface can be defined from knowledge of two statistical functions, the autocorrelation function and the frequency density function.

The probability function is used to point out the spatial characteristics. Some real surfaces, particularly recently grinded surfaces, expose a height repartition close to the normal Gaussian probability function:

$$p(z) = \frac{\sigma}{\sqrt{2 \cdot \pi}} \cdot \exp\left(\frac{-z^2}{2\sigma^2}\right) \tag{1}$$

where σ is standard deviation (r.m.s.) from the average height.

However if the grinded surface is subsequently polished its rugosity is no longer Gaussian. A mathematical representation of the probability function shape can be achieved by using the central moments, defined as:

$$\mu_k = \int_{-\infty}^{+\infty} z^k \cdot p(z) \cdot dz \tag{2}$$

or in discrete form:

$$\mu_k = \frac{1}{n} \cdot \sum_{i=1}^{n} z_i^k \tag{3}$$

Any spatial rugosity parameter presented in the specialized literature can be mathematically established utilizing the first four moments. However, because the first moment equals to zero the arithmetic mean rugosity R_a is defined instead:

$$R_a = \frac{1}{n} \cdot \sum_{i=1}^{n} |z_i| \tag{4}$$

The second moment is the rugosity peaks variance, signifying the root mean square R_q (r.m.s.), or the standard deviation σ, of the surface peaks:

$$R_q = \sigma = \sqrt{\mu_2} \tag{5}$$

The relation from σ and R_a rely, to certain measure, on the surface nature; for a regularly sinusoidal contour $\sigma = \pi/(2 \cdot 2^{1/2}) \cdot R_a \approx 1.11 \cdot R_a$, and for a Gaussian contour $\sigma = (\pi/2)^{1/2} \cdot R_a \approx 1.25 \cdot R_a$.

Rugosity Spectral Characterization. The Autocorrelation Function (ACF). The *ACF*, $R(x, y)$ is the supposed valuables of the multiplication: $z(x, y) \cdot z(x + \lambda_x, y + \lambda_y)$ of the surface peak at point (x, y) and at the point $(x + \lambda_x, y + \lambda_y)$, and λ_x, λ_y are the delay lengths. Under the stationarity assumption this expectation is independent:

$$R(\lambda_x, \lambda_y) = E\langle z(x, y) \cdot z(x + \lambda_x, y + \lambda_y)\rangle, R(0, 0) = \sigma^2. \tag{6}$$

For majority of the fabrication processes the *ACF* attend a negative exponential function. In this presumption the required parameters to provide an *ACF* are the dissolution length in two directions that are normal one to other. These lengths are well-known as the autocorrelation lengths λ_x, λ_y, and are fully described as the length in the x and y directions where the autocorrelation length falls to *10%* from its initial value.

Archard and Whitehouse considered the shape of an arbitrary surface as an arbitrary signal depicted by a peak distribution and an *ACF* and revealed that all characteristics of a surface with Gaussian dispersion of peaks and a negative exponential *ACF* can be depicted by two parameters: σ and λ [27].

Gaussian Surfaces with a Prescribed ACF. A numerical method has been developed to generate a random 3D rough surface, [14]. The surface rugosity peaks were generated corresponding to the next equation:

$$z_{ij} = \sum_{k=1}^{n} \sum_{l=1}^{m} \alpha_{kl} \cdot \eta_{i+k,j+l} \tag{7}$$

where α_{kl} are parameters that give the *ACF* needed and $\eta_{i+k,j+l}$ are individual arbitrarily values that have the variance unit, [13, 14], the following equations are valid:

$$E(\eta_{ij}\eta_{kl}) = \begin{cases} 1 \text{ if } i = k, j = 1 \\ 0 \text{ if } i \neq k, j \neq 1 \end{cases} \tag{8}$$

Using Eqs. (8) the *ACF* is obtained as:

$$R_{pq} = \sum_{k=1}^{n-p} \sum_{l=1}^{m-q} \alpha_{kl} \cdot \alpha_{k+p,l+q} \quad \begin{matrix} p = 0, 1, 2 \ldots, n-1 \\ q = 0, 1, 2, \ldots, m-1 \end{matrix} \tag{9}$$

Equations (9) it is a non-linear system with the unknowns parameters α_{kl}, [13]. An iterative method has been applied to resolve the non-linear system (9) and its solutions were introduced in Eq. (7) to achieve the rugosity peaks with the prescribed *ACF*.

For two coarse surfaces considered in contact an equivalent surface [28] can be defined by adding corresponding values of the rugosity heights:

$$z = z_1 + z_2 \tag{10}$$

Elastic-Plastic Model. An elastic-plastic analytical model has been made to estimate the plastic displacement of the surfaces in rolling contact, that take into account the material cyclical strengthening particularity as depicted by a nonlinear kinematic and an isotropic strengthening law [24, 25]. The plastic flow surface is depicted by the von Mises plastic flow principle:

$$F = f(\sigma - \alpha) - \sigma_Y^0 = 0 \tag{11}$$

where σ_Y^0 is the plastic flow stress and $f(\sigma - \alpha)$ is the von Mises equivalent stress.

The isotropic strengthening part of the model describes the yielding stresses σ_{Y0} as a relationship of the equivalent plastic deformation ε^p:

$$\sigma_Y^0 = \sigma_{Y0} + Q_\infty \cdot \left(1 - e^{-b_\infty \cdot \overline{\varepsilon}^p}\right) \tag{12}$$

where Q_∞ is the limit modifies in the plastic flow surface dimension on the deviatoric level and b_∞ present how fast the limit dimension is achieved [29].

The Ramberg-Osgood's equation has been applied to consider the connection between the strain tensor intensity ε_e and the stress tensor intensity σ_e:

$$\overline{\varepsilon}_e = \frac{\overline{\sigma}_e}{E} + \left(\frac{\overline{\sigma}_e}{B}\right)^N \tag{13}$$

where N is the plastic strengthening index and B is the plastic strengthening parameter [22].

The surfaces plastic displacement u^p was computed by the plastic deformations ε^p gathered in the subsurface stratum:

$$u^p(x, y) = \int_0^z \varepsilon^p(x, y, z) \cdot dz \tag{14}$$

3 Experimental vs. Numerical Results and Discussion

3.1 Experimental Results

The rugosity contours for two discs studied on AMSLER machine are depicted in Fig. 2, before and after the running-in experiment. The original discs rugosity were $R_{q1} = 610$ nm and $R_{q2} = 280$ nm, respectively $R_{a1} = 460$ nm and $R_{a2} = 220$ nm.

The evaluation of discs rugosity before and after experiments revealed modifications for the bottom disc, with a coarser rugosity, for the estimate length of *1* mm in the middle. After the running-in experiment the root mean square of the bottom disc decreases to $R_{q1} = 550$ nm and the arithmetic mean rugosity peaks to $R_{a1} = 390$ nm.

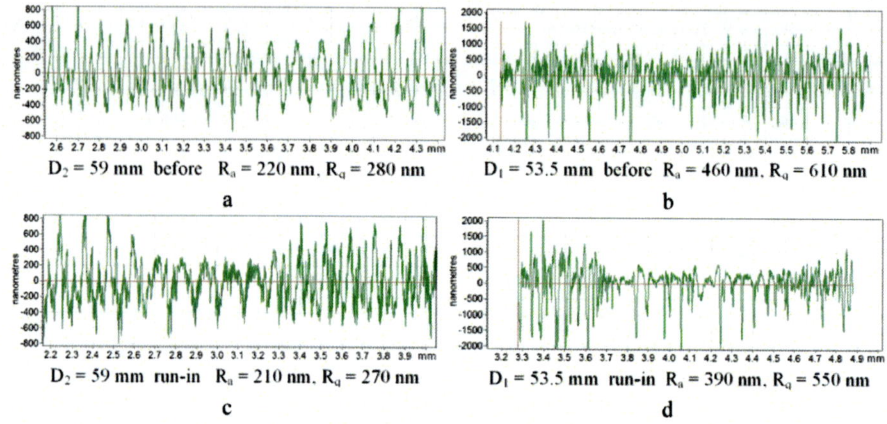

Fig. 2. 2D rugosity contours before and after running-in experiment.

3.2 Experimental vs. Numerical Results Provided by Elastic-Plastic Model

The plastic changes of the surface rugosity are due to the raised peaks in the pressure repartition on the two discs contact area, Fig. 4. These pressure peaks can induce a state stress above the plastic flow limit resulting plastic displacement of the rugosity peaks.

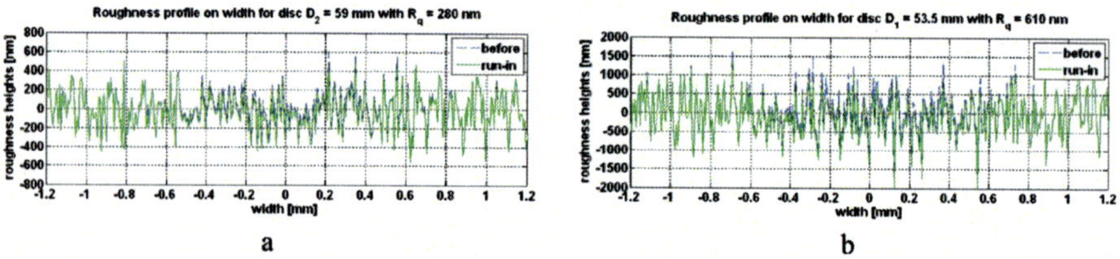

Fig. 3. 2D rugosity contours ahead and further *300* run-in cycles (numerical).

The elastic-plastic running-in model presented in [1, 2] provided numerical results in good agreement with those obtained experimentally. In Fig. 3 are comparatively exemplified the experimental and numerical results obtained after *300* running cycles.

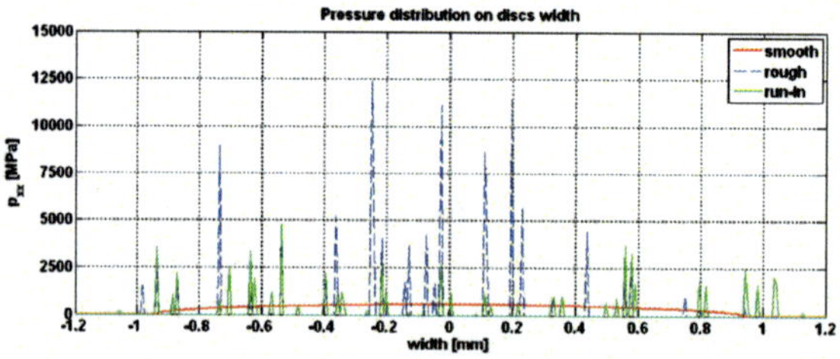

Fig. 4. Pressure distribution on discs width (hypothetic full elastic range).

3.3 Rolling Bearing Application. Numerical Results

The CARB C2318 rolling bearing has been considered for the numerical investigation. The geometry of rolling bearing, Fig. 5a, is depicted by the following parameters: the diameter of outer ring $d_{ce} = 165$ mm, the diameter of inner ring $d_{ci} = 115$ mm, the diameter of roller $d_w = 25$ mm, the length of roller $L_w = 54$ mm, roller crown radius $R_1 = 7000$ mm, inner/outer ring crown radius $R_2 = 7210$ mm, rollers number $Z = 14$, [30]. The measured roughness for rollers, inner and outer raceways are presented in Fig. 5b.

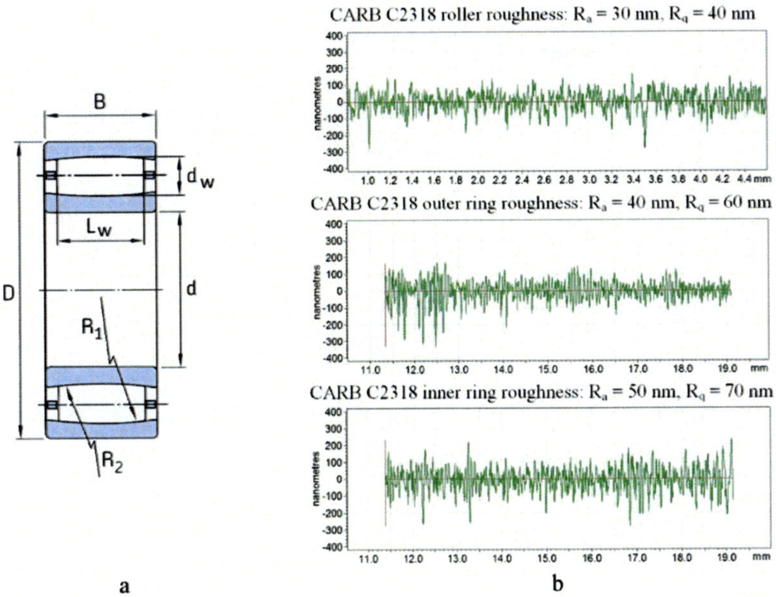

Fig. 5. (a) Toroidal roller bearing geometry [30] and (b) Rugosity of bearing elements.

A moderate radial charging $F_R = .185 \cdot C = 112.5$ kN has been considered in the numerical investigation of the running-in process. Three sizes of the surface rugosity characteristics have been chosen: *i*) a small rugosity with $R_{qw1} = 40$ nm for roller surface and $R_{qi1} = 70$ nm for the raceway surface of the inner ring, like the surface rugosity characteristics determined on Taylor Hobson equipment, Fig. 5b; *ii*) a moderate rugosity with $R_{qw2} = 80$ nm and $R_{qi2} = 140$ nm, and *iii*) a coarse rugosity with $R_{qw3} = 120$ nm and $R_{qi3} = 210$ nm.

For the roller bearing steel loaded in the elastic-plastic field the following parameters were used: the plastic flow stress $\sigma_{Y0} = 1650$ MPa, the coefficient of Poisson $\nu = 0.27$, the modulus of Young $E = 203$ GPa, $b_\infty = 120$ that shows how fast the limit dimension is achieved and $Q_\infty = -100$ MPa, the limit modifies in the plastic flow surface dimension on the deviatoric level [28], the plastic strengthening index $N = 12.6$ and the plastic strengthening parameter $B = 4320$ MPa [22, 23].

The 2D rugosity contours of roller/inner ring ahead and further *300* run-in cycles are presented in Fig. 6. It may be remarked so the running-in process is more pronounced for the inner ring and the rugosity peaks are plastic flattened in the middle: a length of *10* mm for $R_{qi1} = 70$ nm, a length of *20* mm for $R_{qi2} = 140$ nm and a length of *40* mm for $R_{qi3} = 210$ nm.

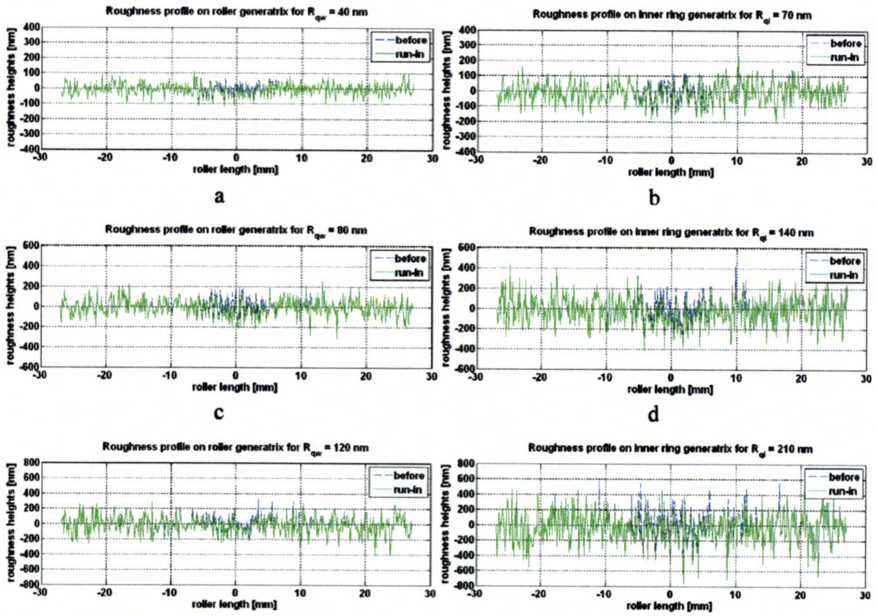

Fig. 6. 2D rugosity contours of roller/inner ring ahead and further *300* run-in cycles.

The higher is the rugosity peaks the larger is the plastic deformation.

The 2D pressures repartitions on roller generatrix for the heavy charged roller/inner ring contact are depicted in Fig. 7. In the center of the contact area the maximum peak pressure for the smooth surface is *2250* MPa. The existence of the rugosity on the working surfaces causes high peaks in pressures repartition. The higher are pressure peaks as the rugosity it is coarser and have the highest magnitude of *2700* MPa for the small rugosity, *3200* MPa for the moderate rugosity and *4000* MPa for the coarse rugosity.

It may be noticed that after *300* run-in cycles the pressure peaks are a little attenuated in the center of the contact area for the small and moderate rugosity and are much attenuated on all contact region for the coarse rugosity.

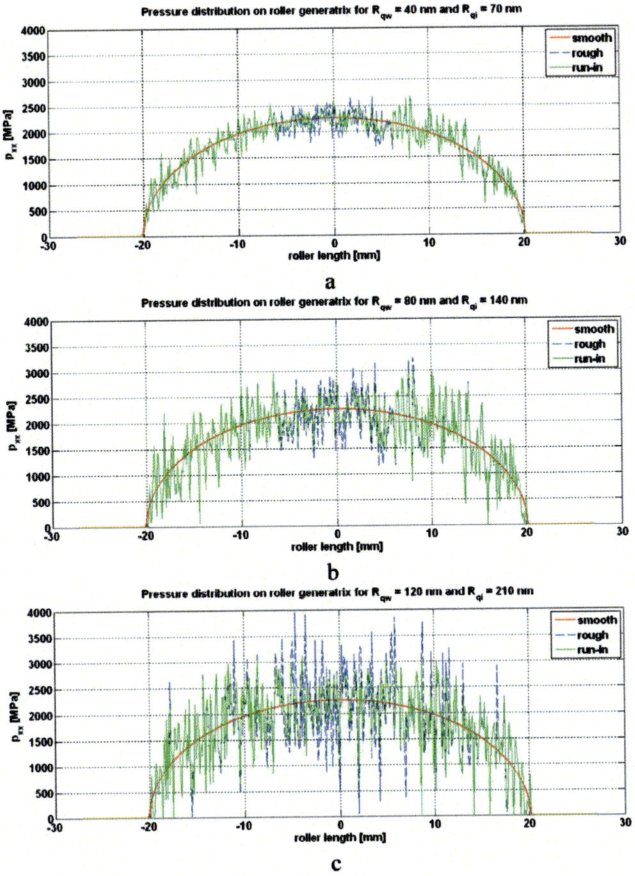

Fig. 7. Pressures repartitions for the heavy charged roller/inner ring contact.

4 Bearing Lives Comparisons

The generally evaluation of the rolling bearings basic rating life take into account just the dynamic capacity charge C, the equivalent charge on the bearing P and an exponent p relying on the shape of rolling element:

$$L_{10} = \left[\frac{C}{P}\right]^{p} \tag{15}$$

where $p = 10/3$ for rollers and $p = 3$ for balls [31, 33].

To consider the changes of microtopography on the active surfaces as result of the running-in process, the basic reference rating life and modified rating life as described in [26, 31, 34] have been evaluated in this article.

4.1 Modified Bearing Life

The κ Parameter Approach. Viscosity Rate and Film Parameter Approach. The magnitude of the dynamic capacity charge C given by the rolling bearing manufacturers and the calculation method depicted in [31] are often utilized for general estimation of the modified rating life, L_{10m}:

$$L_{10m} = a_{ISO} \cdot L_{10} = a_{ISO} \cdot \left[\frac{C}{P}\right]^{p} \tag{16}$$

The modification factor, a_{ISO}, is given by the subsequent relation:

$$a_{ISO} = f\left(\frac{\eta_c \cdot P_u}{P}, \kappa\right) \tag{17}$$

where: f is a function described by ISO 281:2007, (Eqs. (34)–(37) [31]), η_c is the contamination index, and P_u is the fatigue charge threshold.

Corresponding to [31], for lubrication with mineral oil and raceway surfaces of rolling bearing processed with *quality of good manufacture*, the state of lubrication is depicted by a complex lubricate parameter κ , described as the relation between the current kinematic viscosity ν and the reference kinematic viscosity ν_1, [32]:

$$\kappa = \nu / \nu_1 \tag{18}$$

The lubricate parameter λ just rely on two factors: the thickness of lubricating layer and rugosity, although κ, it is just associated to the thickness of lubricating layer, and permits for more manageability the consideration of another specific life of rolling bearing associated to safety factors as inlet shear heating effects and starvation.

In [19] it is remarked that the actual meaning of κ (ISO 281:2007 [31]) still fails to consider important effects as: an actual film height equation and rugosity deformation.

4.2 Rolling Bearing Durability

To evaluate the roller bearing durability with the methodology given by standards ISO 281:2007 [31] and ISO 16281:2008 [26] two levels of loading have been considered: a medium radial loading (for running-in process presented previously) and a light radial loading (for normal operating condition).

Table 1 presents the modified reference rating lives L_{10m} for a medium load (F_R = *112.5* kN), smooth and three values of the surface rugosity parameters. The same running conditions have been considered in all three simulations: working temperature $t = 50$ °C, level of contamination $\eta_c = 0.8$, rotation speed $n = 1000$ rpm.

The modified reference rating life has lower values for rough surfaces like smooth surface, and more lower as the rugosity is coarser. Further the running-in process the modified reference rating life increase with approximate *1000* h for the smaller and moderate rough surfaces and with more than *1250* h for the coarser rough surface.

Table 1. Modified reference rating life for a medium load ($F_R = 0.185 \cdot C = 112.5$ kN).

Rugosity of roller	Rugosity of inner ring	Basic rating life, L_{10}, *hours*	Modified reference rating life, L_{10m}, *hours*	
R_{qw} [*nm*]	R_{qi} [*nm*]		Before	Run-in
Smooth	Smooth	*4668*	*10796*	
40	*70*		*5997*	*6915*
80	*140*		*1567*	*2561*
120	*210*		*246*	*1505*

Table 2. Modified reference rating life for a light load ($F_R = 0.055 \cdot C = 33.5$ kN).

Rugosity of roller	Rugosity of inner ring	Basic rating life, L_{10}, *hours*	Modified reference rating life, L_{10m}, *hours*	
R_{qw} [*nm*]	R_{qi} [*nm*]		Before	Run-in
Smooth	Smooth	*264700*	*688750*	
40	*70*		*218660*	*324140*
80	*140*		*28205*	*74993*
120	*210*		*2297*	*38602*

The modified reference rating lives L_{10m} for a light load ($F_R = 33.5$ kN) are presented in Table 2. As in the previous table, the modified reference rating life has lower values for rough surfaces like smooth surface. For the modified surfaces subjected to a medium load further *300* run-in cycles, the modified reference rating life increase, seemingly a greater enhancement for the coarser rugosity, approximately *16* times higher like ahead run-in, however a importantly enhancement for the smaller rugosity, almost *105500* h furthermore like ahead run-in.

5 Conclusions

The microtopography of the working surfaces subjected on rolling contact rely primarily on the surface machining technologies and secondly on the evolution of the running-in process.

To simulate the running-in phenomenon, random three-dimensional rough surfaces have been generated numerically and further considered into an elastic-plastic numerical model. For the case of CARB rolling bearing the values for modified reference rating life have been considered to estimate the effect of the run-in.

The experimental data, of running-in process, acquired on AMSLER machine present a very good agreement with the numerical data for two discs test and with numerical data obtained with the running-in model exhibited in [1, 35].

The asperities presence on working surfaces in rolling contact causes high peaks in pressure repartitions, and as the rough surfaces are coarser the higher are the pressure peaks.

The modified rating life is seriously reduced for the coarse rough surface, by two size order, right further running-in process, by one size order, like for the fine rough surface or smooth surface.

References

1. Jamari, J.: Running-in of rolling contacts. Ph.D. thesis, Twente (2006)
2. Jamari, J., de Rooij, M.B., Schipper, D.J.: Plastic deterministic contact of rough surfaces. ASME Trans. J. Tribol. **129**, 957–962 (2007)
3. Tasan, Y.C., de Rooij, M.B., Schipper, D.J.: Changes in the micro-geometry of a rolling contact. Tribol. Int. **40**, 672–679 (2007)
4. Robbe-Valloire, F.: Statistical analysis of asperities on a rough surface. Wear **249**, 401–408 (2001)
5. Taylor Hobson Precision: Exploring surface texture. TH Ltd. (2003)
6. Stawicki, T., Sedlak, P., Koniuszy, A.: The testing of the influence of the roughness of the crankshaft journal upon the durability of the crankshaft bearing in engines of agricultural machines. Sci. Probl. Mach. Op. Maint. **4**, 7–17 (2010)
7. Jacobson, S., Hogmark, S.: Surface modifications in tribological contacts. Wear **266**, 370–378 (2009)
8. Andersson, M., Sosa, M., Olofsson, U.: The effect of running-in on the efficiency of superfinished gears. Tribol. Int. **93**, 71–77 (2016)
9. Sjöberg, S., Sosa, M., Andersson, M., Olofsson, U.: Analysis of efficiency of spur ground gears and the influence of running-in. Tribol. Int. **93**, 172–181 (2016)
10. Mallipeddi, D., Norell, M., Sosa, M., Nyborg, L.: Influence of running-in on surface characteristics of efficiency tested ground gears. Tribol. Int. **115**, 45–58 (2017)
11. Allwood, J., Ciftci, H.: An incremental solution method for rough contact problems. Wear **258**, 1601–1615 (2005)
12. Allwood, J.: Survey and performance assessment of solution methods for elastic rough contact problems. ASME Trans. J. Tribol. **127**, 10–23 (2005)
13. Bakolas, V.: Numerical generation of arbitrarily oriented non-gaussian three-dimensional rough surfaces. Wear **254**, 546–554 (2003)
14. Creţu, S.: Random simulation of Gaussian rough surfaces: Part I. Theoretical formulation. Bul. Instit. Pol. Iaşi **1–2**(LII), 1–16 (2006)
15. Akbarzadeh, S., Khonsari, M.M.: Experimental and theoretical investigation of running-in. Tribol. Int. **44**, 92–100 (2011)
16. Clarke, A., Weeks, I.J.J., Snidle, R.W., Evans, H.P.: Running-in and micropitting behavior of steel surfaces under mixed lubrication conditions. Tribol. Int. **101**, 59–68 (2016)
17. Pu, W., Zhu, D., Wang, J., Wang, Q.J.: Rolling-sliding contact fatigue of surfaces with sinusoidal roughness. Int. J. Fatigue **90**, 57–68 (2016)
18. Rycerz, P., Olver, A., Kadiric, A.: Propagation of surface initiated rolling contact fatigue cracks in bearing steel. Int. J. Fatigue **97**, 29–38 (2017)
19. Morales-Espejel, G.E., Gabelli, A., Ioannides, E.: Micro-geometry lubrication and life ratings of rolling bearings. Proc. Inst. Mech. E Part C: J. Mech. Eng. Sci. **224**, 2610–2628 (2010)
20. Morales-Espejel, G.E., Gabelli, A., de Vries, A.J.C.: A model for rolling bearing life with surface and subsurface survival-tribological effects. Tribol. Trans. **58**, 894–906 (2015)

21. Liu, M., Wu, C., Yan, C.: Predicting fatigue life for finite roller contacts based on a mixed EHL model using realistic surface roughness. J. Mech. Sci. Technol. **31**(7), 3419–3428 (2017)
22. Benchea, M., Iovan-Dragomir, A., Creţu, S.: Misalignment effects in cylindrical roller bearings. Appl. Mech. Mater. **658**, 277–282 (2014)
23. Creţu, S., Benchea, M., Iovan-Dragomir, A.: On basic reference rating life of cylindrical roller bearings. Part II - elastic-plastic analysis. J. Balk. Tribol. Assoc. **22**, 272–280 (2016)
24. Chaboche, J.L.: A review of some plasticity and viscoplasticity constitutive theories. Int. J. Plast. **24**, 1642–1693 (2008)
25. Besson, J., Cailletaud, G., Chaboche, J.L., Forest, S., Blétry, M.: Non-Linear Mechanics of Materials. Springer (2010)
26. ISO 16281:2008: Rolling bearings-methods for calculating the modified reference rating life for universally loaded bearings
27. Whitehouse, D.J., Archard, J.F.: The proprieties of random surface of significance in their contact. Proc. R. Soc. Lond. A **316**, 97–121 (1970)
28. Creţu, S.: The influence of the correlation length on pressure distribution and stress state in concentrated rough contacts. In: Proceedings of ASME/ASLE IJTC-2006, San Antonio, USA (2006)
29. Linares Arregui, I., Alfredsson, B.: Elastic-plastic characterization of a high strength bainitic roller bearing steel - experiments and modelling. Int. J. Mech. Sci. **52**, 1254–1268 (2010)
30. SKF Group: Rolling bearings catalogue (2013)
31. ISO 281:2007: Rolling bearings-dynamic load ratings and rating life
32. Ioannides, E., Bergling, G., Gabelli, A.: An analytical formulation for the life rating of rolling bearings. Acta Polytech. Scand. Mech. **137**, 1–80 (1999)
33. Creţu, S., Benchea, M., Creţu, O.: Compressive residual stresses effect on fatigue life of rolling bearings. In: Proceedings of IMECE-07, Seattle, paper 43561 (2007)
34. Benchea, M., Creţu, S.: Profile evolution in cylindrical roller bearings. II. Rating lives evaluation. Bull. Inst. Polit. Iaşi **62/66**(3), 35–42 (2016)
35. Jamari, J., Schipper, D.J.: Deformation due to contact between a rough surface and a smooth ball. Wear **262**, 138–145 (2007)

Author Index

J. Machado et al. (Eds.): icieng 2021, LNME, pp. 473–475, 2022.
https://doi.org/10.1007/978-3-030-79165-0